14

Grundlehren der mathematischen Wissenschaften 241

A Series of Comprehensive Studies in Mathematics

Walter Rudin

Function Theory in the Unit Ball of $\mathbb{C}^n$

Springer-Verlag

New York Heidelberg Berlin

Walter Rudin
University of Wisconsin
Department of Mathematics
Madison, WI 53706
USA

AMS Subject Classifications: 32-02, 31Bxx, 31Cxx, 32Axx, 32Fxx, 32Hxx

Library of Congress Cataloging in Publication Data

Rudin, Walter, 1921–
    Function theory in the unit ball of $\mathbb{C}^n$

    (Grundlehren der mathematischen Wissenschaften; 241)
    Bibliography: p.
    Includes index.
    1. Holomorphic functions.   2. Unit ball.   I. Title.
    II. Series.
QA331.R863   515   80-19990

Printed in the United States of America.

9 8 7 6 5 4 3 2 1

ISBN 0-387-90514-6 Springer-Verlag New York
ISBN 3-540-90514-6 Springer-Verlag Berlin Heidelberg

Preface

Around 1970, an abrupt change occurred in the study of holomorphic functions of several complex variables. Sheaves vanished into the background, and attention was focused on integral formulas and on the "hard analysis" problems that could be attacked with them: boundary behavior, complex-tangential phenomena, solutions of the $\bar{\partial}$-problem with control over growth and smoothness, quantitative theorems about zero-varieties, and so on. The present book describes some of these developments in the simple setting of the unit ball of $\mathbb{C}^n$.

There are several reasons for choosing the ball for our principal stage. The ball is the prototype of two important classes of regions that have been studied in depth, namely the strictly pseudoconvex domains and the bounded symmetric ones. The presence of the second structure (i.e., the existence of a transitive group of automorphisms) makes it possible to develop the basic machinery with a minimum of fuss and bother. The principal ideas can be presented quite concretely and explicitly in the ball, and one can quickly arrive at specific theorems of obvious interest. Once one has seen these in this simple context, it should be much easier to learn the more complicated machinery (developed largely by Henkin and his co-workers) that extends them to arbitrary strictly pseudoconvex domains.

In some parts of the book (for instance, in Chapters 14–16) it would, however, have been unnatural to confine our attention exclusively to the ball, and no significant simplifications would have resulted from such a restriction.

Since the Contents lists the topics that are covered, this may be the place to mention some that might have been included but were not:

The fact that the automorphisms of the ball form a Lie group has been totally ignored.

There is no discussion of concepts such as curvature or geodesics with respect to the geometry that has these automorphisms as isometries.

The Heisenberg group is only mentioned in passing, although it is an active field of investigation in which harmonic analysis interacts with several complex variables.

Most of the refined estimates that allow one to control solutions of the $\bar{\partial}$-problem have been omitted. I have included what was needed to present the

v

Henkin–Skoda theorem that characterizes the zeros of functions of the Nevanlinna class.

Functions of bounded mean oscillation are not mentioned, although they have entered the field of several complex variables and will certainly play an important role there in the future.

To some extent, these omissions are due to considerations of space—I wanted to write a book of reasonable size—but primarily they are of course a matter of personal choice.

As regards prerequisites, they consist of advanced calculus, the basic facts about holomorphic functions of one complex variable, the Lebesgue theory of measure and integration, and a little functional analysis. The existence of Haar measure on the group of unitary matrices is the most sophisticated fact assumed from harmonic analysis. Everything that refers specifically to several complex variables is proved.

I have included a collection of open problems, in the hope that this may be one way to get them solved. Some of these look very simple. The fact that they are still unsolved shows quite clearly that we have barely begun to understand what really goes on in this area of analysis, in spite of the considerable progress that has been made.

I have tried to be as accurate as possible with regard to credits and priorities. The literature grows so rapidly, however, that I may have overlooked some important contributions. If this happened, I offer my sincere apologies to their authors.

Several friends have helped me to learn the material that is presented here—in conversations, by correspondence, and in writing joint papers. Among these, I especially thank Pat Ahern, Frank Forelli, John Fornaess, Alex Nagel, and Lee Stout.

Finally, I take this opportunity to express my appreciation to the National Science Foundation for supporting my work over a period of many years, to the William F. Vilas Trust Estate for one of its Research Professorships, and to the Mathematics Department of the University of Wisconsin for being such a friendly and stimulating place to work in.

Madison, Wisconsin   Walter Rudin
March 1980

Contents

List of Symbols and Notations   xi

Chapter 1

Preliminaries   1
1.1 Some Terminology   1
1.2 The Cauchy Formula in Polydiscs   3
1.3 Differentiation   7
1.4 Integrals over Spheres   12
1.5 Homogeneous Expansions   19

Chapter 2

The Automorphisms of B 23
2.1 Cartan's Uniqueness Theorem   23
2.2 The Automorphisms   25
2.3 The Cayley Transform   31
2.4 Fixed Points and Affine Sets   32

Chapter 3

Integral Representations   36
3.1 The Bergman Integral in B 36
3.2 The Cauchy Integral in B 38
3.3 The Invariant Poisson Integral in B 50

Chapter 4

The Invariant Laplacian   47
4.1 The Operator $\tilde{\Delta}$ 47
4.2 Eigenfunctions of $\tilde{\Delta}$ 49
4.3 $\mathscr{M}$-Harmonic Functions   55
4.4 Pluriharmonic Functions   59

                                                               vii

Chapter 5
Boundary Behavior of Poisson Integrals   65

5.1 A Nonisotropic Metric on S 65
5.2 The Maximal Function of a Measure on S 67
5.3 Differentiation of Measures on S 70
5.4 K-Limits of Poisson Integrals   72
5.5 Theorems of Calderón, Privalov, Plessner   79
5.6 The Spaces $N(B)$ and $H^p(B)$ 83
5.7 Appendix: Marcinkiewicz Interpolation   88

Chapter 6
Boundary Behavior of Cauchy Integrals   91

6.1 An Inequality   92
6.2 Cauchy Integrals of Measures   94
6.3 Cauchy Integrals of L^p-Functions   99
6.4 Cauchy Integrals of Lipschitz Functions   101
6.5 Toeplitz Operators   110
6.6 Gleason's Problem   114

Chapter 7
Some L^p-Topics   120

7.1 Projections of Bergman Type   120
7.2 Relations between H^p and $L^p \cap H$ 126
7.3 Zero-Varieties   133
7.4 Pluriharmonic Majorants   145
7.5 The Isometries of $H^p(B)$ 152

Chapter 8
Consequences of the Schwarz Lemma   161

8.1 The Schwarz Lemma in B 161
8.2 Fixed-Point Sets in B 165
8.3 An Extension Problem   166
8.4 The Lindelöf–Čirka Theorem   168
8.5 The Julia-Carathéodory Theorem   174

Chapter 9
Measures Related to the Ball Algebra   185

9.1 Introduction   185
9.2 Valskii's Decomposition   187
9.3 Henkin's Theorem   189
9.4 A General Lebesgue Decomposition   191
9.5 A General F. and M. Riesz Theorem   195

9.6 The Cole–Range Theorem   198
9.7 Pluriharmonic Majorants   198
9.8 The Dual Space of $A(B)$ 202

Chapter 10

Interpolation Sets for the Ball Algebra   204

10.1 Some Equivalences   204
10.2 A Theorem of Varopoulos   207
10.3 A Theorem of Bishop   209
10.4 The Davie–Øksendal Theorem   211
10.5 Smooth Interpolation Sets   214
10.6 Determining Sets   222
10.7 Peak Sets for Smooth Functions   229

Chapter 11

Boundary Behavior of H^∞-Functions   234

11.1 A Fatou Theorem in One Variable   234
11.2 Boundary Values on Curves in S 237
11.3 Weak*-Convergence   244
11.4 A Problem on Extreme Values   247

Chapter 12

Unitarily Invariant Function Spaces   253

12.1 Spherical Harmonics   253
12.2 The Spaces $H(p, q)$ 255
12.3 $\mathscr{U}$-Invariant Spaces on S 259
12.4 $\mathscr{U}$-Invariant Subalgebras of $C(S)$ 264
12.5 The Case $n = 2$ 270

Chapter 13

Moebius-Invariant Function Spaces   278

13.1 $\mathscr{M}$-Invariant Spaces on S 278
13.2 $\mathscr{M}$-Invariant Subalgebras of $C_0(B)$ 280
13.3 $\mathscr{M}$-Invariant Subspaces of $C(\bar{B})$ 283
13.4 Some Applications   285

Chapter 14

Analytic Varieties   288

14.1 The Weierstrass Preparation Theorem   288
14.2 Projections of Varieties   291
14.3 Compact Varieties in $\mathbb{C}^n$ 294
14.4 Hausdorff Measures   295

Chapter 15
Proper Holomorphic Maps   300
15.1 The Structure of Proper Maps   300
15.2 Balls vs. Polydiscs   305
15.3 Local Theorems   309
15.4 Proper Maps from B to B 314
15.5 A Characterization of B 319

Chapter 16
The $\bar{\partial}$-Problem   330
16.1 Differential Forms   330
16.2 Differential Forms in $\mathbb{C}^n$ 335
16.3 The $\bar{\partial}$-Problem with Compact Support   338
16.4 Some Computations   341
16.5 Koppelman's Cauchy Formula   346
16.6 The $\bar{\partial}$-Problem in Convex Regions   350
16.7 An Explicit Solution in B 357

Chapter 17
The Zeros of Nevanlinna Functions   364
17.1 The Henkin–Skoda Theorem   364
17.2 Plurisubharmonic Functions   366
17.3 Areas of Zero-Varieties   381

Chapter 18
Tangential Cauchy–Riemann Operators   387
18.1 Extensions from the Boundary   387
18.2 Unsolvable Differential Equations   395
18.3 Boundary Values of Pluriharmonic Functions   397

Chapter 19
Open Problems   403
19.1 The Inner Function Conjecture   403
19.2 RP-Measures   409
19.3 Miscellaneous Problems   413

Bibliography   419

Index   431

List of Symbols and Notations

The numbers that follow the symbols indicate the paragraphs in which their meanings are explained. For example, 10.4.2 means Chapter 10, Section 4, paragraph 2.

Sets

$\mathbb{C}, \mathbb{C}^n$	1.1.1	$(Z), (P)$	10.1.1
B_n, B	1.1.2	$(I), (PI)$	10.1.1
$S = \partial B_n$	1.1.2	$(N), (TN)$	10.1.1
U, T	1.1.2	$V(\zeta, \delta)$	10.4.2
$D(a; r)$	1.1.5	(D)	10.6.1
U^n, T^n	1.1.5	$E_1(f), \ldots, E_3(f)$	11.4.2
$E(a, \varepsilon)$	2.2.7	Q	12.3.1
$Q(\zeta, \delta)$	5.1.1	D_k	12.4.3
$D_\alpha(\zeta)$	5.4.1	$\Sigma(\Omega)$	12.4.3
$\Omega(E, \alpha)$	5.5.1	Δ, Δ'	14.1.1
$Z(f)$	7.3.1	D_z	15.3.1
E_c	8.5.3	Δ	16.6.1

Function Spaces

$L^p, C^k, C(X)$	1.1.1	M_z	9.1.2
$H(\Omega)$	1.1.4	$A^\perp$	9.1.4
$(L^p \cap H)(B)$	3.1.1	A^*	9.2.1
$A(B)$	3.2.3	$\mathrm{Re}\, A$	9.5.2
X_λ	4.2.1	$C_R(X)$	9.5.2
$C_0(B)$	4.2.6	HM, TS	9.8.1
$RP(\Omega)$	4.4.1	H	10.6.4
$C^\infty(\{0\})$	4.4.3	$A(\Omega)$	10.6.7
$H_\varphi(B), H^p(B)$	5.6.1	$A^m(B)$	10.7.1
$N(B)$	5.6.1	$A^\infty(B)$	10.7.1
$A(S), H^p(S)$	5.6.7	$\mathscr{P}_k, \mathscr{H}_k$	12.1.1
$L \log L$	6.3.2	$H(p, q)$	12.2.1
$H_E^\infty(B)$	6.6.2	E_Ω, X_Ω	12.3.1
$A(B, E, \{\alpha\})$	6.6.2	$\mathrm{conj}\, A(S)$	13.1.3

$(LH)^p(\Omega)$	7.4.1	plh(S)	13.1.3
l^∞, c_0	7.4.5	$P(B)$	13.3.1
$C_0(\mathbb{C})$	7.5.2	plh(B)	13.3.1
$C(X)^*$	9.1.2	conj $A(B)$	13.3.1
$M(X)$	9.1.2	$W, \tilde{W}$	19.1.6
		$N_*(B)$	19.1.11

Maximal Functions

$M\mu$	5.5.2		
$M_\alpha F$	5.4.4	$M_{\text{rad}} F$	5.4.11

Kernels and Transforms

$K(z, w)$	3.1.1	$K_s(z, w)$	7.1.1
$K[f]$	3.1.1	$T_s f$	7.1.1
$C(z, \zeta)$	3.2.1	$K_z(w)$	12.2.5
$C[f], C[\mu]$	3.2.1	$K_s(z, \zeta)$	16.5.1
$P(z, \zeta)$	3.3.1	$K_b(z, \zeta)$	16.5.2
$P[f], P[\mu]$	3.3.1	Tf	16.7.2

Derivatives

$D_j, \bar{D}_j$	1.2.2	$\mathscr{R}f$	6.4.4
D^α	1.2.2	d	16.1.3
$\partial/\partial z_j, \partial/\partial \bar{z}_j$	1.3.1	$\partial, \bar{\partial}$	16.2.2
Δ	1.3.4	Δ_{rad}	17.2.2
F'	1.3.6	Δ_{tan}	17.2.2
$\tilde{\Delta}$	4.1.1	$L_{ij}, \bar{L}_{ij}$	18.3.1
$\mathscr{D}\mu$	5.3.3		

Differential Forms

$\wedge$	16.1.1	$dz_i, d\bar{z}_i, dz_I, d\bar{z}_J$	16.2.1
dx_I	16.1.1	$\omega(z), \omega_j(z), \omega'(z)$	16.4.1
α_T	16.1.4		

Measures

ν	1.4.1	$|\mu|, \|\mu\|$	5.2.1
σ	1.4.1	$\mu \ll \sigma, \mu \perp \sigma$	5.2.1
τ	2.7.6	μ_a, μ_s	12.2.4

Other Symbols

$\langle z, w \rangle$	1.1.2	$\|f\|_p$	5.6.1
$|z|$	1.1.2	$\Delta(\zeta, \omega, \alpha, \delta)$	6.1.2
$|\alpha|, \alpha!$	1.1.6	T_φ	6.5.1
z^α	1.1.6	$\omega_\varphi(t)$	6.5.1
f_ζ	1.2.5	V_φ	6.5.4
$JF, J_R F$	1.3.6	ρf	7.2.3
$O(2n)$	1.4.1	Eg	7.2.3
$\mathscr{U}$	1.4.6	n_f, N_f	7.3.2
I	2.1.1	$\|\|\|f\|\|\|_p$	7.4.3
φ_a	2.2.1	F_x, F^y	9.4.1
$\|f\|_\infty$	3.2.3	π_{pq}	12.2.4
u_r	3.3.4	$[f, g]$	12.2.4
$\mathscr{M}$	3.3.6	$\mu(p, q; r, s)$	12.4.3
$f^\#$	4.2.1	$g_\alpha(z, w)$	12.5.1
$g_\alpha(z)$	4.2.2	$\#(w)$	15.1.3
$d(a, b)$	5.1.1	ρ	15.5.1
A_3	5.2.2	$N(w)$	15.5.1
$z \cdot w$	5.4.2	H_w, P_w, Q_w	15.5.1
$T_\zeta\ T_\zeta^C$	5.4.2	$\partial\Phi$	16.1.5
K-lim	5.4.6	$M(u)$	17.2.3
g^*	5.5.8	$A(E)$	17.3.1
f_r	5.6.1	$A(V)$	17.3.3
		$(\#_i f)(w)$	17.3.3

Chapter 1

Preliminaries

1.1. Some Terminology

1.1.1. Throughout this book, $\mathbb{C}$ will denote the complex field, and $\mathbb{C}^n$ will be the cartesian product of n copies of $\mathbb{C}$; here n is any positive integer. The points of $\mathbb{C}^n$ are thus ordered n-tuples $z = (z_1, \ldots, z_n)$, where each $z_i \in \mathbb{C}$. Algebraically, $\mathbb{C}^n$ is an n-dimensional vector space over $\mathbb{C}$. Topologically, $\mathbb{C}^n$ is the euclidean space R^{2n} of real dimension $2n$.

The usual vector space notations

$$(1) \qquad\qquad \lambda A = \{\lambda a : \lambda \in \mathbb{C}, a \in A\},$$

$$(2) \qquad\qquad A + B = \{a + b : a \in A, b \in B\}$$

will be freely used (for $A \subset \mathbb{C}^n, B \subset \mathbb{C}^n, \lambda \in \mathbb{C}$), as will the customary symbols for the Lebesgue spaces $L^p(\mu)$ (consisting of measurable complex functions f such that $|f|^p$ is integrable with respect to whatever measure μ is under consideration) and for the spaces C^k (consisting of complex functions whose kth-order partial derivatives are continuous).

The symbol

$$(3) \qquad\qquad f : X \to Y$$

means that f is a map with domain X, whose range lies in Y.

As usual, $C(X)$ is the space of all continuous functions $f : X \to \mathbb{C}$, where X is any topological space.

1.1.2. The inner product

$$(1) \qquad\qquad \langle z, w \rangle = \sum_{j=1}^{n} z_j \bar{w}_j \qquad (z, w \in \mathbb{C}^n)$$

and the associated norm

$$(2) \qquad\qquad |z| = \langle z, z \rangle^{1/2} \qquad (z \in \mathbb{C}^n)$$

1

make $\mathbb{C}^n$ into an n-dimensional Hilbert space whose *open unit ball* will be denoted by B_n, or simply by B when it seems unnecessary to mention the dimension explicitly.

Thus B consists of all $z \in \mathbb{C}^n$ with $|z| < 1$.

The *boundary* of B is the sphere S, the set of all *unit vectors* in $\mathbb{C}^n$, i.e., the set of all $z \in \mathbb{C}^n$ with $|z| = 1$.

To emphasize the special role played by the case $n = 1$, the open unit disc in $\mathbb{C}$ will be denoted by U in place of B_1, and T will be the circle that bounds U.

The *standard orthonormal basis* for $\mathbb{C}^n$ consists of the vectors $e_1, \ldots, e_n$, where e_k is the ordered n-tuple that has 1 in the kth spot and 0 everywhere else.

1.1.3. Holomorphic Functions. If $\Omega \subset \mathbb{C}^n$ is open, a function $f: \Omega \to \mathbb{C}$ is said to be *holomorphic in Ω* provided

(a) $f \in C(\Omega)$ and
(b) f is holomorphic in each variable separately.

More explicitly, (b) requires the following: If $a \in \Omega$, $1 \le i \le n$, and

$$g(\lambda) = f(a + \lambda e_i),$$

then g is to be a holomorphic function of λ is some neighborhood of 0 in $\mathbb{C}$.

(Throughout this book, the term "neighborhood of p" refers to an *open* set that contains the point p.)

This definition of "holomorphic" seems to depend on the particular basis $\{e_1, \ldots, e_n\}$; but it is actually independent of any choice of basis; this will be shown in §1.2.4.

It is a remarkable fact that (a) (the continuity of f) is really a consequence of (b). This is an old theorem of Hartogs [1; p. 12] which even now is not quite easy. There are reasonably simple proofs on pp. 28–29 of Hörmander [2] and on pp. 43–47 of Narasimhan [1]. Since we shall not need Hartog's theorem, we adopt the above (somewhat redundant) definition.

1.1.4. The class of all homomorphic functions with domain Ω will be denoted by $H(\Omega)$. It is clear that $H(\Omega)$ is an algebra over $\mathbb{C}$, with respect to pointwise addition and multiplication, and that $H(\Omega)$ is closed relative to uniform convergence on compact subsets of Ω.

1.1.5. Polydiscs. The principal topic of this book is a detailed study of $H(B)$. But the easiest approach to the most fundamental facts about holomorphic functions of several complex variables is based on *polydiscs* rather than balls.

If $a \in \mathbb{C}^n$ and $r = (r_1, \ldots, r_n)$ is an n-tuple of positive numbers, the polydisc $D(a; r)$ is defined to be the set of all $z \in \mathbb{C}^n$ whose coordinates satisfy the inequalities $|z_i - a_i| < r_i$ for $i = 1, \ldots, n$. Polydiscs in $\mathbb{C}^n$ are thus cartesian products of n discs in $\mathbb{C}$.

Taking $a = 0$ and $r_i = 1$ for $i = 1, \ldots, n$ gives the *unit polydisc* U^n, consisting of all $z \in \mathbb{C}^n$ with $|z_i| < 1$ for $1 \le i \le n$. Part of the boundary of U^n is the torus T^n, the set of all $z \in \mathbb{C}^n$ with $|z_i| = 1$ for $1 \le i \le n$.

T^n is a compact abelian group (with componentwise multiplication as group operation) and as such carries a Haar measure λ_n. This is nothing but ordinary Lebesgue measure, divided by $(2\pi)^n$ in order to have $\lambda_n(T^n) = 1$.

1.1.6. Multi-indices. In any discussion of functions of n variables, the term *multi-index* refers to an ordered n-tuple

$$\alpha = (\alpha_1, \ldots, \alpha_n) \tag{1}$$

of nonnegative integers α_i. The following abbreviated notations will be used:

$$\alpha_1 + \cdots + \alpha_n = |\alpha|, \tag{2}$$

$$\alpha_1! \ldots \alpha_n! = \alpha!, \tag{3}$$

$$z_1^{\alpha_1} \ldots z_n^{\alpha_n} = z^{\alpha}. \tag{4}$$

In (4), $z = (z_1, \ldots, z_n)$. We shall refer to z^{α} as a *holomorphic monomial*.

1.2. The Cauchy Formula in Polydiscs

1.2.1. To begin with, suppose Ω is open in $\mathbb{C}^n$, $f \in H(\Omega)$, and $\Omega \supset \overline{U}^n$, the closure of the unit polydisc U^n. Write $z \in \mathbb{C}^n$ in the form $z = (z', z_n)$, where $z' \in \mathbb{C}^{n-1}$. Cauchy's familiar formula for holomorphic functions of one complex variable can be applied to $f(z', \cdot)$. In our present notation this formula is

$$f(z', z_n) = \int_T f(z', w_n)(1 - \overline{w}_n z_n)^{-1} \, d\lambda_1(w_n)$$

for $(z', z_n) \in U^n$. Similarly, $f(z', w_n)$ is equal to

$$\int_T f(z_1, \ldots, z_{n-2}, w_{n-1}, w_n)(1 - \overline{w}_{n-1} z_{n-1})^{-1} \, d\lambda_1(w_{n-1})$$

for $z' \in U^{n-1}$, $w_n \in T$. Insert this into the preceding formula and repeat the process. The final result is the *Cauchy formula*

$$f(z) = \int_{T^n} f(w) \prod_{j=1}^{n} (1 - \overline{w}_j z_j)^{-1} \, d\lambda_n(w), \tag{1}$$

valid for $z \in U^n$.

[Note that Fubini's theorem was applied in the proof of (1). The assumed continuity of f makes this application legitimate. Continuity could of course be replaced by local boundedness. In fact, the main point in Hartog's theorem is the proof that every function that is holomorphic in each variable separately is bounded on compact sets.]

The product that appears in the integrand is the *Cauchy kernel* for U^n. For $z \in U^n$ and $w \in T^n$, this kernel can be expanded in the series

$$
(2) \qquad \prod_{j=1}^{n} (1 - \bar{w}_j z_j)^{-1} = \sum_{\alpha} \bar{w}^{\alpha} z^{\alpha}
$$

in which α ranges over the multi-indices. The series is dominated by $\sum |z_1|^{\alpha_1} \dots |z_n|^{\alpha_n}$, which converges uniformly on every compact subset of U^n. Substitution of (2) into (1) shows therefore that f is the sum of the *multiple power series*

$$
(3) \qquad f(z) = \sum_{\alpha} c(\alpha) z^{\alpha} \qquad (z \in U^n)
$$

whose coefficients are given by

$$
(4) \qquad c(\alpha) = \int_{T^n} f(w) \bar{w}^{\alpha} \, d\lambda_n(w),
$$

and which converges absolutely and uniformly on every compact subset of U^n (no matter how the terms are ordered).

1.2.2. Suppose again that $f \in H(\Omega)$, and suppose that Ω contains the closure of some polydisc $D(p; r)$. Repetition of the preceding argument, or application of an obvious change of variables to 1.2.1(1), shows that there is a Cauchy formula for f in $D(p; r)$, and that there is a power series

$$
(1) \qquad \sum_{\alpha} \gamma(\alpha)(z - p)^{\alpha}
$$

which converges absolutely and uniformly to $f(z)$ on every compact subset of $D(p; r)$.

Thus every holomorphic function is locally the sum of a convergent power series.

From this, and from the existence of the local Cauchy formula (in $D(p; r)$), one can derive a number of important properties of holomorphic functions, by proofs that are exactly the same as in the familiar case of one variable. Here are some of these:

(i)   Every $f \in H(\Omega)$ has mixed partial derivatives (with respect to $z_1, \dots, z_n$) of all orders, and these derivatives are themselves holomorphic.

In fact, if $f(z)$ is represented by (1), then

(2) $$(D^\alpha f)(p) = \alpha! \gamma(\alpha)$$

where $D^\alpha = D_1^{\alpha_1} \ldots D_n^{\alpha_n}$, $D_j = \partial/\partial z_j$.

(ii)   By (2), f determines the coefficients $\gamma(\alpha)$; the expansion (1) is thus unique.

(iii) If $f(z) = 0$ for all z in some nonempty open set in Ω, and Ω is connected, then $f \equiv 0$.

(iv)   The maximum modulus theorem holds: If $f \in H(\Omega)$, Ω is connected, and $|f|$ has a local maximum in Ω, then f is constant.

(v)   If K is a compact subset of Ω, there are constants $A(\Omega, K, \alpha) < \infty$ such that

$$\|D^\alpha f\|_K < A(\Omega, K, \alpha)\|f\|_\Omega$$

for every $f \in H(\Omega)$. (Here $\|f\|_E$ denotes the supremum of $|f(z)|$ for $z \in E$.)

Consequently, if $f_j \in H(\Omega)$ and $f_j \to f$ uniformly, then $D^\alpha f_j \to D^\alpha f$, uniformly on compact subsets of Ω.

(vi)   If Γ is a uniformly bounded family of holomorphic functions in Ω, then Γ is equicontinuous on every compact subset of Ω. In other words, Γ is a normal family.

1.2.3. Compositions of Holomorphic Maps. Suppose

(a) Ω_1 and Ω_2 are open sets in $\mathbb{C}^k$ and $\mathbb{C}^m$, respectively,

(b) $f_1, \ldots, f_m \in H(\Omega_1)$,

(c) $F = (f_1, \ldots, f_m)$ maps Ω_1 into Ω_2 (such an F is called a *holomorphic map*),

(d) $G: \Omega_2 \to \mathbb{C}^n$ is a holomorphic map,

(e) $H = G \circ F$ is their composition.

Then $H: \Omega_1 \to \mathbb{C}^n$ is holomorphic.

To prove this, it is sufficient to take the case $n = 1$. Also, we may assume that $0 \in \Omega_1$ and that $F(0) = 0$. Then $G(w) = \sum c(\alpha)w^\alpha$ in some polydisc D_2 centered at 0. There is a polydisc D_1 in Ω_1, centered at 0, such that $F(D_1) \subset D_2$. For $z \in D_1$,

$$H(z) = G(F(z)) = \sum_\alpha c(\alpha) f_1(z)^{\alpha_1} \ldots f_m(z)^{\alpha_m}.$$

This series converges absolutely and uniformly on every compact subset of D_1. Thus H is holomorphic in D_1, a neighborhood of 0. The same argument applies at every other point of Ω_1.

1.2.4. Here is an application of the preceding result: Let Ω be open in $\mathbb{C}^n$, pick $a \in \Omega$, $b \in \mathbb{C}^n$. There is then a neighborhood V of 0 in $\mathbb{C}$ such that $a + \lambda b \in \Omega$ for all $\lambda \in V$. Clearly, $\lambda \to a + \lambda b$ is a holomorphic map. If $f \in H(\Omega)$, *the result proved in §1.2.3 shows that*

$$\lambda \to f(a + \lambda b)$$

is holomorphic in V.

Note that the definition of "holomorphic" given in §1.1.3 postulated this behavior only for the vectors $e_1, \ldots, e_n$ in place of b.

1.2.5. As a further specialization, let $f \in H(B)$, and associate with each $\zeta \in S$ (the sphere that bounds B) the function f_ζ defined for $\lambda \in U$ by

$$f_\zeta(\lambda) = f(\lambda\zeta).$$

Then f_ζ is holomorphic in U.

We call the functions f_ζ (one for each $\zeta \in S$) the *slice functions* of f.

If L_ζ is the 1-dimensional subspace of $\mathbb{C}^n$ generated by ζ (the "complex line" through 0 and ζ) then f_ζ may be thought of as the restriction of f to the disc $B \cap L_\zeta$.

Slice functions will be a useful tool that will allow us to apply facts from the function theory of U to questions in B.

1.2.6. In certain situations, the local power series representation of a holomorphic function is actually global:

Proposition. *Suppose Ω is open in $\mathbb{C}^n$, Ω is a union of polydiscs centered at 0, and $f \in H(\Omega)$. Then there is a power series $\sum c(\alpha)z^\alpha$ that converges to $f(z)$ at every point $z \in \Omega$.*

Proof. Let D_1 and D_2 be polydiscs, centered at 0, that lie in Ω. The argument given in §1.2.2 shows that there are power series Σ_1 and Σ_2 that converge to f in D_1 and in D_2, respectively. Since they have the same sum in $D_1 \cap D_2$, they must be identical, by (ii) of §1.2.2. The proposition follows from this.

Remarks. (i) The requirement that Ω be a union of polydiscs centered at 0 can be stated differently: *If $z \in \Omega$, and if the coordinates of a point $w \in \mathbb{C}^n$ satisfy $|w_i| \le |z_i|$ for $1 \le i \le n$, then $w \in \Omega$.* This obviously holds, for example, when $\Omega = B$, the unit ball. *We conclude that every $f \in H(B)$ has a global power series representation.*

(ii) If Ω is as in the Proposition, it may happen that there is a larger open set $\tilde{\Omega}$ such that every power series that converges in Ω actually converges in $\tilde{\Omega}$.

To see this, note first that convergence of $\sum c(\alpha)z^\alpha$ for *all* $z \in \Omega$ implies absolute convergence. For if $z \in \Omega$ then $(1 + \varepsilon)z \in \Omega$ for some $\varepsilon > 0$, so that

$$\sup_\alpha (1 + \varepsilon)^{|\alpha|} |c(\alpha)z^\alpha| = \sup_\alpha |c(\alpha)((1 + \varepsilon)z)^\alpha| = M < \infty.$$

Thus

$$\sum_\alpha |c(\alpha)z^\alpha| \le M \sum_\alpha (1 + \varepsilon)^{-|\alpha|} = M(1 + \varepsilon)^n \varepsilon^{-n}.$$

Next, if $z \in \Omega$, $w \in \Omega$, $0 < t < 1$, and the coordinates p_i of $p \in \mathbb{C}^n$ satisfy $|p_i| \le |z_i|^t |w_i|^{1-t}$, then

$$|p^\alpha| \le |z^\alpha|^t |w^\alpha|^{1-t} \le t|z^\alpha| + (1 - t)|w^\alpha|,$$

so that $\sum |c(\alpha)p^\alpha| < \infty$.

The point is that this can happen even if p is not in Ω.

For a simple example, take $n = 2$, let $D_{a,b}$ be the set of all $z = (z_1, z_2)$ with $|z_1| < a$, $|z_2| < b$, and put $\Omega = D_{1,2} \cup D_{2,1}$. If $\tilde{\Omega}$ is the set defined by

$$|z_1| < 2, \qquad |z_2| < 2, \qquad |z_1 z_2| < 2$$

then the preceding reasoning shows that every power series $\sum c(\alpha)z^\alpha$ that converges in Ω actually converges in all of $\tilde{\Omega}$. Hence, by the Proposition, every $f \in H(\Omega)$ *extends to a holomorphic function in the larger region* $\tilde{\Omega}$.

We have arrived at a point where it would seem natural to introduce the concept of "domain of holomorphy" (the above Ω is *not* one) and to characterize these domains. However, this lies outside the program of the present book. We refer instead to Chap. II of Hörmander [2].

1.3. Differentiation

1.3.1. Writing the coordinates z_j of a point $z \in \mathbb{C}^n$ in the form $z_j = x_j + iy_j$, with x_j and y_j real, the differential operators

$$\frac{\partial}{\partial z_j} = \frac{1}{2}\left(\frac{\partial}{\partial x_j} - i\frac{\partial}{\partial y_j}\right) \quad \text{and} \quad \frac{\partial}{\partial \bar{z}_j} = \frac{1}{2}\left(\frac{\partial}{\partial x_j} + i\frac{\partial}{\partial y_j}\right)$$

$(1 \le j \le n)$ offer certain advantages over just $\partial/\partial x_j$ and $\partial/\partial y_j$ in problems related to holomorphic functions. We shall often use the simpler notation

$$D_j = \frac{\partial}{\partial z_j}, \qquad \bar{D}_j = \frac{\partial}{\partial \bar{z}_j}.$$

Let us note explicitly that D_j and $\bar{D}_j$ can be applied to arbitrary differentiable (not necessarily holomorphic) functions of the $2n$ real variables $x_1, \ldots, x_n$, $y_1, \ldots, y_n$.

1.3.2. The Cauchy–Riemann Equations. Let Ω be open in $\mathbb{C}^n$. Then $C^1(\Omega)$ is the class of all complex functions with domain Ω whose derivatives $\partial f/\partial x_j$ and $\partial f/\partial y_j$ are continuous in Ω. Let $f \in C^1(\Omega)$.

   Then $f \in H(\Omega)$ if and only if the Cauchy–Riemann equations

$$\bar{D}_j f = 0 \qquad (1 \le j \le n)$$

are satisfied.

   This follows immediately from Definition 1.1.3 and the corresponding characterization of holomorphic functions of one variable.

1.3.3. The Chain Rule. Suppose Ω is open in $\mathbb{C}^k$, $F = (f_1, \ldots, f_n)$ maps Ω into $\mathbb{C}^n$, g maps the range of F into $\mathbb{C}$, and $f_1, \ldots, f_n, g$ are of class C^1. If

$$h = g \circ F = g(f_1, \ldots, f_n)$$

then, for $1 \le j \le k$ and $z \in \Omega$,

$$(D_j h)(z) = \sum_{i=1}^{n} \{(D_i g)(w)(D_j f_i)(z) + (\bar{D}_i g)(w)(D_j \bar{f}_i)(z)\},$$

$$(\bar{D}_j h)(z) = \sum_{i=1}^{n} \{(D_i g)(w)(\bar{D}_j f_i)(z) + (\bar{D}_i g)(w)(\bar{D}_j \bar{f}_i)(z)\},$$

where $w = F(z)$.

   These formulas are derived from the usual real variable form of the chain rule by straightforward manipulations; it is of course enough to verify them for real-linear functions.

   When F is a holomorphic map, then $\bar{D}_j f_i = 0$, hence $D_j \bar{f}_i = 0$, and the formulas simplify to

$$D_j h = \sum_{i=1}^{n} D_i g \cdot D_j f_i$$

$$\bar{D}_j h = \sum_{i=1}^{n} \bar{D}_i g \cdot \bar{D}_j \bar{f}_i.$$

1.3.4. The Laplacian. If Ω is open in $\mathbb{C}^n$ and $f \in C^2(\Omega)$, its *Laplacian* is defined to be

(1)
$$\Delta f = \sum_{j=1}^{n} \left(\frac{\partial^2 f}{\partial x_j^2} + \frac{\partial^2 f}{\partial y_j^2} \right).$$

In terms of D_j and $\bar{D}_j$, this can be rewritten in the form

$$(2) \qquad \Delta f = 4 \sum_{j=1}^{n} D_j \bar{D}_j f.$$

To give an application of the chain rule, let f be of class C^2 in a neighborhood of a point $a \in \mathbb{C}^n$, pick $b \in \mathbb{C}^n$, and define

$$(3) \qquad g(\lambda) = g_{a,b}(\lambda) = f(a + \lambda b)$$

for λ in a sufficiently small neighborhood of the origin in $\mathbb{C}$. We wish to express $(\Delta g)(0)$ in terms of f.

Referring to §1.3.3, we now have $k = 1$, $F(\lambda) = a + \lambda b$, so F is holomorphic, and two applications of the chain rule to $g = f \circ F$ yield

$$(4) \qquad (\Delta g)(0) = 4 \sum_{j,k=1}^{n} (D_j \bar{D}_k f)(a) b_j \bar{b}_k.$$

Let $H_f(a)$ be the $n \times n$ matrix with $(D_j \bar{D}_k f)(a)$ in the kth row and jth column. (This is the so-called *complex Hessian* of f at a.) Then (4) takes the form

$$(5) \qquad (\Delta g_{a,b})(0) = 4\langle H_f(a)b, b\rangle.$$

This leads to the conclusion that *every $g_{a,b}$ is harmonic if and only if $H_f(a) = 0$ for all $a \in \Omega$*, i.e., if and only if

$$(6) \qquad D_j \bar{D}_k f = 0 \qquad (j, k = 1, \ldots, n).$$

The functions that satisfy (6) are called *pluriharmonic*.

1.3.5. A Lemma about Determinants. Let $A = (a_{jk})$ be an $n \times n$ matrix with complex entries

$$a_{jk} = b_{jk} + i c_{jk} \qquad (j, k = 1, \ldots, n)$$

where b_{jk}, c_{jk} are real. Put $B = (b_{jk})$, $C = (c_{jk})$, and

$$\tilde{A} = \begin{pmatrix} B & -C \\ C & B \end{pmatrix},$$

so that $\tilde{A}$ is a real $(2n) \times (2n)$ matrix.

Lemma. $\det \tilde{A} = |\det A|^2$.

This will be useful when dealing with Jacobians of holomorphic maps.

Proof. Both det $\tilde{A}$ and $|\det A|^2$ are polynomials in the $N = 2n^2$ real variables b_{jk}, c_{jk}. These polynomials will be identical if they coincide on some nonempty open set $V \subset R^N$.

Let V be a small neighborhood of the point of R^N that has $b_{jk} = 0$ for all $j, k, c_{jk} = 0$ when $j \neq k, c_{kk} = k$ for $1 \leq k \leq n$. The corresponding matrix A is diagonal and has eigenvalues $i, 2i, \ldots, ni$. If V is small enough, every A that corresponds to a point of V has n *distinct* eigenvalues, all with *positive imaginary part*.

Consider such an A. Let λ be one of its eigenvalues, with corresponding eigenvector z. Since $A = B + iC$, it follows that

$$\tilde{A}\begin{pmatrix} z \\ -iz \end{pmatrix} = \begin{pmatrix} B & -C \\ C & B \end{pmatrix}\begin{pmatrix} z \\ -iz \end{pmatrix} = \begin{pmatrix} Az \\ -iAz \end{pmatrix} = \lambda\begin{pmatrix} z \\ -iz \end{pmatrix}.$$

Thus λ is an eigenvalue of $\tilde{A}$, and so is $\bar{\lambda}$, since $\tilde{A}$ is a real matrix.

If now $\lambda_1, \ldots, \lambda_n$ are the (distinct) eigenvalues of A, then $\lambda_1, \ldots, \lambda_n$, $\bar{\lambda}_1, \ldots, \bar{\lambda}_n$ are $2n$ distinct numbers that are eigenvalues of $\tilde{A}$. Hence there are no others, and since the determinant of any square matrix over $\mathbb{C}$ is the product of its eigenvalues, the lemma is proved.

1.3.6. The Derivative of a Holomorphic Map. Let m, n be positive integers, let $\Omega \subset \mathbb{C}^n$ be open, and suppose that $F: \Omega \to \mathbb{C}^m$ is holomorphic. With each point $z \in \Omega$ is then associated a unique linear operator $F'(z): \mathbb{C}^n \to \mathbb{C}^m$ (called the *derivative* of F at z) that satisfies

$$(1) \qquad\qquad F(z + h) = F(z) + F'(z)h + O(|h|^2)$$

for h near the origin of $\mathbb{C}^n$.

In fact, if the components of $F(z + h)$ are expanded in power series about the point $h = 0$, then $F'(z)h$ is just made up from the linear terms in these series.

For $1 \leq k \leq n$, let $h = \lambda e_k$ in (1), and let $\lambda \to 0$. It follows that

$$(2) \qquad\qquad F'(z)e_k = (D_k F)(z),$$

a vector in $\mathbb{C}^m$ with components $(D_k f_j)(z), 1 \leq j \leq m$, where $F = (f_1, \ldots, f_m)$. The linear operator $F'(z)$ is thus represented by a matrix $A = (a_{jk})$, with

$$(3) \qquad\qquad a_{jk} = (D_k f_j)(z) \qquad (1 \leq j \leq m, 1 \leq k \leq n).$$

Here are some special cases that illustrate these concepts and that will occur later.

(i) If $n = 1$ then F has domain $\Omega \subset \mathbb{C}$, and $F'(z)$ is the linear map that takes $h \in \mathbb{C}$ to the vector

(4) $$F'(z)h = (f'_1(z)h, \ldots, f'_m(z)h)$$

in $\mathbb{C}^m$.

(ii) If $m = 1$, then F maps Ω into $\mathbb{C}$, and $F'(z)$ is the linear functional that takes $h = (h_1, \ldots, h_n) \in \mathbb{C}^n$ to the complex number

(5) $$F'(z)h = \sum_{k=1}^{n} (D_k F)(z)h_k.$$

(iii) If $m = n$, then the matrix A defined by (3) is square. The (complex) *Jacobian* of F at z can thus be defined by

(6) $$(JF)(z) = \det A.$$

The point of calling this the *complex* Jacobian is that $F = (f_1, \ldots, f_n)$, with $f_j = u_j + iv_j$, can also be regarded as a map that takes the real variables $x_1, \ldots, x_n, y_1, \ldots, y_n$ to the real variables $u_1, \ldots, u_n$, $v_1, \ldots, v_n$, and as such it has a real Jacobian which we shall denote by $(J_R F)(z)$. By the Cauchy–Riemann equations,

(7) $$D_k f_j = \frac{\partial u_j}{\partial x_k} - i \frac{\partial u_j}{\partial y_k} = \frac{\partial v_j}{\partial y_k} + i \frac{\partial v_j}{\partial x_k}$$

and therefore $J_R F = \det \tilde{A}$, where $\tilde{A}$ is related to A as in Lemma 1.3.5.

We conclude from all this that

(8) $$J_R F = |JF|^2.$$

One consequence of (8) is that volumes of images under F can be computed directly from the complex Jacobian JF.

For another consequence, assume $F'(z)$ is invertible. Then $JF \neq 0$, hence $J_R F > 0$. *This says that F preserves orientation* and implies the fact (which we merely mention in passing) that complex manifolds are necessarily orientable.

1.3.7. The Inverse Function Theorem. *Let Ω be open in $\mathbb{C}^n$, let $F: \Omega \to \mathbb{C}^n$ be holomorphic, and suppose that $F'(p)$ is invertible at some point $p \in \Omega$.*

Then there are neighborhoods V of p and W of $F(p)$ such that F is a one-to-one map of V onto W whose inverse is holomorphic in W.

(Holomorphic maps with holomorphic inverses are usually called *bi-holomorphic*.)

Proof. Since $F'(p)$ is invertible as a linear operator on $\mathbb{C}^n$, it is also invertible when regarded as a real-linear operator on R^{2n}. The familiar real-variable version of the inverse function theorem shows therefore that F is a one-to-one map of some neighborhood V of p onto a neighborhood W of $F(p)$, whose inverse G is of class C^1; furthermore, $F'(z)$ is invertible for all $z \in V$. It remains to be shown that $G = (g_1, \ldots, g_n)$ is a holomorphic map.

For $z \in V$, $G(F(z)) = z$, hence, for $1 \le i \le n$,

$$g_i(F(z)) = z_i,$$

and if we apply $\bar{D}_k$ to this equation, we obtain

$$\sum_{j=1}^{n} (\bar{D}_j g_i)(F(z))(\bar{D}_k \bar{f}_j)(z) = 0,$$

by the chain rule. Since $\bar{D}_k \bar{f}_j = \overline{D_k f_j}$, the matrix $(\bar{D}_k \bar{f}_j)$ is invertible, and we conclude that $\bar{D}_j g_i = 0$ in W. Hence $g_i \in H(W)$.

Note: This theorem has a converse: *If $\Omega \subset \mathbb{C}^n$ is open and $F:\Omega \to \mathbb{C}^n$ is holomorphic and one-to-one, then $F'(p)$ is invertible for all $p \in \Omega$.*

This will be proved later, using information about proper holomorphic maps, in Theorem 15.1.8.

1.4. Integrals over Spheres

1.4.1. The Measures v and σ. Let n be a fixed positive integer. We let v be Lebesgue measure on $\mathbb{C}^n = R^{2n}$, so normalized that

$$v(B) = 1.$$

If m_{2n} is ordinary Lebesgue measure on R^{2n} (normalized so that the unit cube has measure 1) then $c_n v = m_{2n}$ for some constant c_n. The exact value of c_n is not very important for our purposes; it will fall out of a computation in §1.4.9.

We let σ be the *rotation-invariant* positive Borel measure on S for which

$$\sigma(S) = 1.$$

(Positive measures of total mass 1 are often called *probability measures.*) The term "rotation-invariant" refers to the orthogonal group $O(2n)$, the group of all isometries of R^{2n} that fix the origin; these isometries are called *rotations*, and the invariance property of σ is: $\sigma(\rho E) = \sigma(E)$ for every Borel set $E \subset S$ and for every rotation ρ.

Of course, v is also rotation-invariant.

As is the case with B and S, we usually suppress the dimension n in the notations v and σ. This should cause no confusion.

A remark made at the end of §1.4.7 establishes the uniqueness of σ.

1.4.2. Lemma. *If G is a compact subgroup of $O(2n)$ then*

(1)
$$\int_S f \, d\sigma = \int_S d\sigma(\zeta) \int_G f(g\zeta) dg$$

and

(2)
$$\int_{\mathbb{C}^n} f \, dv = \int_{\mathbb{C}^n} dv(z) \int_G f(gz) dg.$$

Here dg denotes the Haar measure on G. This means, we recall, that dg is a positive Borel measure on G, of total mass 1, which is *G-invariant* in the sense that

$$\int_G F(g_1 g) dg = \int_G F(g) dg = \int_G F(gg_1) dg$$

for all $F \in C(G)$, $g_1 \in G$. The simplest proof of its existence and uniqueness is probably due to von Neumann. (See Rudin [2], pp. 123, 377.) The point of the lemma is that the integrals on the left are not changed if the integrand is first "averaged" over G.

We have stated no assumptions on f. It is of course understood that the integrals have to make sense. We shall prove (1) for continuous f, (2) for continuous f with compact support. It is clear that these identities are then true for (say) all nonnegative Borel functions, and for arbitrary Lebesgue integrable functions. The same convention will be made with regard to other integral identities that occur in this section.

Proof. When $f \in C(S)$, then the function that takes (g, ζ) to $f(g\zeta)$ is continuous on $G \times S$. Fubini's theorem shows that the right side of (1) is

(3)
$$\int_G dg \int_S f(g\zeta) d\sigma(\zeta).$$

Since σ is G-invariant, the inner integral in (3) is independent of g. Hence (3) equals $\int_S f \, d\sigma$. This proves (1). The same argument proves (2).

1.4.3. Integration in Polar Coordinates. *The measures v and σ are related by the formula*

$$\int_{\mathbb{C}^n} f \, dv = 2n \int_0^\infty r^{2n-1} \, dr \int_S f(r\zeta) d\sigma(\zeta).$$

Proof. Apply Lemma 1.4.2 to the integrals over $\mathbb{C}^n$ and S, with the full group $O(2n)$ in place of G. It follows that it is enough to prove the formula for *radial* functions f, i.e., for f such that $f(z)$ depends only on $|z|$. Hence it is enough to prove it for radial step functions, i.e., for finite linear combinations of characteristic functions of balls tB, $0 < t < \infty$ (where tB is the set of all $z \in \mathbb{C}^n$ with $|z| < t$). For such a characteristic function f, each side of the formula equals t^{2n}. This completes the proof.

1.4.4. Functions of Fewer Variables. Suppose $1 \leq k < n$, and f is a function on $S = \partial B_n$ that depends only on $z_1, \ldots, z_k$. Then f may be regarded as defined on $\bar{B}_k$, and $\int_S f \, d\sigma$ can be expressed as an integral over B_k, as follows:

Let P be the orthogonal projection of $\mathbb{C}^n$ onto $\mathbb{C}^k$. *Then*

$$(1) \qquad \int_{\partial B_n} (f \circ P) d\sigma = \binom{n-1}{k} \int_{B_k} (1 - |w|^2)^{n-k-1} f(w) dv_k(w)$$

for every $f \in L^1(v_k)$ *on* B_k. (Forelli [4], p. 381.)

It is enough to prove (1) for continuous functions f whose support lies in $r_0 B_k$ for some $r_0 < 1$. Fix such an f, and consider the integral

$$(2) \qquad I(r) = \int_{rB_n} (f \circ P) dv_n \qquad (0 < r < \infty).$$

In polar coordinates,

$$(3) \qquad I(r) = 2n \int_0^r t^{2n-1} \, dt \int_S (f \circ P)(t\zeta) d\sigma(\zeta),$$

and if we differentiate this we obtain

$$(4) \qquad I'(1) = 2n \int_S (f \circ P) d\sigma.$$

On the other hand, there is a constant $c(n, k)$ (depending on the normalizations of v_n and v_k) such that an application of Fubini's theorem converts (2) into

$$(5) \qquad I(r) = c(n, k) \int_{B_k} (r^2 - |w|^2)^{n-k} f(w) dv_k(w)$$

if $r > r_0$. (Note that we can then integrate over B_k in place of rB_k, since f vanishes outside $r_0 B_k$.) Differentiation of (5) gives

$$(6) \qquad I'(1) = 2c(n, k)(n - k) \int_{B_k} (1 - |w|^2)^{n-k-1} f(w) dv_k(w).$$

Comparison of (4) and (6) gives (1), except for the multiplicative constant. To determine the latter, take $f = 1$, and compute the right side of (1) in polar coordinates.

1.4.5. Two Special Cases. When $k = n - 1$, the binomial coefficient in 1.4.4(1) is 1, and the formula simplifies to

$$(1) \qquad \int_{\partial B_n} (f \circ P) d\sigma = \int_{B_{n-1}} f \, dv_{n-1}.$$

At the other extreme, let $k = 1$, let f be a function of one complex variable. Then, for any $\eta \in S = \partial B_n$, we have

$$(2) \qquad \int_S f(\langle \zeta, \eta \rangle) d\sigma(\zeta) = \frac{n-1}{\pi} \iint_U (1 - r^2)^{n-2} f(re^{i\theta}) r \, dr \, d\theta.$$

The point is that we can choose our coordinate system in $\mathbb{C}^n$ so that $\eta = e_1$. Then the left side of (2) is $\int_S f(\zeta_1) d\sigma(\zeta)$, and 1.4.4(1) applies; recall that $B_1 = U$, the unit disc in $\mathbb{C}$.

1.4.6. The Unitary Group. We let $\mathcal{U} = \mathcal{U}(n)$ be the group of all unitary operators on the Hilbert space $\mathbb{C}^n$. These are the linear operators U that preserve inner products:

$$\langle Uz, Uw \rangle = \langle z, w \rangle$$

for $z, w \in \mathbb{C}^n$, $U \in \mathcal{U}$. Clearly, $\mathcal{U}$ is a compact subgroup of $O(2n)$.

1.4.7. Proposition. *The identities*

$$(1) \qquad \int_S f \, d\sigma = \int_S d\sigma(\zeta) \frac{1}{2\pi} \int_{-\pi}^{\pi} f(e^{i\theta}\zeta) d\theta$$

$$(2) \qquad \int_S f \, d\sigma = \int_{B_{n-1}} dv(\zeta') \frac{1}{2\pi} \int_{-\pi}^{\pi} f(\zeta', e^{i\theta}\zeta_n) d\theta$$

$$(3) \qquad \int_S f \, d\sigma = \int_{\mathcal{U}} f(U\eta) dU$$

hold.

(1) may be called integration by slices; see §1.2.5. It was apparently first used by Bochner [3].

In (2), $\zeta' = (\zeta_1, \ldots, \zeta_{n-1})$. Of course, (2) only makes sense when $n > 1$.

In (3), η is an arbitrary point of S.

Proof. (1) is a special case of Lemma 1.4.2, G being the circle group acting on S by sending ζ to $e^{i\theta}\zeta$.

Another application of Lemma 1.4.2, with G the circle group acting on S by sending (ζ', ζ_n) to $(\zeta', e^{i\theta}\zeta_n)$, shows that

(4)
$$\int_S f\, d\sigma = \int_S d\sigma(\zeta) \frac{1}{2\pi} \int_{-\pi}^{\pi} f(\zeta', e^{i\theta}\zeta_n)d\theta.$$

The inner integral on the right of (4) is independent of ζ_n (once ζ' is picked), since $|\zeta_n|^2 = 1 - |\zeta'|^2$. Hence (2) follows from (4), by 1.4.5(1).

A third application of Lemma 1.4.2, with $G = \mathscr{U}$, gives

(5)
$$\int_S f\, d\sigma = \int_S d\sigma(\zeta) \int_{\mathscr{U}} f(U\zeta)dU.$$

Note that $\mathscr{U}$ acts *transitively* on S: If $\zeta \in S$ and $\eta \in S$, then $\zeta = U_1\eta$ for some $U_1 \in \mathscr{U}$. The inner integral in (5) is thus equal to

(6)
$$\int_{\mathscr{U}} f(UU_1\eta)dU = \int_{\mathscr{U}} f(U\eta)dU,$$

by the invariance of the Haar measure dU. The last integral is independent of ζ. Hence (5) and (6) give (3).

Remark. The proof of (3) did not use the full rotation-invariance of σ, but only the fact that σ is $\mathscr{U}$-invariant. Hence (3) implies that σ is the *only* $\mathscr{U}$-invariant Borel probability measure on S.

We now show that the holomorphic monomials z^α are orthogonal to each other in $L^2(\sigma)$:

1.4.8. Proposition. *If α and β are multi-indices and $\alpha \neq \beta$, then*

$$\int_S z^\alpha \bar{z}^\beta\, d\sigma(z) = 0.$$

Proof. Without loss of generality, assume $\alpha_n \neq \beta_n$. Put $f(z) = z^\alpha \bar{z}^\beta$. The inner integral in 1.4.7(2) is then 0.

1.4.9. Proposition. *For every multi-index α,*

(1)
$$\int_S |\zeta^\alpha|^2\, d\sigma(\zeta) = \frac{(n-1)!\alpha!}{(n-1+|\alpha|)!}$$

and

(2)
$$\int_B |z^\alpha|^2 \, dv(z) = \frac{n!\alpha!}{(n + |\alpha|)!}.$$

Proof. To prove (1), Bungart [1], Folland [1], and Fefferman [1] use the integral

$$I = \int_{\mathbb{C}^n} |z^\alpha|^2 \exp(-|z|^2) dm_{2n}(z).$$

The integrand is $\prod |z_j|^{2\alpha_j} \exp(-|z_j|^2)$. Hence Fubini's theorem shows that

$$I = \prod_{j=1}^n \int_{\mathbb{C}} |\lambda|^{2\alpha_j} \exp(-|\lambda|^2) dm_2(\lambda).$$

These integrals over $\mathbb{C}$ are easily computed, and one finds that $I = \pi^n \alpha!$
    If we now apply integration in polar coordinates (§1.4.3) to I, we see that

$$\pi^n \alpha! = c_n \cdot 2n \int_0^\infty r^{2|\alpha| + 2n - 1} e^{-r^2} \, dr \int_S |\zeta^\alpha|^2 \, d\sigma(\zeta)$$

where c_n is the constant that occurs in §1.4.1. Hence

$$\int_S |\zeta^\alpha|^2 \, d\sigma(\zeta) = \frac{\pi^n \alpha!}{(n - 1 + |\alpha|)! n c_n}.$$

Taking $\alpha = 0$ gives $c_n = \pi^n/n!$.
    This proves (1). Another integration in polar coordinates leads from (1) to (2).
    We shall now use Propositions 1.4.8 and 1.4.9 to obtain the asymptotic behavior of certain integrals that will occur later.

1.4.10. Proposition. *For $z \in B$, c real, $t > -1$, define*

$$I_c(z) = \int_S \frac{d\sigma(\zeta)}{|1 - \langle z, \zeta \rangle|^{n+c}}$$

and

$$J_{c,t}(z) = \int_B \frac{(1 - |w|^2)^t \, dv(w)}{|1 - \langle z, w \rangle|^{n+1+t+c}}.$$

When $c < 0$, then I_c and $J_{c,t}$ are bounded in B.

When $c > 0$, then

$$I_c(z) \approx (1 - |z|^2)^{-c} \approx J_{c,t}(z).$$

Finally,

$$I_0(z) \approx \log \frac{1}{1 - |z|^2} \approx J_{0,t}(z).$$

The notation $a(z) \approx b(z)$ means that the ratio $a(z)/b(z)$ has a positive finite limit as $|z| \to 1$.

Proof. The binomial series

$$(1 - x)^{-\lambda} = \sum_{k=0}^{\infty} \frac{\Gamma(k + \lambda)}{k!\Gamma(\lambda)} x^k$$

holds for $|x| < 1$ when λ is not a negative integer. Choose λ so that $2\lambda = n + c$. Then

$$I_c(z) = \int_S \left| \sum_{k=0}^{\infty} \frac{\Gamma(k + \lambda)}{k!\Gamma(\lambda)} \langle z, \zeta \rangle^k \right|^2 d\sigma(\zeta).$$

When $k \neq m$, Proposition 1.4.8 implies that $\langle z, \zeta \rangle^k$ and $\langle z, \zeta \rangle^m$ are orthogonal in $L^2(\sigma)$. Thus

$$I_c(z) = \sum_{k=0}^{\infty} \left| \frac{\Gamma(k + \lambda)}{k!\Gamma(\lambda)} \right|^2 \int_S |\langle z, \zeta \rangle|^{2k} d\sigma(\zeta).$$

Assuming, without loss of generality, that $z = |z|e_1$, the last integral is

$$|z|^{2k} \int_S |\zeta_1^k|^2 d\sigma(\zeta) = \frac{(n - 1)!k!}{(n - 1 + k)!} |z|^{2k},$$

by Proposition 1.4.9. Hence we obtain

$$I_c(z) = \frac{\Gamma(n)}{\Gamma^2(\lambda)} \sum_{k=0}^{\infty} \frac{\Gamma^2(k + \lambda)}{\Gamma(k + 1)\Gamma(k + n)} |z|^{2k}.$$

If we recall that $2\lambda = n + c$, we see that the coefficients in this last series are of order k^{c-1}, as $k \to \infty$.

This proves the assertions made about $I_c(z)$.

To handle $J_{c,t}(z)$, we use polar coordinates ($w = r\zeta$), and note that

$$J_{c,t}(z) = 2n \int_0^1 (1 - r^2)^t I_{1+t+c}(rz) r^{2n-1} \, dr.$$

We insert the series for I_{1+t+c} into this integral, and integrate term by term. The final result is

$$J_{c,t}(z) = \frac{\Gamma(n + 1)\Gamma(t + 1)}{\Gamma^2(\lambda_1)} \sum_{k=0}^{\infty} \frac{\Gamma^2(k + \lambda_1)|z|^{2k}}{\Gamma(k + 1)\Gamma(n + 1 + t + k)},$$

where $2\lambda_1 = n + 1 + t + c$. Again, the coefficients are of order k^{c-1}, by Stirling's formula that describes the asymptotic behavior of the gamma function.

1.5. Homogeneous Expansions

1.5.1. A polynomial P in $\mathbb{C}^n$ is said to be *homogeneous of degree s* if

$$(1) \qquad\qquad P(\lambda z) = \lambda^s P(z) \qquad (\lambda \in \mathbb{C}, z \in \mathbb{C}^n).$$

Assume that $f(z) = \sum c(\alpha)z^\alpha$ in some neighborhood of the origin in $\mathbb{C}^n$. For $s = 0, 1, 2, \ldots$, let $F_s(z)$ be the sum of those terms $c(\alpha)z^\alpha$ in the power series for which $|\alpha| = s$. Then F_s is homogeneous of degree s, and the power series can be written in the grouped form

$$(2) \qquad\qquad f(z) = \sum_{s=0}^{\infty} F_s(z).$$

This is the *homogeneous expansion* of f.

There are several reasons for considering this expansion. One is that it is invariant under linear changes of variables: If L is a linear transformation on $\mathbb{C}^n$, then $F_s \circ L$ is homogeneous of degree s, and the homogeneous expansion of $f \circ L$ is $\sum (F_s \circ L)(z)$.

Another reason is related to slice functions (see §1.2.5): If some $f \in H(B)$ has (2) for its homogeneous expansion, then

$$(3) \qquad\qquad f_\zeta(\lambda) = \sum_{s=1}^{\infty} F_s(\zeta)\lambda^s \qquad (\zeta \in S, \lambda \in U),$$

so that the numbers $F_s(\zeta)$ are the coefficients in the power series expansion of f_ζ.

The main result of this section (Theorem 1.5.6) shows that if a series of homogeneous polynomials (with increasing degrees) converges at every point of an open set $\Omega \subset \mathbb{C}^n$, then the convergence is uniform on every compact subset of Ω. This was proved by Hartogs [1; p. 72]. For a recent extension, see Theorem 5.1 of Bochnak and Siciak [1].

We begin with some lemmas.

1.5.2. Lemma. *If P is a polynomial of one variable, of degree k, and $|P(\lambda)| \leq 1$ for all $\lambda \in U$, then $|P(\lambda)| \leq |\lambda|^k$ for $1 \leq |\lambda| < \infty$.*

Proof. Let $P(\lambda) = a_0 + a_1\lambda + \cdots + a_k\lambda^k$. Since $|P(e^{i\theta})| \leq 1$, we have $|a_k| \leq 1$. The function $P(\lambda)/\lambda^k$ is holomorphic outside $\overline{U}$, it tends to a_k as $\lambda \to \infty$, and its absolute value is at most 1 on T. By the maximum modulus theorem, $|P(\lambda)/\lambda^k| \leq 1$ outside $\overline{U}$.

1.5.3. Subharmonic Functions. Let Ω be open in $\mathbb{C}^n$, and let u be an upper semicontinuous function in Ω, $-\infty \leq u < \infty$. Then u is said to be *subharmonic* if

$$(1) \qquad u(a) \leq \int_S u(a + r\zeta)d\sigma(\zeta)$$

for all $r > 0$ such that $a + r\overline{B} \subset \Omega$.

If we apply (1) with tr in place of r, where $0 \leq t \leq 1$, multiply by $2nt^{2n-1}$, and integrate over $0 \leq t \leq 1$, we see that

$$(2) \qquad u(a) \leq \int_B u(a + rw)dv(w)$$

if u is subharmonic and $a + r\overline{B} \subset \Omega$.

If f is a holomorphic function of one variable, it is well known that $\log|f|$ is subharmonic. This extends to several variables:

1.5.4. Proposition. *Let $f \in H(\Omega)$, where Ω is an open set in $\mathbb{C}^n$. Then $\log|f|$ is subharmonic in Ω, and so is $|f|^c$ for every positive constant c.*

Proof. Put $u = \log|f|$. Fix $a \in \Omega$, $\zeta \in S$. The function $\lambda \to u(a + \lambda\zeta)$ is then subharmonic in a neighborhood of 0 in $\mathbb{C}$, by the one-variable theorem that we just quoted. Hence

$$u(a) \leq \frac{1}{2\pi}\int_{-\pi}^{\pi} u(a + re^{i\theta}\zeta)d\theta$$

for all sufficiently small $r > 0$. If we apply the slice-integration formula 1.4.7(1) to this integral, we obtain the inequality 1.5.3(1). Hence $\log|f|$ is subharmonic.

Since $|f|^c = \exp(c \log |f|)$ and since $\exp(cx)$ is an increasing convex function of x, the subharmonicity of $\log|f|$ implies that of $|f|^c$.

1.5.5. Lemma. *Suppose $\{u_i\}$ is a sequence of subharmonic functions in Ω, and α, β are real numbers such that $u_i(z) \leq \beta$ and*

(1) $$\limsup_{i \to \infty} u_i(z) \leq \alpha$$

for every $z \in \Omega$.

If $K \subset \Omega$ is compact and $\varepsilon > 0$, then there exists i_0 such that

(2) $$u_i(z) < \alpha + \varepsilon$$

for all $z \in K$ and all $i > i_0$.

Proof. The case $\alpha \geq \beta$ is trivial, so assume $\alpha < \beta$. Pick $r > 0$ so that $K + r\bar{B} \subset \Omega$. Put $v_i = \max(\alpha, u_i)$, put $g_j = \sup\{v_i : i \geq j\}$, and define h_j on K by

$$h_j(z) = \int_B g_j(z + rw)dv(w).$$

Each v_i is then subharmonic, and $\alpha \leq v_i \leq \beta$. By (1), $\lim v_i(z) = \alpha$ for all $z \in \Omega$. Hence $g_j(z)$ decreases monotonically to α as $j \to \infty$. The same is true of $h_j(z)$, by the dominated convergence theorem. Since $\alpha \leq g_j \leq \beta$, each h_j is continuous. But if a sequence of continuous functions converges monotonically to a continuous limit, on a compact set, then the convergence is uniform. Thus $h_j(z) \to \alpha$ uniformly on K.

Now the subharmonicity of v_j, combined with the obvious inequality $v_j \leq g_j$, shows that $\alpha \leq v_j \leq h_j$. Hence $v_j \to \alpha$ uniformly on K, and since $u_j \leq v_j$, the lemma is proved.

1.5.6. Theorem. *Suppose*

  (a) Ω *is an open set in $\mathbb{C}^n$,*
  (b) *for $s = 0, 1, 2, \ldots$, F_s is a homogeneous polynomial of degree s, and*
  (c) $\sup_s |F_s(z)| < \infty$ *for every $z \in \Omega$.*

The series $\sum_0^\infty F_s(z)$ converges then uniformly on every compact subset of Ω.

Corollary. *The sum of the series is holomorphic.*

Proof. Pick a compact $K \subset \Omega$, and pick $t > 1$, so close to 1 that $t^2 K \subset \Omega$. By (c) and Baire's theorem, there exist $a \in \Omega, r > 0, M < \infty$, such that

(1) $$|F_s(a + z)| \leq M \qquad (|z| \leq r, s = 0, 1, 2, \ldots).$$

Lemma 1.5.2, applied to the polynomial

(2) $$P(\lambda) = M^{-1}F_s(a + \lambda r\zeta),$$

where $\zeta \in S$, shows that

(3) $$|F_s(a + z)| \le Mr^{-s}|z|^s \qquad (|z| > r).$$

Put $u_s = |F_s|^{1/s}$. By Proposition 1.5.4, each u_s is subharmonic. By (1) and (3), $\{u_s\}$ is uniformly bounded on every compact subset of $\mathbb{C}^n$. By assumption (c),

(4) $$\limsup_{s \to \infty} u_s(z) \le 1 \qquad (z \in \Omega).$$

Lemma 1.5.5 implies therefore that there is an s_0 such that $u_s < t$ on t^2K for all $s > s_0$. Thus

(5) $$|F_s(z)| < t^s \qquad (z \in t^2K, s > s_0),$$

or

(6) $$|F_s(t^2z)| < t^s \qquad (z \in K, s > s_0).$$

Finally, $F_s(t^2z) = t^{2s}F_s(z)$, by the homogeneity of F_s. Thus (6) becomes

(7) $$|F_s(z)| < t^{-s} \qquad (z \in K, s > s_0).$$

Since $t > 1$, $\sum_s t^{-s} < \infty$. This completes the proof.

Chapter 2

The Automorphisms of B

The main purpose of this chapter is the description of the biholomorphic maps of B onto B. These will simply be called the automorphisms of B. They form a group, $\text{Aut}(B)$, which may also be called the *Moebius group* of B, since the automorphisms turn out to be linear fractional transformations.

The chapter begins, however, with some general results about biholomorphic maps between circular regions.

2.1. Cartan's Uniqueness Theorem

2.1.1. Theorem (H. Cartan [1]). *Suppose*

(a) Ω *is a bounded region in* $\mathbb{C}^n$,
(b) $F: \Omega \to \Omega$ *is holomorphic,*
(c) *for some* $p \in \Omega$, $F(p) = p$ *and* $F'(p) = I$. *Then* $F(z) = z$ *for all* $z \in \Omega$.

The derivative $F'(p)$ is defined in §1.3.6; as usual, I denotes the identity operator. A *region* is a *connected* open set.

Proof. Without loss of generality, $p = 0$. Then there exists $r_1 > 0$, $r_2 < \infty$, such that $r_1 B \subset \Omega \subset r_2 B$. For $|z| < r_1$, F has a homogeneous expansion

$$F(z) = z + \sum_{s=2}^{\infty} F_s(z)$$

in which each F_s is a map from $\mathbb{C}^n$ to $\mathbb{C}^n$ whose *components* are homogeneous polynomials of degree s.

Let F^k be the kth iterate of F; explicitly, $F^1 = F$, $F^k = F^{k-1} \circ F$. For $m \geq 2$, make the induction hypothesis (which is vacuously true when $m = 2$) that $F_s = 0$ for $2 \leq s < m$. Then F^k has the homogeneous expansion

$$F^k(z) = z + kF_m(z) + \cdots$$

23

in $r_1 B$, as is easily proved by induction on k. The homogeneity of the maps F_s implies that

$$kF_m(z) = \frac{1}{2\pi} \int_{-\pi}^{\pi} F^k(e^{i\theta}z)e^{-im\theta}\, d\theta \qquad (|z| < r_1).$$

(This is a $\mathbb{C}^n$-valued integral.) Since F^k maps Ω into Ω, we have $|F^k(e^{i\theta}z)| < r_2$ for all $z \in r_1 B$ and for all θ. Thus

$$k|F_m(z)| < r_2$$

for $k = 1, 2, 3, \ldots, z \in r_1 B$. Hence $F_m = 0$, and our induction hypothesis holds with $m + 1$ in place of m.

Thus $F(z) = z$ for all $z \in r_1 B$. Since Ω is connected, the proof is complete.

2.1.2. Definition. A set $E \subset \mathbb{C}^n$ is said to be *circular* if $e^{i\theta}z \in E$ whenever $z \in E$ and θ is real.

2.1.3. Theorem (H. Cartan [1]). *Suppose*

   (a) Ω_1 and Ω_2 are circular regions in $\mathbb{C}^n$, $0 \in \Omega_1$, $0 \in \Omega_2$,
   (b) F is a biholomorphic map of Ω_1 onto Ω_2, with $F(0) = 0$, and
   (c) Ω_1 is bounded.

Then F is a linear transformation.

Note that the conclusion implies that Ω_2 is also bounded.

Proof. Let $G = F^{-1}$, let $A = F'(0)$. Since $G(F(z)) = z$, $G'(0)A = I$, so $G'(0) = A^{-1}$. Fix a real θ, and define

$$H(z) = G(e^{-i\theta}F(e^{i\theta}z)) \qquad (z \in \Omega_1).$$

Since Ω_1 and Ω_2 are circular, $H(z)$ is well defined, and H is a holomorphic map of Ω_1 into Ω_1 that satisfies $H(0) = 0$, $H'(0) = I$. By Theorem 2.1.1, $H(z) = z$. If we apply F to this and multiply by $e^{i\theta}$, we obtain

$$F(e^{i\theta}z) = e^{i\theta}F(z),$$

for all $z \in \Omega_1$, and for every real θ. The linear term in the homogeneous expansion of F is therefore the only one that is different from 0.

2.1.4. Examples. Theorem 2.1.3 fails if boundedness is omitted from its hypotheses, even when $\Omega_1 = \Omega_2$ (except in the case $n = 1$, where the supply of circular regions is rather limited).

For example, let $\Omega \subset \mathbb{C}^2$ consist of all (z, w) with $|zw| < 1$. [When $n = 2$, we shall often write (z, w) in place of (z_1, z_2).] Let $h: U \to \mathbb{C}$ be any zero-free holomorphic function, and put

$$F_h(z, w) = \left(zh(zw), \frac{w}{h(zw)} \right).$$

If $\zeta = zh(zw)$, $\eta = w/h(zw)$, then $\zeta\eta = zw$. It follows that F_h is a biholomorphic map of Ω onto Ω, with inverse $F_{1/h}$, and $F_h(0) = 0$. Also, $F'_h(0) = I$ if $h(0) = 1$. But F_h is obviously not linear, except when h is constant.

Another simple example is furnished by the map of $\mathbb{C}^2$ onto $\mathbb{C}^2$ that takes (z, w) to $(z, w + f(z))$, where $f: \mathbb{C} \to \mathbb{C}$ is an arbitrary entire function with $f(0) = f'(0) = 0$. This can be done again: let g be entire in $\mathbb{C}$, with $g(0) = g'(0) = 0$, and define

$$F(z, w) = (z + g(w + f(z)), w + f(z)).$$

In this fashion one can construct more and more complicated biholomorphic maps F of $\mathbb{C}^2$ onto $\mathbb{C}^2$ that fix the origin, that have $F'(0) = I$, and whose Jacobian is 1 at every point of $\mathbb{C}^2$.

Similar constructions can of course be made in any $\mathbb{C}^n$ if $n > 1$. There is a well-known open question related to this (Magnus [1]):

Suppose F is a holomorphic map of $\mathbb{C}^n$ into $\mathbb{C}^n$ whose components are polynomials and whose Jacobian is 1 at every point. Does it follow that the range of F is all of $\mathbb{C}^n$?

If the question is asked with entire functions in place of polynomials, the answer is negative (for every $n > 1$). This was shown by Fatou and Bieberbach; see Bochner–Martin [1], pp. 45–48.

2.2. The Automorphisms

2.2.1. Recall that to every α in the unit disc of $\mathbb{C}$ corresponds an automorphism φ_α of the disc that interchanges α and 0, namely $\varphi_\alpha(\lambda) = (\alpha - \lambda)/(1 - \bar{\alpha}\lambda)$.

The same can be done in the unit ball B of $\mathbb{C}^n$. Fix $a \in B$. Let P_a be the orthogonal projection of $\mathbb{C}^n$ onto the subspace $[a]$ generated by a, and let $Q_a = I - P_a$ be the projection onto the orthogonal complement of $[a]$. To be quite explicit, $P_0 = 0$ and

(1)
$$P_a z = \frac{\langle z, a \rangle}{\langle a, a \rangle} a \qquad \text{if } a \neq 0.$$

Put $s_a = (1 - |a|^2)^{1/2}$ and define

(2)
$$\varphi_a(z) = \frac{a - P_a z - s_a Q_a z}{1 - \langle z, a \rangle}.$$

If $\Omega = \{z \in \mathbb{C}^n: \langle z, a \rangle \neq 1\}$ then $\varphi_a: \Omega \to \mathbb{C}^n$ is holomorphic. Clearly, $\Omega \supset \bar{B}$, since $|a| < 1$.

When $n = 1$, then $P_a = I, Q_a = 0$, and (2) reduces to the above-mentioned automorphism of the unit disc.

Note that $\varphi_0(z) = -z$. Note also that the map $a \to \varphi_a(z)$ is continuous on B, for every fixed $z \in B$, even at $a = 0$; this will not be important in the sequel, and we omit the easy verification.

When there is no danger of confusion, we shall write P, Q, s in place of P_a, Q_a, s_a.

Here are the main properties of the maps φ_a, listed in an order in which they are easy to prove.

2.2.2. Theorem. *For every $a \in B$, φ_a has the following properties:*

(i) $\varphi_a(0) = a$ *and* $\varphi_a(a) = 0$.
(ii) $\varphi_a'(0) = -s^2 P - sQ$ *and* $\varphi_a'(a) = -P/s^2 - Q/s$.
(iii) *The identity*

$$1 - \langle \varphi_a(z), \varphi_a(w) \rangle = \frac{(1 - \langle a, a \rangle)(1 - \langle z, w \rangle)}{(1 - \langle z, a \rangle)(1 - \langle a, w \rangle)}$$

holds for all $z \in \bar{B}$, $w \in \bar{B}$.
(iv) *The identity*

$$1 - |\varphi_a(z)|^2 = \frac{(1 - |a|^2)(1 - |z|^2)}{|1 - \langle z, a \rangle|^2}$$

holds for every $z \in \bar{B}$.
(v) φ_a *is an involution:* $\varphi_a(\varphi_a(z)) = z$.
(vi) φ_a *is a homeomorphism of $\bar{B}$ onto $\bar{B}$, and $\varphi_a \in \mathrm{Aut}(B)$.*

Proof. (i) is obvious from 2.2.1(2). For any $z \in \bar{B}$, 2.2.1(2) can be rewritten in the form

$$\varphi_a(z) = [1 + \langle z, a \rangle + \langle z, a \rangle^2 + \cdots][a - (P + sQ)z]$$
$$= \varphi_a(0) + \langle z, a \rangle a - (P + sQ)z + O(|z|^2),$$

and since $\langle z, a \rangle a = |a|^2 Pz$, we obtain the first formula in (ii); the second one follows from

$$\varphi_a(a + h) = \frac{-Ph - sQh}{s^2 - \langle h, a \rangle}.$$

To prove (iii), note that 2.2.1(2) splits $\varphi_a(z)$ nicely into its component in $[a]$ and the one that is orthogonal to $[a]$. The left side of (iii) is thus equal to

$$1 - \frac{\langle a - Pz, a - Pw \rangle + s^2 \langle Qz, Qw \rangle}{(1 - \langle z, a \rangle)(1 - \langle a, w \rangle)}.$$

In this expression, Pw and Qw can be replaced by w, since P and Q are self-adjoint projections. If we note also that

$$\langle z, a \rangle \langle a, w \rangle = \langle \langle z, a \rangle a, w \rangle = |a|^2 \langle Pz, w \rangle$$

then it is easy to derive (iii).

Taking $z = w$ in (iii) gives (iv). As an obvious consequence of (iv), note that $|\varphi_a(z)| < 1$ if and only if $|z| < 1$ and that $|\varphi_a(z)| = 1$ if and only if $|z| = 1$. Thus φ_a maps B into B, S into S.

To prove (v), put $\psi = \varphi_a \circ \varphi_a$. Then ψ is a holomorphic map of B into B, with $\psi(0) = 0$, and (ii) shows that

$$\psi'(0) = \varphi_a'(a) \varphi_a'(0) = P + Q = I$$

(since $P^2 = P$, $Q^2 = Q$, $PQ = QP = 0$). By Theorem 2.1.1, $\psi(z) = z$, and (v) is proved.

It follows from (v) that φ_a is a one-to-one map of $\bar{B}$ onto $\bar{B}$, and that $\varphi_a^{-1} = \varphi_a$. This proves (vi).

2.2.3. Theorem. $\mathrm{Aut}(B)$ *acts transitively on* B.

Proof. If $a \in B$ and $b \in B$ then $\varphi_b \circ \varphi_a$ is an automorphism of B that takes a to b.

By combining this with Cartan's uniqueness theorem, one can show that the Riemann mapping theorem fails quite spectacularly in $\mathbb{C}^n$ when $n > 1$:

2.2.4. Theorem. *Suppose* Ω *is a circular region in* $\mathbb{C}^n$, $0 \in \Omega$, *and some biholomorphic* F *maps* B *onto* Ω. *Then there is a linear transformation of* $\mathbb{C}^n$ *that maps* B *onto* Ω.

Proof. Let $a = F^{-1}(0)$. Then $F \circ \varphi_a$ is a biholomorphic map of B onto Ω that fixes 0. By Theorem 2.1.3, $F \circ \varphi_a$ is linear.

Corollary. *When* $n > 1$, *there is no biholomorphic map of* B *onto the polydisc* U^n.

Proof. Invertible linear transformations map balls onto ellipsoids.

Note that Theorem 2.2.4 holds (with the same proof) if B is replaced by any bounded circular region that contains 0 and that has a transitive group of automorphisms.

We shall encounter further variations on this theme in the context of proper holomorphic maps.

The automorphisms of B that we have met so far are the maps φ_a and the unitary transformations. The next theorem shows that all automorphisms are obtained from these:

2.2.5. Theorem. *If $\psi \in \mathrm{Aut}(B)$ and $a = \psi^{-1}(0)$ then there is a unique $U \in \mathcal{U}$ such that*

$$(1) \qquad\qquad\qquad\qquad \psi = U\varphi_a.$$

The identity

$$(2) \qquad\qquad 1 - \langle \psi(z), \psi(w) \rangle = \frac{(1 - \langle a, a \rangle)(1 - \langle z, w \rangle)}{(1 - \langle z, a \rangle)(1 - \langle a, w \rangle)}$$

holds for all $z \in \bar{B}$, $w \in \bar{B}$.

Proof. The map $\psi \circ \varphi_a$ is an automorphism of B that fixes 0, hence is linear, by Theorem 2.1.3. Since the unitary transformations are the only linear ones on $\mathbb{C}^n$ that preserve B, there is a $U \in \mathcal{U}$ such that $\psi \circ \varphi_a = U$, hence $\psi = U\varphi_a$. The uniqueness of U is trivial. The rest follows from Theorem 2.2.2(iii), since $\langle U\varphi_a(z), U\varphi_a(w) \rangle = \langle \varphi_a(z), \varphi_a(w) \rangle$.

2.2.6. Theorem. (i) *If $\psi \in \mathrm{Aut}(B)$ and $a = \psi^{-1}(0)$, then the real Jacobian of ψ at $z \in B$ is*

$$(1) \qquad\qquad (J_R\psi)(z) = \left(\frac{1 - |a|^2}{|1 - \langle z, a \rangle|^2} \right)^{n+1}.$$

(ii) *If τ is the measure defined on B by*

$$(2) \qquad\qquad d\tau(z) = (1 - |z|^2)^{-n-1}\, dv(z)$$

then

$$(3) \qquad\qquad \int_B f\, d\tau = \int_B (f \circ \psi)\, d\tau$$

for every $f \in L^1(\tau)$ and every $\psi \in \mathrm{Aut}(B)$.

This τ is thus invariant under $\mathrm{Aut}(B)$.

Proof. Put $\psi(z) = w$. Then $\varphi_w \circ \psi \circ \varphi_z$ is an automorphism of B that fixes 0, hence it is unitary, by another application of Theorem 2.1.3, as in the preceding proof. Thus $\psi = \varphi_w U \varphi_z$ for some $U \in \mathcal{U}$. Hence

(4) $$\psi'(z) = \varphi'_w(0) U \varphi'_z(z).$$

By Theorem 2.2.2(ii), the linear operator $\varphi'_w(0)$ has a 1-dimensional eigenspace with eigenvalue $-s^2$ (where $s^2 = 1 - |w|^2$) and an $(n-1)$-dimensional eigenspace with eigenvalue $-s$, so that its determinant is $(-1)^n s^{n+1}$. Hence (see §1.3.6)

(5) $$(J_R \varphi_w)(0) = |\det \varphi'_w(0)|^2 = (1 - |w|^2)^{n+1}.$$

Similarly, $(J_R \varphi_z)(z) = (1 - |z|^2)^{-n-1}$, and now (4) implies that

(6) $$(J_R \psi)(z) = \left(\frac{1 - |\psi(z)|^2}{1 - |z|^2} \right)^{n+1}.$$

The identity 2.2.5 (with $w = z$) shows that (6) is the same as (1). Next,

$$\int_B f \, d\tau = \int_B f(w)(1 - |w|^2)^{-n-1} \, dv(w)$$

$$= \int_B f(\psi(z))(1 - |\psi(z)|^2)^{-n-1}(J_R \psi)(z) dv(z).$$

By (6), the last integral is

$$\int_B f(\psi(z))(1 - |z|^2)^{-n-1} \, dv(z) = \int_B (f \circ \psi) d\tau.$$

2.2.7. We now describe the sets onto which φ_a maps balls centered at 0. Fix $a \in B$, fix ε, $0 < \varepsilon < 1$, and define

(1) $$E(a, \varepsilon) = \varphi_a(\varepsilon B).$$

Since φ_a is an involution, $z \in E(a, \varepsilon)$ if and only if $|\varphi_a(z)| < \varepsilon$. If one squares this and uses the definition 2.2.1(2) of $\varphi_a(z)$, a little manipulation shows that $E(a, \varepsilon)$ consists of all z that satisfy

(2) $$\frac{|Pz - c|^2}{\varepsilon^2 \rho^2} + \frac{|Qz|^2}{\varepsilon^2 \rho} < 1,$$

where $P = P_a$, $Q = Q_a$, and

(3)
$$c = \frac{(1 - \varepsilon^2)a}{1 - \varepsilon^2 |a|^2}, \qquad \rho = \frac{1 - |a|^2}{1 - \varepsilon^2 |a|^2}.$$

Thus $E(a, \varepsilon)$ is an ellipsoid with center at c; this is close to a when ε is small. The intersection of $E(a, \varepsilon)$ with $[a]$ is a disc of radius $\varepsilon\rho$, which is roughly εs^2 when ε is small; its intersection with the real $(2n - 2)$-dimensional space perpendicular to $[a]$ at c is a ball of the much larger radius $\varepsilon\sqrt{\rho} \sim \varepsilon s$.

Note that this agrees with Theorem 2.2.2(ii), i.e., with $\varphi'_a(0) = -s^2 P - sQ$. Finally, note that $v(\varepsilon B) = \varepsilon^{2n}$. Thus

(4)
$$\frac{v(E(a, \varepsilon))}{v(\varepsilon B)} = \rho^2 (\sqrt{\rho})^{2n-2} = \rho^{n+1}.$$

As $\varepsilon \to 0$, the quotient converges to $(J_R \varphi_a)(0)$. By (3), it follows that

(5)
$$(J_R \varphi_a)(0) = (1 - |a|^2)^{n+1}.$$

This is another derivation of 2.2.6(5).

2.2.8. Extensions of Automorphisms. Suppose $1 \le n < N$. Corresponding to the orthogonal direct sum decomposition $\mathbb{C}^N = \mathbb{C}^n \oplus \mathbb{C}^{N-n}$, each $z \in \mathbb{C}^N$ decomposes into $z = z' + z''$, where $z' \in \mathbb{C}^n$, $z'' \in \mathbb{C}^{N-n}$. Let B_n and B_N be the corresponding unit balls.

Every $\psi \in \text{Aut}(B_n)$ extends then to a $\Psi \in \text{Aut}(B_N)$.

It is enough to prove this for $\psi = \varphi_a$ ($a \in B_n$), since the extension of unitary operators presents no problem. Given φ_a, put $s = (1 - |a|^2)^{1/2}$, and define

$$\Phi_a(z) = \varphi_a(z') - \frac{sz''}{1 - \langle z', a \rangle}.$$

Since $\langle z', a \rangle = \langle z, a \rangle$, Φ_a is precisely the involution defined in §2.2.1, but with B_N in place of B_n. Thus $\Phi_a \in \text{Aut}(B_N)$.

2.2.9. The automorphisms of B_2 occur in Poincaré [1]. I do not know when they were first written down for $n > 2$. As the present description of the involutions φ_a in terms of inner products and projections suggests, many properties of $\text{Aut}(B)$ extend from B_n to the unit ball of an arbitrary Hilbert space. See Hayden–Suffridge [1], and, for even more far-reaching extensions, Harris [2], [3], [4].

2.3. The Cayley Transform

2.3.1. One often transfers problems from the unit disc to the upper half-plane by means of the correspondence $w = i(1 + z)/(1 - z)$. A similar device is available in $\mathbb{C}^n$.

The appropriate "upper half-plane" in $\mathbb{C}^n$ is the region Ω consisting of all $w = (w_1, w')$ such that

$$(1) \qquad\qquad \mathrm{Im}\, w_1 > |w'|^2$$

where $w' = (w_2, \ldots, w_n)$, $|w'|^2 = |w_2|^2 + \cdots + |w_n|^2$.

The Cayley transform is the map Φ that sends $z \in \mathbb{C}^n$ ($z_1 \neq 1$) to $w \in \mathbb{C}^n$, by

$$(2) \qquad\qquad w = i\,\frac{e_1 + z}{1 - z_1}$$

where, as usual, $e_1 = (1, 0')$. Simple computations show that (2) implies

$$(3) \qquad\qquad \mathrm{Im}\, w_1 - |w'|^2 = \frac{1 - |z|^2}{|1 - z_1|^2}$$

and

$$(4) \qquad\qquad z = \frac{2w}{i + w_1} - e_1.$$

Hence Φ is a biholomorphic map of B onto Ω.

Let $\bar{\Omega} = \Omega \cup \partial\Omega$ be the closure of Ω in $\mathbb{C}^n$; clearly, $w \in \partial\Omega$ if and only if

$$(5) \qquad\qquad \mathrm{Im}\, w_1 = |w'|^2.$$

Let $\bar{\Omega} \cup \{\infty\}$ be the one-point compactification of $\bar{\Omega}$, and let $\Phi(e_1) = \infty$. Then Φ is a homeomorphism of $\bar{B}$ onto $\bar{\Omega} \cup \{\infty\}$, and it is clear that Φ induces an isomorphism between $\mathrm{Aut}(B)$ and $\mathrm{Aut}(\Omega)$.

2.3.2. $\mathrm{Aut}(\Omega)$ has a subgroup that is isomorphic to the multiplicative group of the positive real numbers and that consists of the so-called "non-isotropic" dilations δ_t, defined by

$$(1) \qquad\qquad \delta_t(w) = (t^2 w_1, tw') \qquad (0 < t < \infty).$$

When $t \neq 1$, δ_t fixes only 0 and ∞.

Hence $\Phi^{-1} \circ \delta_t \circ \Phi$ is an automorphism of B whose only fixed points on $\bar{B}$ are e_1 and $-e_1$.

2.3.3. Another subgroup of $\mathrm{Aut}(\Omega)$ consists of the "translations" h_a (one for each $a \in \partial\Omega$) that are defined by

(1) $$h_a(w) = (w_1 + a_1 + 2i\langle w', a' \rangle, w' + a').$$

To see this, put $h_a(w) = \zeta = (\zeta_1, \zeta')$. Using the fact that $\mathrm{Im}\, a_1 = |a'|^2$, it is easy to verify that

(2) $$\mathrm{Im}\, \zeta_1 - |\zeta'|^2 = \mathrm{Im}\, w_1 - |w'|^2.$$

Every h_a thus maps Ω into Ω, $\partial\Omega$ into $\partial\Omega$. Moreover, if $a, b \in \partial\Omega$ and $c = h_a(b)$, then $h_a \circ h_b = h_c$. If $b = (-\bar{a}_1, -a')$ then $h_a(b) = 0$, hence $h_b = h_a^{-1}$, since h_0 is the identity map.

Thus $\{h_a : a \in \partial\Omega\}$ is indeed a subgroup of $\mathrm{Aut}(\Omega)$. It induces a binary operation $\#$ on $\partial\Omega$, defined by

(3) $$a \mathbin{\#} b = (h_a \circ h_b)(0) \qquad (a, b \in \partial\Omega),$$

or simply by $a \mathbin{\#} b = h_a(b)$. This makes $\partial\Omega$ into a group that has been called the Heisenberg group and that has been the setting for a great deal of recent research in harmonic analysis and partial differential equations. See, for instance Greiner and Stein [1], Rothschild and Stein [1].

Note that h_a fixes no point of $\bar{\Omega}$ when $a \neq 0$.

Hence $\Phi^{-1} \circ h_a \circ \Phi$ is an automorphism of B whose only fixed point on $\bar{B}$ is e_1.

2.4. Fixed Points and Affine Sets

2.4.1. Affine Sets. In place of $\mathbb{C}^n$, let us for the moment consider arbitrary vector spaces X and Y over a field F. A nonempty set $E \subset X$ is said to be affine if the following is true: whenever $x_1, \ldots, x_k \in E$, $c_1, \ldots, c_k \in F$, and $\Sigma c_i = 1$, then $\Sigma c_i x_i \in E$.

One sees easily that E is affine if and only if $E = X_0 + p$ for some $p \in X$ and some vector space $X_0 \subset X$.

Now fix $b \in Y$, $\beta \in F$, let $L: X \to Y$ and $\Lambda: X \to F$ be linear, let $D = \{x \in X : \Lambda x \neq \beta\}$, and define a "Moebius transformation" $\psi : D \to Y$ by

(1) $$\psi(x) = \frac{b - Lx}{\beta - \Lambda x}.$$

If $x_i \in D$, $c_i \in F$, $\Sigma c_i = 1$, and $\Sigma c_i x_i = x \in D$, it follows that

(2) $$\psi(x) = \Sigma \gamma_i \psi(x_i)$$

where $\gamma_i = c_i(\beta - \Lambda x_i)/(\beta - \Lambda x)$. The linearity of Λ shows that $\Sigma \gamma_i = 1$.

Somewhat imprecisely (because of the possible vanishing of denominators), this says that ψ maps the affine set generated by $x_1, \ldots, x_k$ into the affine set generated by $\psi(x_1), \ldots, \psi(x_k)$.

Let us now return to the case $X = Y = \mathbb{C}^n$. If $\psi \in \text{Aut}(B)$, then $\psi = U\varphi_a$ (Theorem 2.2.5). Setting $b = Ua$, $\beta = 1$, $L = U(P + sQ)$, $\Lambda z = \langle z, a \rangle$, we see that ψ has the form (1), and that $\langle z, a \rangle \neq 1$ for $z \in B$. If we use the term "affine subset of B" for any intersection $E \cap B$, where E is affine in $\mathbb{C}^n$, then the preceding reasoning (applied to ψ and ψ^{-1}) establishes the following result:

2.4.2. Proposition. *If $\psi \in \text{Aut}(B)$ and E is an affine subset of B, then so is $\psi(E)$.*

One sees in the same way that affine sets are preserved by the Cayley transform.

2.4.3. Definition. *If f is a map of a set X into X, the fixed-point set of f is the set of all $x \in X$ for which $f(x) = x$.*

As we shall now see, the automorphisms of B have very simple fixed-point sets.

2.4.4. Theorem. *If $\psi \in \text{Aut}(B)$ fixes a point of B, then the fixed-point set of ψ is affine. Conversely, every affine subset of B is the fixed-point set of some $\psi \in \text{Aut}(B)$.*

Proof. Suppose $\psi(a) = a$ for some $a \in B$. Then $\varphi_a \circ \psi \circ \varphi_a$ fixes 0, hence is unitary, hence its fixed point set E (in $\mathbb{C}^n$) is either $\{0\}$ or it is a subspace Y of $\mathbb{C}^n$ corresponding to the eigenvalue 1. The fixed-point set of ψ is $\varphi_a(E \cap \bar{B})$. The first assertion follows now from 2.4.2.

For the converse, let E be an affine subset of B. If $a \in E$, Proposition 2.4.2 implies that $\mathbb{C}^n$ has a subspace Y such that $\varphi_a(E) = B \cap Y$. Choose $U \in \mathcal{U}$ so that Y is the eigenspace of U with eigenvalue 1. If $\psi = \varphi_a U \varphi_a$, then ψ has E as its fixed-point set.

2.4.5. As a consequence of the Brouwer fixed point theorem, every $\psi \in \text{Aut}(B)$ fixes at least one point of $\bar{B}$. In Section 2.3 we saw examples of automorphisms that fix no point of B and whose fixed-point set on S consists of exactly one or two points. This is as far as one can go:

2.4.6. Theorem (Hayden–Suffridge [1]). *If $\psi \in \text{Aut}(B)$ and if ψ fixes three distinct points of S, then ψ fixes a point of B.*

Consequently (by Theorem 2.4.4) the fixed-point set of ψ is then affine. The following proof is due to David Ullrich.

Proof. Let z_1, z_2, z_3 be distinct fixed points of ψ on S. The identity 2.2.5(2) becomes now

$$1 - \langle z_i, z_k \rangle = \frac{(1 - \langle a, a \rangle)(1 - \langle z_i, z_k \rangle)}{(1 - \langle z_i, a \rangle)(1 - \langle a, z_k \rangle)}$$

where $a = \psi^{-1}(0)$. If $i \neq k$ then $\langle z_i, z_k \rangle \neq 1$, hence

$$1 - \langle z_i, a \rangle = \frac{(1 - \langle a, a \rangle)}{(1 - \langle a, z_k \rangle)}.$$

With $k = 3$ and $i = 1$ or 2, this implies that

(1) $$\langle z_1, a \rangle = \langle z_2, a \rangle.$$

Put $z = \frac{1}{2}(z_1 + z_2)$. Then $z \in B$. Since (1) holds, the definition of φ_a implies

(2) $$\varphi_a(z) = \tfrac{1}{2}\varphi_a(z_1) + \tfrac{1}{2}\varphi_a(z_2).$$

Since $\psi = U\varphi_a$ for some $U \in \mathcal{U}$, we conclude that

$$\psi(z) = \tfrac{1}{2}\psi(z_1) + \tfrac{1}{2}\psi(z_2) = \tfrac{1}{2}(z_1 + z_2) = z.$$

The following theorem characterizes the automorphisms φ_a as precisely those involutions that fix only one point of B:

2.4.7. Theorem. (i) *Each φ_a fixes exactly one point of B, and no point of S.*

(ii) *If $b \in B$ and $a = 2b/(1 + |b|^2)$, then φ_a is the only involution in $\mathrm{Aut}(B)$ that has b as its only fixed point.*

Proof. Suppose $z \in \bar{B}$ and $\varphi_a(z) = z$. Then $Q\varphi_a(z) = Qz$, where $Q = Q_a$, and 2.2.1(2) implies

(1) $$-sQz = (1 - \langle z, a \rangle)Qz.$$

This forces $Qz = 0$, since $\mathrm{Re}(1 - \langle z, a \rangle) > 0$. Thus $z = \lambda a$ for some $\lambda \in \mathbb{C}$. The quadratic equation $\varphi_a(\lambda a) = \lambda a$ has $\lambda = (1 \pm s)^{-1}$ as roots, where $s = (1 - |a|^2)^{1/2}$. Since $a/(1 - s)$ is not in $\bar{B}$, the point

(2) $$b = \frac{a}{(1 + s)}$$

is the only fixed point of φ_a in $\bar{B}$. This proves (i).

If $b \in B$ and $a = 2b/(1 + |b|^2)$, then (2) holds, as is easily verified. Thus $\varphi_a(b) = b$.

Assume now that $\psi \in \text{Aut}(B)$ is an involution with b as its only fixed point. Then $\varphi_b \circ \psi \circ \varphi_b$ is an involution with 0 as its only fixed point. But if a unitary transformation is an involution, then 1 and -1 are its only possible eigenvalues, and if it fixes only 0, then 1 is excluded from the eigenvalues. Hence $\varphi_b \circ \psi \circ \varphi_b = \varphi_0$. This establishes the uniqueness of ψ and shows, incidentally, that

$$(3) \qquad\qquad \varphi_a = \varphi_b \circ \varphi_0 \circ \varphi_b$$

if $a = 2b/(1 + |b|^2)$.

Chapter 3

Integral Representations

3.1. The Bergman Integral in B

3.1.1. The Spaces $(L^p \cap H)(B)$. As usual, n is a fixed positive integer, $n \geq 1$, $B = B_n$ is the open unit ball of $\mathbb{C}^n$, and v is the normalized Lebesgue measure on $\mathbb{C}^n$ defined in §1.4.1.

For $1 \leq p < \infty$, $(L^p \cap H)(B)$ is the space of all *holomorphic* $f: B \to \mathbb{C}$ that belong to $L^p(v)$.

With respect to the usual norm

$$(1) \qquad \qquad \|f\|_p = \left\{ \int_B |f|^p \, dv \right\}^{1/p},$$

$L^p \cap H$ turns out to be a *closed* subspace of L^p. This follows quickly from the mean value formula

$$(2) \qquad \qquad f(z) = r^{-2n} \int_{z+rB} f \, dv \qquad (r < 1 - |z|),$$

valid for every $f \in H(B)$, since Hölder's inequality leads from (2) to

$$(3) \qquad \qquad |f(z)| \leq (1 - |z|)^{-2n/p} \|f\|_p \qquad (z \in B).$$

[Formula (2) can be proved by the argument that was applied to subharmonic functions in §1.5.3.]

If now $\{f_j\}$ is a Cauchy sequence in $L^p \cap H$, (3) implies that $\{f_j\}$ converges uniformly on every compact subset of B. Hence $L^p \cap H$ is closed in L^p.

3.1.2. The Bergman Kernel. This is the function K defined on $B \times B$ (or, if preferred, on all of $\mathbb{C}^n \times \mathbb{C}^n$ except where $\langle z, w \rangle = 1$) by

$$(1) \qquad \qquad K(z, w) = (1 - \langle z, w \rangle)^{-n-1}.$$

Note that $K(z, w)$ is holomorphic in z, conjugate holomorphic in w, and that

$$(2) \qquad \qquad K(w, z) = \overline{K(z, w)}.$$

36

This kernel associates to every $f \in L^1(v)$ a function $K[f]$, defined in B by the integral

$$K[f](z) = \int_B K(z, w) f(w) dv(w).$$

(3)

This is well-defined, since $K(z, w)$ is a bounded function of w, for each $z \in B$. It is clear that $K[f]$ is always holomorphic. The most important property of K is that it *reproduces* holomorphic functions:

3.1.3. Theorem. (a) $f = K[f]$ *for every* $f \in (L^1 \cap H)(B)$.
  (b) *On* $L^2(v)$, $h \to K[h]$ *is the orthogonal projection whose range is* $(L^2 \cap H)(B)$.

A more explicit statement of (a) is that every $f \in (L^1 \cap H)(B)$ satisfies

$$f(z) = \int_B \frac{f(w) dv(w)}{(1 - \langle z, w \rangle)^{n+1}}$$

(1)

for every $z \in B$.

Proof. Fix $f \in (L^1 \cap H)(B)$, fix $a \in B$, and define

$$g(z) = \frac{K(a, a)}{K(z, a)} f(z) \qquad (z \in B).$$

(2)

Since $1/K(z, a)$ is bounded, $g \in (L^1 \cap H)(B)$, and the same is true of $g \circ \varphi_a$. The mean value formula

$$(g \circ \varphi_a)(0) = \int_B g(\varphi_a(u)) dv(u)$$

(3)

holds therefore. (See §3.1.1.) Since $f(a) = g(a)$, the change of variables $u = \varphi_a(w)$ turns (3) into

$$f(a) = \int_B g(w)(J_R \varphi_a)(w) dv(w).$$

(4)

By Theorem 2.2.6 and the definition of $K(z, w)$,

$$(J_R \varphi_a)(w) = \left\{ \frac{1 - |a|^2}{|1 - \langle w, a \rangle|^2} \right\}^{n+1} = \frac{K(a, w) K(w, a)}{K(a, a)}.$$

(5)

If we insert (2) and (5) into (4) we obtain (1).

To prove (b), it is now enough to show that $K[h] = 0$ for every $h \in L^2(v)$ that is orthogonal to $(L^2 \cap H)(B)$. Fix $a \in B$, define $K_a(w) = K(w, a)$. Then $K_a \in (L^2 \cap H)(B)$, and therefore

$$K[h](a) = \int_B K(a, w)h(w)dv(w) = \int_B h\overline{K}_a \, dv = 0.$$

3.1.4. Remark. There is a Bergman kernel associated to every region $\Omega \subset \mathbb{C}^n$, as follows: For every $z \in \Omega$, the evaluation functional $f \to f(z)$ is continuous on $(L^2 \cap H)(\Omega)$, essentially by 3.1.1(3). Hence there corresponds to every $z \in \Omega$ a function $K_z \in L^2 \cap H$ such that $f(z) = [f, K_z]$ for all $f \in L^2 \cap H$, where $[\ ,\ ]$ is the inner product in L^2. If $\{u_i\}$ is an orthonormal basis for $L^2 \cap H$, then

$$K_z = \sum_i [K_z, u_i]u_i = \sum_i \overline{u_i(z)}u_i.$$

By Parseval's theorem, $\sum |u_i(z)|^2 = \|K_z\|_2^2 < \infty$, and

$$f(z) = \sum_i [f, u_i]u_i(z) = \int_\Omega f(w) \sum_i u_i(z)\overline{u_i(w)}dv(w)$$

for every $f \in L^2 \cap H, z \in \Omega$.

The sum $\sum u_i(z)\overline{u_i(w)}$ is the Bergman kernel $K(z, w)$ of Ω. The associated integral transform projects $L^2(\Omega)$ orthogonally onto $(L^2 \cap H)(\Omega)$. The uniqueness of this projection implies that $K(z, w)$ is independent of the choice of $\{u_i\}$; any orthonormal basis of $(L^2 \cap H)(\Omega)$ will do. (Bergman [1].)

In the ball, the normalized holomorphic monomials form such a basis (see Proposition 1.4.9), and the kernel 3.1.2(1) can be obtained by summing the resulting series.

3.2. The Cauchy Integral in B

3.2.1. Definition. The *Cauchy kernel* for B is the function C defined by

(1) $$C(z, \zeta) = (1 - \langle z, \zeta \rangle)^{-n}$$

for all $(z, \zeta) \in \mathbb{C}^n \times \mathbb{C}^n$ that satisfy $\langle z, \zeta \rangle \neq 1$.

If $f \in L^1(\sigma)$, or, more generally, if μ is a (complex Borel) measure on S, define

(2) $$C[f](z) = \int_S C(z, \zeta)f(\zeta)d\sigma(\zeta) \qquad (z \in B)$$

and

$$(3) \qquad C[\mu](z) = \int_S C(z, \zeta)d\mu(\zeta) \qquad (z \in B).$$

We call $C[f]$ and $C[\mu]$ the *Cauchy integrals* of f and μ, respectively. It is clear that these are holomorphic functions in B.

The operator that takes f to $C[f]$, or μ to $C[\mu]$, is called the *Cauchy transform*.

3.2.2. Lemma. *The Cauchy transform commutes with the action of the unitary group $\mathscr{U}$.*

Proof. The conclusion is, more explicitly, that

$$(1) \qquad C[f \circ U] = (C[f]) \circ U \qquad (f \in L^1(\sigma), U \in \mathscr{U}).$$

Since $C(z, U^{-1}\zeta) = C(Uz, \zeta)$ and σ is $\mathscr{U}$-invariant, (1) follows from

$$\int_S C(z, \zeta)f(U\zeta)d\sigma(\zeta) = \int_S C(z, U^{-1}\zeta)f(\zeta)d\sigma(\zeta) = \int_S C(Uz, \zeta)f(\zeta)d\sigma(\zeta).$$

3.2.3. Definition. The *ball algebra* $A(B)$ is the class of all $f: B \to \mathbb{C}$ that are continuous on the closed ball $\bar{B}$ and that are holomorphic in its interior B.

Equipped with the supremum norm $\|f\|_\infty$, $A(B)$ is a Banach algebra. When $n = 1$, it is the well-known disc algebra.

3.2.4. The Cauchy Formula in B. *If $f \in A(B)$ and $z \in B$, then*

$$(1) \qquad f(z) = \int_S \frac{f(\zeta)}{(1 - \langle z, \zeta \rangle)^n} \, d\sigma(\zeta).$$

This will later be extended to larger classes of holomorphic functions. We begin with $A(B)$ since here there is no problem concerning the existence of boundary values.

When $n = 1$ there is of course no need to prove this. When $n > 1$, the following proof shows that the Cauchy formula in B_n is a consequence of the Bergman formula in B_{n-1}.

Proof. Fix $f \in A(B)$, $z \in B = B_n$. Write points $w \in \mathbb{C}^n$ in the form $w = (w', w_n)$, where $w' = (w_1, \ldots, w_{n-1})$. By Lemma 3.2.2 we may assume, without loss of generality, that $z_n = 0$, i.e., that $z = (z', 0)$. Define

$$(2) \qquad g(w) = C(z, w)f(w) \qquad (w \in \bar{B}).$$

Since $z_n = 0$, $C(z, w) = K(z', w')$, the Bergman kernel for B_{n-1}.

Next, $g(\zeta', w_n)$ is a holomorphic function of w_n, for every $\zeta \in S$, in the disc $|w_n| < |\zeta_n|$, and is continuous on the closure of this disc. Hence

(3)
$$g(\zeta', 0) = \frac{1}{2\pi} \int_{-2\pi}^{\pi} g(\zeta', e^{i\theta}\zeta_n)d\theta.$$

We now integrate (3) over B_{n-1}. On the right, Proposition 1.4.7(2) shows that we obtain

(4)
$$\int_S g \, d\sigma = C[f](z).$$

Since $g(\zeta', 0) = K(z', \zeta')f(\zeta', 0)$, Theorem 3.1.3 shows that the integral of the left side of (4) is $f(z', 0) = f(z)$. This proves (1).

It is curious that this simple formula (1) is not older than it appears to be. It occurs in Hua's 1958 book [1; p. 93], for more general symmetric domains, and was rediscovered in 1964 by Bungart [1], who computed the kernel $C(z, \zeta)$ from the normalized holomorphic monomials in $L^2(\sigma)$, following Bergman's method (see §3.1.4).

As in one variable, there is a Cauchy formula in B that expresses f in terms of its real part on S:

3.2.5. Theorem. *If $f \in A(B)$, $u = \operatorname{Re} f$, and $f(0)$ is real, then, for $z \in B$,*

(1)
$$f(z) = \int_S [2C(z, \zeta) - 1]u(\zeta)d\sigma(\zeta).$$

Proof. Assume $f(0) = 0$, without loss of generality. Then $\int f \, d\sigma = 0$, hence $\int u \, d\sigma = 0$, and the right side of (1) is $C[f] + C[\bar{f}]$. Put

(2)
$$g(w) = C(w, z)f(w) \qquad (w \in \bar{B}).$$

Then $g \in A(B)$, $g(0) = 0$, hence $\int g \, d\sigma = 0$. Since $C[\bar{f}]$ is the complex conjugate of $\int g \, d\sigma$, the right side of (1) is $C[f](z)$.

3.3. The Invariant Poisson Integral in B

3.3.1. Definition. The kernel

(1)
$$P(z, \zeta) = \frac{(1 - |z|^2)^n}{|1 - \langle z, \zeta \rangle|^{2n}} \qquad (z \in B, \zeta \in S)$$

is called the *invariant Poisson kernel* in $B = B_n$. Theorem 3.3.8 justifies this terminology. We shall usually omit the adjective "invariant" when there is no danger of confusing P with the ordinary Poisson kernel that will be briefly discussed in §3.3.10.

The definition (1) shows that the Poisson kernel is related to the Cauchy kernel by the formula

$$(2) \qquad P(z, \zeta) = \frac{C(z, \zeta)C(\zeta, z)}{C(z, z)}.$$

The (invariant) *Poisson integral* $P[f]$ of a function $f \in L^1(\sigma)$ is defined, for $z \in B$, by

$$(3) \qquad P[f](z) = \int_S P(z, \zeta)f(\zeta)d\sigma(\zeta).$$

Similarly,

$$(4) \qquad P[\mu](z) = \int_S P(z, \zeta)d\mu(\zeta)$$

is the Poisson integral of the measure μ on S.

3.3.2. Theorem. *If $f \in A(B)$ then $f(z) = P[f](z)$ for $z \in B$.*

Proof. Fix $z \in B$, put

$$(1) \qquad g(w) = \frac{C(w, z)}{C(z, z)}f(w) \qquad (w \in \bar{B}).$$

Then $g \in A(B)$, and $f(z) = g(z)$. Hence

$$f(z) = \int_S C(z, \zeta)g(\zeta)d\sigma(\zeta) = \int_S P(z, \zeta)f(\zeta)d\sigma(\zeta).$$

The last equality used (1) and 3.3.1(2).

3.3.3. Proposition. *If $0 \le r < 1, \zeta \in S, \eta \in S$, then*

$$(1) \qquad P(r\eta, \zeta) = P(r\zeta, \eta).$$

Also

$$(2) \qquad \int_S P(r\eta, \zeta)d\sigma(\zeta) = 1 = \int_S P(r\eta, \zeta)d\sigma(\eta).$$

Proof. (1) is clear from 3.3.1(1). The first equality in (2) is the special case $f = 1$ of Theorem 3.3.2; the second follows from (1).

3.3.4. Theorem. (a) *If $f \in C(S)$ and F is defined on $\bar{B}$ so that $F = f$ on S and $F = P[f]$ in B, then $F \in C(\bar{B})$ and $\|F\|_\infty = \|f\|_\infty$.*
   (b) *If $1 \leq p \leq \infty$, $f \in L^p(\sigma)$, $u = P[f]$, and*

$$u_r(\zeta) = u(r\zeta) \qquad (0 \leq r < 1, \zeta \in S)$$

then $\|u_r\|_p \leq \|f\|_p$. If also $p < \infty$ then

$$\lim_{r \to 1} \|u_r - f\|_p = 0.$$

   (c) *If μ is a complex Borel measure on S, with total variation $\|\mu\|$, and $u = P[\mu]$, then $\|u_r\|_1 \leq \|\mu\|$, for $0 \leq r < 1$, and*

$$\lim_{r \to 1} u_r \, d\sigma = d\mu$$

in the weak*-topology of the dual space of $C(S)$.
   (d) *If $P[\mu](z) = 0$ for all $z \in B$, then $\mu = 0$.*

The proofs are so similar to the classical case of the Poisson integral in the disc that a brief sketch will suffice.

By 3.3.1(1), $P(z, \zeta) > 0$ and $P(r\eta, \zeta) \to 0$ uniformly, as $r \to 1$, if $\eta \in S$ lies outside any prescribed neighborhood of ζ. Part (a) follows now exactly as in the disc, because of 3.3.3(2).

Part (b) follows from Hölder's inequality, and the fact that $C(S)$ is dense in $L^p(\sigma)$ when $p < \infty$.

The first assertion of (c) is obvious. With the aid of 3.3.3(1), Fubini's theorem gives

$$\int_S u_r f \, d\sigma = \int_S P[f]_r \, d\mu$$

if $u = P[\mu]$ and $f \in C(S)$. By (a), $P[f]_r \to f$ uniformly. Thus $\int u_r f \, d\sigma \to \int f \, d\mu$, as $r \to 1$.

Finally, (d) is a corollary of the second assertion in (c).

The following theorem shows how the Poisson kernel is related to the automorphisms of B.

3.3.5. Theorem. *If $\psi \in Aut(B)$, $z \in B$, and $\zeta \in S$, then*

(1) $P(\psi(z), \zeta) = P(z, \psi^{-1}(\zeta))P(\psi(0), \zeta).$

Proof. Put $a = \psi^{-1}(0)$. The identity 2.2.5(2) is equivalent to

$$(2) \qquad\qquad C(\psi(z), \psi(\zeta)) = \frac{C(a, a)C(z, \zeta)}{C(z, a)C(a, \zeta)},$$

and 3.3.1(2) converts this to

$$(3) \qquad\qquad P(\psi(z), \psi(\zeta)) = \frac{P(z, \zeta)}{P(a, \zeta)}.$$

In particular, $P(\psi(0), \psi(\zeta)) = 1/P(a, \zeta)$, since $P(0, \zeta) = 1$. Thus

$$(4) \qquad\qquad P(\psi(z), \psi(\zeta)) = P(z, \zeta)P(\psi(0), \psi(\zeta)).$$

If we replace ζ by $\psi^{-1}(\zeta)$ in (4), we obtain (1).

3.3.6. Definition. A function $f \in C(B)$ is said to have the *invariant mean value property* if

$$(1) \qquad\qquad f(\psi(0)) = \int_S f(\psi(r\zeta))d\sigma(\zeta)$$

for every $\psi \in \mathrm{Aut}(B), 0 < r < 1$.

If we set $\psi(0) = a$, then $\psi = \varphi_a U$ for some $U \in \mathcal{U}$, and the $\mathcal{U}$-invariance of σ shows that

$$(2) \qquad\qquad f(a) = \int_S f(\varphi_a(r\zeta))d\sigma(\zeta)$$

if (1) holds. The integral in (2) is an average of f over the boundary of the ellipsoid $E(a, r)$; see §2.2.7.

A class X of functions with domain B or S or $\bar{B}$ is said to be *Moebius-invariant*, or simply $\mathcal{M}$-*invariant*, if $f \circ \psi \in X$ for every $f \in X$ and for every $\psi \in \mathrm{Aut}(B)$.

As an example of this terminology, observe that *the class of all f with the invariant mean value property is $\mathcal{M}$-invariant.* For if f satisfies (1), $\psi_1 \in \mathrm{Aut}(B)$, and $g = f \circ \psi_1$, then g satisfies (1), simply because $g \circ \psi = f \circ (\psi_1 \circ \psi)$, and $\psi_1 \circ \psi \in \mathrm{Aut}(B)$.

3.3.7. Theorem. *If μ is a complex Borel measure on S, then $P[\mu]$ has the invariant mean value property.*

Proof. Put $u = P[\mu]$. If z is replaced by $r\eta$ in Theorem 3.3.5 (with $\eta \in S$), and if we integrate the result with respect to $d\sigma(\eta)$, we obtain, for any $\psi \in \mathrm{Aut}(B)$,

(1)
$$\int_S P(\psi(r\eta), \zeta)d\sigma(\eta) = P(\psi(0), \zeta)$$

by Proposition 3.3.3. Multiply (1) by $d\mu(\zeta)$ and use Fubini's theorem to get the desired conclusion

(2)
$$\int_S (u(\psi(r\eta))d\sigma(\eta) = u(\psi(0)).$$

3.3.8. Theorem. *If $f \in L^1(\sigma)$ then*

(1)
$$P[f \circ \psi] = P[f] \circ \psi$$

for every $\psi \in \mathrm{Aut}(B)$.

In other words, the Poisson transform commutes with the action of $\mathrm{Aut}(B)$. The theorem also implies that the class of all Poisson integrals of L^1-functions is $\mathscr{M}$-invariant.

Proof. Since $C(S)$ is dense in $L^1(\sigma)$, it is enough to prove the theorem for $f \in C(S)$. By Theorem 3.3.7, with $d\mu = f\, d\sigma$,

(2)
$$P[f](\psi(0)) = \int_S P[f](\psi(r\eta))d\sigma(\eta).$$

When $r \to 1$, the integrand converges uniformly to $f(\psi(\eta))$, by 3.3.4(a). Hence

(3)
$$P[f](\psi(0)) = P[f \circ \psi](0),$$

which establishes (1) at the origin.

For any $z \in B$, two applications of (3) give

$$P[f](\psi \circ \varphi_z)(0) = P[f \circ \psi \circ \varphi_z](0) = P[f \circ \psi](\varphi_z(0)).$$

Thus $P[f](\psi(z)) = P[f \circ \psi](z)$.

Remark. Since $P(0, \zeta) = 1$, (3) is the same as

(4)
$$\int_S P(\psi(0), \zeta)f(\zeta)d\sigma(\zeta) = \int_S f(\psi(w))d\sigma(w).$$

Replace f by $f \circ \psi^{-1}$, then replace ψ by ψ^{-1}, and (4) turns into a change-of-variables formula

(5)
$$\int_S P(\psi^{-1}(0), \zeta) f(\psi(\zeta)) d\sigma(\zeta) = \int_S f \, d\sigma$$

in which the Poisson kernel plays the role of a Jacobian.

3.3.9. Example. Lemma 3.2.2 showed that the Cauchy transform commutes with $\mathscr{U}$. But it does not have the stronger invariance property of Theorem 3.3.8, for any $n \geq 1$.

For instance, let $f(\zeta) = \bar\zeta_1$. Then $C[f](z) = 0$ for every $z \in B$, but if $a = re_1, 0 < r < 1$, then

$$(f \circ \varphi_a)(\zeta) = \frac{r - \bar\zeta_1}{1 - r\bar\zeta_1}$$

so that

$$C[f \circ \varphi_a](0) = \int_S (f \circ \varphi_a) d\sigma = r \neq 0.$$

Thus $C[f \circ \varphi_a] \neq C[f] \circ \varphi_a$.

3.3.10. Remark. The kernel $P(z, \zeta)$ is intimately related to the Cauchy kernel of B and to the automorphisms of B. The functions that it produces (i.e., the Poisson integrals) are annihilated by a differential operator, the so-called *invariant Laplacian*, which will be the topic of Chapter 4. But these Poisson integrals need not be harmonic functions in the ordinary sense; i.e., their ordinary Laplacian need not be 0, except when $n = 1$.

Let us note, very briefly, that there is a Poisson kernel $Q(x, y)$ associated to the unit ball of any euclidean space R^N:

$$Q(x, y) = \frac{1 - |x|^2}{|x - y|^N} \qquad (|x| < 1, |y| = 1).$$

Straightforward (but tedious) differentiations show that Q is a harmonic function of x, for every y with $|y| = 1$. Also

$$\int Q(x, y) d\sigma(y) = 1$$

because this integral is a *radial* harmonic function in the ball, hence is constant (by Gauss' mean value theorem), hence can be evaluated by putting $x = 0$. Thus Q solves the Dirichlet problem for harmonic functions.

When $N = 2n$, and R^{2n} is identified with $\mathbb{C}^n$, then

$$Q(z, \zeta) = \frac{1 - |z|^2}{|\zeta - z|^{2n}} \qquad (z \in B, \zeta \in S).$$

Thus $Q(z, \zeta) = P(z, \zeta)$ *only* when $n = 1$.

Chapter 4

The Invariant Laplacian

4.1. The Operator $\tilde{\Delta}$

4.1.1. Definition. Suppose Ω is an open subset of B, $f \in C^2(\Omega)$, and $a \in \Omega$. We define

$$(1) \qquad\qquad (\tilde{\Delta}f)(a) = \Delta(f \circ \varphi_a)(0),$$

where φ_a is the involution defined in §2.2.1, and Δ is the ordinary Laplacian given by

$$(2) \qquad\qquad \Delta g = 4 \sum_{i=1}^{n} D_i \bar{D}_i g$$

as in §1.3.4.

This operator $\tilde{\Delta}$ is called the *invariant* Laplacian because it commutes with the automorphisms of B:

4.1.2. Theorem. *If $f \in C^2(\Omega)$ and $\psi \in \mathrm{Aut}(B)$, then*

$$\tilde{\Delta}(f \circ \psi) = (\tilde{\Delta}f) \circ \psi$$

in $\psi^{-1}(\Omega)$.

Proof. Pick $z \in \psi^{-1}(\Omega)$, put $w = \psi(z)$. Then $\varphi_w \circ \psi \circ \varphi_z$ fixes 0, hence is unitary, so that $\psi \circ \varphi_z = \varphi_w \circ U$ for some $U \in \mathcal{U}$. Hence

$$\tilde{\Delta}(f \circ \psi)(z) = \Delta(f \circ \psi \circ \varphi_z)(0) = \Delta(f \circ \varphi_w \circ U)(0).$$

But g and $g \circ U$ have the same Laplacian at 0 (for any $g \in C^2$), so that $\tilde{\Delta}(f \circ \psi)(z)$ equals

$$\Delta(f \circ \varphi_w)(0) = (\tilde{\Delta}f)(w) = (\tilde{\Delta}f)(\psi(z)).$$

The next theorem lists some other expressions for $\tilde{\Delta}f$ that will be used later. It is curious that they totally hide the invariance that was just proved.

4.1.3. Theorem. *If $f \in C^2(\Omega)$ and $a \in \Omega$, then*

(i) *in terms of integral averages,*

$$(\tilde{\Delta}f)(a) = \lim_{r \to 0} \frac{4n}{r^2} \int_S \{f(\varphi_a(r\zeta)) - f(a)\}d\sigma(\zeta);$$

(ii) *in terms of partial derivatives,*

$$(\tilde{\Delta}f)(a) = 4(1 - |a|^2) \sum_{i,k=1}^{n} (\delta_{ik} - a_i\bar{a}_k)(D_i\bar{D}_k f)(a),$$

where $\delta_{ik} = 0$ when $i \neq k$, $\delta_{ii} = 1$;

(iii) *in terms of the slice function f_a defined by $f_a(\lambda) = f(\lambda a)$,*

$$(\tilde{\Delta}f)(a) = (1 - |a|^2)[(\Delta f)(a) - (\Delta f_a)(1)].$$

Proof. If $h = f \circ \varphi_a$ then the formula

$$(\Delta h)(0) = \lim_{r \to 0} \frac{4n}{r^2} \int_S \{h(r\zeta) - h(0)\}d\sigma(\zeta)$$

is an easy consequence of the Taylor expansion of h about 0, since $h \in C^2$. By 4.1.1(1), this proves (i).

The discussion in §1.3.4 shows that (iii) is just another way of writing (ii).

We turn to the proof of (ii). Put $h = f \circ \varphi_a$, as above. Denote the components of φ_a by $\varphi_1, \ldots, \varphi_n$. Since these are holomorphic, the chain rule gives

$$(\Delta h)(0) = 4 \sum_{i,k} (D_i\bar{D}_k f)(a) \sum_m (D_m\varphi_i)(0)\overline{(D_m\varphi_k)(0)}.$$

Let us compute $(D_m\varphi_k)(0)$. The definition of φ_a (§2.2.1) shows that

$$\varphi_a(z) = \{1 + \langle z, a \rangle + \cdots\}\{a - Pz - s(z - Pz)\}$$

$$= a - sz + \frac{s}{1+s}\langle z, a \rangle a + \cdots.$$

The missing terms have z-degree 2 or more. Hence

$$(D_m\varphi_k)(0) = -s\delta_{mk} + \frac{s}{1+s}\bar{a}_m a_k.$$

Insert this into the expression for $(\Delta h)(0)$. After some computation, the inner sum (over m) simplifies to

$$s^2(\delta_{ik} - a_i \bar{a}_k).$$

Since $s^2 = 1 - |a|^2$ (this was of course used in the preceding computation), and since $(\tilde{\Delta}f)(a) = (\Delta h)(0)$, the proof is complete.

4.1.4. Remark. When $n = 1$, formula 4.1.3(ii) simplifies to

$$(\tilde{\Delta}f)(a) = (1 - |a|^2)^2(\Delta f)(a).$$

Thus, although Δ and $\tilde{\Delta}$ are not the same, they do annihilate the same functions. This fails to be true if $n > 1$.

4.1.5. Remark. It is an immediate consequence of Theorem 4.1.3(i) that $\tilde{\Delta}f = 0$ for every $f \in C^2(B)$ that has the invariant mean value property. Since $P[\mu]$ has this property (Theorem 3.3.7) for every complex Borel measure μ on S, it follows that $\tilde{\Delta}P[\mu] = 0$. Taking for μ a point mass at a fixed $\zeta \in S$, and putting $u(z) = P(z, \zeta)$, we see, in particular, that $\tilde{\Delta}u = 0$.

4.2. Eigenfunctions of $\tilde{\Delta}$

4.2.1. Definition. For every $\lambda \in \mathbb{C}$, we define X_λ to be the space of all $f \in C^2(B)$ that satisfy the equation

$$(1) \qquad\qquad\qquad \tilde{\Delta}f = \lambda f.$$

Theorem 4.1.2 shows that every X_λ is $\mathcal{M}$-invariant. The case $\lambda = 0$ will be the most interesting; the members of X_0 will be called $\mathcal{M}$-harmonic functions.

Recall that a function f with domain B is said to be *radial* if $f \circ U = f$ for every $U \in \mathcal{U}$. We associate to each $f \in C(B)$ its socalled "radialization" $f^{\#}$ by the formula

$$(2) \qquad\qquad f^{\#}(z) = \int_{\mathcal{U}} f(Uz)dU \qquad (z \in B).$$

The $\mathcal{U}$-invariance of dU shows that $f^{\#}$ is indeed a radial function.

Each X_λ is an eigenspace of $\tilde{\Delta}$, since none of them reduce to 0. This is part of the content of the following theorem.

4.2.2. Theorem. *If α and λ are related by*

$$(1) \qquad\qquad\qquad \lambda = -4n^2\alpha(1 - \alpha)$$

then X_λ contains every f of the form

(2) $$f(z) = \int_S P^\alpha(z, \zeta)d\mu(\zeta) \qquad (z \in B),$$

where μ is a complex Borel measure on S. In particular, X_λ contains the radial function g_α defined by

(3) $$g_\alpha(z) = \int_S P^\alpha(z, \zeta)d\sigma(\zeta) \qquad (z \in B).$$

Note that $P > 0$, so that the complex powers P^α are defined, as usual, by

(4) $$P^\alpha(z, \zeta) = \exp\{\alpha \log P(z, \zeta)\},$$

where the logarithm is real.

Proof. Fix $\zeta \in S$, and write (as temporary notation) $P(z)$ in place of $P(z, \zeta)$. Then $\tilde{\Delta}P = 0$ (Remark 4.1.5), hence $(\Delta P)(0) = 0$. Since $P(0) = 1$, it follows that

(5) $$(\Delta P^\alpha)(0) = 4\alpha(\alpha - 1) \sum_{j=1}^{n} (D_j P)(0)(\bar{D}_j P)(0).$$

But $(\bar{D}_j P)(0) = n\zeta_j$. Thus (5) becomes

(6) $$(\Delta P^\alpha)(0) = -4n^2\alpha(1 - \alpha) = \lambda.$$

Now apply Theorem 3.3.5 with φ_w in place of ψ:

(7) $$P^\alpha(\varphi_w(z), \zeta) = P^\alpha(z, \varphi_w(\zeta))P^\alpha(w, \zeta).$$

Take the Laplacian, with respect to z, at $z = 0$; by (6), the result is

(8) $$(\tilde{\Delta}P^\alpha)(w) = \lambda P^\alpha(w).$$

Hence $\tilde{\Delta}f = \lambda f$ for every f of the form (2).

Since $P(Uz, U\zeta) = P(z, \zeta)$ for every $U \in \mathcal{U}$, the $\mathcal{U}$-invariance of σ shows that g_α is radial.

In the remainder of this section, α and λ will always be related by (1).

4.2.3. Theorem. *If $f \in X_\lambda$ and f is radial, then $f = f(0)g_\alpha$.*

Corollary. $g_{1-\alpha} = g_\alpha$.

Proof. Every radial $f \in C^2(B)$ has the form $f(z) = u(|z|^2)$, and a little computation converts $\tilde{\Delta}f = \lambda f$ to

(1) $$Lu = \lambda u$$

where $Lu = au'' + bu'$, with

(2) $$a(t) = 4t(1 - t)^2, \qquad b(t) = 4(1 - t)(n - t).$$

The equation $Lu = 0$ has a solution

(3) $$u_0(t) = \int_t^1 s^{-n}(1 - s)^{n-1}\, ds \qquad (0 < t < 1)$$

which is unbounded as $t \to 0$.

If u_λ is a solution of (1), with $u_\lambda(0) = 1$, if v satisfies $u_\lambda^2 v' = u_0'$ (for small $t > 0$, where $u_\lambda(t) \neq 0$) and if $w = u_\lambda v$, one finds that $Lw = \lambda w$ and that $w(t)$ is unbounded as $t \to 0$. Since the solution space of (1) is 2-dimensional, the solutions of (1) that stay bounded as $t \to 0$ form a space of dimension 1. This proves the uniqueness asserted by the theorem.

The corollary follows, since 4.2.2(1) is unchanged if α is replaced by $1 - \alpha$.

4.2.4. Theorem. *Every $f \in X_\lambda$ satisfies*

(1) $$\int_S f(\psi(r\zeta))d\sigma(\zeta) = g_\alpha(r\eta)f(\psi(0))$$

for every $\psi \in \mathrm{Aut}(B)$, $0 \le r < 1$, $\eta \in S$. Equivalently

(2) $$\int_{\mathscr{U}} f(\varphi_z Uw)dU = g_\alpha(w)f(z) \qquad (z, w \in B).$$

Conversely, if $f \in C(B)$ and f satisfies (1), then $f \in C^\infty(B)$ and $f \in X_\lambda$.

Proof. If $f \in X_\lambda$, so is its radialization

(3) $$f^\# = \int_{\mathscr{U}} (f \circ U)dU.$$

By Theorem 4.2.3, $f^\# = f(0)g_\alpha$. This is (1) with f in place of $f \circ \psi$. The general case of (1) follows from the $\mathscr{M}$-invariance of X_λ.

If we take $\psi = \varphi_z$ in (1) and put $r = |w|$, Proposition 1.4.7(3) gives (2).

For the converse, pick a radial function $h \in C^\infty(B)$, with compact support in B, such that $\int_B h g_\alpha \, d\tau = 1$, where τ is the $\mathcal{M}$-invariant measure of Theorem 2.2.6. Multiply (2) by $h(w) d\tau(w)$, integrate, invert the order of integration, and use the $\mathcal{M}$-invariance of τ to obtain

(4) $$f(z) = \int_{\mathcal{U}} dU \int_B f(w) h(U^{-1} \varphi_z(w)) d\tau(w).$$

Since h is radial, the inner integral is independent of U, and (4) simplifies to

(5) $$f(z) = \int_B f(w) h(\varphi_z(w)) d\tau(w).$$

This makes it evident that $f \in C^\infty(B)$.

We can now differentiate (2) with respect to w, and find that

(6) $$\tilde{\Delta} f(\varphi_z(0)) = \lambda f(z),$$

i.e., that $\tilde{\Delta} f = \lambda f$. Thus $f \in X_\lambda$.

Corollary 1. *Every X_λ is a closed subspace of $C(B)$, in the topology of uniform convergence on compact sets.*

Proof. (1) is preserved if $f_j \to f$ uniformly on compact sets.

Corollary 2. *Every $\mathcal{M}$-harmonic function in B has the invariant mean value property, and every $f \in C(B)$ with the invariant mean value property is $\mathcal{M}$-harmonic.*

Proof. This is the case $\lambda = 0$ of the theorem. Note that $g_0 = g_1 = 1$.

Corollary 3. *For every $\lambda \in \mathbb{C}$, $X_\lambda \subset C^\infty(B)$.*

Actually, more is true:

4.2.5. Theorem. *Every $f \in X_\lambda$ is real-analytic in B.*

The term *real-analytic* refers of course to analyticity relative to R^{2n} rather than $\mathbb{C}^n$. A real-analytic function is thus one that is locally the sum of a convergent power series in the $2n$ real variables $x_1, y_1, \ldots, x_n, y_n$.

Proof. We will see that $\tilde{\Delta}$ meets the requirements of the following theorem about partial differential equations (Hörmander [1], Theorem 7.5.1): If L is an elliptic differential operator with real-analytic coefficients, then every solution u of $Lu = 0$ is real-analytic.

Theorem 4.1.3(ii) allows us to represent $\tilde{\Delta}$ in the form

$$(\tilde{\Delta}u)(z) = p(z, D)u$$

where (replacing $\partial/\partial x_j$ by u_j, $\partial/\partial y_j$ by v_j, $w_j = u_j + iv_j$)

$$p(z, w) = (1 - |z|^2)\left\{\sum_i w_i \bar{w}_i - \sum_{j,k} z_j \bar{z}_k \bar{w}_j w_k\right\}$$

$$= (1 - |z|^2)\{|w|^2 - |\langle z, w\rangle|^2\}.$$

This is a polynomial in the variables w_j, $\bar{w}_k$, whose coefficients are evidently real-analytic functions of the z-variables. For $z \in B$, $p(z, w) = 0$ only when $w = 0$. By the definition of "elliptic" (Hörmander [1], p. 177) this proves the ellipticity of $\tilde{\Delta}$.

4.2.6. Definition. $C_0(B)$ is the space of all $f \in C(B)$ such that $f(z) \to 0$ as $|z| \to 1$.

4.2.7. Theorem. (i) $X_\lambda \cap C_0(B) \neq \{0\}$ if and only if

$$(1) \qquad\qquad 4n^2 \operatorname{Re} \lambda + (\operatorname{Im} \lambda)^2 < 0.$$

(ii) X_λ contains a bounded function (not identically 0) if and only if

$$(2) \qquad\qquad 4n^2 \operatorname{Re} \lambda + (\operatorname{Im} \lambda)^2 \leq 0.$$

Proof. Let $\lambda = -4n^2\alpha(1 - \alpha)$, as before. Then (1) and (2) are equivalent to $0 < \operatorname{Re}\alpha < 1$ and $0 \leq \operatorname{Re}\alpha \leq 1$, respectively. Suppose $f \in X_\lambda$. If $f \in C_0(B)$, or f is bounded, the same is true of $f^\# = f(0)g_\alpha$ (Theorem 4.2.3). The following three assertions about g_α therefore prove the theorem. (Recall that $g_{1-\alpha} = g_\alpha$.)

   (a)   If $0 < \operatorname{Re}\alpha \leq \frac{1}{2}$ then $g_\alpha \in C_0(B)$.
   (b)   If $\operatorname{Re}\alpha = 0$ then $|g_\alpha| \leq 1$ but $g_\alpha \notin C_0(B)$.
   (c)   If $\operatorname{Re}\alpha < 0$ then g_α is unbounded in B.

To prove (a), write formula 4.2.2(3) more explicitly:

$$(3) \qquad\qquad g_\alpha(z) = (1 - |z|^2)^{n\alpha} \int_S \frac{d\sigma(\zeta)}{|1 - \langle z, \zeta\rangle|^{2n\alpha}}.$$

Proposition 1.4.10 shows that the integral in (3) is bounded if $\operatorname{Re}\alpha < \frac{1}{2}$ and that it is $O(-\log(1 - |z|))$ if $\operatorname{Re}\alpha = \frac{1}{2}$. In either case, (a) follows from (3), as does part of (b), namely that $|g_\alpha(z)| \leq 1$ if α is pure imaginary.

Put $\beta = -n\alpha$. Then (c) and the second part of (b) will be proved once we show that the integral

(4) $$\int_S |1 - \langle z, \zeta \rangle|^{2\beta} \, d\sigma(\zeta)$$

has a *non-zero* limit, as $|z| \to 1$, if $\text{Re } \beta \geq 0$.

Since (4) is a radial function of z, and since the absolute value of the integrand is bounded by a constant that depends only on β, it suffices to show that

(5) $$\int_S |1 - \zeta_1|^{2\beta} \, d\sigma(\zeta) \neq 0 \qquad (\text{Re } \beta \geq 0).$$

The details of the computation that establishes (5) are not very enlightening, and we only sketch the proof. Formula 1.4.5(2) converts (5) to an integral over U. Replace $\lambda \in U$ by $(z - 1)/(z + 1)$, $z = x + iy$, $x > 0$, to get an integral over the right half-plane, which, after replacing y by $(1 + x)t$, becomes a product of two integrals, one involving only x, the other only t. Each of these turns out to be a beta-function. This proves that the integral in (5) is not 0. Actually, the result can be further simplified, by means of Legendre's duplication formula for the gamma function, to

(6) $$\int_S |1 - \zeta_1|^{2\beta} \, d\sigma(\zeta) = \frac{\Gamma(n)\Gamma(n + 2\beta)}{\Gamma^2(n + \beta)},$$

valid for $\text{Re } \beta > -n/2$, by analytic continuation.

4.2.8. A Digression. Every function f with domain B "lifts" to a function f^* on the group $\text{Aut}(B)$, by setting

(1) $$f^*(\psi) = f(\psi(0)) \qquad (\psi \in \text{Aut}(B)).$$

Clearly, $f^*(\psi U) = f^*(\psi)$, for every $U \in \mathcal{U}$. If f is radial, then f^* has the bi-invariance property

(2) $$f^*(U_1 \psi U_2) = f^*(\psi) \qquad (U_1, U_2 \in \mathcal{U});$$

conversely, it is easy to see that every bi-invariant function on $\text{Aut}(B)$ arises in this way, from a radial function on B.

Formula 4.2.4(2) holds with g_α in place of f, and gives

(3) $$g_\alpha^*(\varphi_z) g_\alpha^*(\varphi_w) = \int_{\mathcal{U}} g_\alpha^*(\varphi_z U \varphi_w) \, dU.$$

If $\psi_1, \psi_2 \in \text{Aut}(B)$, then $\psi_1 = U_1\varphi_z$, $\psi_2 = \varphi_w U_2$, for some $z, w \in B$, $U_1, U_2 \in \mathcal{U}$, and the bi-invariance of g_α^* converts (3) to

$$(4) \qquad g_\alpha^*(\psi_1)g_\alpha^*(\psi_2) = \int_{\mathcal{U}} g_\alpha^*(\psi_1 U \psi_2)dU.$$

This says that g_α^* satisfies the functional equation that characterizes the so-called "spherical functions" on the group $\text{Aut}(B)$. We refer to Helgason [1], [2] and Lang [1] for more information on this subject.

4.3. $\mathcal{M}$-Harmonic Functions

4.3.1. Corollary 2 of Theorem 4.2.4 shows that the $\mathcal{M}$-harmonic functions are precisely those continuous function on B that have the invariant mean value property. Since Poisson integrals satisfy this (§4.1.5), every Poisson integral is $\mathcal{M}$-harmonic. The main purpose of the present section is to prove the converse (Theorem 4.3.3), subject, of course, to the obviously necessary growth restrictions.

But first we state a maximum principle:

4.3.2. Theorem. *Suppose Ω is an open subset of B, $u \in C(\overline{\Omega})$, $\tilde{\Delta}u = 0$ in Ω, and $u \le 0$ on $\partial\Omega$. Then $u \le 0$ in Ω.*

(Boundary and closure of Ω are understood to be taken with respect to $\mathbb{C}^n$, not just B.)

Proof. Put $h(z) = u(z) + \varepsilon z_1\bar{z}_1$, for some $\varepsilon > 0$. Then $h \le \varepsilon$ on $\partial\Omega$, and

$$(\tilde{\Delta}h)(z) = 4\varepsilon(1 - |z|^2)(1 - z_1\bar{z}_1) > 0$$

for all $z \in \Omega$. If h had a local maximum at some $z \in \Omega$, then $h \circ \varphi_z$ would have a local maximum at 0, which is impossible since

$$\Delta(h \circ \varphi_z)(0) = (\tilde{\Delta}h)(z) > 0.$$

Thus $h < \varepsilon$ in Ω. Since $u \le h$, the proof is completed by letting ε tend to 0.

4.3.3. Theorem. *Suppose $F: B \to \mathbb{C}$ is an $\mathcal{M}$-harmonic function that satisfies the growth condition*

$$(1) \qquad \sup_{0 < r < 1} \|F_r\|_p = M_p < \infty$$

for some p, $1 \le p \le \infty$.

If (1) *holds for some $p > 1$, then there is an $f \in L^p(\sigma)$ such that $F = P[f]$.*
If (1) *holds for $p = 1$, then there is a measure μ on S such that $F = P[\mu]$.*

Note that this is a converse of parts (b) and (c) of Theorem 3.3.4; part (d) of the same theorem establishes the uniqueness of the f and μ that occur in the present conclusion.

The norm in (1) is of course

$$
(2) \qquad \|F_r\|_p = \left\{ \int_S |F(r\zeta)|^p \, d\sigma(\zeta) \right\}^{1/p}
$$

when $1 \leq p < \infty$. When $p = \infty$, (1) simply says that F is bounded in B.

For $\mathcal{M}$-harmonic functions F that have continuous extensions to $\bar{B}$, the theorem is trivial: Let $u = P[F]$ be the Poisson integral of the restriction of F to S. Then $u - F$ is continuous on $\bar{B}$, $\mathcal{M}$-harmonic in B, 0 on S, hence 0 in B. (We have used Theorems 3.3.4(a) and 4.3.2.)

When dealing with harmonic functions in the disc (or in the unit ball of R^N), one can deduce the L^p-result by applying the preceding paragraph to the dilates F_r of F. This device is not available in the present context, since dilates of $\mathcal{M}$-harmonic functions need not be $\mathcal{M}$-harmonic. (See Theorem 4.4.10.)

For bounded F, Theorem 4.3.3 is contained in a much more general result proved by Fürstenberg [1] that depends on a fairly heavy dose of Lie group machinery. The L^p statements were added by Korányi [2]. The elementary proof that follows is due to David Ullrich. Its main ingredient is a simple equicontinuity argument.

Proof. Let $h: \mathcal{U} \to [0, \infty)$ be a continuous function that satisfies

$$
(3) \qquad \int_{\mathcal{U}} h(U)dU = 1,
$$

where dU is the Haar measure of $\mathcal{U}$. Define

$$
(4) \qquad G(z) = \int_{\mathcal{U}} F(Uz)h(U)dU \qquad (z \in B).
$$

For any $\zeta \in S$, $0 \leq r < 1$,

$$
(5) \qquad \int_{\mathcal{U}} |F(Ur\zeta)|^p \, dU = \int_S |F_r|^p \, d\sigma.
$$

(See Proposition 1.4.7.) Hence Hölder's inequality, applied to (4), gives

$$
(6) \qquad |G(z)| \leq M_p \|h\|_q \qquad (z \in B),
$$

where q is the exponent conjugate to p, and

(7) $\|G_r\|_p \le M_p \quad (0 < r < 1).$

We claim that $\{G_r : 0 < r < 1\}$ is an *equicontinuous* family of functions on S: Pick $\varepsilon > 0$. There is a neighborhood N of the identity in $\mathcal{U}$, such that

(8) $|h(U) - h(UU_0^{-1})| < \varepsilon \quad (U \in \mathcal{U}, U_0 \in N)$

and there is a $\delta > 0$ with the following property: If $\zeta \in S$, $w \in S$, $|\zeta - w| < \delta$, then $w = U_0 \zeta$ for some $U_0 \in N$.

The invariance of the Haar measure dU leads from (4) to

(9) $$G(U_0 z) = \int_{\mathcal{U}} F(Uz)h(UU_0^{-1})dU.$$

If we combine (4) and (9), and use (8), we obtain the desired equicontinuity:

(10) $|G_r(\zeta) - G_r(w)| \le M_1 \varepsilon \quad (0 < r < 1),$

whenever $|\zeta - w| < \delta$; note that $M_1 \le M_p$.

Since $\{G_r\}$ is equicontinuous and uniformly bounded (see (6)), there is a sequence $r_i \to 1$ such that $\{G_{r_i}\}$ converges uniformly to a function $g \in C(S)$. Let

(11) $\varepsilon_i = \sup_{\zeta} |G(r_i\zeta) - P[g](r_i\zeta)|.$

Since $|G(r_i\zeta) - g(\zeta)|$ and $|g(\zeta) - P[g](r_i\zeta)|$ tend to 0 uniformly, as $i \to \infty$, we see that $\varepsilon_i \to 0$.

We now use the fact that $\tilde{\Delta}F = 0$. By (4), $\tilde{\Delta}G = 0$. Thus $G - P[g]$ is $\mathcal{M}$-harmonic, and the maximum principle (Theorem 4.4.2), combined with (11), shows, for every i, that

(12) $|G(z) - P[g](z)| \le \varepsilon_i \quad (|z| \le r_i).$

Hence $G(z) = P[g](z)$ for all $z \in B$.

Note also that $\|g\|_p \le M_p$, by (7) and Fatou's lemma.

To finish, we do the preceding for a sequence $\{h_j\}$ in place of h, in which the supports of h_j shrink to the identity element of $\mathcal{U}$. The corresponding functions G_j, defined by (4) with h_j in place of h, converge then pointwise to $F(z)$. They are Poisson integrals of functions $g_j \in L^p(\sigma)$, with $\|g_j\|_p \le M_p$. If $p > 1$, some subsequence of $\{g_j\}$ converges, in the weak*-topology of $L^p(\sigma)$, to some $f \in L^p(\sigma)$. In particular, $P[g_j](z) \to P[f](z)$. Thus $F = P[f]$.

When $p = 1$, the weak*-convergence takes place in the dual of $C(S)$, and results in a measure μ that satisfies $F = P[\mu]$.

Corollary. *If F is a positive $\mathcal{M}$-harmonic function in B then there is a positive measure μ on S such that $F = P[\mu]$.*

Proof. By the mean-value property (Corollary 2 to Theorem 4.2.4) $\|F_r\|_1 = F(0)$ for all r, $0 < r < 1$. Thus, by the theorem, $F = P[\mu]$ for some μ, and μ is positive because μ is the weak*-limit of the positive functions F_r. (See Theorem 3.3.4.)

We shall now show, for continuous functions on the closed ball, that a much weaker mean value property implies $\mathcal{M}$-harmonicity:

4.3.4. The One-Radius Theorem. *Suppose $u \in C(\bar{B})$, and suppose that there corresponds to every $z \in B$ just one radius $r(z)(0 < r(z) < 1)$ such that*

$$(1) \qquad\qquad u(z) = \int_S u(\varphi_z(r(z)\zeta))d\sigma(\zeta).$$

Then u is $\mathcal{M}$-harmonic in B.

If $C(\bar{B})$ is replaced by $C(B)$, the theorem becomes false, as we shall see in §4.3.5. It is not known whether the theorem is true for bounded continuous functions in B (or, for that matter, for bounded real-analytic functions in B), even in the case $n = 1$.

See Zalcman [1] and Berenstein–Zalcman [1] for further information about such mean-value problems.

Proof. Let u be real, without loss of generality. Let v be the Poisson integral of the restriction of u to S, put $f = u - v$. We have to prove that $f = 0$.

Note that $f \in C(\bar{B})$, $f = 0$ on S, and (1) holds with f in place of u, since Poisson integrals have the invariant mean value property. Assume $f > 0$ at some point of B, let E be the set of all $z \in B$ at which f attains its maximum. Then E is a compact subset of B, and E therefore contains a point z which is nearest to S. The integral (1) (with f in place of u) is an average of f over the boundary of an ellipsoid around z; this boundary cannot lie in E; the integral is therefore smaller than $f(z)$. This contradiction proves the theorem.

4.3.5. Theorem. *To every r $(0 < r < 1)$ correspond infinitely many $\lambda \in \mathbb{C}$ such that*

$$u(z) = \int_S u(\varphi_z(r\zeta))d\sigma(\zeta)$$

for every $u \in X_\lambda$ and for every $z \in B$.

Proof. Fix $w \in B$ with $|w| = r$. By Theorem 4.2.4, it is enough to show that $g_\alpha(w) = 1$ for infinitely many $\alpha \in \mathbb{C}$.

Put $F(\alpha) = g_\alpha(w) - 1$. The definition of g_α (Theorem 4.2.2) shows that F is an entire function of order 1. Since $g_{1-\alpha} = g_\alpha$, we have $F(\alpha) = F(1 - \alpha)$. If $G(\lambda) = F(\frac{1}{2} + \lambda)$, it follows that G is even, and therefore $G(\lambda) = H(\lambda^2)$, where H is an entire function of order $\frac{1}{2}$. Since $F(\alpha)$ grows exponentially as $\alpha \to \infty$ through positive values, H is not a polynomial, and Hadamard's factorization theorem (Titchmarsh [1], Theorem 8.24) shows that H has infinitely many zeros in $\mathbb{C}$. The same is then true of F.

4.4. Pluriharmonic Functions

4.4.1. Definitions. Let Ω be open in $\mathbb{C}^n$. A function $u \in C^2(\Omega)$ is said to be *pluriharmonic* if it satisfies the n^2 differential equations

$$(1) \qquad\qquad D_j \bar{D}_k u = 0 \qquad (j, k = 1, \ldots, n).$$

The class of all real parts of holomorphic functions in Ω will be denoted by $RP(\Omega)$.

4.4.2. Remarks. (a) If $f \in H(\Omega)$ then $\bar{D}_k f = 0$ and $D_j \bar{f} = 0$. Hence f and $\bar{f}$ are pluriharmonic. So is $f + \bar{f}$. *Every* $u \in RP$ *is thus pluriharmonic.*

(b)   By Theorem 4.1.3(ii), every pluriharmonic u in B satisfies $\tilde{\Delta}u = 0$. *The pluriharmonic functions in B are thus $\mathcal{M}$-harmonic as well as (trivially) harmonic.*

(c)   If $u \in C^2(\Omega)$, $a \in \Omega$, $b \in \mathbb{C}^n$, put

$$u_{a,b}(\lambda) = u(a + \lambda b)$$

for all $\lambda \in \mathbb{C}$ for which $a + \lambda b \in \Omega$. As we saw in §1.3.4, *u is then pluriharmonic in Ω if and only if every $u_{a,b}$ is harmonic.*

Theorem 4.4.9 will show that (a) and (b) have converses. But first we present a recent result of Forelli which is related to (c) but uses far fewer complex lines.

4.4.3. Definitions. A complex function u, defined in a neighborhood of the origin in $\mathbb{C}^n$, is said to be of class $C^\infty(\{0\})$ if there corresponds to every positive integer k a neighborhood of 0 in which u is of class C^k.

A set $E \subset \mathbb{C}^n$ is said to be *balanced* if $\lambda z \in E$ whenever $z \in E$, $\lambda \in \mathbb{C}$, and $|\lambda| \le 1$.

If u is a function with domain E, and if $z \in E$, then u_z is defined by $u_z(\lambda) = u(\lambda z)$.

4.4.4. Theorem (Forelli [7]). *Let Ω be a balanced region in $\mathbb{C}^n$. Suppose $u: \Omega \to R$ has the following properties:*

(a) $u \in C^\infty(\{0\})$.

(b) *For every $z \in \Omega$ the slice function u_z is harmonic in the disc $\{\lambda \in \mathbb{C}: \lambda z \in \Omega\}$.*

Then $u \in RP(\Omega)$.

Note that the continuity of u in Ω is not part of the hypothesis, nor is it even assumed that u is bounded on compact subsets of Ω.

Proof. By (b), every u_z is harmonic in a neighborhood of the closed unit disc $\overline{U}$ in $\mathbb{C}$. Hence there corresponds to every $z \in \Omega$ a sequence of coefficients $F_k(z)$ such that

$$(1) \qquad u(re^{i\theta}z) = \sum_{k=-\infty}^{\infty} F_k(z) r^{|k|} e^{ik\theta}$$

for $0 \leq r \leq 1, 0 \leq \theta \leq 2\pi$. The series converges absolutely. In particular,

$$(2) \qquad \sum_{-\infty}^{\infty} |F_k(z)| < \infty \qquad (z \in \Omega).$$

Define

$$(3) \qquad f(z) = \sum_{k=1}^{\infty} F_k(z) \qquad (z \in \Omega).$$

Then

$$(4) \qquad u(z) = u(0) + f(z) + \overline{f(z)} \qquad (z \in \Omega)$$

and (1) implies, for $\lambda \in U, \mu \in U$, that

$$\sum_{k=1}^{\infty} F_k(\lambda z)\mu^k = f(\lambda\mu z) = \sum_{k=1}^{\infty} F_k(z)\lambda^k\mu^k.$$

Consequently, for $z \in \Omega, |\lambda| \leq 1, k = 1, 2, 3, \ldots$,

$$(5) \qquad F_k(\lambda z) = \lambda^k F_k(z).$$

Another consequence of (1) is

$$(6) \qquad F_k(z) = \frac{1}{2\pi} \int_{-\pi}^{\pi} u(e^{i\theta}z) e^{-ik\theta} \, d\theta.$$

Now consider a fixed positive integer k. By (a), there is a ball, centered at 0, in which $u \in C^k$. By (6), $F_k \in C^k$ in this ball. In particular, F_k is bounded there, and hence (5) implies that $F_k(z) = O(|z|^k)$ as $z \to 0$. The Taylor expansion of F_k thus has the simple form

$$(7) \qquad F_k(z) = \sum_{r+s=k} P_{rs}(z) + |z|^k \gamma(z)$$

where $P_{rs}(z)$ has total degree r in the variables $z_1, \ldots, z_n$, total degree s in $\bar{z}_1, \ldots, \bar{z}_n$, and $\gamma(z) \to 0$ as $z \to 0$. If we combine (5) and (7), we see that

$$(8) \qquad \sum_{r+s=k} \lambda^r \bar{\lambda}^s P_{rs}(z) + |\lambda z|^k \gamma(\lambda z)$$

is equal to

$$(9) \qquad \sum_{r+s=k} \lambda^k P_{rs}(z) + \lambda^k |z|^k \gamma(z).$$

When $\lambda > 0$, the two sums are equal, so that $\gamma(\lambda z) = \gamma(z)$. Letting $\lambda \to 0$, we see that $\gamma = 0$. The two sums in (8) and (9) are thus equal for all $\lambda \in U$. Hence $P_{rs}(z) = 0$ except when $r = k$, $s = 0$.

Returning to (7), we have now proved that F_k is a *homogeneous (holomorphic) polynomial of degree k*.

By (2) and (3), Theorem 1.5.6 tells us now that $f \in H(\Omega)$. By (4), this completes the proof.

4.4.5. Theorem. *Let* Ω *be a balanced region in* $\mathbb{C}^n$, *and suppose* $f: \Omega \to \mathbb{C}$ *satisfies*

(a) $f \in C^\infty(\{0\})$, *and*
(b) *all slice functions* f_z *are holomorphic.*

Then $f \in H(\Omega)$.

Proof. Forelli's theorem applies to $u = \operatorname{Re} f$, and shows that there exists $F \in H(\Omega)$ such that $u = \operatorname{Re} F$. For each $z \in \Omega$, $F_z - f_z$ is holomorphic, pure imaginary, hence equal to a constant $c(z)$. But

$$c(z) = F_z(0) - f_z(0) = F(0) - f(0),$$

independent of z. Thus $f = c + F$.

This characterization of $H(\Omega)$ leads to a very simple formulation and proof of a theorem concerning removable singularities (Theorem 4.4.7).

4.4.6. Definition. Let Ω be a region in $\mathbb{C}^n$. A (relatively) closed subset E of Ω is said to be H^∞-*removable in* Ω if every bounded $f \in H(\Omega \backslash E)$ has an extension $F \in H(\Omega)$.

For example, when $n = 1$, then every discrete set $E \subset \Omega$ is H^∞-removable. The same is true if the 1-dimensional Hausdorff measure of E is 0. (Ahlfors–Beurling [1].)

Let us say that E is 1-*dimensionally* H^∞-*removable* in Ω if (informally) $L \cap E$ is H^∞-removable in $L \cap \Omega$, for every complex line L that contains a point of $\Omega \backslash E$.

More precisely: If $a \in \Omega \backslash E$, $b \in \mathbb{C}^n$,

$$V = \{\lambda \in \mathbb{C} : a + \lambda b \in \Omega\},$$

and V_0 is the component of V that contains 0, then

$$\{\lambda \in V_0 : a + \lambda b \in E\}$$

is to be H^∞-removable in V_0.

4.4.7. Theorem. *If E is a relatively closed subset of a region $\Omega \subset \mathbb{C}^n$, and if E is 1-dimensionally H^∞-removable in Ω, then E is H^∞-removable in Ω.*

Proof. Let $f \in H^\infty(\Omega \backslash E)$. Pick $w \in E$. Since E has empty interior, w lies in a ball $p + rB \subset \Omega$, with $p \notin E$. Let $p = 0$, without loss of generality. Our hypothesis on E says then that every slice function f_z (with $z \in rS$) extends to a holomorphic function in U. This extends f to a function in rB which, by Theorem 4.4.5, is holomorphic in rB. A neighborhood of w has thus been removed from the set of singularities of f. Continuing in this way, all of E can be wiped out.

The following special case is a standard result that is frequently encountered:

Corollary. *If Ω is a region in $\mathbb{C}^n$, $g \in H(\Omega)$, $g \not\equiv 0$, and*

$$E = \{z \in \Omega : g(z) = 0\}$$

then E is H^∞-removable in Ω.

Proof. If $a \in \Omega$, $g(a) \neq 0$, $b \in \mathbb{C}^n$, and V and V_0 are as in Definition 4.4.6, then

$$\{\lambda \in V_0 : g(a + \lambda b) = 0\}$$

is a discrete subset of V_0. Hence E is 1-dimensionally H^∞-removable in Ω, and the corollary follows from the theorem.

4.4.8. Let Ω again be a balanced region in $\mathbb{C}^n$, and suppose $f \in C^2(\Omega)$. Forelli's theorem asserts then that if f satisfies the smoothness condition

(1) $f \in C^\infty(\{0\})$

as well as the differential equation

(2) $\sum_{j,k=1}^{n} z_j \bar{z}_k (D_j \bar{D}_k f)(z) = 0 \qquad (z \in \Omega),$

then f is pluriharmonic in Ω. Thus, in the presence of (1), the single equation (2) implies that the n^2 equations $D_j \bar{D}_k f = 0$ hold.

Similarly, Theorem 4.4.5 asserts that $f \in H(\Omega)$ provided that f satisfies (1) and

(3) $\sum_{k=1}^{n} \bar{z}_k (\bar{D}_k f)(z) = 0 \qquad (z \in \Omega).$

It is a very curious fact that these results fail completely if the assumption (1) is replaced by the existence of any *finite* number of derivatives in a neighborhood of 0, even if f is assumed to be real-analytic in the complement of the origin.

To see a simple example of this phenomenon, fix $n > 1$, let $p \geq 2$ be an integer, and define f on $\mathbb{C}^n$ by

(4) $f(z) = |z|^{-2} z_1^{2+p} \bar{z}_2 \qquad (z \neq 0),$

and $f(0) = 0$. Then $f \in C^p(\mathbb{C}^n)$, f is real-analytic except at 0, and all slice functions f_z are entire; in fact

(5) $f_z(\lambda) = f(\lambda z) = f(z)\lambda^{p+1} \qquad (z \in \mathbb{C}^n, \lambda \in \mathbb{C}).$

Thus f satisfies equations (2) and (3). Nevertheless, f is not even a harmonic function of z_2 (for fixed $z_1 \neq 0$), since $f \to 0$ as $z_2 \to \infty$.

Multiplication of (4) by suitable functions that are constant on each slice (except at 0) leads to solutions of (2) and (3) that are in $C^p(\mathbb{C}^n)$ but are not real-analytic anywhere.

4.4.9. Theorem. *If a function* $u \colon B \to R$ *has one of the following five properties, then it has the other four.*

    (a) $u \in RP(B)$.
    (b)   *u is pluriharmonic in* B.
    (c) $\Delta u = 0$ *and* $\tilde{\Delta} u = 0$ *in* B.
    (d) $u \in C^\infty(\{0\})$ *and* u_ζ *is harmonic in* U, *for every* $\zeta \in S$.
    (e) $u \circ \psi$ *is harmonic in* B, *for every* $\psi \in \mathrm{Aut}(B)$.

Proof. The implications (a) → (b) → (c) were noted in §4.4.2. That (c) → (d) follows from Theorem 4.1.3(iii), and Theorem 4.4.4 shows that (d) → (a).

It is trivial that (a) → (e). If (e) holds, then

$$(\tilde{\Delta}u)(a) = \Delta(u \circ \varphi_a)(0) = 0$$

for all $a \in B$, since $u \circ \varphi_a$ is harmonic. Thus (e) → (c), and the proof is complete.

The implications (e) → (a) and (c) → (a) occur in Nagel–Rudin [1], p. 865, and in Rudin [10], respectively. As is clear from the above proof, Forelli's theorem is now the only nontrivial step in these equivalences.

If B is replaced by any region $\Omega \subset \mathbb{C}^n$, the implication (b) → (a) holds in every ball in Ω, hence it holds globally whenever Ω is simply connected.

We conclude this section by clarifying a remark made prior to the proof of Theorem 4.3.3, to the effect that dilates of $\mathcal{M}$-harmonic functions need not be $\mathcal{M}$-harmonic:

4.4.10. Theorem. *If f is $\mathcal{M}$-harmonic in B and if there is one r, $0 < r < 1$, such that f_r is also $\mathcal{M}$-harmonic in B, then f is pluriharmonic in B.*

Proof. Put $g(z) = f(rz)$, for $z \in B$. Since $\tilde{\Delta}f = 0$ and $\tilde{\Delta}g = 0$, both f and g are solutions of the equation $(\Delta u)(z) = (\Delta u_z)(1)$, by 4.1.3(iii). Since $g_z(\lambda) = f_{rz}(\lambda)$, it follows that

$$\begin{aligned}
0 &= (\Delta g)(z) - (\Delta g_z)(1) \\
&= r^2(\Delta f)(rz) - (\Delta f_{rz})(1) = -(1 - r^2)(\Delta f)(rz).
\end{aligned}$$

Thus $\Delta f = 0$ in rB.

Being $\mathcal{M}$-harmonic, f is real-analytic in B (Theorem 4.2.5), hence so is Δf, and it follows that $\Delta f = 0$ in all of B. Thus f has property (c) of Theorem 4.4.9.

Chapter 5

Boundary Behavior of Poisson Integrals

The principal result of the present chapter is Korányi's Theorem 5.4.5—the fact that certain maximal functions associated to invariant Poisson integrals are of weak type $(1, 1)$. The existence of what we call "K-limits" at almost all points of S follows easily from this.

The chapter begins with the definition of a certain $\mathscr{U}$-invariant metric and the associated maximal function $M\mu$ of a measure μ on S. The corresponding maximal theorem 5.2.4 is true for many other choices of metrics. As the proof will show, the finiteness of the constant A_3 in Lemma 5.2.3 is basically all that is needed. The reason for the particular choice made is that $M\mu$ dominates $P[\mu]$ in a way that Theorem 5.4.5 makes precise.

The chapter contains some applications to holomorphic functions that do not depend on the more delicate theorems concerning the boundary behavior of Cauchy integrals. The latter will be studied in Chapter 6.

5.1. A Nonisotropic Metric on S

5.1.1. Definitions. For $a \in \bar{B}$, $b \in \bar{B}$,

(1)
$$d(a, b) = |1 - \langle a, b \rangle|^{1/2}.$$

For $\zeta \in S, \delta > 0$,

(2)
$$Q(\zeta, \delta) = \{\eta \in S : d(\zeta, \eta) < \delta\}.$$

Note the $\mathscr{U}$-invariance: for every $U \in \mathscr{U}$,

(3)
$$d(Ua, Ub) = d(a, b), \qquad UQ(\zeta, \delta) = Q(U\zeta, \delta).$$

The letter Q will always denote a "ball" $Q(\zeta, \delta)$. Occasionally, when the "center" ζ is of no particular importance, we shall write Q_δ in place of $Q(\zeta, \delta)$. If Q and Q' are balls with the same center ζ but with "radii" δ and

65

$t\delta$, we shall sometimes write $Q' = tQ$. Mathematically, this notation makes little sense, but it is quite convenient.

Note that $Q_\delta = S$ when $\delta > \sqrt{2}$.

Throughout this chapter, the letter A will denote a positive finite constant that depends only on the dimension n. Any dependence on other parameters will be explicitly indicated.

5.1.2. Proposition. (i) *The triangle inequality*

$$d(a, c) \le d(a, b) + d(b, c)$$

holds for all $a, b, c \in \bar{B}$.

(ii)   *On* S, *d is a metric; the sets* $Q(\zeta, \delta)$ *are the corresponding balls.*

Proof. Since d is $\mathcal{U}$-invariant, one may take $b = re_1 (0 \le r \le 1)$, and one then has to prove that

(1) $$|1 - \langle a, c \rangle| \le \{|1 - ra_1|^{1/2} + |1 - rc_1|^{1/2}\}^2.$$

Put $a' = a - a_1 e_1$, $c' = c - c_1 e_1$. The left side of (1) is then

$$|1 - a_1 \bar{c}_1 - \langle a', c' \rangle| \le |1 - a_1 \bar{c}_1| + |a'||c'|.$$

Since

$$|1 - a_1 \bar{c}_1| = |1 - ra_1 + a_1(r - \bar{c}_1)| \le |1 - ra_1| + |1 - r\bar{c}_1|$$

and

$$|a'|^2 \le 1 - |a_1|^2 \le 1 - |ra_1|^2 \le 2|1 - ra_1|,$$

with a similar estimate for $|c'|^2$, (1) holds.

This proves (i), and (ii) is an immediate consequence, since (on S), $d(\zeta, \eta) = 0$ if and only if $\zeta = \eta$. (This fails in B: if $|a| < 1$ then $d(a, a) > 0$.)

Note: When $n = 1$, our metric d is the square root of the euclidean metric of the unit circle as a subset of the plane.

5.1.3. Let $Q_\delta = Q(e_1, \delta)$, for small δ. Then Q_δ is close to being a $(2n - 1)$-dimensional ellipsoid (hence the term "nonisotropic" for our metric): Recall that $\zeta \in Q_\delta$ provided that $|1 - \zeta_1| < \delta^2$. In the y_1-direction (i.e., if $\zeta = \zeta_1 e_1$, $|\zeta_1| = 1$) the "thickness" (in the euclidean metric) of Q_δ is thus about δ^2. But if $\zeta_1 = 1 - \delta^2$, then

$$|\zeta'|^2 = 1 - |\zeta_1|^2 = 2\delta^2 - \delta^4,$$

so that ζ' ranges over a $(2n - 2)$-dimensional ball of radius about $2^{1/2}\delta$, which is much larger than δ^2.

This indicates that $\sigma(Q_\delta)$ is roughly proportional to $\delta^2 . \delta^{2n-2} = \delta^{2n}$, where δ is small.

We shall now prove this conclusion by an accurate computation, instead of relying on the imprecise argument which led to it.

5.1.4. Proposition. *When $n > 1$, the ratio $\sigma(Q_\delta)/\delta^{2n}$ increases from 2^{-n} to a finite limit A_0 as δ decreases from $\sqrt{2}$ to 0.*

When $n = 1$, this ratio decreases from $\frac{1}{2}$ to $1/\pi$.

Proof. The case $n = 1$ is elementary. When $n > 1$, apply formula 1.4.5(2) to the characteristic function of $Q(e_1, \delta)$, to get

$$\sigma(Q_\delta) = \frac{n-1}{\pi} \int_{E(\delta)} (1 - |\lambda|^2)^{n-2} \, dm_2(\lambda)$$

where $E(\delta) = \{\lambda : |\lambda| < 1 \text{ and } |1 - \lambda| < \delta^2\}$. The change of variables $1 - \lambda = \delta^2/z$ turns this into

$$\frac{\sigma(Q_\delta)}{\delta^{2n}} = \frac{n-1}{\pi} \int_{E'(\delta)} (2x - \delta^2)^{n-2} |z|^{-2n} dx \, dy$$

where $E'(\delta) = \{z = x + iy : 2x > \delta^2 \text{ and } |z| > 1\}$. As δ decreases, both the integrand and the domain of integration increase. The monotone convergence theorem shows therefore that the limit of $\sigma(Q_\delta)/\delta^{2n}$, as $\delta \to 0$, exists and equals

$$A_0 = \frac{(n-1)2^{n-2}}{\pi} \int_1^\infty \frac{dr}{r^{n+1}} \int_{-\pi/2}^{\pi/2} (\cos \theta)^{n-2} \, d\theta.$$

This proves the proposition. Computation of these integrals gives the more explicit result

$$A_0 = \frac{\frac{1}{4}\Gamma(n + 1)}{\Gamma^2(n/2 + 1)}.$$

5.2. The Maximal Function of a Measure on S

5.2.1. Terminology and Notation. If μ is a complex measure on S ("measure" will always mean "Borel measure") then, as usual,

    (i) $|\mu|$ is the total variation measure of μ,

    (ii) $\|\mu\| = |\mu|(S)$,

(iii) $\mu \ll \sigma$ means that μ is *absolutely continuous* with respect to σ, i.e., that $\mu(E) = 0$ for every Borel set $E \subset S$ with $\sigma(E) = 0$, and

(iv) $\mu \perp \sigma$ means that μ is *singular* with respect to σ, i.e., that there is a Borel set $E \subset S$ with $\sigma(E) = 0$, $|\mu|(E) = \|\mu\|$.

We recall the theorem of Lebesgue and Radon–Nikodym: $d\mu = f\, d\sigma + d\mu_s$, uniquely, with $f \in L^1(\sigma)$, $\mu_s \perp \sigma$.

5.2.2. Definition. The *maximal function* of a complex measure μ on S is the function $M\mu: S \to [0, \infty]$ defined by

$$(M\mu)(\zeta) = \sup_{\delta > 0} \frac{|\mu|(Q(\zeta, \delta))}{\sigma(Q_\delta)}.$$

Note that $M\mu = M|\mu|$.

For each fixed $\delta > 0$, the above quotient is easily seen to be a lower semi-continuous function of ζ. Hence $M\mu$ is lower semicontinuous.

The proof of Theorem 5.2.4 will use the following simple covering lemma, that involves the constant

$$A_3 = \sup_Q \frac{\sigma(3Q)}{\sigma(Q)}.$$

Proposition 5.1.4 implies that $A_3 < \infty$.

5.2.3. Lemma. *If E is the union of a finite collection Φ of balls $Q \subset S$, then Φ has a disjoint subcollection Γ such that*

(1)
$$E \subset \bigcup_\Gamma 3Q$$

and

(2)
$$\sigma(E) \le A_3 \sum_\Gamma \sigma(Q).$$

Proof. Order the members $Q_i = Q(\zeta_i, \delta_i)$ of Φ so that $\delta_i \ge \delta_{i+1}$. Put $i_1 = 1$. Suppose $k \ge 1$ and i_k is chosen. If Q_{i_k} intersects Q_i for every $i > i_k$, stop. If not, let i_{k+1} be the first index such that $Q_{i_{k+1}}$ is disjoint from Q_{i_k}. Since Φ is finite, this process stops, say at $k = m$. Put $\Gamma = \{Q_{i_1}, \ldots, Q_{i_m}\}$.

To every $Q_i \in \Phi$ corresponds some k such that $i_k \le i < i_{k+1}$ (except when $i \ge k_m$, since there is no i_{m+1}). Since $\delta_i \le \delta_{i_k}$ and Q_i intersects Q_{i_k}, we have $Q_i \subset 3Q_{i_k}$. This proves (1), and (2) follows from (1) and the definition of A_3.

5.2.4. Theorem. *If μ is a complex measure on S, then*

(1) $$\sigma\{M\mu > t\} \le A_3 t^{-1}\|\mu\|$$

for every $t > 0$.

The left side of (1) replaces the more cumbersome

(2) $$\sigma(\{\zeta \in S: (M\mu)(\zeta) > t\}).$$

We shall often simplify notation in this way.

Proof. Fix μ and t. Let K be a compact subset of the open set $\{M\mu > t\}$. Each $\zeta \in K$ is the center of a ball Q such that $|\mu(Q)| > t\sigma(Q)$. Some finite collection Φ of these Q's covers K. If Γ is as in Lemma 5.2.3, then

$$\sigma(K) \le A_3 \sum_\Gamma \sigma(Q) < A_3 t^{-1} \sum_\Gamma |\mu|(Q) \le A_3 t^{-1}\|\mu\|.$$

The disjointness of Γ was used in the last inequality. Now (1) follows by taking the supremum over all compact $K \subset \{M\mu > t\}$.

5.2.5. Weak L^1. If $f \in L^1(\sigma)$ and $t > 0$, then the inequality

(1) $$\sigma\{|f| > t\} \le t^{-1}\|f\|_1$$

is obvious (and holds equally well for any positive measure in place of σ).

Any measurable function f such that

(2) $$t\sigma\{|f| > t\}$$

is a bounded function of t on $(0, \infty)$ is said to belong to *weak $L^1(\sigma)$*.

Every $f \in L^1(\sigma)$ can be identified with the absolutely continuous measure $f \, d\sigma$. Its maximal function Mf is thus, in accordance with Definition 5.2.2, given by

(3) $$(Mf)(\zeta) = \sup_{\delta > 0} \frac{1}{\sigma(Q_\delta)} \int_{Q(\zeta, \delta)} |f| \, d\sigma.$$

Theorem 5.2.4 (restricted to L^1) can now be restated by saying that the "maximal operator" M sends $L^1(\sigma)$ to weak $L^1(\sigma)$, with constant A_3, i.e., that M is "of weak type (1, 1)" in the customary terminology. Since M is subadditive ($M(f + g) \le Mf + Mg$) and since the inequality $\|Mf\|_\infty \le \|f\|_\infty$ is trivial, the following L^p result is a consequence of the Marcinkiewicz interpolation theorem (Section 5.7):

5.2.6. Theorem. *For* $1 < p < \infty$ *there are constants* $A(p) < \infty$ *such that*

$$\int_S |Mf|^p \, d\sigma \le A(p) \int_S |f|^p \, d\sigma$$

for every $f \in L^p(\sigma)$.

5.2.7. Theorem. *If* μ_s *is the singular part of* μ *and if* $E = \{M\mu < \infty\}$, *then* $|\mu_s|(E) = 0$.

In other words, μ_s is concentrated on the set $\{M\mu = \infty\}$.

Proof. If $E_t = \{M\mu \le t\}$ then $E = \cup E_t$, and it is enough to prove that $|\mu_s|(E_t) = 0$ for every positive integer t. This, in turn, follows once we show that $\mu(K) = 0$ for every compact $K \subset E_t$ with $\sigma(K) = 0$.

Fix such a K. Choose $\varepsilon > 0$. There is an open set $\Omega \supset K$, having $\sigma(\Omega) < \varepsilon$. There is a finite collection Φ of balls $Q \subset \Omega$, with centers in K, whose union covers K. Choose Γ as in Lemma 5.2.3. If $Q \in \Gamma$, then $3Q$ is a ball with center in $K \subset E_t$, so that $|\mu|(3Q) \le t\sigma(3Q)$. Lemma 5.2.3 implies therefore that

$$|\mu|(K) \le \sum_\Gamma |\mu|(3Q) \le t \sum_\Gamma \sigma(3Q) \le A_3 t \sum_\Gamma \sigma(Q) \le A_3 t\sigma(\Omega) < A_3 t\varepsilon$$

by the disjointness of Γ. Thus $|\mu|(K) = 0$.

5.3. Differentiation of Measures on S

Theorem 5.2.4 gives easy access to differentiation theorems. We shall deal separately with the absolutely continuous case and with the singular case.

5.3.1. Theorem. *If* $f \in L^1(\sigma)$ *then*

(1)
$$\lim_{\delta \to 0} \frac{1}{\sigma(Q_\delta)} \int_{Q(\zeta, \delta)} |f - f(\zeta)| d\sigma = 0$$

for almost every $\zeta \in S$. *Hence*

(2)
$$f(\zeta) = \lim_{\delta \to 0} \frac{1}{\sigma(Q_\delta)} \int_{Q(\zeta, \delta)} f \, d\sigma \qquad \text{(a.e.)}$$

Note: The points ζ for which (1) holds are called the *Lebesgue* points of f. If $E \subset S$ is measurable and f is the characteristic function of E, then every point of E that is a Lebesgue point of f is called a *point of density* of E. The theorem implies that almost all points of E are points of density of E.

Proof. Define $T_f(\zeta)$ to be the left side of (1), but with lim sup in place of lim. Choose $t > 0$ and $\varepsilon > 0$. Choose $g \in C(S)$ with $\|f - g\|_1 < \varepsilon$. Put $h = f - g$. Then

    (i) $T_f \le T_g + T_h$,
    (ii) $T_g = 0$ since g is continuous, and
    (iii) $T_h \le |h| + Mh$.

Thus $T_f \le |h| + Mh$, so that $\{Tf > t\}$ is a subset of

$$E_t = \left\{|h| > \frac{t}{2}\right\} \cup \left\{Mh > \frac{t}{2}\right\}.$$

By Theorem 5.2.4, and 5.2.5(1),

$$\sigma(E_t) \le At^{-1}\|h\|_1 < At^{-1}\varepsilon.$$

Since ε was arbitrary, $\{Tf > t\}$ is a subset of Borel sets of arbitrarily small measure. Thus $\sigma\{Tf > t\} = 0$ for every $t > 0$. This implies that $(Tf)(\zeta) = 0$ (a.e.).

Corollary. *If $f \in L^1(\sigma)$ then $|f(\zeta)| \le (Mf)(\zeta)$ at every Lebesgue point of f.*

This follows directly from (2).

5.3.2. Theorem. *If μ is a complex measure on S and $\mu \perp \sigma$, then*

(1) $$\lim_{\delta \to 0} \frac{\mu}{\sigma}(Q(\zeta, \delta)) = 0 \qquad (\text{a.e. } [\sigma]).$$

Proof. Assume $\mu \ge 0$, with loss of generality. Pick $t > 0$, $\varepsilon > 0$. There is a decomposition $\mu = \mu_1 + \mu_2$, where μ_1 is the restriction of μ to some compact K with $\sigma(K) = 0$, and $\|\mu_2\| < \varepsilon$.

Write $(\mathscr{D}\mu)(\zeta)$ for the limit in (1), and $(\bar{\mathscr{D}}\mu)(\zeta)$ for the corresponding lim sup. Off K, $\bar{\mathscr{D}}\mu_1 = 0$, hence $\bar{\mathscr{D}}\mu = \bar{\mathscr{D}}\mu_2$. Thus

$$K \cup \{\bar{\mathscr{D}}\mu > t\} = K \cup \{\bar{\mathscr{D}}\mu_2 > t\} \subset K \cup \{M\mu_2 > t\},$$

and σ of this last set is at most

$$A_3 t^{-1}\|\mu_2\| < A_3 t^{-1}\varepsilon,$$

by Theorem 5.2.4. Letting $\varepsilon \to 0$, it follows that

$$\sigma\{\bar{\mathscr{D}}\mu > t\} = 0$$

for every $t > 0$. Letting $t \to 0$, we see that $\mathscr{D}\mu = 0$ a.e. $[\sigma]$.

5.3.3. The two preceding theorems can be combined: *If μ is any complex Borel measure on S, then its derivative*

$$(\mathscr{D}\mu)(\zeta) = \lim_{\delta \to 0} \frac{\mu}{\sigma}(Q(\zeta, \delta))$$

exists a.e. $[\sigma]$; *if* $d\mu = f\,d\sigma + d\mu_s$ *is the Lebesgue decomposition of μ, then*

$$(\mathscr{D}\mu)(\zeta) = f(\zeta) \qquad (\text{a.e. } [\sigma]).$$

5.4. K-Limits of Poisson Integrals

5.4.1. Approach Regions. For $\alpha > 1$ and $\zeta \in S$, we let $D_\alpha(\zeta)$ be the set of all $z \in \mathbb{C}^n$ such that

(1) $$|1 - \langle z, \zeta \rangle| < \frac{\alpha}{2}(1 - |z|^2).$$

It is clear that $D_\alpha(\zeta) \subset B$. When $\alpha \le 1$, (1) defines the empty set. As $\alpha \to \infty$, the regions $D_\alpha(\zeta)$ fill B, for every fixed $\zeta \in S$.

Note that every $U \in \mathscr{U}$ permutes the D_α's: the relation

(2) $$U(D_\alpha(\zeta)) = D_\alpha(U\zeta)$$

follows obviously from (1).

[The notational similarity between the regions $D_\alpha(\zeta)$ and the differential operators D^α should cause no confusion.]

In one variable, the classical theorem of Fatou states that Poisson integrals of measures on the unit circle have nontangential limits almost everywhere. The theorem extends, with the same proof, to harmonic functions in balls in R^N, where "nontangential" means approach within a cone. However, Korányi [2] proved that the invariant Poisson integrals in B have limits at almost all $\zeta \in S$, as z approaches ζ within any $D_\alpha(\zeta)$, although—and this is the remarkable point—approach to ζ within $D_\alpha(\zeta)$ is not restricted to be nontangential when $n > 1$.

To examine the shape of $D_\alpha(\zeta)$ more closely, let us take $\zeta = e_1$ [by (2), this involves no loss of generality] and write D_α in place of $D_\alpha(e_1)$. Thus $z \in D_\alpha$ if and only if

(3) $$|1 - z_1| < \frac{\alpha}{2}(1 - |z|^2).$$

The intersection of D_α with the complex line through 0 and e_1 is the familiar angular region

(4)
$$|1 - z_1| < \frac{\alpha}{2}(1 - |z_1|^2)$$

in the unit disc. However, the intersection of D_α with the copy of R^{2n-1} obtained by setting $y_1 = 0$ in (3) is the ball

(5)
$$\left(x_1 - \frac{1}{\alpha}\right)^2 + |z'|^2 < \left(1 - \frac{1}{\alpha}\right)^2,$$

where $z' = (z_2, \ldots, z_n)$. This ball is *tangent* to S at e_1.

Although they are not really needed for the boundary theorems that are our present concern, this is a natural place to introduce the so-called *complex tangents*. They will play an important role later on.

5.4.2. Complex Tangent Spaces. Any point $z = (z_1, \ldots, z_n) \in \mathbb{C}^n$ is also a point of R^{2n}, and can thus be written in the form

(1)
$$z = (x_1, y_1, \ldots, x_n, y_n)$$

where $z_k = x_k + iy_k$. Note that then

(2)
$$iz = (-y_1, x_1, \ldots, -y_n, x_n).$$

If $w = (w_1, \ldots, w_n)$, $w_k = u_k + iv_k$, then, in addition to the Hilbert space inner product $\langle z, w \rangle$ on $\mathbb{C}^n$ there is the dot-product

(3)
$$z \cdot w = \sum_{k=1}^{n} (x_k u_k + y_k v_k)$$

on R^{2n}. The two are related by

(4)
$$z \cdot w = \text{Re}\langle z, w \rangle.$$

Thus

(5)
$$z \cdot (iw) = -(iz) \cdot w.$$

In particular, $w \cdot (iw) = 0$.

Let Y now be an R-subspace of $\mathbb{C}^n = R^{2n}$, of real codimension 1. Pick $w \in \mathbb{C}^n$, $|w| = 1$, perpendicular to Y, i.e., $w \cdot z = 0$ for all $z \in Y$. Let Y_0 be the set of all $z \in Y$ that are perpendicular to both w and iw. It follows from (5) that Y_0 is a $\mathbb{C}$-subspace of $\mathbb{C}^n$, of complex codimension 1.

This maximal $\mathbb{C}$-subspace Y_0 of Y could also have been defined by $Y_0 = Y \cap (iY)$.

Now let $\zeta \in S$. The *tangent space* $T_\zeta = T_\zeta(S)$ to S at ζ consists then of all vectors $w \in \mathbb{C}^n$ that are perpendicular to the radius of B which ends at ζ. Thus $w \in T_\zeta$ if and only if $w \cdot \zeta = 0$. The *complex tangent space* $T_\zeta^{\mathbb{C}}$ is defined to be $T_\zeta \cap (iT_\zeta)$, and consists of all w that satisfy $\langle w, \zeta \rangle = 0$.

The preceding discussion shows that the vector $i\zeta$ lies in T_ζ but that $i\zeta$ is perpendicular to $T_\zeta^{\mathbb{C}}$.

To return to our regions $D_\alpha(\zeta)$, they are tangential to S in the direction of $T_\zeta^{\mathbb{C}}$. In formula 5.4.1(1), the left side is the euclidean distance from z to the affine set $\zeta + T_\zeta^{\mathbb{C}}$; the easiest way to see this is to refer to 5.4.1(3). The right side is essentially proportional to $1 - |z|$, the euclidean distance from z to S. This interpretation of 5.4.1(1) suggests generalizations which led E. M. Stein [2] to extend Korányi's theorem to a much wider class of regions.

When $n = 1$, this discussion is vacuous, since $\{0\}$ is then the only complex subspace of codimension 1.

5.4.3. Lemma. *If* $\eta \in S$, $\zeta \in S$, $z \in D_\alpha(\zeta)$, $|z| = r$, *then*

(1)
$$|1 - \langle \zeta, \eta \rangle| < 4\alpha |1 - \langle z, \eta \rangle|$$

and

(2)
$$P(z, \eta) < \frac{[32\alpha^2(1 - r)]^n}{|1 - \langle \zeta, \eta \rangle|^{2n}}.$$

Proof. $|1 - \langle z, \zeta \rangle| < \alpha(1 - |z|) \leq \alpha |1 - \langle z, \eta \rangle|$, so that $d(z, \zeta) < \sqrt{\alpha}\, d(z, \eta)$. By the triangle inequality (Proposition 5.1.2),

$$d(\zeta, \eta) \leq d(\zeta, z) + d(z, \eta) < (1 + \sqrt{\alpha})d(z, \eta).$$

This gives (1), since $\alpha > 1$. The estimate (2) follows from (1) and the definition of the Poisson kernel.

5.4.4. Definition. If $F \in C(B)$ and $\alpha > 1$, the maximal function $M_\alpha F : S \to [0, \infty]$ is defined by

(1)
$$(M_\alpha F)(\zeta) = \sup\{|F(z)| : z \in D_\alpha(\zeta)\}.$$

Note that $\{M_\alpha F \leq t\}$ is closed in S, for every real t, since F is continuous. Thus $M_\alpha F$ is lower semicontinuous.

5.4.5. Theorem (Korányi [2]). *To every $\alpha > 1$ corresponds a constant $A(\alpha) < \infty$ such that the inequality*

(1) $$M_\alpha P[\mu] \leq A(\alpha) M\mu$$

holds for every complex measure μ on S.

In conjunction with Therem 5.2.4 this says that the map $\mu \to M_\alpha P[\mu]$ carries measures into weak L^1.

Proof. Since $M\mu = M|\mu|$ and $|P[\mu]| \leq P[|\mu|]$, it is enough to prove (1) for positive μ. We shall work at the point e_1, and assume $(M\mu)(e_1) = 1$.

By 5.1.4 there is a constant $A_0 < \infty$ such that $\sigma(Q_\delta) \leq A_0 \delta^{2n}$ for all $\delta > 0$. (When $n = 1$, replace A_0 by $\frac{1}{2}$.)

We shall prove that

(2) $$P[\mu](z) \leq 2A_0(16\alpha)^n$$

for every $z \in D_\alpha(e_1)$. This establishes (1), with the right side of (2) for $A(\alpha)$.

Fix $z \in D_\alpha(e_1)$, put $r = |z|$, put $t = 8\alpha(1 - r)$, define

(3) $$V_0 = \{\omega \in S : |1 - \omega_1| < t\},$$

(4) $$V_k = \{\omega \in S : 2^{k-1}t \leq |1 - \omega_1| < 2^k t\},$$

for $k = 1, 2, \ldots$, until $2^k t > 2$. Since $(M\mu)(e_1) = 1$,

(5) $$\mu(V_k) \leq A_0(2^k t)^n \qquad (k = 0, 1, 2, \ldots).$$

Decompose $P[\mu](z)$ into a sum

(6) $$\int_S P(z, \omega) d\mu(\omega) = \int_{V_0} + \sum_{k \geq 1} \int_{V_k}.$$

For any $\omega \in S$, $P(z, \omega) < 2^n(1 - r)^{-n}$. Thus (5), with $k = 0$, shows that

(7) $$\int_{V_0} \leq A_0 \left(\frac{2t}{1 - r}\right)^n = A_0(16\alpha)^n.$$

When $k \geq 1$ and $\omega \in V_k$, Lemma 5.4.3 gives

(8) $$P(z, \omega) < \frac{(4\alpha t)^n}{|1 - \omega_1|^{2n}} \leq \left(\frac{16\alpha}{4^k t}\right)^n,$$

by (4). Hence (5) implies that

$$(9) \qquad \int_{V_k} \le A_0(16\alpha)^n \cdot 2^{-nk} \qquad (k = 1, 2, 3, \ldots).$$

The inequalities (9), added to (7), give (2).

5.4.6. K-Limits. Suppose $\zeta \in S$, Ω is open in B, and to every $\alpha > 1$ corresponds an $r < 1$ such that

$$(1) \qquad \{|z| > r\} \cap D_\alpha(\zeta) \subset \Omega.$$

($\Omega = B$ is the simplest and most important example.)

We say that a function $F: \Omega \to \mathbb{C}$ has K-limit λ at ζ, and write

$$(2) \qquad (K\text{-lim } F)(\zeta) = \lambda$$

if the following is true: For every $\alpha > 1$ and for every sequence $\{z_i\}$ in $D_\alpha(\zeta) \cap \Omega$ that converges to ζ, $F(z_i) \to \lambda$ as $i \to \infty$.

The case $\lambda = \infty$ is not excluded. But usually we will of course be interested in finite K-limits.

Korányi [2] used the terms "admissible limit" and "admissible convergence" in this context.

The next three theorems are consequences of the maximal theorems 5.4.5 and 5.2.4.

5.4.7. Theorem. *If μ is a positive measure on S and if $(\mathcal{D}\mu)(\zeta) = 0$ for some $\zeta \in S$, then*

$$(1) \qquad (K\text{-lim } P[\mu])(\zeta) = 0.$$

Proof. Pick $\varepsilon > 0$. Choose $Q_0 = Q(\zeta, \delta_0)$ so that

$$(2) \qquad \mu(Q(\zeta, \delta)) < \varepsilon\sigma(Q(\zeta, \delta)) \qquad (0 < \delta < \delta_0).$$

Let μ_0 be the restriction of μ to Q_0, put $\mu_1 = \mu - \mu_0$. Then $P[\mu_1]$ has K-limit 0 at ζ, simply because ζ is not in the support of μ_1, and $(M\mu_0)(\zeta) \le \varepsilon$, by (2). Theorem 5.4.5 implies therefore that

$$(3) \qquad \limsup_{i \to \infty} P[\mu](z_i) \le A(\alpha)\varepsilon$$

if $z_i \to \zeta$ within $D_\alpha(\zeta)$. This proves (1).

5.4.8. Theorem. *If* $f \in L^1(\sigma)$ *then*

$$(K\text{-}\lim P[f])(\zeta) = f(\zeta)$$

at every Lebesgue point ζ of f.

Proof. Fix such a ζ and apply Theorem 5.4.7 to the measure μ defined for all Borel sets $E \subset S$ by

$$\mu(E) = \int_E |f - f(\zeta)| d\sigma.$$

5.4.9. Theorem. *If μ is a complex measure on S, then*

$$(K\text{-}\lim P[\mu])(\zeta) = (\mathscr{D}\mu)(\zeta) \qquad \text{a.e. } [\sigma].$$

Proof. Let $d\mu = f d\sigma + d\mu_s$ be the Lebesgue decomposition of μ. By Theorems 5.4.8 and 5.3.3, it is enough to show that $P[\mu_s]$ has K-limit 0 a.e. $[\sigma]$. Since $|\mu_s| \perp \sigma$, this follows from 5.3.2 and 5.4.7.

5.4.10. Theorem. *If $1 < p < \infty$ and $f \in L^p(\sigma)$, then*

$$\int_S |M_\alpha P[f]|^p \, d\sigma \le A(\alpha, p) \int_S |f|^p \, d\sigma.$$

Proof. Combine Theorems 5.4.5 and 5.2.6. (The measurability of $M_\alpha F$ was established in §5.4.4, for any $F \in C(B)$.)

5.4.11. The Radial Maximal Function. For any $F: B \to \mathbb{C}$, define

(1) $$(M_{\text{rad}} F)(\zeta) = \sup_{0 \le r < 1} |F(r\zeta)|.$$

Since $r\zeta \in D_\alpha(\zeta)$ for $0 \le r < 1$ if $\alpha > 2$, Theorem 5.4.5 implies that

(2) $$M_{\text{rad}} P[\mu] \le A M\mu.$$

For *positive* measures, the opposite inequality holds too:

5.4.12. Theorem. *If μ is a positive measure on S, then*

(1) $$M\mu \le 9^n M_{\text{rad}} P[\mu].$$

Consequently, if $\mu \ge 0$ and one of $M\mu$, $M_{\text{rad}} P[\mu]$, $M_\alpha P[\mu]$ is finite at some $\zeta \in S$, so are the others.

Proof. For $0 < \delta \leq \sqrt{2}$, let $r = 1 - \frac{1}{2}\delta^2$. To cover the point $-e_1$, we work with the closed balls $\bar{Q}(e_1, \delta)$. If $\zeta \in \bar{Q}(e_1, \delta)$, then $|1 - \zeta_1| \leq 2(1 - r)$, hence $|1 - r\zeta_1| \leq 3(1 - r)$, so that

$$P(re_1, \zeta) \geq \left\{\frac{1 - r}{9(1 - r)^2}\right\}^n = 9^{-n}(1 - r)^{-n} > \frac{9^{-n}}{\sigma(Q_\delta)},$$

by Proposition 5.1.4. It follows that

$$\left(\frac{\mu}{\sigma}\right)(Q(e_1, \delta)) \leq 9^n P[\mu](re_1) \leq 9^n M_{\mathrm{rad}} P[\mu](e_1).$$

This proves (1) at the point e_1. The same estimate holds at every other point of S.

5.4.13. Examples. When $n = 1$, it is a classical fact that $P[\mu](re^{i\theta}) \to (\mathscr{D}\mu)(e^{i\theta})$ at *every* point of T where the symmetric derivative $\mathscr{D}\mu$ exists. Theorem 5.4.9 shows, for arbitrary n, that this limit relation still holds for almost all $\zeta \in S$.

But, as we shall now see, the result can fail at individual points. In the first two examples that follow we shall take $n = 2$.

The first example gives *a positive measure μ on S with $(\mathscr{D}\mu)(e_1) = 2$ but* $\lim_{r\to 1} P[\mu](re_1) = 4$.

We define μ by requiring that

$$\int_S f \, d\mu = \int_{-\pi}^{\pi} f(e^{i\theta}, 0)|\sin\theta| \, d\theta$$

for every $f \in C(S)$. If $0 < x < 1$, $c \in \mathbb{C}$, $|c| < 1$, then

$$P[\mu](x, c\sqrt{1 - x^2}) = \int_{-\pi}^{\pi} \frac{(1 - |c|^2)^2(1 - x^2)^2|\sin\theta|}{|1 - xe^{i\theta}|^4} \, d\theta$$

$$= 4(1 - |c|^2)^2.$$

With $c = 0$ this gives 4 for the radial limit. We see also that $P[\mu]$ has no K-limit at e_1. Next,

$$\mu(Q(e_1, \delta)) \sim \int_{-\delta^2}^{\delta^2} |\sin\theta| \, d\theta \sim \delta^4$$

whereas $\sigma(Q(e_1, \delta)) \sim \frac{1}{2}\delta^4$. (When $n = 2$, the computation used in the proof of Proposition 5.1.4 can be replaced by a direct estimate of the first integral that occurs in that proof.) Thus $(\mathscr{D}\mu)(e_1) = 2$.

Our second example gives *a real measure μ on S, with $(\mathscr{D}\mu)(e_1) = 0$ but*

$$\limsup_{r \to 1} P[\mu](re_1) = \infty.$$

Choose $m_k > 0$, $0 < t_k < 1$, for $k = 1, 2, 3, \ldots$, so that $\sum m_k < \infty$ but $m_k/(1 - t_k)^2 \to \infty$. Choose θ_k so that $|1 - e^{i\theta_k}| = 1 - t_k$. Let μ be the sum of the point masses

$$m_k \quad \text{at} \quad (e^{i\theta}, 0), \qquad -m_k \quad \text{at} \quad (t_k, \sqrt{1 - t_k^2}).$$

Then $\mu(Q(e_1, \delta)) = 0$ for every $\delta > 0$, hence $(\mathscr{D}\mu)(e_1) = 0$. Next,

$$P[\mu](re_1) = \sum_k \left\{ \frac{(1 - r^2)^2}{|1 - re^{i\theta_k}|^4} - \frac{(1 - r^2)^2}{(1 - rt_k)^4} \right\} m_k.$$

For $0 < r < 1$, each term in this sum is positive. When $r = t_k$, the kth term is

$$\frac{(2t_k + t_k^2)m_k}{(1 - t_k^2)^2}$$

which tends to ∞ as $k \to \infty$.

The proof of Theorem 5.4.5 shows that the constant $A(\alpha)$ is at most $A\alpha^n$. Our third example shows that *this is the correct order of magnitude, as far as its dependence on α is concerned.*

Let μ be the unit point mass at e_1. Then $P[\mu](z) = P(z, e_1)$. Pick $\zeta \in S$, with $0 < \zeta_1 < 1$, put $t = 1 - \zeta_1$. By 5.1.4, $(M\mu)(\zeta) \leq A/t^n$. If $\alpha \geq 3$ and $r = (\alpha - t)/(\alpha + t)$ then $re_1 \in D_\alpha(\zeta)$ (a computation shows this), hence

$$M_\alpha P[\mu](\zeta) \geq P(re_1, e_1) = \left(\frac{\alpha}{t}\right)^n \geq A^{-1}\alpha^n(M\mu)(\zeta).$$

5.5. Theorems of Calderón, Privalov, Plessner

This section extends two classical theorems of Privalov [1] and Plessner [1], concerning the nontangential behavior of harmonic and holomorphic functions in the unit disc, to analogous theorems about $\mathscr{M}$-harmonic and holomorphic functions in the approach regions $D_\alpha(\zeta)$. (Theorems 5.5.7 and 5.5.8.) Their first extension to several variables was made by Calderón [1]—to harmonic and holomorphic functions in half spaces and their behavior on cones with vertices on the boundary. The material of the present section is directly based on Calderón's ideas, although many details are different.

5.5.1. Definition. If $E \subset S$ and $\alpha > 1$, then

$$\Omega(E, \alpha) = \bigcup_{\zeta \in E} D_\alpha(\zeta).$$

For any set E, χ_E denotes its characteristic function: $\chi_E(x) = 1$ if $x \in E$, 0 if $x \notin E$.

5.5.2. Lemma. *If $E \subset S$, $\sigma(E) = m < 1$, and $\alpha > 1$, there is a constant $c = c(\alpha, m) > 0$ such that*

$$P[\chi_{S \setminus E}](z) \geq c$$

for every z outside $\Omega(E, \alpha)$.

Proof. Put $V = S \setminus E$, $f = \chi_V$. Pick $z \in B$, $z \notin \Omega(E, \alpha)$. Without loss of generality, $z = re_1$, $0 \leq r < 1$.

If $r \leq 1/\alpha$, a trivial lower bound for the Poisson kernel shows that

$$(1) \qquad P[f](z) \geq \left(\frac{1 - r}{1 + r}\right)^n \sigma(V) \geq \left(\frac{\alpha - 1}{\alpha + 1}\right)^n (1 - m).$$

So assume $1/\alpha < r < 1$. Let $\varphi_z \in \mathrm{Aut}(B)$ be as in §2.2.1. Then $f \circ \varphi_z$ is the characteristic function of $\varphi_z(V)$, and the $\mathcal{M}$-invariance of Poisson integrals (Theorem 3.3.8) implies that

$$(2) \qquad P[f](z) = P[f \circ \varphi_z](0) = \int_S (f \circ \varphi_z) d\sigma = \sigma(\varphi_z(V)).$$

Put $G_\alpha = \{\eta \in S : 1 + 1/\alpha < |1 - \eta_1| \leq 2\}$.
Pick $\eta \in G_\alpha$, put $\zeta = \varphi_z(\eta)$. Since $r > 1/\alpha$,

$$|1 - r\eta_1| \geq |1 - \eta_1| - (1 - r) > \frac{2}{\alpha}.$$

The definition of $\varphi_z(\eta)$ implies therefore that

$$|1 - r\zeta_1| = \frac{1 - r^2}{|1 - r\eta_1|} < \frac{\alpha}{2}(1 - r^2).$$

Thus $z = re_1 \in D_\alpha(\zeta)$. Since $z \notin \Omega(E, \alpha)$, $\zeta \in V$.

We have shown that $\varphi_z(G_\alpha) \subset V$. Hence $G_\alpha \subset \varphi_z(V)$, $\sigma(\varphi_z(V)) \geq \sigma(G_\alpha) > 0$, so that (2) yields

$$(3) \qquad P[f](z) \geq \sigma(G_\alpha) \qquad \left(|z| > \frac{1}{\alpha}\right).$$

The lemma follows from (1) and (3).

5.5.3. Lemma. *Suppose $E \subset S$ is measurable, ζ is a point of density of E, $\alpha > 1$, $\beta > 1$. Then there exists $r = r(E, \zeta, \alpha, \beta) < 1$ such that*

$$\{|z| > r\} \cap D_\beta(\zeta) \subset \Omega(E, \alpha).$$

(The interesting case is $\beta > \alpha$.)

Proof. Without loss of generality, $\sigma(E) < 1$. By Theorem 5.4.8, the Poisson integral in Lemma 5.5.2 has K-limit 0 at ζ. Since this integral is $\geq c$ outside $\Omega(E, \alpha)$, it follows that every sequence $\{z_i\}$ that converges to ζ within $D_\beta(\zeta)$ must eventually be in $\Omega(E, \alpha)$.

5.5.4. Lemma. *Fix r, $0 < r < 1$. Every $g \in C(rS)$ extends then to a function $G \in C(r\bar{B})$ that satisfies $\tilde{\Delta}G = 0$ in rB. Every uniformly bounded collection of these G's is equicontinuous on every compact subset of rB.*

(The first part of the lemma asserts the solvability of a certain Dirichlet problem.)

Proof. Let μ be a complex measure on S that satisfies

$$(1) \qquad \int_S P[f](r\zeta)d\mu(\zeta) = 0$$

for every $f \in C(S)$. By Fubini's theorem,

$$(2) \qquad \int_S P[\mu](r\omega)f(\omega)d\sigma(\omega) = 0$$

for every f. Thus $P[\mu] \equiv 0$ on rS, hence in rB (by the maximum principle for $\mathcal{M}$-harmonic functions), hence in B by real-analyticity. Thus $\mu = 0$, by Theorem 3.3.4(d).

It follows now from the Hahn–Banach theorem that there is a sequence $\{f_i\} \subset C(S)$ whose Poisson integrals converge to g, uniformly on rS. By the maximum principle, $\{P[f_i]\}$ converges uniformly on $r\bar{B}$ to $G \in C(r\bar{B})$.

If $z \in rB$ and $t > 0$ is so small that $\varphi_z(t\bar{B}) \subset rB$, then G inherits the mean value property

$$(3) \qquad G(z) = \int_S G(\varphi_z(t\zeta))d\sigma(\zeta)$$

from $P[f_i]$. To finish the proof, confine z to a compact set $K \subset rB$, let τ be the invariant measure of Theorem 2.2.6, and pick $h \in C^\infty(B)$ whose support is a very small neighborhood of 0, such that $\int_B h \, d\tau = 1$. Then (3) implies

(4) $$G(z) = \int_B (G \circ \varphi_z) h \, d\tau = \int_B (h \circ \varphi_z) G \, d\tau.$$

The last integral shows that $G \in C^\infty(K)$; by Theorem 4.1.3(i), (3) implies now that $(\tilde{\Delta} G)(z) = 0$. The equicontinuity assertion follows from the representation of G by the last integral in (4).

5.5.5. Theorem. *Suppose* $E \subset S$ *is measurable,* $\alpha > 1$, *and* u *is a bounded function in* $\Omega(E, \alpha)$ *such that* $\tilde{\Delta} u = 0$. *Then* u *has* K-*limits at almost every point of* E.

[Although it is not obvious that $\Omega(E, \alpha)$ satisfies the geometric condition that is needed before it makes sense to talk about K-limits at points of E, Lemma 5.5.3 shows that the condition does in fact hold at every point of density of E.]

Proof. Suppose $0 \leq u \leq 1$, without loss of generality. If we replace α by a slightly smaller value, and write Ω for $\Omega(E, \alpha)$, then u is continuous on $\overline{\Omega} \cap B$, hence u extends to a continuous function on B, with $0 \leq u \leq 1$.

Let $r_j \nearrow 1$, $r_j < r_{j+1}$. By Lemma 5.5.4 there are functions $g_j \in C(r_j \overline{B})$, such that $g_j = u$ on $r_j S$, $\tilde{\Delta} g_j = 0$ in $r_j B$, $0 \leq g_j \leq 1$. By equicontinuity, some subsequence, again denoted by $\{g_j\}$, converges uniformly on compact subsets of B, to an $\mathcal{M}$-harmonic function g in B.

Let v be the Poisson integral in Lemma 5.5.2. (We may of course assume, without loss of generality, that $\sigma(E) < 1$.) Then $v \geq c > 0$ on $B \cap \partial\Omega$, and $|u - g_j| \leq 1$ in $r_j B$. Also, $u - g_j = 0$ on $r_j S$. Thus

(1) $$|u - g_j| \leq c^{-1} v$$

on $\partial(\Omega \cap r_j B)$. By the maximum principle, (1) holds at every point of $\Omega \cap r_j B$. Letting $j \to \infty$, it follows that

(2) $$|u - g| \leq c^{-1} v \quad \text{in} \quad \Omega.$$

By Theorems 4.3.3 and 5.4.8, the K-limit of g exists at almost all points of S. If ζ is a point of density of E, Theorem 5.4.8 and the definition of v show that v has K-limit 0 at ζ. Hence (2) implies that u and g have the same K-limits at almost all points of E.

5.5.6. Definition. A function $F: B \to \mathbb{C}$ is said to be (weakly) K-*bounded* at a point $\zeta \in S$ if there exist $\alpha > 1$, $M < \infty$, such that $|F(z)| \leq M$ for every $z \in D_\alpha(\zeta)$.

The word "weakly" is to draw attention to the fact that boundedness of F on $D_\alpha(\zeta)$ is required only for a *single* $\alpha > 1$, whereas *every* $\alpha > 1$ is involved in the definition of K-limits.

5.5.7. Theorem. *If u is $\mathcal{M}$-harmonic in B, E is a measurable subset of S, and u is weakly K-bounded at every $\zeta \in E$, then $(K\text{-}\lim u)(\zeta)$ exists for almost all $\zeta \in E$.*

Proof. For $i = 1, 2, 3, \ldots$, let E_i be the set of all $\zeta \in S$ such that $|u| \leq i$ in $D_{1+1/i}(\zeta)$. Since u is continuous in B, each E_i is closed. By Theorem 5.5.5, u has K-limits at almost every point of each E_i. Since $E \subset \bigcup_i E_i$, the theorem is proved.

5.5.8. Theorem. *Every $f \in H(B)$ decomposes S into three measurable sets $E_K, E_{\mathbb{C}}, E_N$, such that*

(i) $\sigma(E_N) = 0$,

(ii) f *has finite K-limit at every $\zeta \in E_K$,*

(iii) $f(D_\alpha(\zeta))$ *is dense in $\mathbb{C}$, for every $\alpha > 1$ and for every $\zeta \in E_{\mathbb{C}}$.*

Proof. Let $\{V_j\}$ be a countable base of $\mathbb{C}$, consisting of open discs with centers c_j, radii r_j. Let E_{ij} be the set of all $\zeta \in S$ such that $f(D_{1+1/i}(\zeta))$ does not intersect V_j. Each E_{ij} is closed, and their union is the complement of $E_{\mathbb{C}}$. (Thus $E_{\mathbb{C}}$ is a set of type G_δ.)

Now fix (i, j) and let $\Omega = \Omega(E_{ij}, 1 + 1/i)$, as in Definition 5.5.1. The function $g = 1/(f - c_j)$ is holomorphic in Ω, and $|g| \leq 1/r_j$ in Ω. By Theorem 5.5.5, g has K-limits $g^*(\zeta)$ at almost all $\zeta \in E_{ij}$. As we shall see presently, $g^*(\zeta) \neq 0$ for almost every $\zeta \in E_{ij}$. Thus f has finite K-limits a.e. in each E_{ij}, hence at almost all points of the complement of $E_{\mathbb{C}}$. This is what the theorem asserts.

If $g^*(\zeta)$ were 0 on a set of positive measure, some slice function $g_\zeta = 1/(f_\zeta - c_j)$ would be a meromorphic function in the unit disc U with nontangential limit 0 on a subset of the unit circle that has positive measure. This forces $g_\zeta(\lambda) = 0$ for all $\lambda \in U$ (Zygmund [3], vol. II, p. 203); but g has no zero in B, by its very definition.

For later reference, here is a formal statement of a uniqueness theorem that the preceding argument proves:

5.5.9. Theorem. *If $g \in H(B)$ and if g has K-limit 0 on a set $E \subset S$ with $\sigma(E) > 0$, then $g \equiv 0$.*

5.6. The Spaces $N(B)$ and $H^p(B)$

5.6.1. Definitions. If f is any function with domain B, and $0 < r < 1$, then f_r denotes the dilated function defined for $|z| < 1/r$ by $f_r(z) = f(rz)$.

A function $f \in H(B)$ is in the *Nevanlinna class* $N(B)$ provided that

(1) $$\sup_{0 < r < 1} \int_S \log^+ |f_r| d\sigma < \infty,$$

and is in the *Hardy space* $H^p(B)$ $(0 < p < \infty)$ provided that

(2)
$$\sup_{0<r<1} \int_S |f_r|^p \, d\sigma < \infty.$$

In the latter case, we write $\|f\|_p$ for the pth root of the left side of (2). This is evidently a norm when $p \geq 1$, and $\|f - g\|_p^p$ defines a metric when $0 < p < 1$.

More generally, let $\varphi: [-\infty, \infty) \to [0, \infty)$ be a nondecreasing convex function, not identically 0, and let $H_\varphi(B)$ be the class of all $f \in H(B)$ whose growth is restricted by the requirement

(3)
$$\sup_{0<r<1} \int_S \varphi(\log |f_r|) d\sigma < \infty.$$

If $\varphi(x) = x^+ = \max(0, x)$, then $H_\varphi(B) = N(B)$.
If $\varphi(x) = e^{px}$, then $H_\varphi(B) = H^p(B)$.
Clearly, $H^{p_1} \subset H^{p_2} \subset N$ if $p_1 > p_2$.
Suppose g is real-valued function in B. If there is an $\mathcal{M}$-harmonic u in B such that $g \leq u$, we call u an *$\mathcal{M}$-harmonic majorant* of g. If, furthermore, every other $\mathcal{M}$-harmonic majorant u_1 of g satisfies the inequality $u(z) < u_1(z)$ at every $z \in B$, then u is said to be the *least $\mathcal{M}$-harmonic majorant of g in B.*

5.6.2. Theorem. *Let $f \in H(B)$, let φ be as above, put $v = \varphi(\log |f|)$, and define*

(1)
$$I_p = \sup_{0<r<1} \left\{ \int_S v_r^p \, d\sigma \right\}^{1/p}.$$

   (a) *If $I_1 < \infty$, then v has a least $\mathcal{M}$-harmonic majorant u in B. There is a positive measure μ on S such that $u = P[\mu]$, and $\|\mu\| = I_1$.*
   (b) *If $1 < p < \infty$ and $I_p < \infty$, then the least $\mathcal{M}$-harmonic majorant u of v has the form $u = P[h]$ for some $h \in L^p(\sigma)$ with $\|h\|_p = I_p$.*

Proof. Pick $z \in B$ and $\psi \in \text{Aut}(B)$ so that $\psi(0) = z$. If $0 < r < 1$ then $f_r \circ \psi$ is in the ball algebra, hence $v_r \circ \psi$ is subharmonic, so that

$$v(rz) = (v_r \circ \psi)(0) \leq \int_S (v_r \circ \psi) d\sigma = P[v_r \circ \psi](0).$$

By Theorem 3.3.8, this is the same as

(2)
$$v(rz) \leq P[v_r](z).$$

If $I_1 < \infty$, there is a sequence $r_i \nearrow 1$ such that $\{v_{r_i}\}$ converges to a positive measure μ, in the weak*-topology of the dual of $C(S)$. Clearly, $\|\mu\| \leq I_1$ and $v \leq P[\mu]$. This last inequality implies $I_1 \leq \|\mu\|$. Thus $\|\mu\| = I_1$.

To show that $u = P[\mu]$ is the *least* $\mathcal{M}$-harmonic majorant of v, suppose $\tilde{u} \geq v$ and $\tilde{u}$ is $\mathcal{M}$-harmonic in B. Then $\tilde{u} = P[\tilde{\mu}]$ for some positive measure $\tilde{\mu}$ on S. Since $\tilde{u}_r \geq v_r$ for $0 < r < 1$ and since $\tilde{u}_r \to \tilde{\mu}$ and $v_r \to \mu$, in the weak*-topology, it follows that

$$\int_S g \, d\tilde{\mu} \geq \int_S g \, d\mu$$

for every $g \in C(S)$. In particular, if $z \in B$, then

$$\tilde{u}(z) = P[\tilde{\mu}](z) \geq P[\mu](z) = u(z).$$

This proves (a).

To deduce (b), one proceeds as above, using weak convergence in $L^p(\sigma)$.

5.6.3 Theorem. *A function $f \in H(B)$ is in $H_\varphi(B)$ if and only if $\varphi(\log|f|)$ has an $\mathcal{M}$-harmonic majorant.*

Proof. If $f \in H_\varphi(B)$, then $I_1 < \infty$, and Theorem 5.6.2 furnishes the desired majorant. The converse is trivial.

Corollary. *Every $H_\varphi(B)$ is an $\mathcal{M}$-invariant function space.*

Note: The spaces $H_\varphi(B)$ are also characterized by the existence of harmonic (rather than $\mathcal{M}$-harmonic) majorants of $\varphi(\log|f|)$. This is not difficult to prove, but does not lead to their $\mathcal{M}$-invariance.

The existence of a *pluriharmonic* majorant is a much more restrictive requirement, as we shall see in Chapter 7.

5.6.4. Theorem. *If $f \in N(B)$, $f \not\equiv 0$, then*

(a) *f has finite K-limits f^* a.e. on S,*
(b) *$\log|f^*| \in L^1(\sigma)$,*
(c) *the least $\mathcal{M}$-harmonic majorant u of $\log|f|$ satisfies $u^* = \log|f^*|$ a.e. on S, and*
(d) *there is a singular measure τ on S such that*

$$u = P[\log|f^*| \, d\sigma + d\tau].$$

(Part (a) was proved by Korányi [3] and Stein [2].)

Proof. Theorem 5.6.2, with $\varphi(x) = x^+$ and $p = 1$, shows that $\log^+|f| \leq P[\lambda]$ for some measure λ. The maximal theorems 5.2.4 and 5.4.5 show that

$P[\lambda]$ is K-bounded at almost every $\zeta \in S$. The same is then true of $\log^+ |f|$, hence also of $|f|$ itself. This proves (a), by Theorem 5.5.7.

To prove (b), put

$$A_r = \int_S \log|f_r|\,d\sigma, \qquad B_r = \int_S \log^+ |f_r|\,d\sigma, \qquad C_r = \int_S |\log|f_r||\,d\sigma,$$

for $0 < r < 1$. Since $\log|f|$ and $\log^+ |f|$ are subharmonic, A_r and B_r do not decrease as r increases; B_r is bounded, by hypothesis; $C_r = 2B_r - A_r$; and

$$\log|f(0)| \le A_r \le B_r.$$

If $f(0) \neq 0$, it follows that A_r and B_r have finite limits, as $r \nearrow 1$. Hence

$$\int_S |\log|f^*||\,d\sigma \le \lim_{r \to 1}(2B_r - A_r) < \infty.$$

If $f(0) = 0$, choose $\psi \in \mathrm{Aut}(B)$ so that $(f \circ \psi)(0) \neq 0$, and note that the integrability of $\log|f^* \circ \psi|$ implies that of $\log|f^*|$. This proves (b).

This argument shows also that the functions $\log|f_r|$ are bounded in $L^1(\sigma)$; a subsequence converges therefore weak* to a measure μ on S such that $u = P[\mu]$ is the least $\mathcal{M}$-harmonic majorant of $\log|f|$, just as in the proof of Theorem 5.6.2. The obvious inequality $u^* \ge \log|f^*|$ allows us to apply Fatou's lemma and to conclude that

$$\int_S (u^* - \log|f^*|)\,d\sigma \le \liminf_{r \to 1} \int_S (u_r - \log|f_r|)\,d\sigma.$$

Since both u_r and $\log|f_r|$ tend to μ in the weak*-topology, the integral of the nonnegative function $u^* - \log|f^*|$ is 0.

This proves (c). Since $u^* = \mathscr{D}\mu$ a.e. on S (Theorem 5.4.9), (d) follows from (c) and §5.3.3.

5.6.5. Theorem. *To every $\alpha > 1$ corresponds a constant $A(\alpha) < \infty$ such that*

$$\int_S |M_\alpha f|^p\,d\sigma \le A(\alpha)\|f\|_p^p$$

for every $f \in H^p(B)$, $0 < p < \infty$.

(The case $n = 1$, with M_{rad} in place of M_α, is the original maximal theorem of Hardy and Littlewood [1].)

Proof. Fix p, $0 < p < \infty$, and choose $f \in H^p(B)$. Put $v = |f|^{p/2}$. Then $v^2 = |f|^p$, so that

$$\sup_r \|v_r\|_2^2 = \sup_r \|f_r\|_p^p = \|f\|_p^p.$$

Theorem 5.6.2, with $\varphi(x) = e^{px/2}$, shows therefore that $v \leq P[h]$ for some $h \in L^2(\sigma)$, with $\|h\|_2 = \|f\|_p^{p/2}$. Hence

$$|M_\alpha f|^p = |M_\alpha v|^2 \leq |M_\alpha P[h]|^2,$$

and we conclude from Theorem 5.4.10 (applied to h, with exponent 2) that

$$\int_S |M_\alpha f|^p \, d\sigma \leq A(\alpha) \int_S h^2 \, d\sigma = A(\alpha)\|f\|_p^p.$$

5.6.6. Theorem. *Suppose* $0 < p < \infty$, $f \in H^p(B)$, *and put*

(1) $$f^*(\zeta) = (K\text{-}\lim f)(\zeta) \qquad \text{(a.e. on } S).$$

Then

(2) $$\lim_{r \to 1} \int_S |f^* - f_r|^p \, d\sigma = 0.$$

Note that $f^*(\zeta)$ exists a.e., by Theorem 5.6.4.

Proof. Put $F = M_{\text{rad}} f$. Then $F \leq M_\alpha f$ if $\alpha \geq 2$, and Theorem 5.6.5 shows that $F \in L^p(\sigma)$. The inequalities

$$(x + y)^p \leq \begin{cases} 2^{p-1}(x^p + y^p) & (p \geq 1) \\ x^p + y^p & (0 < p < 1), \end{cases}$$

valid for $x \geq 0$, $y \geq 0$, show that

(3) $$|f^* - f_r|^p \leq \begin{cases} 2^p F^p & (p \geq 1) \\ 2F^p & (0 < p < 1). \end{cases}$$

Since $f_r \to f^*$ a.e. as $r \nearrow 1$, (2) follows from (3) and the dominated convergence theorem.

5.6.7. Definition. We let $A(S)$ denote the class of all $f \in C(S)$ that are restrictions to S of members of the ball algebra $A(B)$. [Note that $A(S)$ is a closed subalgebra of $C(S)$, relative to the sup-norm, by the maximum modulus theorem.]

For $0 < p < \infty$, we let $H^p(S)$ be the L^p-closure of $A(S)$.

5.6.8. Theorem. (a) *If* $f \in H^p(B)$ *then* $f^* \in H^p(S)$ *and* $\|f^*\|_p = \|f\|_p$.

(b) *If* $p \geq 1$ *and* $g \in H^p(S)$, *then* $P[g] = C[g] \in H^p(B)$, *and* $g = C[g]^*$ *a.e.*

Proof. Part (a) is a corollary of Theorem 5.6.6, since $f_r \in A(S)$ for $0 < r < 1$, and since the subharmonicity of $|f|^p$ implies that

$$\lim_{r \to 1} \| f_r \|_p = \sup_{0 < r < 1} \| f_r \|_p = \| f \|_p.$$

Next, $P[g] = C[g]$ when $g \in A(S)$, by Theorem 3.3.2. Hence $P[g] = C[g]$ for every $g \in H^p(S)$. Since $C[g]$ is holomorphic, the rest of (b) follows from Theorems 3.3.4(b) and 5.4.8.

Corollary. *The map $f \to f^*$ is a linear isometry of $H^p(B)$ onto $H^p(S)$.*

5.6.9. Theorem. *The map $f \to C[f]^*$ is the orthogonal projection of $L^2(\sigma)$ onto $H^2(S)$, and*

(1)
$$\int_S |M_\alpha C[f]|^2 \, d\sigma \leq A(\alpha) \int_S |f|^2 \, d\sigma$$

for every $f \in L^2(\sigma)$.

Proof. Every $f \in L^2(\sigma)$ has an orthogonal decomposition $f = g + h$, where $g \in H^2(S)$ and $h \perp H^2(S)$. Fix $z \in B$ and define

$$u_z(\zeta) = \overline{C(z, \zeta)}.$$

Then $u_z \in A(S)$, and therefore

$$C[h](z) = \int_S h \bar{u}_z \, d\sigma = 0.$$

Thus $C[f] = C[g]$, and $C[f]^* = g$ by Theorem 5.6.8. This proves the first assertion.

Next, (1) holds with g in place of f, by the Hardy–Littlewood theorem (5.6.5). Since $C[f] = C[g]$ and $\|g\|_2 \leq \|f\|_2$, (1) holds for f as well.

5.7. Appendix: Marcinkiewicz Interpolation

This appendix contains the special cases of the Marcinkiewicz interpolation theorem that occur in the present book. The theorem is usually stated for a wider variety of function spaces (see, for instance, Zygmund [2], [3; Chap. XII] or Stein and Weiss [1; Chap. 5]), and the proofs are correspondingly more complicated.

We consider two positive measures σ and μ on sets S and X, respectively, and we let T be an operator that associates to every $f \in L^1(\sigma)$ a measurable function $Tf: X \to [0, \infty]$, and that is sub-additive: $T(f + g) \leq Tf + Tg$.

The various maximal functions that occur in Chapters 5 and 6 are examples of such operators.

We let c_r and c_∞ be the smallest constants such that the inequalities

$$\mu\{Tf > t\} \le c_r t^{-r} \int_S |f|^r \, d\sigma,$$

$$\|Tf\|_\infty \le c_\infty \|f\|_\infty$$

hold whenever $f \in L^1(\sigma)$, $1 \le r < \infty$, $0 < t < \infty$.

Theorem. *The following inequalities hold for every* $f \in L^1(\sigma)$:

(a) *If* $1 < p < r \le \infty$, *then*

$$\int_X (Tf)^p \, d\mu \le K_p(c_1, c_r) \int_S |f|^p \, d\sigma.$$

(b) *If* $r > 1$, *then*

$$\int_X (Tf) d\mu \le K'(\|\sigma\|, \|\mu\|, c_r) + K''(c_1, c_r) \int_S |f| \log^+ |f| \, d\sigma.$$

The constants K_p, K', K'' are finite whenever the indicated parameters are finite (and are $+\infty$ otherwise).

Proof. The formula

(1) $$\int_X F^p \, d\mu = p \int_0^\infty \mu\{F > t\} t^{p-1} \, dt$$

holds for every measurable $F: X \to [0, \infty]$ and for every $p \in (0, \infty)$. To prove (1), define φ on $X \times [0, \infty)$ by $\varphi(x, t) = p t^{p-1}$ if $0 < t < F(x)$, 0 otherwise, and apply Fubini's theorem to the integral of φ over $X \times [0, \infty)$ with respect to the product measure $d\mu \times dt$.

Fix $f \in L^1(\sigma)$. For every $t \in (0, \infty)$ split f into $f = g_t + h_t$, where $g_t(\zeta) = f(\zeta)$ if $|f(\zeta)| \ge t$, $g_t(\zeta) = 0$ if $|f(\zeta)| < t$. Put

(2) $$G(t) = \mu\left\{ Tg_t > \frac{t}{2} \right\}, \qquad H(t) = \mu\left\{ Th_t > \frac{t}{2} \right\},$$

and

(3) $$G_1(t) = \frac{2c_1}{t} \int_{|f| \ge t} |f| \, d\sigma, \qquad H_1(t) = \frac{2^r c_r}{t^r} \int_{|f| < t} |f|^r \, d\sigma.$$

The definitions of c_1 and c_r (for $r < \infty$) show that

(4)
$$G(t) \le 2c_1 t^{-1} \int_S |g_t| d\sigma = G_1(t),$$

(5)
$$H(t) \le 2^r c_r t^{-r} \int_S |h_t|^r d\sigma = H_1(t).$$

Integration of (3) yields

(6)
$$\int_1^\infty G_1(t)dt = 2c_1 \int_S |f| \log^+ |f| d\sigma$$

(7)
$$p \int_0^\infty G_1(t)t^{p-1} dt = \frac{2pc_1}{p-1} \int_S |f|^p d\sigma \qquad (1 < p < \infty),$$

(8)
$$p \int_0^\infty H_1(t)t^{p-1} dt = \frac{2^r pc_r}{r-p} \int_S |f|^p d\sigma \qquad (0 < p < r).$$

Since $Tf \le Tg_t + Th_t$, (4) and (5) give

(9)
$$\mu\{Tf > t\} \le G(t) + H(t) \le G_1(t) + H_1(t).$$

Hence (1) implies: $\int_X (Tf)^p d\mu \le$ (7) + (8) if $1 < p < r$. This proves (a) in the case $r < \infty$.

If $r = \infty$ in (a), assume $c_\infty \le \frac{1}{2}$, without loss of generality. (Replace T by $T/2c_\infty$ if necessary.) Since $\|h_t\|_\infty \le t$, we have $\|Th_t\|_\infty \le t/2$, hence $H(t) = 0$, and (9) becomes $\mu\{Tf > t\} \le G_1(t)$. Now (a) follows from (1) and (7).

Since $\mu\{Tf > t\} \le \|\mu\|$, (9) yields

$$\int_0^\infty \mu\{Tf > t\}dt \le \|\mu\| + \int_1^\infty G_1(t)dt + \int_0^\infty H_1(t)dt,$$

so that $\int_X (Tf)d\mu \le \|\mu\| +$ (6) + (8) (with $p = 1$). Since

$$\int_S |f| d\sigma \le e\|\sigma\| + \int_S |f| \log^+ |f| d\sigma,$$

(b) follows.

Chapter 6

Boundary Behavior of Cauchy Integrals

One major difference between the Poisson kernel and the Cauchy kernel is. that the former is positive and the latter is not. The positivity of $P(z, \zeta)$ made it possible to be rather unsubtle in the proof of the basic maximal theorem 5.4.5: μ was replaced by $|\mu|$, and it was then just a matter of making size estimates without giving too much away. No cancellation effects came into play.

No such approach can succeed with the Cauchy integral, for the simple reason that

$$\int_S |C(z, \zeta)| \, d\sigma(\zeta) \to \infty$$

as z approaches any boundary point of B. (This is the case $c = 0$ of 1.4.10.) Nevertheless, Cauchy integrals of measures do have finite K-limits at almost all points of S (Theorem 6.2.3), and a weak-type $(1, 1)$ maximal theorem (6.2.2) does exist, even though $M_\alpha C[\mu]$ is not pointwise dominated by $M\mu$.

The proofs of these facts proceed along a route that has by now become standard when dealing with singular integrals, but they are quite self-contained and rely on no previous acquaintance with singular integrals. (These methods were originated by Calderón and Zygmund [1]; there are now many good expositions, for instance by Rivière [1], Coifman–Weiss [1], Stein [1].)

There is one feature that should perhaps be emphasized. Usually, singular integrals involve a kernel $K(x, y)$ in which both x and y live on the same set; the singularity occurs when $x = y$, and one deals with it by truncating K. In the Cauchy kernel $C(z, \zeta)$, ζ lives on S, z lives in B, and *no truncations will be made*. In this respect, the present treatment differs from the one used by Korányi–Vagi [1; Theorem 7.1]. They introduce kernels $C_\varepsilon(z, \zeta)$ (the notation is different) in which both z and ζ are on S, by

$$C_\varepsilon(z, \zeta) = \begin{cases} (1 - \langle z, \zeta \rangle)^{-n} & \text{if } d(z, \zeta) > \varepsilon \\ 0 & \text{otherwise,} \end{cases}$$

and show, for $f \in L^p(\sigma)$, $1 < p < \infty$, that the functions $C_\varepsilon[f]$ defined on S by

$$C_\varepsilon[f](z) = \int_S C_\varepsilon(z, \zeta) f(\zeta) d\sigma(\zeta)$$

converge in the norm topology of $L^p(\sigma)$, as $\varepsilon \searrow 0$, thus obtaining a generalization of the Marcel Riesz projection theorem. This will be obtained more easily in Theorem 6.3.1 below.

Among the other topics taken up in this chapter, let us just mention that the complex tangent spaces play an essential role in Section 6.4.

6.1. An Inequality

6.1.1. Lemma. *Suppose* ζ, η, $\omega \in S$, $d(\omega, \eta) < \delta$, $d(\omega, \zeta) > 2\delta$, *and* $z \in D_\alpha(\zeta)$. *Then*

$$|C(z, \eta) - C(z, \omega)| < (16\alpha)^{n+1} \delta |1 - \langle \zeta, \omega \rangle|^{-n-1/2}.$$

(As in §5.1.1, $d(\omega, \eta) = |1 - \langle \omega, \eta \rangle|^{1/2}$.)

Proof. The proof will be based on the identity

(1) $$C(z, \eta) - C(z, \omega) = \sum_{j=0}^{n-1} \frac{\langle z, \eta \rangle - \langle z, \omega \rangle}{(1 - \langle z, \eta \rangle)^{j+1}(1 - \langle z, \omega \rangle)^{n-j}} \, .$$

By 5.4.3(1),

(2) $$|1 - \langle z, \omega \rangle|^{-1} < 4\alpha |1 - \langle \zeta, \omega \rangle|^{-1}$$

and

(3) $$|1 - \langle z, \eta \rangle|^{-1} < 4\alpha |1 - \langle \zeta, \eta \rangle|^{-1} < 16\alpha |1 - \langle \zeta, \omega \rangle|^{-1};$$

the last inequality amounts to $d(\zeta, \omega) < 2\, d(\zeta, \eta)$, which is a consequence of the triangle inequality and the hypotheses of the lemma.

We claim that

(4) $$|\langle z, \eta \rangle - \langle z, \omega \rangle| \leq 3\delta \alpha^{1/2} |1 - \langle z, \omega \rangle|^{1/2}.$$

In the proof of (4), take $\omega = e_1$, without loss of generality, and let z', η' be the components of z, η orthogonal to e_1. The left side of (4) is then

$$|\langle z', \eta' \rangle + z_1(\bar{\eta}_1 - 1)| \leq |z'||\eta'| + |1 - \eta_1|$$
$$\leq 2|1 - z_1|^{1/2}|1 - \eta_1|^{1/2} + |1 - \eta_1|$$
$$\leq \delta\{2|1 - z_1|^{1/2} + |1 - \eta_1|^{1/2}\}.$$

Since $|1 - \eta_1| < \delta^2 < \frac{1}{4}|1 - \zeta_1| < \alpha|1 - z_1|$, by (2) and the hypotheses, one obtains

$$|\langle z, \eta \rangle - \langle z, \omega \rangle| \leq \delta(2 + \alpha^{1/2})|1 - z_1|^{1/2},$$

which proves (4), since $\alpha > 1$.

Returning to (1), we first apply (4), then (2) and (3), and conclude that $|C(z, \eta) - C(z, \omega)|$ is at most

$$3\delta\alpha^{1/2}|1 - \langle \zeta, \omega \rangle|^{-n-1/2} \sum_{j=0}^{n-1} (16\alpha)^{j+1}(4\alpha)^{n-j-1/2}.$$

The lemma follows.

6.1.2. Definition. For $\zeta, \omega \in S$, $\alpha > 1$, $\delta > 0$, we define a *maximal difference*,

(1) $$\Delta(\zeta, \omega, \alpha, \delta) = \sup|C(z, \eta) - C(z, \omega)|,$$

the supremum being taken over all $\eta \in Q(\omega, \delta)$ and over all $z \in D_\alpha(\zeta)$.

Thus Lemma 6.1.1 says that

(2) $$\Delta(\zeta, \omega, \alpha, \delta) < (16\alpha)^{n+1}\delta|1 - \langle \zeta, \omega \rangle|^{-n-1/2}$$

if $d(\zeta, \omega) > 2\delta$, i.e., if $\zeta \notin Q(\omega, 2\delta)$. If we integrate (2), we obtain an inequality that will be needed in the proof of Theorem 6.2.2:

6.1.3. Theorem. *If $\omega \in S$, $\alpha > 1$, $\delta > 0$, then*

(1) $$\int_{R(\omega, \delta)} \Delta(\zeta, \omega, \alpha, \delta)d\sigma(\zeta) < A(\alpha),$$

where $R(\omega, \delta) = S \backslash Q(\omega, 2\delta)$.

Here, as in the rest of this chapter, $A(\alpha)$ denotes a finite constant, that may also depend on the dimension n, and that is not necessarily the same at each occurrence.

The integral in (1) is clearly independent of ω. The significant assertion of the theorem is that there is a bound that is independent of δ.

Proof. By 6.1.2(2), it is enough to show that

(2) $$\int_{R(\omega, \delta)} |1 - \langle \zeta, \omega \rangle|^{-n-1/2} d\sigma(\zeta) < \frac{A}{\delta}.$$

When $n > 1$, formula 1.4.5(2) shows that this integral equals

(3)
$$\frac{n-1}{\pi} \int_{E(\delta)} \frac{(1 - |\lambda|^2)^{n-2}}{|1 - \lambda|^{n+1/2}} \, dm_2(\lambda),$$

where $E(\delta) = \{\lambda : |\lambda| < 1, |1 - \lambda| > 4\delta^2\}$. Replace $1 - |\lambda|^2$ by $2|1 - \lambda|$, then put $1 - \lambda = re^{i\theta}$, to see that (3) is less than

$$2^{n-2}(n - 1) \int_{4\delta^2}^\infty r^{-3/2} \, dr = \frac{2^{n-2}(n-1)}{\delta}.$$

This proves (2).

If $n = 1$, (2) still holds, since the integral is then

$$2 \int_{4\delta^2}^\pi |1 - e^{i\theta}|^{-3/2} \, d\theta < 2 \left(\frac{\pi}{2}\right)^{3/2} \delta^{-1}.$$

6.2. Cauchy Integrals of Measures

We need a covering lemma that is more elaborate than 5.2.3, one that is specifically adjusted to the maximal function of a given measure. In place of the constant A_3 that occurred in 5.2.3, we now use

$$A_4 = \sup \frac{\sigma(4Q)}{\sigma(Q)}.$$

By Proposition 5.1.4, $A_4 < \infty$.

6.2.1. Lemma. *Let μ be a complex measure on S. If $t > \|\mu\|$, then there are balls Q_i and pairwise disjoint Borel sets $V_i \subset Q_i$ such that*

(i) $\{M\mu > t\} \subset \bigcup_i Q_i = \bigcup_i V_i$,

(ii) $\sigma(Q_i) \leq A_4 t^{-1} |\mu|(Q_i)$,

(iii) $\sum_i \sigma(Q_i) \leq A_4 t^{-1} \|\mu\|$,

(iv) $|\mu|(V_i) < A_4 t\sigma(V_i)$.

Proof. Put $E_t = \{M\mu > t\}$. Assume $\mu \geq 0$, without loss of generality. To each $\zeta \in E_t$ corresponds then a *largest* δ such that the ball $q = Q(\zeta, \delta)$ satisfies

(1)
$$\mu(q) \geq t\sigma(q).$$

Since $t > \|\mu\|$, we have $q \neq S$; hence

$$(2) \qquad \mu(4q) < t\sigma(4q).$$

E_t is thus covered by a collection Γ_1 of balls q that satisfy (1) and (2).

Let r_1 be the supremum of the radii of the members of Γ_1, and choose $q_1 \in \Gamma_1$ with radius $> 3r_1/4$. Discard all members of Γ_1 that intersect q_1. Call the remaining collection Γ_2, let r_2 be the supremum of the radii of the members of Γ_2, and choose $q_2 \in \Gamma_2$ with radius $> 3r_2/4$. Discard all members of Γ_2 that intersect q_2, let Γ_3 be the remaining collection, etc. (The process stops if some Γ_i is empty, otherwise it continues through the natural numbers.)

We thus get disjoint balls q_i. Put $Q_i = 4q_i$. It is then easy to find disjoint Borel sets V_i such that $q_i \subset V_i \subset Q_i$ for each i, and $\bigcup_i V_i = \bigcup_i Q_i$.

If some $q \in \Gamma_1$ was discarded at the ith stage, then q intersects q_i, and $r(q) < 4r(q_i)/3$, where $r(q)$ is the radius of q. Since $1 + \frac{4}{3} + \frac{4}{3} < 4$, we have $q \subset 4q_i = Q_i$. Thus $E_t \subset \bigcup_i Q_i$, and (i) is proved. Next,

$$(3) \qquad \sigma(Q_i) \leq A_4\sigma(q_i) \leq A_4 t^{-1}\mu(q_i),$$

by (1). This proves (ii). If we add the inequalities (3), the disjointness of $\{q_i\}$ proves (iii). Finally, $\mu(Q_i) < t\sigma(Q_i)$, by (2), so that (iv) follows from

$$\mu(V_i) \leq \mu(Q_i) < t\sigma(Q_i) \leq A_4 t\sigma(q_i).$$

6.2.2. Theorem. *To every $\alpha > 1$ corresponds a constant $A(\alpha) < \infty$ such that*

$$(1) \qquad \sigma\{M_\alpha C[\mu] > t\} \leq A(\alpha)t^{-1}\|\mu\|$$

for all complex measures μ on S and for all $t > 0$.

Let us recall that

$$M_\alpha C[\mu](\zeta) = \sup\{|C[\mu](z): z \in D_\alpha(\zeta)\},$$

in accordance with Definition 5.4.4.

Proof. Fix μ and t. If $t \leq \|\mu\|$, then (1) holds provided that $A(\alpha) \geq 1$. So assume $t > \|\mu\|$, put $E_t = \{M\mu > t\}$, and choose $\{Q_i\}$, $\{V_i\}$ in accordance with Lemma 6.2.1. Let

$$(2) \qquad d\mu = f\,d\sigma + d\mu_s$$

be the Lebesgue decomposition of μ. We construct another decomposition

$$(3) \qquad d\mu = g\,d\sigma + d\beta$$

in which g is a "good" function (one that is not too large) and $\beta = \Sigma \beta_i$ is the "bad" part of the measure μ, by defining

(4)
$$c_i = \left(\frac{\mu}{\sigma}\right)(V_i)$$

(5)
$$g(\zeta) = \begin{cases} c_i & \text{if } \zeta \in V_i \\ f(\zeta) & \text{if } \zeta \in S \setminus \bigcup_i V_i \end{cases}$$

(6)
$$\beta_i(E) = (\mu - c_i \sigma)(V_i \cap E).$$

Since μ_s is concentrated in $E_t \subset \bigcup_i V_i$, it is not hard to verify that (3) holds, i.e., that

(7)
$$\mu(E) = \int_E g \, d\sigma + \beta(E)$$

for every Borel set $E \subset S$, by considering separately the cases $E \subset V_i$ and $E \cap \bigcup V_i = \varnothing$. (In the latter case, $\mu_s(E) = 0$.)

We shall first deal with g. Since $Mf \leq M\mu$, the corollary to Theorem 5.3.1 implies that $|f(\zeta)| \leq t$ a.e. outside E_t, hence $|g|^2 \leq t|f|$, so that

$$\int_{S \setminus E_t} |g|^2 \, d\sigma \leq t \int_S |f| \, d\sigma \leq t \|\mu\|.$$

Part (iv) of Lemma 6.2.1 shows that $|c_i| \leq At$ (where $A = A_4$), and part (iii) implies therefore that

$$\sum_i \int_{V_i} |g|^2 \, d\sigma = \sum_i |c_i|^2 \sigma(V_i) \leq A^2 t^2 \sum_i \sigma(Q_i) \leq A^3 \|\mu\| t.$$

Thus

(8)
$$\int_S |g|^2 \, d\sigma \leq (1 + A^3) \|\mu\| t.$$

Setting $G = M_\alpha C[g]$, Theorem 5.6.9 (the Hardy–Littlewood maximal theorem) gives

$$\int_S G^2 \, d\sigma \leq A(\alpha) \|\mu\| t.$$

Since $\sigma\{G > t\} \leq t^{-2} \int G^2 \, d\sigma$, we conclude that

$$(9) \qquad \sigma\left\{M_\alpha C[g] > \frac{t}{2}\right\} \leq \frac{A(\alpha)\|\mu\|}{t}.$$

Our next objective is the analogous estimate

$$(10) \qquad \sigma\left\{M_\alpha C[\beta] > \frac{t}{2}\right\} \leq \frac{A(\alpha)\|\mu\|}{t}.$$

This will complete the proof, because

$$M_\alpha C[\mu] \leq M_\alpha C[g] + M_\alpha C[\beta],$$

so that (1) follows from (9) and (10).

By (4), $c_i \sigma(V_i) = \mu(V_i)$. Hence $\|\beta_i\| \leq 2|\mu|(V_i)$, by (6), and thus

$$(11) \qquad \sum_i \|\beta_i\| \leq 2\|\mu\|.$$

Let Ω be the set that occurs in (10). Put $W = \bigcup_i (2Q_i)$. Then

$$\Omega \subset W \cup (\Omega\backslash W) \subset W \cup \bigcup_i (\Omega\backslash 2Q_i).$$

By part (iii) of Lemma 6.2.1,

$$(12) \qquad \sigma(W) \leq \Sigma\sigma(2Q_i) \leq A\Sigma\sigma(Q_i) \leq \frac{A^2\|\mu\|}{t}.$$

Note that β_i is concentrated on $V_i \subset Q_i$, and that $\beta_i(Q_i) = 0$. This is a crucial feature of the proof, for it enables us to write the Cauchy integral of β_i in the form

$$(13) \qquad C[\beta_i](z) = \int_{Q_i} \{C(z, \eta) - C(z, \omega_i)\} d\beta_i(\eta),$$

where ω_i is the center of $Q_i = Q(\omega_i, \delta_i)$. Using the notation introduced in §6.1.2, it follows that

$$(14) \qquad |C[\beta_i](z)| \leq \Delta(\zeta, \omega_i, \alpha, \delta_i)\|\beta_i\|$$

for all $z \in D_\alpha(\zeta)$. This inequality persists if its left side is replaced by its supremum as z ranges over $D_\alpha(\zeta)$, i.e., by $M_\alpha C[\beta_i](\zeta)$. If we then integrate, and appeal to Theorem 6.1.3, we see that

$$(15) \qquad \int_{S \setminus 2Q_i} M_\alpha C[\beta_i] d\sigma \leq A(\alpha) \|\beta_i\|.$$

In (15), we can replace $S \setminus 2Q_i$ by its subset $\Omega \setminus W$, and then add, to obtain

$$(16) \qquad \int_{\Omega \setminus W} M_\alpha C[\beta] d\sigma \leq A(\alpha) \sum_i \|\beta_i\| \leq 2A(\alpha) \|\mu\|,$$

by (11). The integrand in (16) exceeds $t/2$ at every point of Ω. Hence (16) shows that

$$(17) \qquad \sigma(\Omega \setminus W) \leq \frac{4A(\alpha) \|\mu\|}{t}.$$

This completes the proof, since (12) and (17) imply (10).

6.2.3. Theorem. *If μ is a complex Borel measure on S, then $C[\mu] \in H^p(B)$ for all $p < 1$.*

Corollary. *$C[\mu]$ has finite K-limits $C[\mu]^*$ at almost all points of S.*

Proof. Fix $\alpha > 2$, put $F = M_\alpha C[\mu]$. Theorem 6.2.2 shows that $\sigma\{F > t\} \leq x/t$, where $x = A(\alpha) \|\mu\|$. Hence, for $0 < p < 1$,

$$\int_S F^p \, d\sigma = p \int_0^\infty \sigma\{F > t\} t^{p-1} \, dt$$

$$\leq p \int_0^x t^{p-1} \, dt + px \int_x^\infty t^{p-2} \, dt = \frac{x^p}{(1-p)}.$$

Since $|C[\mu](r\zeta)| \leq F(\zeta)$, the theorem is proved.

The corollary follows from Theorem 5.6.4.

6.2.4. Remark. There exist bounded function u on S whose Cauchy integrals tend to ∞ along some radius of B. For example, let u be the real part of $i \log(1 - z_1)$. This shows that the precise analogue of Theorem 5.4.5 fails for Cauchy integrals, since $(M\mu)(\zeta)$ may be finite at a point where $M_\alpha C[\mu]$, or even $M_{\text{rad}} C[\mu]$ is infinite.

The conclusion of Theorem 6.2.3 cannot be strengthened to $C[\mu] \in H^1(B)$. Theorem 6.3.5 shows this.

6.3. Cauchy Integrals of L^p-Functions

6.3.1. Theorem (Korànyi–Vagi [1]). *If* $1 < p < \infty$ *and* $\alpha > 1$, *there exists* $A(\alpha, p) < \infty$ *such that*

$$(1) \qquad \int_S |M_\alpha C[f]|^p \, d\sigma \leq A(\alpha, p) \int_S |f|^p \, d\sigma$$

for every $f \in L^p(\sigma)$.

Corollary. *If* $1 < p < \infty$ *and* $f \in L^p(\sigma)$, *then* $C[f] \in H^p(B)$. *The map* $f \to C[f]^*$ *is a bounded linear projection of* $L^p(\sigma)$ *onto* $H^p(S)$.

Proof. The case $p = 2$ is Theorem 5.6.9. Since

$$f \to M_\alpha C[f]$$

is subadditive, the Marcinkiewicz interpolation theorem, combined with Theorem 6.2.2, proves (1) if $1 < p < 2$.

A standard duality argument completes the proof: Fix p, $2 < p < \infty$, let q be the conjugate exponent, pick $f \in L^p(\sigma)$, $h \in L^q(\sigma)$. For $0 < r < 1$,

$$\int_S h(\zeta)\overline{C[f](r\zeta)}d\sigma(\zeta) = \int_S C[h](r\eta)\overline{f(\eta)}d\sigma(\eta).$$

Since $1 < q < 2$, $\|C[h]_r\|_q \leq A(q)\|h\|_q$, so that

$$\left| \int_S h \cdot \overline{C[f]_r} \, d\sigma \right| \leq A(q)\|h\|_q \|f\|_p.$$

This holds for every $h \in L^q(\sigma)$. Hence

$$\|C[f]_r\|_p \leq A(q)\|f\|_p$$

for all $r \in (0, 1)$. This says that $C[f] \in H^p(B)$ and that $\|C[f]\|_p \leq A(q)\|f\|_p$. Now (1) follows from the Hardy–Littlewood maximal theorem.

For the corollary, see Theorem 5.6.8.

6.3.2. Definition. The class of all measurable functions f on S for which

$$\int_S |f| \log^+ |f| d\sigma < \infty$$

is called $L \log L$.

6.3.3. Theorem. *If $f \in L \log L$ then $C[f] \in H^1(B)$.*

Proof. By Marcinkiewicz interpolation, this follows from Theorems 6.2.2 and 6.3.1.

$L \log L$ is actually the largest class in which this conclusion holds. Theorem 6.3.5 will show this. But first we show that there is a close connection between the maximal function $M\mu$ and the class $L \log L$.

6.3.4. Theorem. *If μ is a complex measure on S for which $M\mu \in L^1(\sigma)$, then there is an $f \in L \log L$ such that $d\mu = f \, d\sigma$.*

Stein [4] proved this for functions rather than measures, but the proof is essentially the same.

Proof. Since $M\mu = M|\mu|$ we may assume that $\mu \geq 0$, and that $\|\mu\| = 1$. Our first objective is the inequality

(1) $\mu\{M\mu > t\} \leq At\varphi(t) \qquad (t > 1)$

where $\varphi(t) = \sigma\{A^2 M\mu > t\}$, and A is the constant A_4 in Lemma 6.2.1. Choose Q_i, V_i in accordance with that lemma, and put $E_t = \{M\mu > t\}$.

Pick some $Q_i = Q(\zeta, \delta)$. If $\omega \in Q_i$, it follows that $Q_i \subset Q(\omega, 2\delta)$. Hence

$$(M\mu)(\omega) \geq \frac{\mu}{\sigma}(Q(\omega, 2\delta)) \geq \frac{\mu(Q_i)}{A\sigma(Q_i)} \geq \frac{t}{A^2},$$

by 6.2.1(ii).

Hence $\bigcup_i V_i \subset \{M\mu \geq t/A^2\}$. By 6.2.1(iv),

$$\mu(E_t) \leq \sum_i \mu(V_i) < At \sum_i \sigma(V_i) \leq At\sigma\left\{M\mu \geq \frac{t}{A^2}\right\}.$$

This is (1).

Observe next that $\varphi(t)$ decreases when t increases, and that

(2) $\displaystyle\int_0^\infty \varphi(t)dt = A^2 \int_S (M\mu)d\sigma < \infty,$

by formula 5.7(1). It follows that $t\varphi(t) \to 0$ as $t \to \infty$. Hence $\mu(E_t) \to 0$ as $t \to \infty$, by (1). Since μ_s is concentrated on $\cap E_t$ (Theorem 5.2.7), we conclude that $\mu_s = 0$, hence $d\mu = f \, d\sigma$ for some $f \in L^1(\sigma)$, and $Mf = M\mu$.

At every Lebesgue point of f, $f(\zeta) \leq (Mf)(\zeta)$. Thus $f \leq t$ a.e. outside E_t, or $\{f > t\} \subset E_t$. Hence (1) shows that

(3) $\displaystyle\int_{f>t} f \, d\sigma \leq \mu(E_t) \leq At\varphi(t) \qquad (t > 1)$

so that finally (by Fubini's theorem)

$$A \int_1^\infty \varphi(t)dt \geq \int_1^\infty \frac{dt}{t} \int_{f>t} f \, d\sigma = \int_S f \log^+ f \, d\sigma,$$

which proves the theorem, because of (2).

6.3.5. Theorem. *If $f \in H^1(B)$ and $u = \mathrm{Re}\, f > 0$, then $u^* \in L \log L$.*

Proof. By the Hardy–Littlewood maximal theorem, $M_\alpha f \in L^1(\sigma)$, hence $M_\alpha u \in L^1(\sigma)$ (for any $\alpha > 1$). There is a measure $\mu \geq 0$ such that $u = P[\mu]$. Since $M_\alpha u \in L^1(\sigma)$, it follows from Theorem 5.4.12 that $M\mu \in L^1(\sigma)$, and Theorem 6.3.4 gives the desired conclusion.

6.3.6. When $n = 1$, Theorems 6.2.3, 6.3.1, 6.3.3, and 6.3.4 are due to Kolmogorov, M. Riesz, Zygmund, and M. Riesz, respectively. See Zygmund [3], Chap. VII, §2, and p. 381, for detailed references.

6.4. Cauchy Integrals of Lipschitz Functions

6.4.1. Definitions. If $0 < \alpha \leq 1$, and if f is a complex function with domain S, or B, or $\bar{B}$, we say that f satisfies a Lipschitz condition of order α (briefly: $f \in \mathrm{Lip}\, \alpha$) provided that

(1) $$|f(z) - f(w)| \leq c|z - w|^\alpha$$

for some $c = c_f < \infty$ and all z, w in the domain of f.

Note that this definition of Lip α is based on the euclidean metric, even when the domain of f is S, rather than on the nonisotropic metric introduced in §5.1.1.

A *holomorphic Lipschitz function* is one that lies in $A(B) \cap \mathrm{Lip}\, \alpha$.

A *complex-tangential curve* is a C^1-map $\gamma: I \to S$, where I is some interval on the real line, whose derivative $\gamma'(t)$ lies in the complex tangent space to S at $\gamma(t)$, for all $t \in I$. (See §5.4.2.) Analytically, the requirement is that

(2) $$\langle \gamma'(t), \gamma(t) \rangle = 0 \qquad (t \in I).$$

Stein [3] showed that holomorphic Lipschitz functions of order α are, roughly speaking, twice as smooth on complex-tangential curves. The same conclusion holds if the Lipschitz condition is only imposed on the slices of a holomorphic function (Rudin [12]). Ahern and Schneider [5] went even further, and showed that it holds for Cauchy integrals of arbitrary L^1-functions (not necessarily boundary values of holomorphic functions) whose slices are in Lip α. Theorems 6.4.9 and 6.4.10 summarize these results.

But first it seems advisable to describe some geometric features of the curves that are involved.

6.4.2. Complex-Tangential Curves. Let $\gamma: I \to S$ be a C^1-curve. Then $\langle \gamma, \gamma \rangle = 1$, and if we differentiate this, we obtain $\operatorname{Re}\langle \gamma', \gamma \rangle = 0$. The requirement $\langle \gamma', \gamma \rangle = 0$ which characterizes the complex-tangential curves is thus equivalent to

$$(1) \qquad\qquad\qquad \operatorname{Im}\langle \gamma', \gamma \rangle = 0.$$

Associate to each $\zeta \in S$ the curve Γ_ζ defined by

$$(2) \qquad\qquad\qquad \Gamma_\zeta(\theta) = e^{i\theta}\zeta \qquad (-\pi \le \theta \le \pi).$$

Note that Γ_ζ traces a circle, namely the intersection of S with the complex line through 0 and ζ, and that $\Gamma_\zeta'(0) = i\zeta$. If γ is a C^1-curve in S, and if $\zeta = \gamma(t)$, it follows that

$$(3) \qquad \gamma'(t) \cdot \Gamma_\zeta'(0) = \operatorname{Re}\langle \gamma'(t), i\gamma(t) \rangle = \operatorname{Im}\langle \gamma'(t), \gamma(t) \rangle;$$

the dot-product is as in §5.4.2. Comparison of (1) and (3) leads to the following characterization:

A C^1-curve γ in S is complex-tangential if and only if γ is perpendicular to every circle Γ_ζ that it intersects.

For any $n > 1$ there is a large supply of such curves. To illustrate this, consider the case $n = 2$.

Let $r = r(t)$, $\theta = \theta(t)$ be of class C^1 on I, so that $\gamma_0 = re^{i\theta}$ defines a C^1-curve γ_0 in the complex plane. Choose real functions u, v on I such that

$$(4) \qquad\qquad\qquad v' = \frac{\theta'}{1 + r^2}, \qquad u = v - \theta,$$

and define

$$(5) \qquad\qquad\qquad \gamma = \left(\frac{e^{iu}}{(1 + r^2)^{1/2}}, \frac{re^{iv}}{(1 + r^2)^{1/2}} \right).$$

Then γ is a curve on S.

Consider the map of S onto the Riemann sphere (more appropriately, in the present context, onto the complex projective space of dimension 1) that sends (z, w) to w/z. Its fibers are the circles Γ_ζ. *It carries γ onto our prescribed curve γ_0*, since $v - u = \theta$, by (4).

Differentiation of (5) leads to

$$(6) \qquad \langle \gamma', \gamma \rangle = \frac{i(u' + r^2 v')}{1 + r^2}$$

which is 0, by (4). *Hence* γ *is complex-tangential.*

This construction can be used to connect any point $(a, b) \in S$ to any $(z, w) \in S$ by a complex-tangential curve: Perform a unitary change of variables, if needed, so that none of a, b, z, w are 0. Let $I = [0, 1]$. Find a positive function $r \in C^1(I)$ with $r(0) = |b/a|$, $r(1) = |w/z|$, then find $\theta \in C^1(I)$ so that

$$\theta(0) = \arg\left(\frac{b}{a}\right), \qquad \theta(1) = \arg\left(\frac{w}{z}\right)$$

and

$$\int_0^1 (1 + r^2)^{-1} \theta' \, dt = \arg\left(\frac{w}{b}\right).$$

Put $v(0) = \arg b$, define u and v by (4), and γ by (5). Then $\gamma(0) = (a, b)$, $\gamma(1) = (z, w)$.

6.4.3. A theorem of Hardy and Littlewood (see, for example, Duren [1], p. 74) characterizes the holomorphic function in U that belong to Lip α $(0 < \alpha < 1)$ by the growth condition

$$|f'(z)| = O((1 - |z|)^{\alpha - 1}).$$

We shall reprove this in the n-variable context, but mention it here because it furnishes much of the motivation for the work in this section.

6.4.4. The Radial Derivative $\mathscr{R}f$. Let $f \in H(B)$ have the homogeneous expansion $f = \Sigma F_k$. We define

$$(1) \qquad (\mathscr{R}f)(z) = \sum_{k=0}^{\infty} k F_k(z) \qquad (z \in B).$$

$\mathscr{R}f$ is related to the derivative of the slice functions f_ζ by

$$(2) \qquad (\mathscr{R}f)(\lambda \zeta) = \lambda f_\zeta'(\lambda) \qquad (\zeta \in S, \lambda \in U).$$

The advantage of $\mathscr{R}f$ over f_ζ' is that the former is a holomorphic function in B.

If f is a Cauchy integral, say $f = C[\mu]$ for some measure μ on S, one can use (2) to derive a Cauchy formula for $\mathscr{R}f$, namely

$$(3) \qquad (\mathscr{R}f)(z) = \int_S \frac{n\langle z, \zeta\rangle}{(1 - \langle z, \zeta\rangle)^{n+1}}\, d\mu(\zeta).$$

6.4.5. Lemma. *If $f \in H(B)$ and β is a multi-index, then*

$$(1) \qquad D^\beta \mathscr{R}f - \mathscr{R}D^\beta f = |\beta| D^\beta f$$

and, for $0 < r < 1$, $\zeta \in S$, $\beta \neq 0$,

$$(2) \qquad r^{|\beta|}(D^\beta f)(r\zeta) = \int_0^r (D^\beta \mathscr{R}f)(t\zeta)t^{|\beta|-1}\, dt.$$

Proof. By linearity, it suffices to prove (1) and (2) for $f = F_k$, a homogeneous polynomial of degree k. Then $D^\beta F_k$ has degree $k - |\beta|$ (negative degree indicates the zero-function), so that

$$(D^\beta \mathscr{R} - \mathscr{R}D^\beta)F_k = [kD^\beta - (k - |\beta|)D^\beta]F_k = |\beta| D^\beta F_k,$$

which gives (1). Next,

$$(D^\beta \mathscr{R}F_k)(t\zeta) = kt^{k-|\beta|}(D^\beta F_k)(\zeta)$$

so that the integral in (2) is

$$r^k(D^\beta F_k)(\zeta) = r^{|\beta|}(D^\beta F_k)(r\zeta).$$

6.4.6. Lemma. *If $g \in H(B)$, $c \geq 0$, and*

$$(1) \qquad |g(z)| \leq (1 - |z|)^{-c} \qquad (z \in B)$$

then, for $k = 1, 2, 3, \ldots, 0 < r < 1$,

$$(2) \qquad |(D_2^k g)(re_1)| \leq A_{k,c}(1 - r)^{-c-k/2}.$$

Recall that $D_2 = \partial/\partial z_2$. The derivative on the left of (2) is thus the kth derivative in the direction of e_2, which is orthogonal to e_1. The pair e_1, e_2 could be replaced in the lemma by any pair of orthogonal unit vectors in $\mathbb{C}^n$.

The point of the lemma is the appearance of $k/2$ in the exponent. The growth of $(D_1^k g)(re_1)$ can be much more rapid. For example, if $g(z) = (1 - z_1)^{-c}$ then

$$(D_1^k g)(re_1) = a_{k,c}(1 - r)^{-c-k},$$

where $a_{k,c} = c(c + 1) \dots (c + k - 1)$.

Proof. Fix r and define

(3) $$h(\lambda) = g(re_1 + \lambda e_2)$$

for all $\lambda \in \mathbb{C}$ with $|\lambda|^2 < 1 - r^2$. Then, by (1),

(4) $$|h(\lambda)| < 2^c(1 - r^2 - |\lambda|^2)^{-c},$$

since $2(1 - |z|) > 1 - |z|^2$. Let Γ be the circle on which $|\lambda|^2 = \frac{1}{2}(1 - r^2)$. The Cauchy formula

(5) $$h^{(k)}(0) = \frac{k!}{2\pi i} \int_\Gamma \lambda^{-k-1} h(\lambda) d\lambda$$

gives the estimate

(6) $$|h^k(0)| \le k! \, 2^c \left(\frac{1 - r^2}{2}\right)^{-c-k/2}.$$

Since $(D_2^k g)(re_1) = h^{(k)}(0)$, (2) follows from (6).

6.4.7. Lemma. *Suppose $f \in H(B)$, $0 < \alpha < 1$, and*

(1) $$|(\mathscr{R}f)(z)| \le (1 - |z|)^{\alpha - 1} \qquad (z \in B).$$

Then, for $2 \le j \le n$ and $0 < r < 1$,

(2) $$|(\mathscr{R}D_j f)(re_1)| \le A_\alpha(1 - r)^{\alpha - 3/2},$$

(3) $$|(D_j^2 f)(re_1)| \le A_\alpha(1 - r)^{\alpha - 1},$$

and

(4) $$|(D_j f)(re_1)| \le \begin{cases} A_\alpha(1 - r)^{\alpha - 1/2} & (0 < \alpha < \frac{1}{2}) \\ A_\alpha & (\frac{1}{2} < \alpha < 1). \end{cases}$$

Here, and later, A_α denotes a finite constant, not necessarily the same at each occurrence.

Proof. Lemma 6.4.6, with $\mathscr{R}f$ in place of g, shows that

$$(5) \qquad\qquad |(D_j\mathscr{R}f)(re_1)| \le A_\alpha(1-r)^{\alpha-3/2},$$

$$(6) \qquad\qquad |(D_j^2\mathscr{R}f)(re_1)| \le A_\alpha(1-r)^{\alpha-2}.$$

By 6.4.5(2), (5) gives (4) and (6) gives (3). By 6.4.5(1), (5) and (4) imply (2).

6.4.8. Lemma. *If the gradient of a function* $u: B \to \mathbb{C}$ *satisfies*

$$(1) \qquad\qquad |(\text{grad } u)(z)| \le (1-|z|)^{\alpha-1} \qquad (z \in B)$$

for some $\alpha \in (0, 1)$, *then* $u \in \text{Lip } \alpha$.

(It follows, trivially, that u has a Lip α extension to $\bar{B}$.)

Proof. Choose $a \in B$, $b \in B$, with $0 < |a| \le |b| < 1$. Put $|b - a| = \delta$, and define

$$a' = \frac{1-\delta}{|a|}\, a, \qquad b' = \frac{1-\delta}{|b|}\, b.$$

Three cases occur:

   (i)   When $\delta \le 1 - |b|$ then

$$|\text{grad } u| \le (1-|b|)^{\alpha-1} \le \delta^{\alpha-1}$$

       on the line from a to b. Hence $|u(a) - u(b)| \le \delta^\alpha$.

  (ii) When $1 - |b| < \delta \le 1 - |a|$, then $|a - b'| \le |a - b|$, and

$$|u(a) - u(b)| \le |u(a) - u(b')| + |u(b') - u(b)|.$$

       The first term on the right is estimated as in (i); the second is at most

$$\int_{1-\delta}^{|b|} (1-x)^{\alpha-1}\, dx < \alpha^{-1}\delta^\alpha.$$

  (iii) When $1 - |a| < \delta$, then $|a' - b'| \le |a - b|$, and $|u(a) - u(b)|$ is at most

$$|u(a) - u(a')| + (u(a') - u(b')| + |u(b') - u(b)|.$$

       The first and third term is estimated as in (ii), the second by (i).

6.4.9. Theorem. *Assume* $0 < \alpha < 1, f \in L^1(\sigma)$, *and*

(1) $$|f(e^{i\theta}\zeta) - f(e^{it}\zeta)| \le |e^{i\theta} - e^{it}|^{\alpha}$$

for all $\zeta \in S$ *and all real* θ, t. *The radial derivative of the Cauchy integral* $F = C[f]$ *must then satisfy the growth condition*

(2) $$|(\mathscr{R}F)(z)| \le A_{\alpha}(1 - |z|)^{\alpha - 1}.$$

Proof. If we combine formulas 6.4.4(3) and 1.4.7(1), we see that $(\mathscr{R}F)(z)$ is equal to

(3) $$n \int_S d\sigma(\zeta) \cdot \frac{1}{2\pi} \int_{-\pi}^{\pi} \frac{\langle z, \zeta \rangle e^{-i\theta} f(e^{i\theta}\zeta)}{(1 - \langle z, \zeta \rangle e^{-i\theta})^{n+1}} \, d\theta.$$

Fix z, ζ, for the moment, and denote the inner integral by J. Write $\langle z, \zeta \rangle = re^{it}$, with $0 \le r < 1$, t real. Note that $J = 0$ when f is constant. Hence f can be replaced by $f - c$, for arbitrary $c \in \mathbb{C}$, without changing J. We choose $c = f(e^{it}\zeta)$. By (1), it follows then that

(4) $$|J| \le \frac{r}{2\pi} \int_{-\pi}^{\pi} \frac{|e^{i\theta} - e^{it}|^{\alpha}}{|1 - re^{i(t-\theta)}|^{n+1}} \, d\theta.$$

Note that $|1 - \lambda| < 2|1 - r\lambda|$ if $|\lambda| = 1$. Hence

(5) $$|J| \le \frac{2^{\alpha}}{2\pi} \int_{-\pi}^{\pi} |1 - \langle z, \zeta \rangle e^{-i\theta}|^{-n-1+\alpha} \, d\theta.$$

If we insert this into (3), and use 1.4.7(1) once more, we see that

(6) $$|(\mathscr{R}F)(z)| \le 2n \int_S |1 - \langle z, \zeta \rangle|^{-n-1+\alpha} \, d\sigma(\zeta).$$

Now (2) follows from (6) and Proposition 1.4.10.

Remark. The hypothesis of Theorem 6.4.9 holds for every $f \in \text{Lip } \alpha$, and in particular for every $f \in A(B) \cap \text{Lip } \alpha$, in which case $F = f$. The converse of this remark is part of the following theorem.

6.4.10. Theorem. *Suppose* $0 < \alpha < 1, f \in H(B)$, *and*

(1) $$|(\mathscr{R}f)(z)| \le (1 - |z|)^{\alpha - 1} \qquad (z \in B).$$

Then f has a continuous extension to $\bar{B}$ (still called f) that lies in $A(B) \cap \text{Lip } \alpha$.

If $\gamma: I \to S$ is a complex-tangential C^2-curve and if $g = f \circ \gamma$, then

(a) $g \in \mathrm{Lip}(2\alpha)$ *on I when* $0 < \alpha < \frac{1}{2}$,
(b)   *g is differentiable and $g' \in \mathrm{Lip}(2\alpha - 1)$ on I when* $\frac{1}{2} < \alpha < 1$.

Proof. Since $\int_r^1 (1 - t)^{\alpha-1} dt = \alpha^{-1}(1 - r)^\alpha$, we infer from (1) that $f(r\zeta)$ converges, uniformly on S, to a limit that we call $f(\zeta)$. This extends f, so that $f \in A(B)$. Note also that

$$(2) \qquad\qquad |f(\zeta) - f(r\zeta)| \leq \alpha^{-1}|1 - r|^\alpha$$

for all $\zeta \in S, 0 < r < 1$.

Lemma 6.4.7 shows that the gradient of f satisfies the hypothesis of 6.4.8. Thus $f \in \mathrm{Lip}\,\alpha$.

Assume now, without loss of generality, that our complex-tangential γ has $|\gamma'| = 1$. (This amounts to parametrizing γ by arc length.) Define

$$(3) \qquad\qquad g_r(t) = f(r\gamma(t)) \qquad (t \in I, 0 < r < 1).$$

For any $t \in I$, there is a unitary change of coordinates in $\mathbb{C}^n$ such that $\gamma(t) = e_1$ and $\gamma'(t) = e_2$. The chain rule shows then that

$$(4) \qquad\qquad g_r'(t) = r(D_2 f)(re_1)$$

and

$$(5) \qquad\qquad g_r''(t) = r^2(D_2^2 f)(re_1) + \sum_{j=1}^n r(D_j f)(re_1)\gamma_j''(t).$$

Fix $t_1, t_2 \in I$ so that $t_2 = t_1 + h, 0 < h < 1$.
Put $r = 1 - h^2$.
If $0 < \alpha < \frac{1}{2}$, it follows from (2) that $|g - g_r| \leq A_\alpha h^{2\alpha}$, and it follows from (4) and Lemma 6.4.7 that $|g_r'| \leq A_\alpha h^{2\alpha-1}$, so that

$$(6) \qquad\qquad |g_r(t_2) - g_r(t_1)| \leq (t_2 - t_1)A_\alpha h^{2\alpha-1} = A_\alpha h^{2\alpha}.$$

Hence $|g(t_2) - g(t_1)| \leq A_\alpha h^{2\alpha}$, and (a) is proved.

Finally, assume $\frac{1}{2} < \alpha < 1$. By (5) and Lemma 6.4.7, $|g_r''| \leq A_{\alpha, \gamma} h^{2\alpha-2}$, so that

$$(7) \qquad\qquad |g_r'(t_2) - g_r'(t_1)| \leq A_{\alpha, \gamma} h^{2\alpha-1}.$$

Furthermore, it follows from 6.4.7(2) that

$$(8) \qquad\qquad |g_r'(t) - g_s'(t)| \leq A_\alpha(1 - r)^{\alpha-1/2} = A_\alpha h^{2\alpha-1}$$

if $r < s < 1$. Hence $\{g_s'\}$ converges, uniformly on I, as $s \to 1$, so that g' exists and (8) holds with g' in place of g_s'. Combined with (7), this proves that

$$(9) \qquad |g'(t_2) - g'(t_1)| \leq A_{\alpha, \gamma} h^{2\alpha - 1},$$

which is what (b) asserts.

Remark. In the omitted case $\alpha = \frac{1}{2}$, it is still true that $|g_r''| \leq A_\gamma h^{-1}$. From this and (2) it is easy to deduce that

$$(10) \qquad |g(t + h) + g(t - h) - 2g(t)| \leq A_\gamma h,$$

i.e., that g lies in the space that is often called Λ_1. (Chapter V of Stein [1] describes the spaces Λ_α and their relation to Lip α.) But g need not be in Lip 1.

To see an example (with $n = 2$), take

$$(11) \qquad f(z) = \frac{z_2}{z_1} \log \frac{1}{1 - z_1}.$$

Then $(\mathscr{R}f)(z) = z_2/(1 - z_1)$, and since $|z_2|^2 \leq 1 - |z_1|^2$,

$$(12) \qquad |(\mathscr{R}f)(z)| \leq \left\{ \frac{2}{(1 - |z|)} \right\}^{1/2}.$$

Thus $f \in A(B) \cap \text{Lip } \frac{1}{2}$. If $\gamma(t) = (\cos t, \sin t)$, then γ is complex-tangential, but

$$f(\gamma(t)) = \tan t \cdot \log \frac{1}{1 - \cos t}$$

is not in Lip 1 and is not differentiable at $t = 0$.

The combination of Theorems 6.4.9 and 6.4.10 shows that the Cauchy integral tends to preserve smoothness on slices, while introducing additional smoothness in the complex-tangential directions. An extreme example of this phenomenon is furnished by any $f \in L^1$ that is constant on each slice. The conclusion—probably surprising at first glance—is that $C[f]$ is then constant in B. Actually, the proof follows immediately from the $\mathscr{U}$-invariance of the Cauchy transform: If $Uz = e^{i\theta}z$, then $f \circ U = f$ by assumption, hence $C[f] \circ U = C[f]$, which shows that the holomorphic function $C[f]$ is constant on each slice. Hence it is constant in B, since all slices intersect at the center of B.

6.5. Toeplitz Operators

6.5.1. Introduction. To every $\varphi \in L^{\infty}(\sigma)$ is associated a linear operator T_{φ}—the *Toeplitz operator* with *symbol* φ—that is defined for $f \in L^1(\sigma)$ by

$$(1) \qquad\qquad (T_{\varphi}f)(z) = C[\varphi f](z) \qquad (z \in B).$$

The holomorphic functions $T_{\varphi}f$ have K-limits at almost all points of S (Theorem 6.2.3). This extends the domain of $T_{\varphi}f$ is a natural way to the closed ball $\bar{B}$, with the possible exception of a set of measure zero on S.

The domain of T_{φ} is often restricted to certain subclasses of $L^1(\sigma)$. For example, every T_{φ} is a bounded linear operator on the Hilbert space $H^2(S)$, and the relations between the symbol φ and operator-theoretic properties of T_{φ} provide an interesting topic. This, however, will not be pursued here. (See Coburn [1], Boutet de Monvel [1], Janas [1], McDonald [2], [3].)

Theorems 6.4.9 and 6.4.10 imply that T_{φ} preserves Lip α $(0 < \alpha < 1)$ if $\varphi \in$ Lip α (since Lip α is an algebra).

Theorem 6.3.1 shows that every T_{φ} carries $L^p(\sigma)$ into $H^p(B)$ if $1 < p < \infty$. This is false for $p = 1$, even when $\varphi = 1$: if $f > 0$ on S, Theorem 6.3.5 shows that f must be in $L \log L$ in order for $T_{\varphi}f = C[f]$ to be in $H^1(B)$.

It is also not true that every T_{φ} carries H^{∞} to H^{∞}: Let g be an unbounded holomorphic function in B whose real part is bounded, with $g(0) = 0$. If $\varphi = \operatorname{Re} g^*$, then $T_{\varphi}(2) = g \notin H^{\infty}$. (See Theorem 3.2.5.)

There is a rather weak smoothness assumption which, when imposed on φ, avoids this last difficulty. Theorem 6.5.4 shows this in detail.

We define the *modulus of continuity* ω_{φ} of a function $\varphi : S \to \mathbb{C}$ by

$$(2) \qquad\qquad \omega_{\varphi}(t) = \sup\{|\varphi(\zeta) - \varphi(\eta)| : |\zeta - \eta| \le t\}.$$

We say that φ is a *Dini function* if

$$(3) \qquad\qquad \int_0^2 \omega_{\varphi}(t)\, \frac{dt}{t} < \infty.$$

For example, (3) holds if $\omega_{\varphi}(t) = t^{\alpha}$ for some $\alpha > 0$; every $\varphi \in$ Lip α is thus a Dini function. (The terminology comes from Dini's test for the convergence of the conjugate Fourier series; see Zygmund [3], vol. I., p. 52.)

If φ is a Dini function on S, let us extend φ to a continuous function on $\bar{B}$ (with the same sup-norm), in such a way that $\varphi(z) = \varphi(z/|z|)$ for $\frac{1}{2} \le |z| \le 1$, and let us define

$$(4) \qquad\qquad \Gamma_z(\zeta) = \{\varphi(z) - \varphi(\zeta)\}C(z, \zeta)$$

for $z \in \bar{B}$, $\zeta \in S$, $z \ne \zeta$.

6.5.2. Lemma. *If φ is a Dini function, then $\{\Gamma_z : z \in \bar{B}\}$ is uniformly integrable.*

More explicitly: To every $\varepsilon > 0$ corresponds a $\delta > 0$ such that $\int_E |\Gamma_z| d\sigma$ $< \varepsilon$ for all $z \in \bar{B}$ and for all $E \subset S$ with $\sigma(E) < \delta$.

Proof. $\|\Gamma_z\|_\infty$ is bounded for $|z| \le \frac{1}{2}$. It is thus sufficient to consider z of the form $z = r\eta$, $\frac{1}{2} \le r \le 1$, $\eta \in S$. Since $|1 - \lambda| < 2|1 - r\lambda|$ for $|\lambda| \le 1$, we have

$$(1) \qquad\qquad |C(r\eta, \zeta)| < 2^n |C(\eta, \zeta)|$$

and therefore

$$(2) \qquad\qquad |\Gamma_z(\zeta)| \le 2^n \omega(|\eta - \zeta|)|C(\eta, \zeta)| = 2^n F_\eta(\zeta),$$

say, where $\omega = \omega_\varphi$. Since $\{F_\eta : \eta \in S\}$ is $\mathscr{U}$-invariant, in the sense that $F_{U\eta_*}$ $= F_\eta \circ U^{-1}$, the lemma will follow as soon as we know that

$$(3) \qquad\qquad \int_S F_\eta(\zeta) d\sigma(\zeta) < \infty.$$

When $n = 1$, (3) is elementary. When $n > 1$, we use formula 1.4.5(2). Since ω is nondecreasing and $|\eta - \zeta|^2 = 2\,\mathrm{Re}(1 - \langle \eta, \zeta \rangle)$, the integral in (3) is at most

$$(4) \qquad\qquad \frac{n-1}{\pi} \int_U \frac{(1 - |\lambda|^2)^{n-2}}{|1 - \lambda|^n} \omega(\sqrt{2|1 - \lambda|})dm_2(\lambda).$$

Now use $1 - |\lambda|^2 \le 2|1 - \lambda|$, replace U by the half-disc D with center at $\lambda = 1$ and radius 2 that covers U, write $\lambda = 1 - se^{i\theta}$ ($0 \le s \le 2, |\theta| \le \pi/2$), and then replace $2s$ by t^2. It follows that (4) is less than

$$(5) \qquad\qquad (n-1)2^{n-1} \int_0^2 \omega(t) \frac{dt}{t} < \infty.$$

6.5.3. Lemma. *If φ is a Dini function, then*

$$(1) \qquad\qquad z \to \Gamma_z$$

is a uniformly continuous map of $\bar{B}$ into $L^1(\sigma)$.

Proof. The uniform integrability of $\{\Gamma_z\}$, combined with Egorov's theorem, gives

$$(2) \qquad\qquad \lim_{z \to w} \int_S |\Gamma_z - \Gamma_w| d\sigma = 0$$

for every $w \in \bar{B}$. This shows that (1) is continuous; the compactness of $\bar{B}$ gives uniform continuity.

6.5.4. Theorem. *Let φ be a Dini function.*

(a)   *If V_φ is defined by*

$$V_\varphi f = \varphi f - T_\varphi f$$

   then V_φ carries the closed unit ball of H^∞ into a bounded equicontinuous subset of $C(\bar{B})$.

(b)   *T_φ carries H^∞ into H^∞, and $A(S)$ into $A(B)$.*

(c)   *T_φ carries H^1 into H^1.*

Note. Since bounded equicontinuous subsets of $C(\bar{B})$ have compact closure (Ascoli's theorem), (a) asserts that V_φ is a *compact operator* from H^∞ into $C(\bar{B})$.

Proof. For $z \in B$, $f \in H^1$, the Cauchy formula shows that

(1)
$$(V_\varphi f)(z) = \int_S \{\varphi(z) - \varphi(\zeta)\} f(\zeta) C(z, \zeta) d\sigma(\zeta),$$

i.e., that

(2)
$$(V_\varphi f)(z) = \int_S f \Gamma_z \, d\sigma.$$

If $f \in H^\infty$, it follows that

(3)
$$|(V_\varphi f)(z) - (V_\varphi f)(w)| \le \|f\|_\infty \int_S |\Gamma_z - \Gamma_w| d\sigma$$

for $z \in B$, $w \in B$. By Lemma 6.5.3, (3) implies that $V_\varphi f$ is uniformly continuous in B, hence has a continuous extension to $\bar{B}$. The equicontinuity asserted in (a) is then also a consequence of (3).

Since $T_\varphi f = \varphi f - V_\varphi f$, (a) shows that the holomorphic function $T_\varphi f$ is bounded if $f \in H^\infty$, which proves (b), and (c) will follow from

(4)
$$\sup_{0 < r < 1} \int_S |(V_\varphi f)(r\eta)| d\sigma(\eta) < \infty.$$

By (2) and Fubini's theorem, the estimate

(5)
$$\sup_{0<r<1} \int_S |\Gamma_{rn}(\zeta)| \, d\sigma(\eta) < \infty$$

implies (4). The computation used in the proof of Lemma 6.5.2 establishes (5).

Theorem 6.5.4 will be used in the solution of Gleason's problem (Section 6.6) and in the proof of Henkin's theorem (Chapter 9). Our first application, however, is to the space $H^\infty + C$, the set of all sums $g + h$ with $g \in C(S)$, $h \in H^\infty(S)$:

6.5.5. Theorem (Rudin [7]). $H^\infty(S) + C(S)$ *is a closed sub-algebra of* $L^\infty(\sigma)$.

Proof. If $f \in H^\infty(S)$ and φ is a Dini function, Theorem 6.5.4 shows, after restricting to S, that the product

$$\varphi f = T_\varphi f + V_\varphi f$$

lies in $H^\infty(S) + C(S)$. Since the Dini functions are dense in C (we omit S in the rest of this proof) it remains to be shown that $H^\infty + C$ is closed with respect to convergence in the essential supremum norm.

The following observation is the key:

If $f \in H^\infty + C$, *then* $f = g + h$ *for some* $g \in C$, $h \in H^\infty$, *that satisfy*

$$\|g\|_\infty \le 2\|f\|_\infty, \quad \|h\|_\infty \le 3\|f\|_\infty.$$

Indeed, $f = G + H$, for some $G \in C$, $H \in H^\infty$. The case $\|f\|_\infty = 0$ being trivial, assume $\|f\|_\infty > 0$. Then there exists $r < 1$ such that the Poisson integral $P[G]$ satisfies

$$\|G - P[G]_r\|_\infty < \|f\|_\infty.$$

Define

$$g = G - P[G]_r + P[f]_r,$$
$$h = H - P[H]_r.$$

Then $g \in C$, $h \in H^\infty$, $f = g + h$, $\|g\|_\infty < 2\|f\|_\infty$, and $\|h\|_\infty = \|f - g\|_\infty < 3\|f\|_\infty$.

Finally, if f is in the closure of $H^\infty + C$, there exist $f_i \in H^\infty + C$, $i = 1, 2, 3, \ldots$, such that $\|f_i\|_\infty < 2^{-i}$ for $i \ge 2$, and $f = \sum_1^\infty f_i$. Consequently, $f_i = g_i + h_i$ ($i \ge 1$), where $g_i \in C$, $h_i \in H^\infty$, and (for $i \ge 2$) $\|g_i\|_\infty < 2^{1-i}$, $\|h_i\|_\infty < 2^{2-i}$. Then $g = \sum_1^\infty g_i$ is in C, $h = \sum_1^\infty h_i$ is in H^∞, and hence $f = g + h$ is in $H^\infty + C$.

6.5.6. Remarks. The case $n = 1$ of Theorem 6.5.5 was proved by Sarason [1].

When $n = 1$, then $H^\infty + C$ is the *smallest* closed subalgebra of L^∞ that contains H^∞ and is larger than H^∞. (See Sarason [2].)

This is false when $n > 1$. For example, let E be an arbitrary circular subset of S with $0 < \sigma(E) < 1$; then E is a union of circles Γ_ζ introduced in §6.4.2. Let X_E be the subalgebra of $H^\infty(S) + C(S)$ consisting of all f whose restriction to almost every $\Gamma_\zeta \subset E$ belongs to $H^\infty(T)$. It is clear that X_E is closed, that $H^\infty(S) \subset X_E \subset H^\infty(S) + C(S)$, and that both inclusions are strict.

Aytuna and Chollet [1] have extended Theorem 6.5.5 to strictly pseudo-convex domains.

If B and S are, however, replaced by a polydisc U^n ($n > 1$) and its distinguished boundary T^n, the situation is different: $H^\infty(T^n) + C(T^n)$ is still a closed subspace of $L^\infty(T^n)$, but it fails to be an algebra (Rudin [7]).

6.6. Gleason's Problem

6.6.1. Let X be some class of holomorphic functions in a region $\Omega \subset \mathbb{C}^n$. Gleason's problem for X is the following:

If $a \in \Omega$ and $f \in X$, do there exist functions $g_1, \ldots, g_n \in X$ such that

$$(1) \qquad\qquad f(z) - f(a) = \sum_{k=1}^{n} (z_k - a_k) g_k(z)$$

for all $z \in \Omega$?

Gleason [1] originally asked this for $\Omega = B$, $a = 0$, and $X = A(B)$, the ball algebra. (In other words, do the coordinates functions generate the maximal ideal consisting of all $f \in A(B)$ with $f(0) = 0$?)

The difficulty of the problem depends on Ω. For instance, if B is replaced by a polydisc, Gleason's question becomes trivial: take $n = 2$, for simplicity, and decompose any $f \in H(U^2)$ that is continuous on $\bar{U}^2$ into a sum

$$(2) \qquad\qquad f(z, w) = f(0, 0) + z g_1(z, w) + w g_2(z, w)$$

where

$$(3) \qquad g_1(z, w) = \frac{f(z, 0) - f(0, 0)}{z}, \qquad g_2(z, w) = \frac{f(z, w) - f(z, 0)}{w}.$$

The point is that $z \neq 0$ and $w \neq 0$ on the torus T^2, so that the functions (3) belong to the prescribed class.

Put in different terms, if $f \in H^\infty(U^2)$ and g_2 is as in (3), then $g_2 \in H^\infty(U^2)$. This is not true in B: $w^2/(1 - z)$ is bounded in B_2, but $w/(1 - z)$ is not; to see the latter, take $(z, w) = (\cos t, \sin t)$.

Gleason's original question was solved by Leibenson (see Henkin [4]), who observed that

(4) $$f(z) - f(0) = \int_0^1 \frac{d}{dt} f(tz)dt = \sum_{k=1}^n z_k \int_0^1 (D_k f)(tz)dt$$

and then proved that the functions

(5) $$g_k(z) = \int_0^1 (D_k f)(tz)dt \qquad (1 \le k \le n)$$

lie in $A(B)$ if $f \in A(B)$. (This proof is similar to some of the work in Section 6.4; details may be found on pp. 151–153 of Stout [1].)

Kerzman and Nagel [1] used sheaf-theoretic methods to solve Gleason's problem in smoothly bounded strictly pseudoconvex regions, for functions subject to Lipschitz conditions that could vary from point to point.

The simplest and most far-reaching solution is probably the one found by Ahern and Schneider [4]. They worked in strictly pseudo-convex domains. Here is the special case of the ball:

6.6.2. The Ahern–Schneider Solution. If $f \in H^1(B)$ and $a \in B$, the Cauchy formula gives

(1) $$f(z) - f(a) = \int_S [C(z, \zeta) - C(a, \zeta)]f(\zeta)d\sigma(\zeta) \qquad (z \in B)$$

where $f(\zeta)$ denotes the K-limit of f at $\zeta \in S$. This suggests consideration of the functions

(2) $$g_k(a, z) = \int_S \frac{C(z, \zeta) - C(a, \zeta)}{\langle z - a, \zeta \rangle} \bar{\zeta}_k f(\zeta)d\sigma(\zeta)$$

since they clearly satisfy

(3) $$f(z) - f(a) = \sum_{k=1}^n (z_k - a_k)g_k(a, z).$$

The identity

$$\frac{C(z, \zeta) - C(a, \zeta)}{\langle z - a, \zeta \rangle} = C(z, \zeta) \sum_{j=0}^{n-1} \frac{(1 - \langle z, \zeta \rangle)^j}{(1 - \langle a, \zeta \rangle)^{j+1}}$$

shows that every $g_k(a, z)$ is (for fixed $a \in B$) a finite sum of products, each product being a holomorphic monomial z^α (of degree $|\alpha| \le n - 1$) times a

Toeplitz transform of f whose symbol is in $C^\infty(\bar{B})$; in fact, the symbols are conjugate-holomorphic functions in a neighborhood of $\bar{B}$.

The results of Section 6.5 prove therefore cases (i), (ii), and (iii) (for $0 < \alpha < 1$) of the following conclusion:

The formulas (2) *and* (3) *solve Gleason's problem in any of the following spaces X:*

(i) $X = A(B)$.

(ii) $X = H^p(B), 1 \le p \le \infty$.

(iii) $X = A(B) \cap \text{Lip } \alpha, 0 < \alpha \le 1$.

(iv) $X = H_E^\infty(B)$, the space of all $f \in H^\infty(B)$ that have continuous extensions to $B \cup E$, where $E \subset S$.

(v) $X = A(B, E, \{\alpha\})$.

The spaces (v) are defined as follows: let $E \subset S$ be arbitrary, associate to each $\zeta \in E$ a number $\alpha(\zeta), 0 < \alpha(\zeta) \le \frac{1}{2}$, and let X consist of all $f \in A(B)$ for which

(4) $$\sup_{z \in \bar{B}} \frac{|f(z) - f(\zeta)|}{|z - \zeta|^{\alpha(\zeta)}} = K(f, \zeta) < \infty.$$

(These spaces occur in Kerzman–Nagel [1], with $\alpha(\zeta) < 1/2$.)

Note, furthermore, that the g_k's given by (2) *depend linearly on f when a is fixed, and that they depend holomorphically on a when f is fixed.*

The proofs of the remaining cases (iv), (v), and Lip 1 depend on a more detailed look at the operators V_φ of Theorem 6.5.4, where φ is any of the conjugate-holomorphic symbols that play a role in the definition of g_k, and that are defined on a neighborhood of $\bar{B}$.

Recall that

(5) $$V_\varphi f = \varphi f - T_\varphi f$$

and that

(6) $$(V_\varphi f)(z) = \int_S \{\varphi(z) - \varphi(\zeta)\} C(z, \zeta) f(\zeta) d\sigma(\zeta).$$

If $f \in H^\infty$ then $V_\varphi f \in C(\bar{B})$ (Theorem 6.5.4); by (5), this proves case (iv). Similarly, case (v) will follow from

(7) $$|(\text{grad } V_\varphi f)(z)| < c(1 - |z|)^{-1/2},$$

since $V_\varphi f$ is then in Lip $\frac{1}{2}$, by Lemma 4.6.8, and the case $\alpha = 1$ of (iii) will follow if we show that the above gradient is bounded in B, since then $V_\varphi f \in$ Lip 1.

The gradient of $V_\varphi f$ is obtained by differentiating (6). Since $\varphi \in C^1(\bar{B})$, it suffices to consider

$$(8) \qquad \int_S \{\varphi(z) - \varphi(\zeta)\} \frac{n\bar{\zeta}_j}{(1 - \langle z, \zeta \rangle)^{n+1}} f(\zeta) d\sigma(\zeta)$$

for $1 \le j \le n$.

Since $|\varphi(z) - \varphi(\zeta)| \le c|z - \zeta| \le 2c|1 - \langle z, \zeta \rangle|^{1/2}$, the integrand in (8) is dominated by

$$(9) \qquad C\|f\|_\infty |1 - \langle z, \zeta \rangle|^{-n-1/2}$$

and now (7) follows, for $f \in H^\infty$, by Proposition 1.4.10.

Finally, suppose $f \in A(B) \cap \text{Lip } 1$, and extend f to a bounded Lip 1— function on $\mathbb{C}^n$, for instance by setting $f(z) = f(z/|z|)$ when $|z| > 1$. Put $z = re_1$ in (8), without loss of generality, and put

$$(10) \qquad \psi(\zeta) = \psi_{r,j}(\zeta) = n\bar{\zeta}_j [\varphi(re_1) - \varphi(\zeta)].$$

Writing $\zeta = (\zeta_1, \zeta')$, (8) is then the sum of

$$(11) \qquad \int_S \frac{\psi(\zeta)}{(1 - r\bar{\zeta}_1)^{n+1}} [f(\zeta_1, \zeta') - f(r, \zeta')] d\sigma(\zeta)$$

and (by 1.4.7(2))

$$(12) \qquad \int_{B_{n-1}} f(r, \zeta') dv_{n-1}(\zeta') \frac{1}{2\pi} \int_{-\pi}^{\pi} \frac{\psi(\zeta_1 e^{i\theta}, \zeta')}{(1 - r\bar{\zeta}_1 e^{-i\theta})^{n+1}} d\theta.$$

Since $|f(\zeta_1, \zeta') - f(r, \zeta')| \le c|\zeta_1 - r| \le c|1 - r\bar{\zeta}_1|$ and $|\psi(\zeta)| \le |re_1 - \zeta| \le c|1 - r\bar{\zeta}_1|^{1/2}$, (11) is bounded. Since $\bar{\varphi}$ is holomorphic, the inner integral in (12) is $\psi(0, \zeta')$, so that the absolute value of (12) is at most $2n\|f\|_\infty \|\varphi\|_\infty$.

This completes the proof.

6.6.3. A Multiplier Theorem. We shall now consider the Ahern–Schneider solution in the special case $a = 0$.

Differentiate the Cauchy formula 6.6.2(1) with respect to z_k, then replace z by tz, and integrate over $0 \le t \le 1$. Comparison with 6.6.2(2) shows that

$$(1) \qquad \int_0^1 (D_k f)(tz) dt = g_k(0, z).$$

The Ahern–Schneider solution thus turns out to be identical with Leibenson's, when $a = 0$.

If $f(z) = \sum c(\alpha)z^\alpha$ (the summation extends over all multi-indices α), one computes easily that

(2)
$$z_k \int_0^1 (D_k f)(tz)dt = \sum_{\alpha \neq 0} \frac{\alpha_k}{|\alpha|} c(\alpha)z^\alpha$$

where $|\alpha| = \alpha_1 + \cdots + \alpha_n$, as usual.

In view of the properties of the functions g_k that were established in §6.6.2, the following multiplier theorem emerges:

For $k = 1, \ldots, n$, the transformation

(3)
$$\sum_\alpha c(\alpha)z^\alpha \to \sum_{\alpha \neq 0} \frac{\alpha_k}{|\alpha|} c(\alpha)z^\alpha$$

maps X into X, where X is any of the spaces $A(B)$, $H^p(B)$ ($1 \leq p \leq \infty$), $A(B) \cap \text{Lip } \alpha$ ($0 < \alpha \leq 1$), $H_E^\infty(B)$, or $A(B, E, \{\alpha\})$.

The same is true if the right side of (3) is divided by z_k.

6.6.4. Homomorphisms of $H^\infty(B)$. The fact that Gleason's problem can be solved in $H^\infty(B)$ has implications concerning the possible homomorphisms of the algebra $H^\infty(B)$ (relative to pointwise multiplication) into any algebra of holomorphic functions. This is Theorem 6.6.5 below. Its proof also depends on the following two easy propositions:

(a)   *To each homomorphism Λ of $A(B)$ onto $\mathbb{C}$ corresponds a point $w \in \bar{B}$ such that $\Lambda f = f(w)$ for every $f \in A(B)$.*
(b)   *If Ω is a region in $\mathbb{C}^m$, $F: \Omega \to \mathbb{C}^n$ is a holomorphic map with range in $\bar{B}$, and $F(w_0) \in S$ for some $w_0 \in \Omega$, then F is constant.*

Proof of (a). Let π_i denote the ith coordinate function: $\pi_i(z) = z_i$, $1 \leq i \leq n$. Put $w_i = \Lambda\pi_i$, $w = (w_1, \ldots, w_n)$. Thus $\Lambda\pi_i = \pi_i(w)$. Since Λ is linear and multiplicative, it follows that $\Lambda f = f(w)$ for all polynomials f. Since Λ is a complex homomorphism of a Banach algebra, $\|\Lambda\| = 1$, and the fact that the polynomials are dense in $A(B)$ implies now that $\Lambda f = f(w)$ for all $f \in A(B)$.

Proof of (b). Put $g(w) = \langle F(w), F(w_0) \rangle$. Then $g: \Omega \to \mathbb{C}$ is holomorphic, $|g| \leq 1$, $g(w_0) = 1$. By the maximum modulus theorem, $g(w) = 1$ for all $w \in \Omega$. Since $F(w) \in \bar{B}$, it follows that $F(w) = F(w_0)$ for all $w \in \Omega$.

6.6.5. Theorem. *Suppose B is the unit ball in $\mathbb{C}^n$, Ω is a region in $\mathbb{C}^m$, and T is a multiplicative linear map of $H^\infty(B)$ into $H(\Omega)$.*

If Tf is nonconstant for at least one $f \in A(B)$, then there is a holomorphic map $\psi: \Omega \to B$ such that $Tf = f \circ \psi$ for every $f \in H^\infty(B)$.

Corollary. *The range of T lies in $H^\infty(\Omega)$.*

Proof. The multiplicativity of T implies that $T1 = 1$. If $z \in \Omega$, it follows that $f \to (Tf)(z)$ is a homomorphisms of $A(B)$ onto $\mathbb{C}$. Hence there is a point $\psi(z) \in \bar{B}$ such that $(Tf)(z) = f(\psi(z))$ for all $f \in A(B)$.

In particular, $\pi_i \circ \psi = T\pi_i$, $1 \le i \le n$. Thus ψ is a holomorphic map, and since $Tf \ne$ const. for some $f \in A(B)$, ψ cannot be constant. Hence $\psi(\Omega) \subset B$.

Now let $f \in H^\infty(B)$, pick $z \in \Omega$, and apply the solution of Gleason's problem to f at the point $a = \psi(z)$: there are functions $g_i \in H^\infty(B)$ such that

$$f - f(\psi(z)) = \sum_{i=1}^{n} (\pi_i - \psi_i(z))g_i.$$

Apply T to this equation and evaluate at z:

$$(Tf)(z) - f(\psi(z)) = \sum_{i=1}^{n} [(T\pi_i)(z) - \psi_i(z)](Tg_i)(z).$$

Since $T\pi_i = \psi_i$, the right side is 0, and thus $Tf = f \circ \psi$.

This proof is quite similar to one used by Ahern and Schneider [1].

Corollary. *If T is an automorphism of the algebra $H^\infty(B)$, then $Tf = f \circ \psi$ for some $\psi \in \text{Aut}(B)$.*

This follows if the theorem (with $\Omega = B$) is applied to both T and T^{-1}.

The assumption that Tf is nonconstant for at least one $f \in A(B)$ is needed for the validity of the theorem. In fact, there exist homomorphisms of $H^\infty(U)$ onto $H^\infty(U)$ that send every member of the disc algebra to a constant. For a proof of this (which depends on the existence of analytic embeddings of U in "fibers" of the maximal ideal space of $H^\infty(U)$), we refer to pp. 166–169 Hoffman's book [1].

Chapter 7

Some L^p-Topics

7.1. Projections of Bergman Type

7.1.1. A Class of Reproducing Kernels. Fix $n \geq 1$, write B_n for B, temporarily, let s be a nonnegative integer, and let P be the orthogonal projection of $\mathbb{C}^{n+s}$ onto $\mathbb{C}^n$ (regarding $\mathbb{C}^n$ as the subspace of $\mathbb{C}^{n+s}$ defined by $z_{n+1} = \cdots = z_{n+s} = 0$).

If $f \in (L^1 \cap H)(B_n)$, the Bergman formula can be applied to $f \circ P$ in B_{n+s}:

$$(1) \qquad f(Pz) = \int_{B_{n+s}} \frac{f(P(w))}{(1 - \langle z, w \rangle)^{n+s+1}} \, dv_{n+s}(w).$$

If $z \in B_n$ then $Pz = z$, $\langle z, w \rangle = \langle z, Pw \rangle$, and hence Fubini's theorem converts (1) to

$$(2) \qquad f(z) = c(n, s) \int_{B_n} \frac{(1 - |w|^2)^s}{(1 - \langle z, w \rangle)^{n+s+1}} f(w) dv_n(w).$$

The constants $c(n, s)$ can be computed by taking $f = 1$, $z = 0$ in (1):

$$(3) \qquad 1 = c(n, s) \int_{B_n} (1 - |w|^2)^s \, dv(w).$$

Reverting to our usual notation (B and v in place of B_n and v_n) we obtain the following analogues of the Bergman formula:

$$(4) \qquad f(z) = \binom{n + s}{n} \int_B K_s(z, w) f(w) dv(w) \qquad (z \in B)$$

where

$$(5) \qquad K_s(z, w) = \frac{(1 - |w|^2)^s}{(1 - \langle z, w \rangle)^{n+1+s}} \qquad (z \in B, w \in B)$$

for $s = 0, 1, 2, \ldots$.

120

Both (5) and the right side of (4) make perfectly good sense for arbitrary complex $s = \sigma + it$ (provided $\sigma > -1$, for obvious integrability reasons), if the complex powers are understood to be the usual principal branches: both numerator and denominator in (5) are 1 when $w = 0$. The binomial coefficient in (4) is then

$$\frac{\Gamma(n + s + 1)}{\Gamma(n + 1)\Gamma(s + 1)}.$$

This suggests the introduction of operators T_s defined by

$$(6) \qquad (T_s f)(z) = \binom{n + s}{n} \int_B K_s(z, w) f(w) dv(w)$$

for $z \in B$, for Re $s = \sigma > -1$, and for any $f \in L^1(v)$ (*not necessarily holomorphic*) for which the integrands are in $L^1(v)$.

One can now use a complex interpolation argument to prove that all of these operators have the reproducing property (4):

7.1.2. Proposition. *If $f \in H^\infty(B)$ and $\sigma > -1$, then $T_s f = f$ and $T_s \bar{f} = \overline{f(0)}$.*

(The restriction to H^∞ will be removed in Theorem 7.1.4.)

Proof. The computation that yielded the value of $c(n, s)$ in §7.1.1 was valid for all s with $\sigma > -1$. Hence $T_s 1 = 1$. To prove that $T_s f = f$ amounts therefore to showing that

$$(1) \qquad \int_B \left\{ \frac{1 - |w|^2}{1 - \langle z, w \rangle} \right\}^s \cdot \frac{f(w) - f(z)}{(1 - \langle z, w \rangle)^{n+1}} \, dv(w) = 0$$

for $z \in B$, $\sigma > -1$.

Regard z and f as fixed, and denote the integral (1) by $H(s)$. Then H is holomorphic in the half-plane $\sigma > -1$, and $H(s) = 0$ when s is a nonnegative integer, because of the reproducing formula 7.1.1(4). Since

$$\frac{1 - |w|^2}{|1 - \langle z, w \rangle|} < 2 \quad \text{and} \quad |\arg(1 - \langle z, w \rangle)| < \frac{\pi}{2},$$

there is a constant $\gamma = \gamma(f, z) < \infty$ such that

$$(2) \qquad |H(s)| \leq \gamma 2^\sigma e^{\pi |t|/2} \qquad (s = \sigma + it, \sigma \geq 0).$$

Now define

$$(3) \qquad G(s) = \frac{2^{-s} H(s)}{\sin(\pi s/2)}.$$

Since $2|\sin(\pi s/2)| \geq |e^{\pi t/2} - e^{-\pi t/2}|$, G is a *bounded* holomorphic function in the half-plane $\sigma > 0$. Since H vanishes at every positive integer, G vanishes at every odd positive integer. This zero-set of G violates the Blaschke condition for bounded holomorphic functions in a half-plane. Hence G is identically 0, and so is H. Thus (1) holds, and $T_s f = f$.

To evaluate $T_s \tilde{f}$, it is enough to consider the case $f(0) = 0$. Insert $\tilde{f}$ into 7.1.1(6), use the explicit formula for $K_s(z, w)$, and integrate in polar coordinates (§1.4.3). It is then easy to see that the integrals over S are 0, by the mean-value property of conjugate-holomorphic functions.

7.1.3. The Case $s = n + 1$. This case is particularly simple, because

$$(1) \qquad |K_{n+1}(z, w)| = \left\{\frac{1 - |w|^2}{|1 - \langle z, w\rangle|^2}\right\}^{n+1} = (J_R \varphi_w)(z)$$

(see Theorem 2.2.6), so that

$$(2) \qquad \int_B |K_{n+1}(z, w)|\, dv(z) = 1.$$

Consequently,

$$(3) \qquad \int_B |T_{n+1} f|\, dv \leq \binom{2n+1}{n} \int_B |f|\, dv$$

for every $f \in L^1(v)$. Thus T_{n+1} is a bounded linear operator on $L^1(v)$, with range in $(L^1 \cap H)(B)$. Since $H^\infty(B)$ is dense in $(L^1 \cap H)(B)$, the conclusions of Proposition 7.1.2 hold if $f \in (L^1 \cap H)(B)$.

Hence T_{n+1} is a bounded linear projection of $L^1(v)$ onto $(L^1 \cap H)(B)$.

This is a special case of the next theorem, which says precisely on which L^p a given T_s is bounded, or, equivalently, which T_s are bounded on a given L^p. Since K_0 is the Bergman kernel, the theorem shows that the Bergman transform is not bounded on $L^1(v)$, but that it is bounded on $L^p(v)$ for $1 < p < \infty$.

7.1.4. Theorem (Forelli–Rudin [1])

(a) *For $1 \leq p < \infty$, T_s is a bounded linear operator on $L^p(v)$ if and only if*

$$(1) \qquad \operatorname{Re}(1 + s) > \frac{1}{p}.$$

(b) *If* (1) *holds then* T_s *projects* $L^p(v)$ *onto* $(L^p \cap H)(B)$: *in fact*

(2) $$T_s f = f \quad and \quad T_s \bar{f} = \overline{f(0)}$$

for every $f \in (L^p \cap H)(B)$.

Proof. By Proposition 7.1.2, part (b) is an immediate consequence of (a), since $H^\infty(B)$ is dense in $(L^p \cap H)(B)$.

The estimate

(3) $$e^{-\pi|t|/2}|K_\sigma(z, w)| \le |K_s(z, w)| \le e^{\pi|t|/2}|K_\sigma(z, w)|$$

(where $s = \sigma + it$, as before) will be used in the proof of (a); it holds because $1 - \langle z, w \rangle$ has positive real part, so that its argument lies between $-\pi/2$ and $\pi/2$.

We first take $p = 1$, in which case (1) becomes $\sigma > 0$.

If $\sigma < 0$ and $z \in B$, then $K_s(z, w)$ is not a bounded function in B, and therefore $T_s f$ fails to exist for some $f \in L^1(B)$.

So assume $\sigma \ge 0$. The duality between L^1 and L^∞ shows that T_s is bounded on L^1 if and only if

(4) $$\sup_{w \in B} \int_B |K_s(z, w)|\, dv(z) < \infty.$$

By (3), we can replace s by σ in (4), and now Proposition 1.4.10 implies that (4) fails when $\sigma = 0$ but that (4) holds when $\sigma > 0$.

This proves (a) in the case $p = 1$.

If $1 < p < \infty$ and $1 + \sigma \le 1/p$, then $\sigma q \le -1$, where $1/p + 1/q = 1$. Hence

(5) $$\int_B |K_s(z, w)|^q\, dv(w) = \infty$$

for every $z \in B$, and $T_s f$ fails to exist for some $f \in L^p(v)$.

There remains the case $1 < p < \infty$, $1 + \sigma > 1/p$. Put $c = 1/pq$, and

(6) $$h(z) = (1 - |z|^2)^{-c} \quad (z \in B).$$

Since $\sigma - qc = \sigma - 1/p > -1$, Proposition 1.4.10 implies that

(7) $$\int_B |K_\sigma(z, w)|h(w)^q\, dv(w) = \int_B \frac{(1 - |w|^2)^{\sigma - qc}}{|1 - \langle z, w \rangle|^{n+1+\sigma}}\, dv(w)$$
$$\le [ah(z)]^q$$

for some $a = a(n, \sigma, q) < \infty$. Also, $pc = 1/q < 1$ and $\sigma + pc = \sigma + 1/q > 0$, so that

(8)
$$\int_B |K_\sigma(z, w)| h(z)^p \, dv(z) = (1 - |w|^2)^\sigma \int_B \frac{(1 - |z|^2)^{-pc}}{|1 - \langle z, w \rangle|^{n+1+\sigma}} \, dv(z)$$
$$\leq [bh(w)]^p$$

for some $b = b(n, \sigma, q) < \infty$, by another application of 1.4.10.

Now fix some $f \in L^p(v)$, and put

(9)
$$F(z) = \int_B |K_\sigma(z, w)| |f(w)| \, dv(w) \qquad (z \in B).$$

By Hölder's inequality and (7),

(10)
$$F(z) = \int_B |K_\sigma|^{1/q} h \cdot |K_\sigma|^{1/p} \left| \frac{f}{h} \right| dv$$
$$\leq ah(z) \left\{ \int_B |K_\sigma(z, w)| \left| \frac{f}{h} \right|^p (w) dv(w) \right\}^{1/p}.$$

If we raise (10) to the pth power and apply (8), we obtain

(11)
$$\int_B F^p \, dv \leq a^p \int_B \left| \frac{f}{h} \right|^p (w) dv(w) \int_B |K_\sigma(z, w)| h(z)^p \, dv(z)$$
$$\leq (ab)^p \int_B |f|^p \, dv.$$

If we combine (3), (9), and (11), we see that T_s is bounded on $L^p(v)$ if $1 < p < \infty$ and $1 + \sigma > 1/p$. This completes the proof.

One consequence of the theorem just proved is that the $L^p(v)$-norm of any holomorphic function in B is controlled by the corresponding norm of its real part:

7.1.5. Theorem. *To every p, $1 \leq p < \infty$, corresponds a constant $M_p = M_{p,n} < \infty$ with the following property:*

If $f \in H(B)$, $f(0) = 0$, $u = \mathrm{Re}\, f$, then

(1)
$$\int_B |f|^p \, dv \leq M_p \int_B |u|^p \, dv.$$

Proof. Fix p and choose s so that $1 + \sigma > 1/p$. Then there exists $A < \infty$ such that $\| T_s g \|_p \leq A \| g \|_p$ for every $g \in L^p(v)$. Apply this to $g = f_r$ and $g = \tilde{f}_r$

(where $f_r(z) = f(rz)$, $0 < r < 1$, as usual). By Theorem 7.1.4,

$$f_r = T_s(f_r) = T_s(f_r + \bar{f}_r) = 2T_s(u_r),$$

so that

$$\int_B |f_r|^p \, dv = 2^p \int_B |T_s(u_r)|^p dv \le (2A)^p \int_B |u_r|^p \, dv$$

or

(2) $$\int_{rB} |f|^p \, dv \le (2A)^p \int_{rB} |u|^p \, dv \qquad (0 < r < 1).$$

Now (1) follows as $r \nearrow 1$ in (2).

7.1.6. The real part of the function

$$f(z) = i \log(1 - z_1)$$

is bounded in B, but its imaginary part is not. Theorem 7.1.5 can therefore not be extended to the case $p = \infty$. As in the proof of 7.1.5 shows, Theorem 7.1.4 also fails for $p = \infty$, no matter what s we choose.

7.1.7. The Korányi–Vagi theorem shows that Theorem 7.1.5 has an analogue in which integration over B is replaced by integration over S:
There are constants $M_{p,n} < \infty$, for $1 < p < \infty$, $n = 1, 2, 3, \ldots$ such that

(1) $$\int_S |f|^p \, d\sigma \le M_{p,n} \int_S |u|^p \, d\sigma$$

for every $f = u + iv$ in $H^p(S)$ with $v(0) = 0$.
    In contrast to Theorem 7.1.5 this fails when $p = 1$ and $n = 1$; the functions

(2) $$f_r(z) = \frac{1}{1 - rz} \qquad (0 < r < 1)$$

show this.
    However, it is not known whether there exist constants $M_{1,n} < \infty$ for $n = 2, 3, 4, \ldots$, such that (1) holds. In other words, if S is (for example) the unit sphere in $\mathbb{C}^2$, it is not known whether

(3) $$\sup \frac{\int_S |f| d\sigma}{\int_S |u| d\sigma} < \infty,$$

where the supremum is taken over all holomorphic polynomials $f \not\equiv 0$, with $f(0)$ real, and $u = \operatorname{Re} f$.

It seems quite likely that (3) is true when $n > 1$. This conjecture is closely related to the inner function problem which will be discussed in Chapter 19.

7.2. Relations between H^p and $L^p \cap H$

7.2.1. Plurisubharmonic Functions. Let Ω be a region in $\mathbb{C}^n$. An upper semi-continuous function $u \colon \Omega \to [-\infty, \infty)$ is said to be *plurisubharmonic* if to each $a \in \Omega$ and $b \in \mathbb{C}^n$ corresponds a neighborhood V of 0 in $\mathbb{C}$ such that

$$(1) \qquad\qquad \lambda \to u(a + \lambda b)$$

is subharmonic in V. Less precisely, the restrictions of u to all complex lines should be subharmonic.

The most obvious examples of plurisubharmonic functions are $\log|f|$ and $|f|^c$ $(c > 0)$, for any $f \in H(\Omega)$.

The slice-integration formula shows that every plurisubharmonic function u in Ω is subharmonic in Ω (see the proof of Proposition 1.5.4) and that therefore $\Delta u \geq 0$, provided, of course, that $u \in C^2$. When $\Omega = B$, the class of plurisubharmonic functions is Moebius invariant. It follows that *every plurisubharmonic $u \in C^2(B)$ satisfies*

$$(2) \qquad\qquad \Delta u \geq 0 \quad \text{and} \quad \tilde{\Delta} u \geq 0.$$

Contrary to what one might expect by analogy with Theorem 4.4.9, (the equivalence of (b) and (c)), this last statement does not have a converse (at least not when $n > 2$): The function

$$(3) \qquad\qquad u(z) = z_1 \bar{z}_1 + z_2 \bar{z}_2 - z_3 \bar{z}_3$$

has $\Delta u = 4$ and $\tilde{\Delta} u = 4(1 - |z|^2)(1 - u) > 0$ in B, but $u(0, 0, w) = -|w|^2$ is not subharmonic.

Plurisubharmonic functions will play an important role in Chapter 17. They are also closely related to the characterization of domains of holomorphy (see, for instance, §2.6 in Hörmander [2]). At present, we only need the following inequality. Its statement uses the familiar notation $z = (z', z_n)$ for $z \in B_n$, where $z' = (z_1, \ldots, z_{n-1}) \in B_{n-1}$.

7.2.2. Proposition. *If u is plurisubharmonic in B_n $(n > 1)$ and $u \geq 0$, then*

$$(1) \qquad\qquad \int_{B_{n-1}} u(z', 0) dv_{n-1}(z') \leq \sup_{0 < r < 1} \int_S u(r\zeta) d\sigma(\zeta)$$

where S is the boundary of B_n. If, furthermore, $u(z', z_n)$ is a harmonic function of z_n, for every $z' \in B_{n-1}$, then equality holds in (1).

Notes. (i) The assumption $u \geq 0$ can be removed. It is made because its presence assures the existence of all integrals that occur in the proof and it allows us to apply Fubini's theorem without any argument.

(ii) The integrals on the right of (1) are nondecreasing functions of r, since u is subharmonic. The supremum is therefore equal to the limit, as $r \nearrow 1$.

Proof. To each $z' \in B_{n-1}$ corresponds a $z_n \in \mathbb{C}$ so that $(z', z_n) \in S$. For $0 < r < 1$,

$$(2) \qquad u(rz', 0) \leq \frac{1}{2\pi} \int_{-\pi}^{\pi} u(rz', rz_n e^{i\theta}) d\theta,$$

since u is plurisubharmonic. By formula 1.4.7(2), it follows that

$$(3) \qquad \int_{B_{n-1}} u(rz', 0) dv_{n-1}(z') \leq \int_S u(r\zeta) d\sigma(\zeta).$$

Setting $rz' = w'$, the left side of (3) is seen to be

$$r^{2-2n} \int_{rB_{n-1}} u(w', 0) dv(w').$$

Hence (1) follows from (3) as $r \nearrow 1$. If $u(z', z_n)$ is harmonic in z_n, then equality holds in (2), hence also in (3) and (1).

7.2.3. Restrictions and Extensions. With B_n and B_{n-1} as above, let f and g be functions with domains B_n and B_{n-1}, respectively, and define a restriction operator ρ and an extension operator E by

$$(1) \qquad\qquad (\rho f)(z') = f(z', 0) \qquad (z' \in B_{n-1}),$$

$$(2) \qquad\qquad (Eg)(z', z_n) = g(z') \qquad (z \in B_n).$$

Since $(\rho E g)(z') = (Eg)(z', 0) = g(z')$, ρE is the identity operator.

7.2.4. Theorem. *Assume* $n > 1, 0 < p < \infty$.

(a) *The extension E is a linear isometry of $(L^p \cap H)(B_{n-1})$ into $H^p(B_n)$.*

(b) *The restriction ρ is a linear norm-decreasing map of $H^p(B_n)$ onto $(L^p \cap H)(B_{n-1})$.*

Proof. Apply Proposition 7.2.2 to the plurisubharmonic function $u = |f|^p$, where $f \in H(B)$. This gives (a) and most of (b). The fact that $\rho(H^p(B_n))$ is all of $(L^p \cap H)(B_{n-1})$ follows from (a) and the identity $\rho Eg = g$.

Note: An analogue of this result holds in polydiscs, for restrictions to the diagonal, but it is not quite so easy. See Rudin [1], pp. 53, 69, Duren–Shields [1], Horowitz–Oberlin [1], J. H. Shapiro [1], Moulin–Rosay [1], and Detraz [1].

7.2.5. Theorem. *Suppose $n \geq 1, 0 < p < \infty$.*

(a) *If $f \in H^p(B_n)$ then*

(1) $$|f(z)| \leq 2^{n/p}\|f\|_p(1 - |z|)^{-n/p}$$

and

(2) $$\lim_{|z| \to 1} (1 - |z|)^{n/p}|f(z)| = 0 \qquad (z \in B_n).$$

(b) *If $f \in (L^p \cap H)(B_n)$ then (1) and (2) hold with n replaced by $n + 1$.*

Proof. Assume $f \in H^p(B_n)$. Theorems 5.6.2 and 5.6.3 show that $|f|^p$ has an $\mathscr{M}$-harmonic majorant $u = P[d\mu]$, where μ is a positive measure on S whose total mass is $\|f\|_p^p$. The estimate

(3) $$P(z, \zeta) = \frac{(1 - |z|^2)^n}{|1 - \langle z, \zeta \rangle|^{2n}} \leq \left\{\frac{1 + |z|}{1 - |z|}\right\}^n$$

therefore yields (1).

If we apply (1) to $f - f_r$ in place of f and recall that $\|f - f_r\|_p \to 0$ as $r \nearrow 1$ (Theorem 5.5.6), we obtain (2).

Next, if $f \in (L^p \cap H)(B_n)$, (a) can be applied to the function $Ef \in H^p(B_{n+1})$; see Theorem 7.2.4. This gives (b). $\blacksquare$

Notes: (i) When $p \geq 1$, Theorem 7.2.5 can be proved more directly by applying Hölder's inequality to the Cauchy formula and the Bergman formula, respectively.

(ii) If $c < n/p$ and $f(z) = (1 - z_1)^{-c}$ then $f \in H^p(B_n)$, by Proposition 1.4.10. The exponents that occur in (1) and (2) can therefore not be replaced by smaller ones.

7.2.6. In several complex variables it is much harder to construct functions that exhibit interesting phenomena than it is in one. The required techniques are not sufficiently developed. We shall now present two rather modest examples (7.2.9 and 7.2.10) that illustrate the possible boundary behavior of functions in $H^2(B_n)$ and in $(L^2 \cap H)(B_n)$, for $n = 2$ and $n = 3$. Similar examples undoubtedly exist when $n \geq 3$, but different constructions would be required for them.

The following "scrambling lemma" works, however, for all n. It is basically due to Calderón (see Zygmund [3], Vol. II, p. 165) and to Stein [5; p. 146].

7.2.7. Lemma. *Suppose* $\{f_i\}$ *is a sequence of Borel functions on* S, $0 \le f_i \le 1$, *and*

$$(1) \qquad\qquad \sum_{i=1}^{\infty} \int_S f_i \, d\sigma = \infty.$$

Then there exist unitary U_i *such that*

$$(2) \qquad\qquad \sum_{i=1}^{\infty} f_i(U_i\zeta) = \infty \quad \text{a.e.} \quad \text{on } S.$$

Proof. Let $\mathscr{U} = \mathscr{U}(n)$ be the unitary group on $\mathbb{C}^n$, and let $\mathscr{U}^{\infty}$ be the cartesian product of countably many copies of $\mathscr{U}$, with the Tychonoff topology, so that $\mathscr{U}^{\infty}$ is a compact group whose Haar measure is the usual product of the Haar measures of the copies of $\mathscr{U}$. Consider the function

$$(3) \qquad\qquad g(\zeta, \{U_i\}) = \prod_{i=1}^{\infty} [1 - f_i(U_i\zeta)]$$

on $S \times \mathscr{U}^{\infty}$. The definition of the product measure dU on $\mathscr{U}^{\infty}$ shows, for any $\zeta \in S$, that

$$(4) \qquad\qquad \int_{\mathscr{U}^{\infty}} g(\zeta, \{U_i\}) dU = \prod_{i=1}^{\infty} \int_{\mathscr{U}} [1 - f_i(U_i\zeta)] dU_i.$$

By Proposition 1.4.7(3), (4) is the same as

$$(5) \qquad\qquad \int_{\mathscr{U}^{\infty}} g(\zeta, \{U_i\}) dU = \prod_{i=1}^{\infty} \left[1 - \int_S f_i \, d\sigma \right].$$

By (1), this infinite product is 0. Since $g \ge 0$, Fubini's theorem implies now that

$$(6) \qquad\qquad \int_S g(\zeta, \{U_i\}) d\sigma(\zeta) = 0$$

for almost all $\{U_i\} \in \mathscr{U}^{\infty}$. For any such choice of $\{U_i\}$, $g(\zeta, \{U_i\}) = 0$ for almost all $\zeta \in S$, and therefore (2) holds.

7.2.8. Proposition. *For* $z = (z_1, \ldots, z_n) \in \mathbb{C}^n$, *define*

$$F_k(z) = n^{kn}(z_1 \cdots z_n)^{2k} \qquad (k = 1, 2, 3, \ldots).$$

Then

    (i) $|F_k(z)| \le 1$ *for all* $z \in \bar{B}$,

    (ii) $\int_S |F_k| \, d\sigma \approx k^{-(n-1)/2}$,

    (iii) $\int_S |F_k|^2 \, d\sigma \approx k^{-(n-1)/2}$,

    (iv) $\int_B |F_k|^2 \, dv \approx k^{-(n+1)/2}$.

The symbol $\approx$ means that the ratio of the two terms involved has a positive finite limit as $k \to \infty$.

Proof. $|F_k(z)|$ attains its maximum on $\bar{B}$ when

$$|z_1| = \cdots = |z_n| = n^{-1/2}.$$

This gives (i). By 1.4.9, the integral in (ii) is

$$\frac{n^{kn}(n-1)!(k!)^n}{(n-1+nk)!}$$

Hence (ii) follows from Stirling's formula, and (iii) is just (ii) with $2k$ in place of k. Since F_k is homogeneous of degree $2nk$, an integration in polar coordinates shows that

$$\int_B |F_k|^2 \, dv = \frac{1}{1+2k} \int_S |F_k|^2 \, d\sigma.$$

Thus (iii) implies (iv).

7.2.9. Example. There is an $f \in H^2(B)$ (when $n = 2, 3$) such that almost no slice function f_ζ has a Taylor series that converges absolutely on the closed unit disc:

When $n = 2$ or 3, Proposition 7.2.8 shows that there exists $\{c_k\}, 0 < c_k < 1$, such that

(1)
$$\sum_{k=1}^{\infty} \int_S |c_k F_k| \, d\sigma = \infty$$

but

(2)
$$\sum_{k=1}^{\infty} \int_S |c_k F_k|^2 \, d\sigma < \infty.$$

(Take $c_k = k^{-1/2}$ if $n = 2$, $c_k = 1/\log(k+2)$ if $n = 3$.)

Since $|c_k F_k| \leq 1$, it follows from Lemma 7.2.7 that there exist $U_k \in \mathcal{U}$ such that

$$(3) \qquad \sum_{k=1}^{\infty} |c_k F_k(U_k \zeta)| = \infty \quad \text{a.e.} \quad \text{on } S.$$

If $f(z) = \sum c_k F_k(U_k z)$, the orthogonality of the functions $F_k \circ U_k$ in $L^2(\sigma)$ shows that $f \in H^2(B)$, by (2). Finally, note that

$$(4) \qquad f_\zeta(\lambda) = \sum_{k=1}^{\infty} c_k F_k(U_k \zeta) \lambda^{2nk}.$$

If ζ satisfies (3) then the series (4) does not converge absolutely on the unit circle.

7.2.10. Example. There is an $f \in (L^2 \cap H)(B)$ (when $n = 2, 3$) such that

$$\sup_{0 < r < 1} |f(r\zeta)| = \infty$$

for almost all $\zeta \in S$:

When $n = 2$ or 3, Proposition 7.2.8 shows that

$$(1) \qquad \sum_{k=1}^{\infty} \int_B |F_k|^2 \, dv < \infty$$

whereas

$$(2) \qquad \sum_{k=1}^{\infty} \int_S |F_k|^2 \, d\sigma = \infty.$$

Since $|F_k| \leq 1$, it follows from Lemma 7.2.7 that there exist $U_k \in \mathcal{U}$ such that

$$(3) \qquad \sum_{k=1}^{\infty} |F_k(U_k \zeta)|^2 = \infty \quad \text{a.e.} \quad \text{on } S.$$

Now let $\{\varphi_k(t)\}$ be the Rademacher functions (independent random variables that take the values ± 1 with probability $\frac{1}{2}$) and define

$$(4) \qquad g_t(z) = \sum_{k=1}^{\infty} \varphi_k(t) F_k(U_k z) \qquad (0 \leq t \leq 1).$$

Since the functions $F_k \circ U_k$ are mutually orthogonal in $L^2(v)$, (1) shows that every g_t belongs to $(L^2 \cap H)(B)$.

If $\zeta \in S$ satisfies (3), it follows from a well-known theorem about Rademacher series that

$$(5) \qquad \sup_r |g_t(r\zeta)| = \sup_r \left| \sum_k F_k(U_k\zeta)r^{2nk}\varphi_k(t) \right| = \infty$$

for almost all $t \in [0, 1]$. (See Zygmund [3], vol. II, pp. 212, 205.) By Fubini's theorem, it is therefore true for almost every t that the supremum in (5) is ∞ for almost all $\zeta \in S$. Putting $f = g_t$ for such a value of t gives our example.

7.2.11. Example. If $f \in H^2(B_2)$, Theorem 7.2.4(b) implies that every slice function of f satisfies

$$(1) \qquad \int_U |f_\zeta(\lambda)|^2 \, dv(\lambda) \le \|f\|_2^2,$$

the norm on the right being the H^2-norm of f. In particular, the left side of (1) is a bounded function of ζ on S.

One may ask whether there is a converse to this, i.e., whether H^2 can be characterized by the slice functions forming a bounded set in the Bergman space. The answer is negative:

Put $F_k(z) = (2z_1z_2)^k$, for $k = 1, 2, 3, \ldots$. As in Proposition 7.2.8,

$$(2) \qquad \int_S |F_k|^2 \, d\sigma \approx k^{-1/2}$$

whereas another simple computation gives

$$(3) \qquad \int_U |F_k(\lambda\zeta)|^2 \, dv(\lambda) = \frac{|F_k(\zeta)|^2}{2k + 1} \le \frac{1}{2k + 1}$$

for every $\zeta \in S$. If $c_k \in \mathbb{C}$ is chosen so that

$$(4) \qquad \sum_{k=1}^\infty \frac{|c_k|^2}{2k + 1} = 1 \quad \text{but} \quad \sum_{k=1}^\infty \frac{|c_k|^2}{\sqrt{k}} = \infty,$$

and if $f = \sum c_k F_k$, then (2) shows that f is not in $H^2(B_2)$, although

$$(5) \qquad \int_U |f_\zeta(\lambda)|^2 \, dv(\lambda) \le 1$$

for every $\zeta \in S$.

7.3. Zero-Varieties

7.3.1. Introduction. If Ω is a region in $\mathbb{C}^n$ and $f \in H(\Omega)$, we write $Z(f)$ for the set of all $z \in \Omega$ at which $f(z) = 0$. We call a set $E \subset \Omega$ a *zero-variety in* Ω if $E = Z(f)$ for some $f \in H(\Omega)$, $f \not\equiv 0$. If X is a subspace of $H(\Omega)$ and $E = Z(f)$ for some $f \in X$, $f \not\equiv 0$, then E is said to be a *zero-variety for X.*

At the other extreme, E is said to be a *determining set for X* if the assumptions $f \in X$, $E \subset Z(f)$ force $f \equiv 0$.

The local structure of zero-varieties is described by the Weierstrass preparation theorem (Chapter 14). The present section is devoted to quantitative global properties of zero varieties of certain subspaces of $H(B)$, with special emphasis on the differences that exist between the case $n = 1$ and the cases $n > 1$.

When $n = 1$, the zero-varieties in U are precisely the discrete subsets of U. Such a set E is a zero-variety for $H^\infty(U)$ (hence also for all $H^p(U)$ and for the Nevanlinna class $N(U)$) if $E = \{\alpha_i\}$ satisfies the Blaschke condition

$$\sum (1 - |\alpha_i|) < \infty.$$

Conversely, E is a determining set for $N(U)$ (hence also for all $H^p(U)$, including $H^\infty(U)$) if E violates the Blaschke condition. (Duren [1], Hoffman [1], Rudin [3], etc.)

We turn to $n > 1$. A recent very impressive theorem, proved independently by Henkin and by Skoda, characterizes the zero-varieties of $N(B)$ by an analogue of the Blaschke condition. This forms the subject of Chapter 17. One result of the present section (Theorem 7.3.8) is that to distinct values of p correspond distinct classes of zero-varieties. But no characterizations of any of these classes are obtained. In view of some examples, it seems quite unlikely that any reasonably simple characterizations can exist. Theorem 7.3.3 gives a simple necessary condition, which is however very far from being sufficient.

In one variable, zeros of holomorphic functions can occur with multiplicities greater than 1. The Blaschke condition is usually stated so as to take these multiplicities into account. When $n > 1$, one can also assign multiplicities (positive integers) to the irreducible branches (we shall not define this) of a zero-variety. This leads to the concept of a *divisor*. The introduction of divisors can be avoided by the following simple device which will accomplish the same thing for us:

Two functions $f, g \in H(B)$ are said to *have the same zeros* if both f/g and g/f are holomorphic in B, i.e., if $f = gh_1$ and $g = fh_2$, with $h_1, h_2 \in H(B)$. Having the same zeros is clearly an equivalence relation, and many properties involving zeros (such as the consequences of Jensen's formula discussed in §7.3.2) depend only on the equivalence class to which a function belongs.

Some of our results will be stated for the spaces $H_\varphi(B)$ defined in §5.6.1.

7.3.2. Counting Functions. If $f \in H(U)$, the customary notation for the number of zeros of f (counted with multiplicities) in the disc tU is $n_f(t)$. If $f(0) \neq 0$, there is also the integrated counting function

(1)
$$N_f(r) = \int_0^r n_f(t) \frac{dt}{t} \qquad (0 < r \leq 1).$$

Jensen's formula (see Titchmarsh [1], p. 126) expresses $N_f(r)$ directly in terms of f:

(2)
$$N_f(r) = \frac{1}{2\pi} \int_{-\pi}^{\pi} \log|f(re^{i\theta})| d\theta - \log|f(0)|.$$

Assume now that $f \in H(B)$, and $f(0) = 1$ for simplicity. For each $\zeta \in S$, let $n_f(\zeta, t)$ and $N_f(\zeta, t)$ be associated to the slice function f_ζ in the above manner. Thus

(3)
$$N_f(\zeta, r) = \frac{1}{2\pi} \int_{-\pi}^{\pi} \log|f(re^{i\theta}\zeta)| d\theta \qquad (0 < r < 1).$$

Note that both sides of (3) *are unchanged if* f *is replaced by some* $g \in H(B)$ *which has the same zeros as* f *(provided that* $g(0) = 1$*).*

If φ is any nonnegative increasing convex function on $(-\infty, \infty)$, Jensen's inequality leads from (3) to

(4)
$$\varphi(N_f(\zeta, r)) \leq \frac{1}{2\pi} \int_{-\pi}^{\pi} \varphi(\log|f(re^{i\theta}\zeta)|) d\theta$$

and if (4) is integrated over S one obtains

(5)
$$\int_S \varphi(N_f(\zeta, r)) d\sigma(\zeta) \leq \int_S \varphi(\log|f_r|) d\sigma,$$

by 1.4.7. Since $N_f(\zeta, r)$ increases as r increases, the monotone convergence theorem leads from (5) to the inequality

(6)
$$\int_S \varphi(N_f(\zeta, 1)) d\sigma(\zeta) \leq \sup_{0 < r < 1} \int_S \varphi(\log|f_r|) d\sigma.$$

This provides a necessary condition which the zeros of certain holomorphic functions must satisfy (Weyland [1]):

7.3.3. Theorem. *Suppose* $f \in H(B)$, $f(0) = 1$. *The inequality*

(1)
$$\int_S \varphi(N_f(\zeta, 1)) d\sigma(\zeta) < \infty$$

holds then if $f \in H_\varphi(B)$. In particular,

$$(2) \qquad \int_S N_f(\zeta, 1)d\sigma(\zeta) < \infty$$

if f is in the Nevanlinna class $N(B)$, and

$$(3) \qquad \int_S \exp\{pN_f(\zeta, 1)\}d\sigma(\zeta) < \infty$$

if $f \in H^p(B)$ for some $p, 0 < p < \infty$.

Part (1) was proved above. The case $\varphi(x) = x^+$ gives (2), and $\varphi(x) = e^{px}$ gives (3).

The Henkin–Skoda theorem that was mentioned in §7.3.1 gives a converse of (2): *If f satisfies (2), then f has the same zeros as some $g \in N(B)$.* (Chapter 17.)

On the other hand, (3) does not ensure that f has the same zeros as some $g \in H^p(B)$. We shall see this in §7.3.5, after the following theorem, which (for $n = 1$) occurs in Shapiro–Shields [1]. Points of B will be written in the form

$$z = (z_1, z'), \qquad z' = (z_2, \ldots, z_n).$$

7.3.4. Theorem. *Suppose $\Sigma(1 - x_i) = \infty$, where $0 < x_i < 1$ for all i, and let E be the set of all $z \in B$ that have $z_1 \in \{x_i\}$.*

If $f \in H(B)$ satisfies the growth condition

$$(1) \qquad |f(z)| < \exp\left\{\frac{c}{1 - |z|^2}\right\}^\alpha$$

for some $c < \infty$ and some $\alpha < \frac{1}{2}$, and if $Z(f) \supset E$, then $f \equiv 0$.

Note that E is a countable union of balls of complex dimension $n - 1$, and that E is the zero-variety of some holomorphic function in B that depends only on z_1. The theorem asserts that E is a determining set for the class of functions described by (1).

Proof. Let Ω be the set of all $w \in B$ with

$$(2) \qquad |2w_1 - 1| < 1 \quad \text{and} \quad |w'| < |1 - w_1|.$$

Fix $w \in \Omega$. Since Ω is a nonempty open set, it is enough to prove that $f(w) = 0$. Define

$$(3) \qquad h(\lambda) = \left(\frac{1 + \lambda}{2}, \frac{1 - \lambda}{2(1 - w_1)} w'\right) \qquad (\lambda \in U).$$

Put $\delta = 1 - |w'|^2|1 - w_1|^{-2}$. Then $0 < \delta < 1$. A simple computation shows that

(4) $4(1 - |h(re^{i\theta})|^2) = 2(1 - r^2) + \delta(1 - 2r\cos\theta + r^2) > \delta\sin^2\theta$

for $0 \le r < 1, |\theta| \le \pi$. It follows that h maps U into B, so that one can define

(5) $$g(\lambda) = f(h(\lambda)) \qquad (\lambda \in U).$$

By (1) and (4),

(6) $$\log|g(re^{i\theta})| < \left\{\frac{c}{1 - |h(re^{i\theta})|^2}\right\}^\alpha$$
$$< \left\{\frac{4c}{\delta\sin^2\theta}\right\}^\alpha = c_1|\sin\theta|^{-2\alpha}$$

for $0 \le r < 1, |\theta| \le \pi$. Since $\alpha < \frac{1}{2}$, (6) implies that $g \in N(U)$.

Next, note that

(7) $$g(2x_i - 1) = f\left(x_i, \frac{1 - x_i}{1 - w_1}w'\right) = 0$$

for all i, since $Z(f) \supset E$. The sequence $\{2x_i - 1\}$ violates the Blaschke condition. Since $g \in N(U)$, it follows that $g(\lambda) = 0$ for all $\lambda \in U$. In particular, $2w_1 - 1 \in U$, so that

(8) $$f(w) = f(w_1, w') = g(2w_1 - 1) = 0.$$

7.3.5. Example. If $n > 1$, there is a function $g \in H(B)$ such that

(a) $$\int_S \exp\{pN_g(\zeta, 1)\}d\sigma(\zeta) < \infty$$

for every $p < \infty$, although

(b)   no product of the form $f = gh$ (with $h \in H(B)$) satisfies a growth condition of the form 7.3.4(1), unless $h \equiv 0$.

In particular, no $f \in H^p(B)$ has the same zeros as g.

To construct g, put $x_k = 1 - (k\log k)^{-1}$ for $k = 3, 4, 5, \ldots$, let G be any holomorphic function in U with simple zeros at precisely the points x_k, and put $g(z) = g(z_1, z') = G(z_1)$.

Since $\Sigma(1 - x_k) = \infty$, Theorem 7.3.4 shows that g has property (b). To prove (a) we need an upper estimate for $N_g(\zeta, 1)$.

For $\zeta \in S$, $n_g(\zeta, t)$ is the number of λ's in tU such that $\lambda\zeta_1 \in \{x_k\}$. This is the number of k's such that $x_k < t|\zeta_1|$, i.e., such that

$$k\log k < \frac{1}{1 - t|\zeta_1|}.$$

If $k \log k = A$, then $A < k^2$, hence

$$k = \frac{A}{\log k} < \frac{2A}{\log A}.$$

Taking $A = (1 - t|\zeta_1|)^{-1}$, it follows that

$$n_g(\zeta, t) < \frac{2}{(1 - t|\zeta_1|)\log\{1/(1 - t|\zeta_1|)\}}.$$

Since

$$N_g(\zeta, r) = \int_0^r n_g(\zeta, t)\, \frac{dt}{t}$$

it follows that

$$N_g(\zeta, 1) < c \log \log \frac{1}{1 - |\zeta_1|}$$

for some $c < \infty$ and all $|\zeta_1|$ sufficiently close to 1. Hence

$$\exp\{pN_g(\zeta, 1)\} < \left\{\log \frac{1}{1 - |\zeta_1|}\right\}^{pc}.$$

This implies (a).

7.3.6. Example. Suppose V_1 and V_2 are zero-varieties in B, $V_1 \subset V_2$, and V_2 is a zero-variety for some class $X \subset H(B)$. In some sense this imposes a restriction on the size of V_2, and one might expect that V_1 should therefore also be a zero-variety for X. But it need not be:

Take $n = 2$, write (z, w) in place of (z_1, z_2). Fix $p > 0$, let m be an integer, $m > 4/p$. Put

$$D = \{(z, 0): |z| < 1\},$$

$$r_k = 1 - 2^{-k}, \qquad k = 1, 2, 3, \ldots,$$

$$T_k = \left\{r_k e^{i\theta}: \theta = \frac{2\pi j}{m2^k}, 1 \le j \le m2^k\right\},$$

$$E_k = \{(z, w) \in B: z \in T_k\},$$

$$E = \bigcup_{k=1}^{\infty} E_k.$$

Each E_k is the union of $m2^k$ discs "in the w-direction" and D is a disc orthogonal to those that make up E. This example has the following properties:

(a) E is a zero-variety,
(b) $E \cup D = Z(h)$ for some $h \in A(B)$, but
(c) no $g \in H^p(B)$ has $Z(g) = E$.

The proof of (a) and (b) uses the product

$$
(1) \qquad\qquad P(z) = \prod_{k=1}^{\infty} \left\{ 1 - \left(\frac{z}{r_k}\right)^{m2^k} \right\} \qquad (z \in U).
$$

Since $(r_k)^{2^k} > \frac{1}{4}$, we have

$$
(2) \qquad |P(z)| \le \prod_{k=0}^{\infty} \{1 + 4^m |z|^{m2^k}\} = 1 + \sum_{t=1}^{\infty} 4^{mc(t)} |z|^{mt}
$$

where $c(t)$ is the number of 1's in the binary expansion of t. Thus $c(t) \le \log t + 1/\log 2$, so that

$$
(3) \qquad |P(z)| \ge 1 + \sum_{t=1}^{\infty} (t + 1)^{2m} |z|^{mt} \le \frac{c}{(1 - |z|)^{2m+1}}.
$$

Since $P(z) = 0$ if and only if $z \in T_k$ for some k, the function $Q(z, w) = P(z)$ establishes (a).

Since $|w|^2 \le 1 - |z|^2$ in B, the function

$$
(4) \qquad\qquad h(z, w) = w^{4m+3} P(z)
$$

establishes (b), because of (3).

To prove (c), suppose $g \in H(B)$, $Z(g) \supset E$, but $g(0) = 1$. (Note that $0 \notin E$.) Putting $f(z) = g(z, 0)$, Theorem 7.2.4(b) shows that it suffices to prove that

$$
(5) \qquad\qquad \int_U |f|^p \, dv = \infty.
$$

As in §7.3.2,

$$
(6) \qquad \int_0^r n_f(t) dt \le \int_0^r n_f(t) \frac{dt}{t} = \frac{1}{2\pi} \int_{-\pi}^{\pi} \log |f(re^{i\theta})| \, d\theta.
$$

Multiply (6) by p, exponentiate, apply the geometric vs. arithmetic mean inequality on the right, multiply by $r \, dr$, and integrate:

$$(7) \qquad \int_0^1 \left\{ \exp \int_0^r p n_f(t) dt \right\} r \, dr \leq \frac{1}{2\pi} \int_U |f|^p r \, dr \, d\theta.$$

If $\frac{1}{2} < t < 1$, then $r_k < t \leq r_{k+1}$ for some k. Since f has $m2^k$ zeros in T_k, it follows that

$$(8) \qquad p n_f(t) \geq p m 2^k > \frac{4}{1 - r_k} \geq \frac{2}{1 - t}.$$

The left side of (7) is thus ∞, and (5) is proved.

   Note. Each of the $m2^k$ discs in E_k has radius

$$(1 - r_k^2)^{1/2} \sim 2^{-(k-1)/2}$$

so that the area of E_k is about $2\pi m$, for every k. Thus (b) shows that $E \cup D$ *is a zero-variety for $A(B)$, of infinite area.*

   No such example seems to be known when $n > 2$ (where "area" must of course be replaced by $(2n - 2)$-dimensional volume), even with $H^\infty(B)$ in place of $A(B)$.

   The following lemma will be used in the proof of Theorem 7.3.8.

7.3.7. Lemma. *Suppose*

(a) *μ is a finite positive measure on a set Ω;*

(b) *v is a real measurable function on Ω, with $0 \leq v < 1$ a.e., whose essential supremum is 1;*

(c) *Φ is a continuous nondecreasing real function on $[0, \infty)$, $\Phi(0) = 0$, $\Phi(x) \to \infty$ as $x \to \infty$;*

(d) *$0 < \delta < \infty, 0 < t < \infty$.*

Then there exist constants $c_k \in (0, \infty)$, for $k = 1, 2, 3, \ldots$, such that

$$(1) \qquad \int_\Omega \Phi(c_k v^k) d\mu = \delta.$$

If $|\alpha| < 1$ and if $Y_k = Y_k(t)$ is the set of all $\omega \in \Omega$ at which $c_k v^k(\omega) > t$, then

$$(2) \qquad \lim_{k \to \infty} c_k \alpha^k = 0$$

and

$$(3) \qquad \lim_{k \to \infty} \int_{Y_k} \Phi(c_k v^k) d\mu = \delta.$$

Proof. The monotone convergence theorem shows that

$$(4) \qquad\qquad c \rightarrow \int_\Omega \Phi(cv^k) d\mu$$

is, for each k, a continuous mapping of $[0, \infty)$ onto $[0, \infty)$ which carries 0 to 0. Hence (1) holds for some $c_k \in (0, \infty)$.

Assume $|\alpha| < \beta < 1$, and let E be the set of all $\omega \in \Omega$ at which $v(\omega) > \beta$. Since the essential supremum of v is 1, $\mu(E) > 0$. Also,

$$(5) \qquad\qquad \Phi(c_k \beta^k) \mu(E) \leq \int_E \Phi(c_k v^k) d\mu \leq \delta.$$

This shows that $\{c_k \beta^k\}$ is a bounded sequence, and (2) holds because $|\alpha| < \beta$.

To prove (3), define

$$(6) \qquad\qquad g_k(\omega) = \begin{cases} 0 & \text{if } \omega \in Y_k \\ c_k v^k(\omega) & \text{if } \omega \notin Y_k. \end{cases}$$

Then $0 \leq g_k \leq t$. Since $0 \leq v < 1$ a.e., (2) implies that $g_k(\omega) \rightarrow 0$ a.e. on Ω. Thus

$$(7) \qquad\qquad \int_{\Omega \setminus Y_k} \Phi(c_k v^k) d\mu = \int_\Omega \Phi(g_k) d\mu \rightarrow 0$$

as $k \rightarrow \infty$, by the dominated convergence theorem. Now (3) follows from (1) and (7).

7.3.8. Theorem. *Fix $n > 1$. Assume that φ and ψ are nonconstant, non-negative, nondecreasing convex functions defined on $(-\infty, \infty)$, and that*

$$(1) \qquad\qquad \frac{\psi(t)}{\varphi(t)} \rightarrow +\infty \quad as \quad t \rightarrow +\infty.$$

Suitably chosen points $\zeta_i \in S$, positive integers k_i, and positive numbers a_i produce then a function

$$(2) \qquad\qquad f(z) = \prod_{i=1}^\infty (1 - a_i \langle z, \zeta_i \rangle^{k_i})$$

such that

  (i) $f \in H_\varphi(B)$, *but*
  (ii) *if $b \in H^\infty(B)$, $g \in H(B)$, $g \not\equiv 0$, and*

$$(3) \qquad\qquad h = (f + b)g$$

then some constant multiple of h fails to be in $H_\psi(B)$.

The classes $H_\varphi(B)$ were defined in §5.6.1. Some comments on the theorem will follow its proof, which, for convenience, is split into four steps.

Step 1. We claim that there exist nonempty circular sets K and X_i $(i = 1, 2, 3, \ldots)$ on S such that

(a) K is compact, $\sigma(K) > 0$,
(b) the sets X_i are pairwise disjoint open subsets of $S \backslash K$, and
(c) every circular open subset of S that intersects K contains infinitely many X_i.

The standard projection π of S onto the complex projective space P of dimension $n - 1$ provides an easy way to see this. By definition, $\pi(\zeta) = \pi(\eta)$ if and only if $\eta = e^{i\theta}\zeta$ for some real θ; a set $E \subset P$ is open if and only if $\pi^{-1}(E)$ is open in S; and $\sigma(\pi^{-1}(E))$ defines a Borel measure on P. Now let K' be a compact subset of P, of positive measure, with empty interior, let $\{p_i\}$ be a sequence of distinct points in $P \backslash K'$ whose set of subsequential limits is exactly K', pick pairwise disjoint neighborhoods X_i' of p_i that do not intersect K', and define

$$K = \pi^{-1}(K'), \qquad X_i = \pi^{-1}(X_i').$$

Step 2. We now construct f. It will be convenient to assume that

(4) $$\varphi(t) = 0 \quad \text{if} \quad t \le 1.$$

To see that this involves no loss of generality, put $\varphi_1(t) = \varphi(t) - \varphi(1)$ if $t \ge 1$, and put $\varphi_1(t) = 0$ if $t < 1$. Then $\varphi - \varphi_1$ is bounded, hence $H_\varphi = H_{\varphi_1}$, and φ_1 satisfies (4).

For $t \ge 0$, define

(5) $$\Phi(t) = \varphi(1 + \log(1 + t)), \qquad \Psi(t) = \psi(1 + \log(1 + t)).$$

By (1), there are numbers $t_i > i$ $(i = 1, 2, 3, \ldots)$ such that

(6) $$\Psi(t) > i^3 \Phi(t) \quad \text{if} \quad t \ge t_i.$$

Let the sets X_i be as in Step 1. For each i, pick $\zeta_i \in X_i$.

We now apply Lemma 7.3.7, with (S, σ) in place of (Ω, μ), with $v(z) = |\langle z, \zeta_i \rangle|$, with $\delta_i = 2/i^2$, and with α_i the maximum of $|\langle z, \zeta_i \rangle|$ on $S \backslash X_i$ (so that $\alpha_i < 1$). By the lemma, there exist positive numbers $a_i = c_{k_i}$ (where k_i is a sufficiently large positive integer) such that, setting

(7) $$F_i(z) = a_i \langle z, \zeta_i \rangle^{k_i} \qquad (z \in \mathbb{C}^n)$$

we have

$$\text{(8)} \qquad \int_S \Phi(|F_i|)d\sigma = \frac{2}{i^2},$$

$$\text{(9)} \qquad |F_i(z)| < 2^{-i} \quad \text{if} \quad z \in S \backslash X_i \quad \text{and if} \quad |z| < 1 - \frac{1}{i},$$

$$\text{(10)} \qquad \int_{Y_i} \Phi(|F_i|)d\sigma > \frac{1}{i^2},$$

where $Y_i = \{\zeta \in S: |F_i(\zeta)| > t_i\}$. We define

$$\text{(11)} \qquad f(z) = \prod_{i=1}^{\infty} (1 - F_i(z)) \qquad (z \in B).$$

This product has the form (2). Because of (9), the product converges uniformly on compact subsets of B (thus $f \in H(B)$), and it is 0 only where one of the factors is 0.

Note also that $Y_i \subset X_i$, by (9), and that (6) and (10) imply

$$\text{(12)} \qquad \int_{Y_i} \Psi(|F_i|)d\sigma > i.$$

Step 3. To prove that $f \in H_\varphi(B)$, put

$$\text{(13)} \qquad M_N(z) = \prod_{i=1}^{N} |1 - F_i(z)|, \qquad M(z) = \prod_{i=1}^{\infty} \{1 + |F_i(z)|\}.$$

Note that $\prod(1 + 2^{-i}) < \exp(\sum 2^{-i}) = e$. Since the sets X_i are disjoint, (9) implies therefore that

$$\text{(14)} \qquad M(\zeta) \leq \begin{cases} e & \text{in} \quad S \backslash \bigcup X_i \\ e(1 + |F_j(\zeta)| & \text{in} \quad X_j. \end{cases}$$

By (4) and (5), it follows that

$$\text{(15)} \qquad \varphi(\log M(\zeta)) \leq \begin{cases} 0 & \text{in} \quad S \backslash \bigcup X_i \\ \Phi(|F_j|) & \text{in} \quad X_j. \end{cases}$$

Hence (8) implies

$$\text{(16)} \qquad \int_S \varphi(\log M)d\sigma \leq \sum_{i=1}^{\infty} \frac{2}{i^2} < 4.$$

M_N is the absolute value of a holomorphic function. Hence $\log M_N$ is subharmonic, for each N, and so is $\varphi(\log M_N)$, since φ is convex and non-decreasing. For $0 < r < 1$, it follows that

$$(17) \qquad \int_S \varphi(\log M_N(r\zeta))d\sigma(\zeta) \le \int_S \varphi(\log M_N)d\sigma < 4.$$

If we fix r and let $N \to \infty$, $M_N(r\zeta) \to |f(r\zeta)|$ uniformly on S. Hence (17) gives

$$(18) \qquad \int_S \varphi(\log |f_r|)d\sigma < 4 \qquad (0 < r < 1).$$

Thus $f \in H_\varphi(B)$.

Step 4. We now prove part (ii). Suppose $b \in H^\infty(B)$, $g \in H(B)$, $h = (f + b)g$, and $g \not\equiv 0$. Then $g = G_m + G_{m+1} + \cdots$, where each G_i is a homogeneous polynomial of degree i, and $G_m \not\equiv 0$.

Referring to Step 1, $\sigma(K) > 0$, hence $G_m(\zeta) \ne 0$ for some $\zeta \in K$. Since $|G_m(e^{i\theta}\zeta)| = |G_m(\zeta)|$ for all real θ, $|G_m|$ is bounded from 0 on some circular open subset of S that intersects K. By Step 1, there is a $\delta > 0$ and there is an infinite set J of natural numbers, such that

$$(19) \qquad |G_m(\zeta)| > \delta \quad \text{if} \quad \zeta \in X_i \quad \text{and} \quad i \in J.$$

We shall see that ch is not in $H_\psi(B)$ if $c = 16e/\delta$, by factoring

$$(20) \qquad \text{ch} = 8e(f + b) \cdot 2\delta^{-1}g.$$

Since $t_i \to \infty$ (see (6)) we may, after discarding at most finitely members of J, assume that

$$(21) \qquad t_i > 5 + 8\|b\|_\infty \qquad (i \in J).$$

We now fix $j \in J$ and choose $r = r(j) < 1$ so that

$$(22) \qquad 2r^m > 1 \quad \text{and} \quad (1 - r^{k_j})\|F_j\|_\infty < 1.$$

Fix $\zeta \in Y_j$ and put

$$(23) \qquad \gamma(\lambda) = 2\delta^{-1}\lambda^{-m}g(\lambda\zeta) \qquad (\lambda \in U)$$

Then $\gamma \in H(U)$, $|\gamma(0)| = 2\delta^{-1}|G_m(\zeta)| > 2 > r^{-m}$,

$$2\delta^{-1}|g(re^{i\theta}\zeta)| = r^m|\gamma(re^{i\theta})|,$$

and the subharmonicity of $\log|\gamma|$ implies therefore that

(24)
$$\frac{1}{2\pi} \int_{-\pi}^{\pi} \log|2\delta^{-1}g(re^{i\theta}\zeta)|\,d\theta \geq \log|r^m\gamma(0)| > 0.$$

Next we note that

(25)
$$\prod_{i=1}^{\infty}(1 - 2^{-i}) > \tfrac{1}{4}.$$

If $\zeta \in Y_j$ and $|\lambda| = r$, it follows from (9) and (11) that

$$4|f(\lambda\zeta)| > |F_j(\lambda\zeta)| - 1 = r^{k_j}|F_j(\zeta)| - 1 > |F_j(\zeta)| - 2,$$

by (22). Since $|F_j(\zeta)| > t_j$ on Y_j, (21) implies now that

(26) $8|f(\lambda\zeta) + b(\lambda\zeta)| \geq 2|F_j(\zeta)| - 4 - 8\|b\|_\infty \geq 1 + |F_j(\zeta)|$

By (20), (24), and (26),

(27)
$$\frac{1}{2\pi} \int_{-\pi}^{\pi} \log|\mathrm{ch}(re^{i\theta}\zeta)|\,d\theta > \log\{e(1 + |F_j(\zeta)|)\}.$$

Since ψ is convex and nondecreasing, Jensen's inequality leads from (27) to

(28)
$$\frac{1}{2\pi} \int_{-\pi}^{\pi} \psi(\log|\mathrm{ch}(re^{i\theta}\zeta)|)\,d\theta > \Psi(|F_j(\zeta)|).$$

This holds for every $\zeta \in Y_j$. If we integrate (28) over Y_j and recall the rotation-invariance of σ, (12) shows that

(29)
$$\int_{Y_j} \psi(\log|\mathrm{ch}(r\zeta)|)\,d\sigma(\zeta) > j.$$

This was done for fixed $r = r(j)$. But (29) implies obviously that

(30)
$$\sup_{0<r<1} \int_S \psi(\log|\mathrm{ch}(r\zeta)|)\,d\sigma(\zeta) > j$$

for every $j \in J$. The left side of (30) is therefore infinite, and we have proved that ch is not in H_ψ.

7.3.9. Remarks. (a) The appearance of a "constant multiple" may be a bit puzzling in the conclusion of Theorem 7.3.8. The phrase may be omitted when ψ satisfies the growth condition

(1)
$$\limsup_{t \to \infty} \frac{\psi(t+1)}{\psi(t)} < \infty$$

since $H_\psi(B)$ is then closed under scalar multiplication. In that case, the conclusion is simply that $h \notin H_\psi(B)$.

The case $\varphi(t) = \exp(e^t)$, $\psi(t) = \varphi(t+1)$, shows the need for the "constant multiple": For every $f \in H_\varphi$ we have $e^{-1}f \in H_\psi$. Thus H_φ and H_ψ have the same zero-varieties, although $\psi(t)/\varphi(t) \to \infty$ as $t \to \infty$.

(b)   Theorem 7.3.8 states that no "bounded perturbation" $f + b$ of f has the same zeros as any member of $H_\psi(B)$. Letting b range over the constants, this gives some information about the level sets of f, or, in other words, about the distribution of values of f.

(c)   Fix p, $0 < p < \infty$, and define

(2)
$$\varphi(t) = e^{pt}, \qquad \psi(t) = (2 + p^2 t^2)e^{pt}.$$

Then ψ is convex and increasing, $H_\varphi = H^p$, and $H^q \subset H_\psi$ for all $q > p$. Since ψ satisfies (1), *Theorem 7.3.8 furnishes an $f \in H^p(B)$ whose zero-variety is a determining set for every $H^q(B)$ with $q > p$.*

(d)   With a gap series in place of a product, Theorem 7.3.8 appears in Rudin [9]. An earlier result, for polydiscs, is due to Miles [1]; it strengthened Theorem 4.1.1 of Rudin [1]. Similar theorems have also been proved for Bergman spaces, by Horowitz [1] and J. H. Shapiro [2].

7.4. Pluriharmonic Majorants

7.4.1. Introduction. If Ω is a region in $\mathbb{C}^n$ and $0 < p < \infty$, *Lumer's Hardy space* $(LH)^p(\Omega)$ is defined to consist of all $f \in H(\Omega)$ such that $|f|^p \le u$ for some pluriharmonic u. (Lumer [1].)

When $n = 1$, pluriharmonic is the same as harmonic, so that this definition coincides with the classical one that involves harmonic majorants. But when $n > 1$, then $(LH)^p(\Omega)$ is a proper subclass of what is usually called $H^p(\Omega)$; see Stein [2], Stout [5], for instance. The use of pluriharmonic majorants leads to some appealing function-theoretic properties of $(LH)^p(\Omega)$; from the standpoint of functional analysis, however, $(LH)^p(B)$ has some unexpectedly pathological properties, as we shall see presently; for example, $(LH)^2(B)$ is not even a Hilbert space.

One appealing property is what may be called *holomorphic invariance*: If $\Phi: \Omega_1 \to \Omega_2$ is holomorphic and $f \in (LH)^p(\Omega_2)$ then $f \circ \Phi \in (LH)^p(\Omega_1)$. This is trivial, for if u is pluriharmonic, so is $u \circ \Phi$, simply because u is locally the real part of a holomorphic function (Theorem 4.4.9).

As regards zeros, $(LH)^p(\Omega)$ behaves much better than $H^p(B)$ does: If Ω is *simply connected, then every* $f \in (LH)^p(\Omega)$ *has the same zeros as some* $h \in H^\infty(\Omega)$.

This is a corollary of the following result:

7.4.2. Proposition. *Suppose Ω is a simply connected region in $\mathbb{C}^n$ and $f \in H(\Omega)$. Then f has the same zeros as some $h \in H^\infty(\Omega)$ if and only if there is a $g \in H(\Omega)$ such that*

$$(1) \qquad\qquad\qquad \log|f| \le \operatorname{Re} g.$$

Proof. If (1) holds, put $h = f \cdot \exp(-g)$. Then $|h| \le 1$ and h has the same zeros as f. Conversely, if $h \in H^\infty(\Omega)$ has the same zeros as f, then f/h is a zero-free holomorphic function which has a holomorphic logarithm g, since Ω is simply connected.

7.4.3. For most of the remainder of this section we confine ourselves to the case $\Omega = B$, $n > 1$. In addition to the difference in zero-varieties that we just saw, there are at least two other ways of seeing that $(LH)^p(B)$ is a rather small subclass of $H^p(B)$.

The first of these involves rates of growth. If u is a positive pluriharmonic function in B, then each slice function u_ζ is a positive harmonic function in U, hence $u_\zeta(\lambda) \le 2u(0)/(1 - |\lambda|)$. If $f \in (LH)^p(B)$, it follows that

$$(1) \qquad\qquad |f(z)| \le c(1 - |z|)^{-1/p} \qquad (z \in B)$$

for some $c = c(f) < \infty$.

This is much more restrictive than the corresponding H^p bound (Theorem 7.2.5). For instance, if $1/p < t < n/p$, and

$$g(z) = (1 - z_1)^{-t}$$

then $g \in H^p(B)$ but $g \notin (LH)^p(B)$.

The second difference involves the norm

$$(2) \qquad\qquad\qquad \||f\||_p = \inf u(0)^{1/p},$$

the infimum being taken over all pluriharmonic majorants of $|f|^p$ in B. As pointed out by Lumer [1], it is easy to prove that this norm turns $(LH)^p(B)$ into a Banach space if $p \ge 1$.

If $\|f\|_p$ and $\|f_\zeta\|_p$ denote the norms of f and f_ζ in $H^p(B)$ and $H^p(U)$, respectively, it follows that

$$(3) \qquad\qquad \|f\|_p = \left\{ \int_S \|f_\zeta\|_p^p \, d\sigma(\zeta) \right\}^{1/p}$$

by slice-integration of $|f|^p$, whereas

(4)
$$\|\|f\|\|_p \geq \sup_{\zeta \in S} \|f_\zeta\|_p$$

since u_ζ is a harmonic majorant of $|f_\zeta|^p$ for every u that competes in (2).

There is a variant of (4) in which equality holds, and in which the right side is replaced by an expression that involves *all* representing measures of the origin (not just those that come from slices). This will be taken up in Section 9.7. Equality need not hold in (4). In fact, the left side may be ∞ although the right side is finite:

7.4.4. Example. When $n = 2$, there is an $f \in H^2(B)$ which extends continuously to $\bar{B}$ except for one boundary point, such that

(1)
$$\frac{1}{2\pi} \int_{-\pi}^{\pi} |f(re^{i\theta}\zeta)|^2 \, d\theta \leq 1$$

for all $\zeta \in S, 0 < r < 1$, although f is not in $(LH)^2(B)$:

Writing (z, w) in place of (z_1, z_2), define

(2)
$$g_m(z, w) = (1 - z)^{-m-1}w^{2m+1},$$

let $c_m \geq 0$ satisfy

(3)
$$\sum_{m=1}^{\infty} c_m = 1 \quad \text{but} \quad \sum_{m=1}^{\infty} m^{1/2}c_m^2 = \infty,$$

and put

(4)
$$f(z, w) = \sum_{m=1}^{\infty} c_m \binom{2m}{m}^{-1/2} g_m(z, w).$$

To see an example of (3), put $c_m = 0$ unless m is a power of 16. If $m = 16^k$, put $c_m = 2^{-k}$ $(k = 1, 2, 3, \ldots)$.

The proof that (4) defines a function with the desired properties is as follows.

First, we claim, for $0 < \delta < 1$, that

(5)
$$|g_m(z, w)| < \left(\frac{2}{\delta}\right)^{1/2}(2 - \delta)^m$$

for all $(z, w) \in \bar{B}$ that satisfy $|1 - z| \geq \delta$. Indeed, since $|w|^2 \leq 1 - |z|^2$ in $\bar{B}$,

(6)
$$|g_m(z, w)| \leq |1 - z|^{-m-1}(1 - |z|^2)^{m+1/2}.$$

On the set defined by $|z| \leq 1$, $|1 - z| \geq \delta$, the right side of (6) attains its maximum at the point $z = 1 - \delta$. This gives (5). By Stirling's formula,

$$(7) \qquad \binom{2m}{m} \sim \frac{4^m}{\sqrt{\pi m}}.$$

Hence (5) shows that the mth term in (4) is dominated by a constant times

$$(8) \qquad \left(\frac{2}{\delta}\right)^{1/2} m^{1/4} \left(1 - \frac{\delta}{2}\right)^m$$

if $|1 - z| \geq \delta$. Thus f is continuous on $\bar{B}$, except at the point $(1, 0)$.

Next, we claim that

$$(9) \qquad \frac{1}{2\pi} \int_{-\pi}^{\pi} |g_m(re^{i\theta}\zeta)|^2 \, d\theta \leq \binom{2m}{m}$$

for $\zeta \in S$, $0 \leq r \leq 1$. It is of course enough to do this when $r = 1$. Insert the binomial expansion

$$(10) \qquad (1 - z)^{-m-1} = \sum_{k=0}^{\infty} \binom{k+m}{m} z^k$$

into (2), and apply Parseval's theorem; if $\zeta = (z, w) \in S$ and $r = 1$, the left side of (9) equals

$$(11) \qquad |w|^{4m+2} \sum_{k=0}^{\infty} \binom{k+m}{m}^2 |z|^{2k}.$$

It is easily verified that

$$(12) \qquad \binom{k+m}{m}^2 \leq \binom{2m}{m}\binom{k+2m}{2m}.$$

If (12) is inserted into (11), another application of the binomial theorem shows that (11) is at most

$$(13) \qquad \binom{2m}{m}|w|^{4m+2}(1 - |z|^2)^{-2m-1} = \binom{2m}{m}.$$

This proves (9). Since $\Sigma c_m = 1$, (9) and (4) show that every f_ζ is a convex combination of functions whose H^2-norm is at most 1. Hence f satisfies (1).

It follows of course that $f \in H^2(B)$.

Finally, assume that $|f|^2 \leq u$ for some pluriharmonic u in B. This will lead to a contradiction:

For $0 < x < 1$, $u(x, w)$ is a harmonic function of w in $|w|^2 < 1 - x^2$. For $0 < r < 1$ it follows that

$$(14) \qquad u(x, 0) = \frac{1}{2\pi} \int_{-\pi}^{\pi} u(x, (1 - x^2)^{1/2} re^{i\theta}) d\theta$$

$$\geq \frac{1}{2\pi} \int_{-\pi}^{\pi} |f(x, (1 - x^2)^{1/2} re^{i\theta})|^2 \, d\theta.$$

Let $r \to 1$ in the last integral, and apply Parseval's theorem to (4), regarded as a power series in w. The result is

$$(15) \qquad (1 - x)u(x, 0) \geq \sum_{m=1}^{\infty} c_m^2 \binom{2m}{m}^{-1} (1 + x)^{2m+1}.$$

As $x \nearrow 1$, the sum of this series tends to infinity, by (3) and (7). But $u(z, 0)$ is a positive harmonic function in U, so that $(1 - x)u(x, 0)$ must be bounded.

This contradiction shows that $|f|^2$ has no pluriharmonic majorant in B. Hence f is not in $(LH)^2(B)$.

7.4.5. Notation. As usual, l^{∞} is the Banach space of all bounded complex sequences, with the sup-norm, and c_0 is the subspace of l^{∞} consisting of those sequences that converge to 0.

The following theorem describes some pathological features of $(LH)^p(B)$:

7.4.6. Theorem (Rudin [11]). *Fix $n > 1$, $1 \leq p < \infty$, $\varepsilon > 0$.*

(a) *There exists a linear map of l^{∞} into $(LH)^p(B)$ which assigns to each $\gamma \in l^{\infty}$ a function f_γ that satisfies*

$$(1) \qquad \|\gamma\|_{\infty} \leq \|\|f_\gamma\|\|_p \leq \|f_\gamma\|_{\infty} \leq (1 + \varepsilon)\|\gamma\|_{\infty}.$$

(b) *If γ is not in c_0, then $U \to f_\gamma \circ U$ is a discontinuous map of $\mathcal{U}$ into $(LH)^p(B)$.*

(c) *If γ is not in c_0, then $(f_\gamma)_r$ does not converge to f_γ in the norm topology of $(LH)^p(B)$, as $r \nearrow 1$.*

Before we turn to the proof, let us list some consequences of the theorem; recall that two Banach spaces are said to be *isomorphic* if there is a linear homeomorphism of one onto the other.

(i) $(LH)^p(B)$ *contains a closed subspace that is isomorphic to l^{∞} and lies in $H^{\infty}(B)$.*

(ii) $(LH)^p(B)$ *is not separable.*

(iii) $A(B)$ *is not dense in $(LH)^p(B)$.*

(iv) $(LH)^2(B)$ *is not isomorphic to any Hilbert space.*

Indeed, (i) follows immediately from (a), and (i) obviously implies (ii). Since $A(B)$ is separable in the sup-norm topology, it is *a fortiori* separable in the norm topology of $(LH)^p(B)$; thus (ii) implies (iii). Finally, (iv) follows from (i), since every closed subspace of a Hilbert space is a Hilbert space, but l^∞ (not being reflexive) is not isomorphic to any Hilbert space.

Proof. As in the proof of Theorem 7.3.8, let $\{X_i\}$ be a sequence of pairwise disjoint, nonempty, circular open subsets of S. Enlarge each X_i to a set V_i that is open in $\mathbb{C}^n$, so that $V_i \cap V_j = \emptyset$ if $i \neq j$. Pick $\zeta_i \in X_i$, pick $U_i \in \mathscr{U}$ so that $\{U_i\}$ converges to the identity element of $\mathscr{U}$ as $i \to \infty$, and so that $|\langle U_i\zeta_i, \zeta_i\rangle| < 1$, and pick positive integers n_i such that

(2) $$|\langle z, \zeta_i\rangle|^{n_i} < \frac{\varepsilon}{2^i} \quad \text{if} \quad z \in B\backslash V_i$$

and

(3) $$|\langle U_i\zeta_i, \zeta_i\rangle|^{n_i} < \tfrac{1}{2}.$$

The linear map mentioned in part (a) is the one that assigns to every $\gamma = \{c_i\} \in l^\infty$ the function

(4) $$f_\gamma(z) = \sum_{i=1}^\infty c_i\langle z, \zeta_i\rangle^{n_i} \quad (z \in B).$$

Since no two of the sets V_i intersect, the inequality (2) fails (for any given $z \in B$) for at most one term of the series (4). Thus

$$|f_\gamma(z)| \leq \|\gamma\|_\infty + \varepsilon \sum_{i=1}^\infty |c_i|2^{-i} \leq (1 + \varepsilon)\|\gamma\|_\infty$$

so that $\|f_\gamma\|_\infty \leq (1 + \varepsilon)\|\gamma\|_\infty$. That $\||f_\gamma\||_p \leq \|f_\gamma\|_\infty$ is trivial, and the first inequality in (1) is a consequence of the following, with f_γ in place of g:
If $g = \Sigma G_k$ is the homogeneous expansion of some $g \in (LH)^p(B)$, then

(5) $$|G_k(\zeta)| \leq \||g\||_p \quad (\zeta \in S, k = 1, 2, 3, \ldots).$$

To see this, note that $G_k(\zeta)$ is a coefficient in the Taylor series of the slice function g_ζ, so that $|G_k(\zeta)| \leq \|g_\zeta\|_p \leq \||g\||_p$, by 7.4.3(4).
The proof of (a) is now complete.
Observe next that

$$(f_\gamma - f_\gamma \circ U_k)(z) = \sum_{i=1}^\infty c_i\{\langle z, \zeta_i\rangle^{n_i} - \langle U_kz, \zeta_i\rangle^{n_i}\}.$$

When $z = \zeta_k$, the absolute value of the kth term of this series is

$$|c_k||1 - \langle U_k \zeta_k, \zeta_k \rangle^{n_k}| \geq \tfrac{1}{2}|c_k|,$$

by (3). Another application of (5) shows therefore that

$$\tfrac{1}{2} \limsup_{k \to \infty} |c_k| \leq \limsup_{k \to \infty} \| f_\gamma - f_\gamma \circ U_k \|_p.$$

Hence $f_\gamma \circ U_i$ does not converge to f_γ in $(LH)^p(B)$ if $\{c_i\}$ fails to converge to 0. This proves (b).

The proof of (c) is quite similar. If r_i is chosen so that $(r_i)^{n_i} = \tfrac{1}{2}$, then

$$f_\gamma(z) - f_\gamma(r_k z) = \sum_{i=1}^{\infty} c_i \{1 - (r_k)^{n_i}\} \langle z, \zeta_i \rangle^{n_i}.$$

When $z = \zeta_k$ it follows from (5) that

$$\| f_\gamma - (f_\gamma)_{r_k} \|_p \geq \tfrac{1}{2}|c_k|$$

for $k = 1, 2, 3, \ldots$. Thus

$$\limsup_{r \to 1} \| f_\gamma - (f_\gamma)_r \|_p \geq \tfrac{1}{2} \limsup_{k \to \infty} |c_k|,$$

which proves (c).

We conclude this section with an observation which is just as easily proved for arbitrary regions:

7.4.7. Proposition. *Suppose Ω is a region in $\mathbb{C}^n$, $f \in H(\Omega)$, $|\text{Im } f| < 1$, and $0 < t < \pi/2$. Then there is an $h \in H(\Omega)$ such that*

$$\exp\{t|f|\} < \text{Re } h.$$

Proof. If $f = u + iv$, then $|v| < 1$, and

$$\text{Re}(e^{tf}) = e^{tu} \cos(tv) > e^{tu} \cos t.$$

The same holds with $-f$ in place of f. Since $|f| < 1 + |u|$, it follows that

$$e^{t|f|} < e^t(e^{tu} + e^{-tu}) < \frac{e^t}{\cos t} \text{Re}(e^{tf} + e^{-tf}).$$

Corollary. *Under the same hypotheses, $f \in (LH)^p(\Omega)$ for every $p < \infty$.*

Proof. $|f|^p \le (p/te)^p e^{t|f|}$.

(This corollary occurs in Stout [2], for star-shaped domains, with a different proof.)

Note: The Proposition fails when $t = \pi/2$. The function

$$f(z) = \frac{2}{\pi} \log \frac{1 + z_1}{1 - z_1} \qquad (z \in B)$$

shows this, since

$$\exp\left\{\frac{\pi}{2}|f(z)|\right\} \ge \left|\frac{1 + z_1}{1 - z_1}\right|$$

and $(1 + z_1)/(1 - z_1)$ is not in $H^1(U)$.

7.5. The Isometries of $H^p(B)$

7.5.1. Introduction. The isometries in question are the linear maps T of $H^p(B)$ *onto* $H^p(B)$ that satisfy

(1) $\|Tf\|_p = \|f\|_p$

for every $f \in H^p(B)$. These will be completely described, for all $n \ge 1$ and for all $p \ne 2, 0 < p < \infty$.

It should be clear why the case $p = 2$ is special: $H^2(B)$ is a Hilbert space, isometrically isomorphic to every other separable Hilbert space (of infinite dimension), and its isometries are therefore the unitary operators; the fact that $H^2(B)$ is also a space of holomorphic functions plays no role here.

For any p, the following construction yields isometries:

Choose $\psi \in \mathrm{Aut}(B)$, $c \in \mathbb{C}$, $|c| = 1$, put $\psi^{-1}(0) = a$, and define

(2) $(Tf)(z) = c \dfrac{(1 - |a|^2)^{n/p}}{(1 - \langle z, a \rangle)^{2n/p}} f(\psi(z))$.

It is easily seen that these satisfy (1), since

$$\int_S |Tf|^p \, d\sigma = \int_S P(a, \zeta)|f(\psi(\zeta))|^p \, d\sigma(\zeta)$$

$$= P[|f|^p \circ \psi](a) = P[|f|^p](\psi(a)) = \int_S |f|^p \, d\sigma.$$

This computation made use of the $\mathcal{M}$-invariance of Poisson integrals (Theorem 3.3.8) and the fact that $\psi(a) = 0$; we have identified f and Tf with their boundary values f^* and $(Tf)^*$ and will continue to do so, unless there seems to be some danger of confusion.

It is also easily seen that the range of any T given by (2) is all of $H^p(B)$, for if $g \in H^p(B)$ and

$$f(z) = \bar{c} \frac{(1 - \langle \psi^{-1}(z), a \rangle)^{2n/p}}{(1 - |a|^2)^{n/p}} g(\psi^{-1}(z)),$$

then $f \in H^p(B)$ and $Tf = g$.

The principal result of this section, Theorem 7.5.6, is that *every linear isometry of $H^p(B)$ onto $H^p(B)$ has the form* (2), when $p \neq 2$.

The first case of this ($n = 1, p = 1$) was proved by deLeeuw–Rudin–Wermer [1], using the fact that the extreme points of the closed unit ball of $H^1(U)$ were completely known. Forelli [2] devised a different method that handled $H^p(U)$ for all p, and which was then applied by Schneider [1] to $H^p(U^n)$, and by Forelli [3] to $H^p(B)$ for $p > 2$. The proof of Theorem 7.5.6 (Rudin [8]) is based on these ideas, in combination with some very general facts about L^p-isometries on arbitrary finite measure spaces, namely Theorems 7.5.2 and 7.5.3. As shown by Korányi–Vagi [2], this proof extends without difficulty to arbitrary bounded symmetric domains in place of balls. Stephenson [1] has studied the isometries of the Nevanlinna class.

In the following two theorems, μ_1 and μ_2 will be finite positive measures, on some sets that will play no role at all and that will not even be named.

7.5.2. Theorem. *Assume $0 < p < \infty$, $p \neq 2, 4, 6, \ldots\ldots$ If $f_i \in L^p(\mu_i)$, $i = 1, 2$, and*

$$(1) \qquad\qquad \int |1 + \lambda f_1|^p \, d\mu_1 = \int |1 + \lambda f_2|^p \, d\mu_2$$

for every $\lambda \in \mathbb{C}$, then

$$(2) \qquad\qquad \int (h \circ f_1) d\mu_1 = \int (h \circ f_2) d\mu_2$$

for every bounded Borel function $h: \mathbb{C} \to \mathbb{C}$, and also for every positive Borel function h on $\mathbb{C}$.

Proof. It is enough to prove that (2) holds for every $h \in C_0 = C_0(\mathbb{C})$, the space of all continuous functions on $\mathbb{C}$ that vanish at infinity.

Let X be the set of all $h \in C_0$ such that (1) implies (2). If f_1, f_2 satisfy (1), then so do $c + f_1$, $c + f_2$ and cf_1, cf_2, for any $c \in \mathbb{C}$. It follows that X is a uniformly closed subspace of C_0 which is invariant under translations, dilations, and rotations. We have to prove that $X = C_0$.

Define

$$(3) \qquad u(\lambda) = \frac{1}{2\pi} \int_{-\pi}^{\pi} |1 + \lambda e^{i\theta}|^p \, d\theta \qquad (\lambda \in \mathbb{C}).$$

By Fubini's theorem,

$$(4) \qquad \int (u \circ f_j) d\mu_j = \frac{1}{2\pi} \int_{-\pi}^{\pi} d\theta \int |1 + e^{i\theta} f_j|^p \, d\mu_j$$

for $j = 1, 2$, so that (1) implies

$$(5) \qquad \int (u \circ f_1) d\mu_1 = \int (u \circ f_2) d\mu_2.$$

Equation (5) remains true if u is replaced by u_t, where $u_t(\lambda) = u(t\lambda)$, $t > 0$, $\lambda \in \mathbb{C}$, hence also if u is replaced by any finite linear combination

$$(6) \qquad v(\lambda) = \sum_{i=1}^{N} a_i u(t_i \lambda).$$

As we shall now see, a_i and t_i can be so chosen that $v \in C_0$, hence $v \in X$.

Since p is not an even integer, none of the coefficients b_k in the binomial expansion

$$(7) \qquad (1 + \lambda)^{p/2} = \sum_{k=0}^{\infty} b_k \lambda^k \qquad (|\lambda| < 1)$$

is zero. Replace λ by $\lambda e^{i\theta}$ in (7) and substitute into (3). It follows that

$$(8) \qquad u(\lambda) = \sum_{k=0}^{\infty} b_k^2 |\lambda|^{2k} \qquad (|\lambda| < 1).$$

By (3), $u(\lambda) = |\lambda|^p u(1/\lambda)$. Hence

$$(9) \qquad u(\lambda) = \sum_{k=0}^{\infty} b_k^2 |\lambda|^{p-2k} \qquad (|\lambda| > 1).$$

Now choose t_i, $0 < t_1 < \cdots < t_N < \infty$, where $N > 2 + p$. For $k = 0, 1, 2, \ldots$, define

$$(10) \qquad c_k = \sum_{i=1}^{N} a_i t_i^{2k}, \qquad \gamma_k = \sum_{i=1}^{N} a_i t_i^{p-2k}.$$

Then $\gamma_0 = \gamma_1 = \cdots = \gamma_{N-2} = 0$ is a homogeneous linear system of $N - 1$ equations in N unknowns $a_1, \ldots, a_N$. Pick some nontrivial solution, and use it to define v by (6). Since the t_i are distinct, c_k can be 0 for at most finitely many k. For $|\lambda| < 1/t_N$,

$$(11) \qquad v(\lambda) = \sum_{k=0}^{\infty} c_k b_k^2 |\lambda|^{2k},$$

by (6), (8), (10).

Thus v is not identically 0.

For $|\lambda| > 1/t_1$, one sees similarly that

$$(12) \qquad v(\lambda) = \sum_{k=N-1}^{\infty} \gamma_k b_k^2 |\lambda|^{p-2k} = O(|\lambda|^{-2-p}),$$

since $p - 2(N-1) < -2 - p$.

Thus $v \in C_0$, hence $v \in X$; in fact v is also in L^1, relative to Lebesgue measure of the plane, by (12).

Suppose now that μ is a measure on $\mathbb{C} = R^2$ such that $\int h \, d\mu = 0$ for every $h \in X$. Since X contains v and all its dilates v_t, and since X is translation invariant, it follows that all convolutions $v_t * \mu$ are 0. Taking Fourier transforms, we see that

$$(13) \qquad \hat{v}\left(\frac{x}{t}\right) \hat{\mu}(x) = 0 \qquad (x \in R^2, t > 0),$$

where $\hat{v}$, $\hat{\mu}$ are the Fourier transforms of v, μ.

Note that v is a radial function. Hence so is $\hat{v}$, and if $x \neq 0$, $\hat{v}(x/t) \neq 0$ for some $t > 0$, since $\hat{v} \not\equiv 0$. Thus $\hat{\mu}(x) = 0$ for all $x \neq 0$, hence also for $x = 0$, by continuity.

Therefore $\mu = 0$, $X = C_0$, and the proof is complete.

7.5.3. Theorem. *Assume* $0 < p < \infty$, $p \neq 2$. *Assume that* $M \subset L^\infty(\mu_1)$, *that* $1 \in M$, *and that* M *is an algebra over* $\mathbb{C}$, *relative to pointwise multiplication.*

Let A *be a linear map of* M *into* $L^p(\mu_2)$, *such that* $A1 = 1$ *and*

$$(1) \qquad \int |Af|^p \, d\mu_2 = \int |f|^p \, d\mu_1$$

for every $f \in M$. *Then*

$$(2) \qquad A(fg) = Af \cdot Ag \qquad \text{a.e. } [\mu_2],$$

$$(3) \qquad \int Af \cdot \overline{Ag} \, d\mu_2 = \int f\bar{g} \, d\mu_1$$

and

(4) $$\|Af\|_\infty = \|f\|_\infty$$

for all $f \in M$, $g \in M$. Moreover,

(5) $$\int h(Af_1, \ldots, Af_N)d\mu_2 = \int h(f_1, \ldots, f_N)d\mu_1$$

for all $f_1, \ldots, f_N \in M$ and for every Borel function $h: \mathbb{C}^n \to \mathbb{C}$ that is bounded or positive.

Proof. If p is not an even integer, then Theorem 7.5.2, with f and Af in place of f_1 and f_2, shows that

(6) $$\int |Af|^{2m} \, d\mu_2 = \int |f|^{2m} \, d\mu_1$$

for $m = 1, 2, 3, \ldots$. If p is an even integer, then (1) is the same as (6), for some $m \geq 2$. Thus we may assume (6), for every $f \in M$, and for some integer $m \geq 2$ which is fixed from now on.

Pick $f \in M$, $g \in M$, and put $u = Af$, $v = Ag$, $w = A(fg)$. For any $\alpha, \beta, \gamma \in \mathbb{C}$, the integral

(7) $$\int (1 + \alpha u + \beta v + \gamma w)^m (1 + \overline{\alpha u} + \overline{\beta v} + \overline{\gamma w})^m \, d\mu_2$$

is then equal to

(8) $$\int (1 + \alpha f + \beta g + \gamma fg)^m (1 + \overline{\alpha f} + \overline{\beta g} + \overline{\gamma fg})^m \, d\mu_1.$$

The coefficients of $\alpha\beta\overline{\alpha}\overline{\beta}$, $\alpha\beta\overline{\gamma}$, $\gamma\alpha\overline{\beta}$, $\gamma\overline{\gamma}$ are the same in (7) and (8). (It is here that $m \geq 2$ is used.) Each of the 4 integrals

(9) $$\int u v \overline{u} \overline{v}, \quad \int u v \overline{w}, \quad \int w \overline{u} \overline{v}, \quad \int w \overline{w}$$

is thus equal to $\int |fg|^2$, hence they are equal to each other, and therefore

(10) $$\int |w - uv|^2 \, d\mu_2 = 0.$$

It follows that $w = uv$ a.e. This proves (2).

Comparison of the coefficients of $\alpha\overline{\beta}$ in (7) and (8) leads to (3).

By (2), $(Af)^k = A(f^k)$ for $f \in M$, $k = 1, 2, 3, \ldots$. Hence

(11) $$\int |Af|^{2k} = \int |A(f^k)|^2 = \int |f^k|^2 = \int |f|^{2k};$$

the second of these equalities follows from (3). If we raise (11) to the exponent $1/2k$ and then let $k \to \infty$, we obtain (4).

We turn to the proof of (5). If h is a monomial in $z_1, \ldots, z_N, \bar{z}_1, \ldots, \bar{z}_N$, then (5) holds by repeated application of (2) and (3). Hence (5) holds if h is any polynomial in these variables. The range of $(f_1, \ldots, f_N)$ has compact closure in $\mathbb{C}^N$; the same is true for $(Af_1, \ldots, Af_N)$, by (4). The Stone–Weirstrass theorem shows therefore that (5) holds for every continuous h. The final assertion of the theorem follows now by standard approximation arguments.

7.5.4. Lemma. *Suppose* f, g, $h \in N(B)$, *and their boundary values satisfy* $f^* = g^*h^*$ *a.e. on* S. *Then* $f = gh$ *in* B.

Proof. Since $gh \in N(B)$ and $(gh)^* = g^*h^*$ a.e., we have $(f - gh)^* = 0$ a.e. Apply Theorem 5.6.4(b).

For the next lemma we recall that $H^p(S)$ is the class of all boundary functions of members of $H^p(B)$, as in §5.6.7.

7.5.5. Lemma (Schneider [1]). *Suppose* $0 < p \leq \infty$, $u \in L^\infty(S)$, $g \in H^p(S)$, $g \neq 0$ *a.e., and* $u^k g \in H^p(S)$ *for* $k = 1, 2, 3, \ldots$. *Then* $u \in H^\infty(S)$.

Proof. Without loss of generality, suppose $|u| \leq 1$ and $\|g\|_p \leq 1$. Put $h_k = u^k g$. Denote the holomorphic extensions of h_k and g into B by the same letters. Then

(1) $$h_1^k = (ug)^k = h_k g^{k-1} \quad \text{a.e.} \quad \text{on } S.$$

Let Ω be the set of all $z \in B$ where $g(z) \neq 0$. If $z \in \Omega$, then (1) and Lemma 7.5.4 give

(2) $$\left| \left(\frac{h_1}{g} \right)(z) \right|^k = \left| \left(\frac{h_k}{g} \right)(z) \right|.$$

Since $\|h_k\|_p \leq 1$, Theorem 7.2.5 shows that $\{h_k(z)\}$ is a bounded sequence; the bound depends on z and p. If we take kth roots in (2) and let $k \to \infty$, we conclude that $|h_1/g| \leq 1$ at all points of Ω. The corollary to Theorem 4.4.7 shows therefore that there exists $f \in H^\infty(B)$ such that $h_1 = gf$ in B. Since $h_1 = gu$ on S, we see that $u = f^*$.

We now come to the main result of this section.

7.5.6. Theorem. *Suppose $0 < p < \infty$, $p \neq 2$, $n \geq 1$, and T is a linear isometry of $H^p(B)$ onto $H^p(B)$. Then there is a $\psi \in \mathrm{Aut}(B)$ and a $c \in \mathbb{C}$, $|c| = 1$, such that*

(1) $$(Tf)(z) = c\,\frac{(1 - |a|^2)^{n/p}}{(1 - \langle z, a\rangle)^{2n/p}} \cdot f(\psi(z))$$

for all $f \in H^p(B)$, $z \in B$, where $a = \psi^{-1}(0)$.

The word "linear" refers to complex scalars, naturally. The reason for mentioning this explicitly is that there are other isometries. For instance, when $n = 1$, the map that associates $\overline{f(\bar{z})}$ to $f(z)$ is an isometry of $H^p(U)$ onto $H^p(U)$ is is not $\mathbb{C}$-linear.

Proof. The natural one-to-one correspondence between $f \in H^p(B)$ and its boundary function $f^* \in H^p(S)$ is a linear isometry, and can be used to identify $H^p(B)$ with $H^p(S)$. We may accordingly regard T as acting on $H^p(S)$, and shall make no notational distinction between f and f^*.

Put $g = T1$. Then $\|g\|_p = 1$, hence $g \neq 0$ a.e. on S, and

(2) $$Af = \frac{Tf}{g}$$

is well-defined a.e. on S, for every $f \in H^p(S)$. The isometry hypothesis on T becomes

(3) $$\int_S |Af|^p |g|^p \, d\sigma = \int_S |f|^p \, d\sigma$$

for all $f \in H^p(S)$. Since $A1 = 1$, we can apply Theorem 7.5.3, with $\mu_1 = \sigma$, $d\mu_2 = |g|^p \, d\sigma$, and $M = H^\infty(S)$. Since $|g|^p \in L^1(\sigma)$ and $|g| \neq 0$ a.e. $[\sigma]$, μ_2 and σ are mutually absolutely continuous. By Theorem 7.5.3, A is a multiplicative linear map of $H^\infty(S)$ into $L^\infty(\mu_2) = L^\infty(\sigma)$ that preserves L^∞-norms. The multiplicativity of A shows that

(4) $$g \cdot (Af)^k = g \cdot A(f^k) = T(f^k) \in H^p(S)$$

for $k = 1, 2, 3, \ldots$ and $f \in H^\infty(S)$. By Lemma 7.5.5, $Af \in H^\infty(S)$.

Passing from $H^\infty(S)$ to $H^\infty(B)$, we have now proved that A is a homomorphism of $H^\infty(B)$ into $H^\infty(B)$. Theorem 6.6.5 shows therefore that $(Af)(z) = f(\psi(z))$ for some holomorphic $\psi: B \to B$. Thus

(5) $$(Tf)(z) = g(z)f(\psi(z)) \quad (z \in B)$$

for every $f \in H^\infty(B)$. Since $H^\infty(B)$ is dense in $H^p(B)$ and evaluations at points of B are continuous linear functionals on $H^p(B)$ (Theorem 7.2.5), (5) holds actually for all $f \in H^p(B)$.

We now use the assumption that the range of T is all of $H^p(B)$. Everything proved so far for T is thus equally true for T^{-1}. In particular, the analogue of (5) is

$$(6) \qquad\qquad (T^{-1}f)(z) = h(z)f(\varphi(z))$$

where $Th = 1$ and $\varphi: B \to B$ is holomorphic.

By (5) and (6),

$$(7) \qquad f = T^{-1}Tf = h \cdot [(Tf) \circ \varphi] = h \cdot (g \circ \varphi) \cdot (f \circ \psi \circ \varphi)$$

for every $f \in H^p(B)$. With $f = 1$, (7) becomes

$$(8) \qquad\qquad h \cdot (g \circ \varphi) = 1$$

so that $f = f \circ \psi \circ \varphi$ for every $f \in H^p(B)$. Hence $\psi \circ \varphi$ is the identity map on B. The same argument (interchange T and T^{-1} in (7)) shows that $\varphi \circ \psi$ is the identity map.

We conclude that $\psi \in \text{Aut}(B)$.

It remains to identify g. Put $a = \psi^{-1}(0)$. By (5), $T(f \circ \psi^{-1}) = gf$. Hence

$$(9) \qquad \int_S |fg|^p \, d\sigma = \int_S |f \circ \psi^{-1}|^p \, d\sigma = P[|f|^p](a)$$

or

$$(10) \qquad \int_S |f|^p |g|^p \, d\sigma = \int_S |f|^p P_a \, d\sigma$$

for all $f \in H^\infty(S)$; here $P_a(\zeta) = P(a, \zeta)$.

Another application of Theorem 7.5.3 leads now from (10) to

$$(11) \qquad \int_S f_1 \bar{f}_2 |g|^p \, d\sigma = \int_S f_1 \bar{f}_2 P_a \, d\sigma$$

for, say, all f_1, f_2 in the ball algebra. The linear span of these functions $f_1 \bar{f}_2$ is a self-adjoint subalgebra of $C(S)$. It follows therefore from (11) and the Stone–Weierstrass theorem that

$$(12) \qquad |g(\zeta)| = P(a, \zeta)^{1/p} \qquad \text{a.e.} \quad \text{on } S.$$

Hence $g \in H^\infty(B)$, and (5) implies that T maps $H^\infty(B)$ into $H^\infty(B)$. The same is true of T^{-1}; thus $h \in H^\infty(B)$, and (8) shows that $1/g \in H^\infty(B)$.

Finally, put

$$(13) \qquad k(z) = \frac{(1 - \langle z, a \rangle)^{2n/p}}{(1 - |a|^2)^{n/p}} \qquad (z \in \bar{B}).$$

By (12) and (13), $|gk| = 1$ a.e. on S. Also, both gk and $1/gk$ are in $H^\infty(B)$. Thus $gk = c$, a constant of absolute value 1, and (1) follows from (5) and (13).

7.5.7. As regards the sup-norm isometries of $H^\infty(B)$ onto $H^\infty(B)$, and of $A(B)$ onto $A(B)$, their characterization depends on the following general theorem:

Every linear isometry T of a sup-norm algebra A onto A has the form

$$Tf = \alpha \cdot T_1 f,$$

where $\alpha \in A$, $1/\alpha \in A$, $|\alpha| = 1$, and T_1 is an automorphism of A.

A proof of this may be found on pp. 144–147 of Hoffman [1], and in deLeeuw–Rudin–Wermer [1]. Since the automorphisms of $H^\infty(B)$, and likewise of $A(B)$, have the form $f \to f \circ \psi$ for some $\psi \in \mathrm{Aut}(B)$ (Theorem 6.6.5), one obtains the following description of their surjective isometries:

If T is a linear isometry of $H^\infty(B)$ onto $H^\infty(B)$, or of $A(B)$ onto $A(B)$, then there is a $\psi \in \mathrm{Aut}(B)$ and a $c \in \mathbb{C}$, $|c| = 1$, such that

$$Tf = cf \circ \psi.$$

Note that formally this is a limiting case of 7.5.6(1), obtained by letting p tend to ∞.

Chapter 8

Consequences of the Schwarz Lemma

8.1. The Schwarz Lemma in B

8.1.1. The familiar classical Schwarz lemma deals with functions defined in the open unit disc $U \subset \mathbb{C}$, and asserts the following:

(a) If $f: U \to U$ is holomorphic, then $|f'(0)| < 1$, except when $f(\lambda) = c\lambda$ for some $c \in \mathbb{C}$ with $|c| = 1$.

(b) If also $f(0) = 0$, then $|f(\lambda)| < |\lambda|$ for every $\lambda \in U \setminus \{0\}$, except when $f(\lambda) = c\lambda$, as in (a).

As we shall see, this implies a variety of analogous results in several variables. Our first example concerns holomorphic maps of one *balanced* region into another; a set $E \subset \mathbb{C}^n$ is said to be balanced if $\lambda z \in E$ whenever $z \in E$ and $\lambda \in \mathbb{C}, |\lambda| \leq 1$. This terminology is customary in functional analysis. Balanced open sets in $\mathbb{C}^n$ are also known as *star-shaped circular regions*. Note that every balanced region is a neighborhood of the origin.

8.1.2. Theorem. *Suppose that*

(i) Ω_1 *and* Ω_2 *are balanced regions in* $\mathbb{C}^n$ *and* $\mathbb{C}^m$ *respectively,*

(ii) Ω_2 *is convex and bounded,*

(iii) $F: \Omega_1 \to \Omega_2$ *is holomorphic.*

Then

(a) $F'(0)$ *maps* Ω_1 *into* Ω_2, *and*

(b) $F(r\Omega_1) \subset r\Omega_2$ $(0 < r \leq 1)$ *if* $F(0) = 0$.

Recall that $F'(0)$ is a linear operator carrying $\mathbb{C}^n$ into $\mathbb{C}^m$; see §1.3.6.

Proof. The assumptions made on Ω_2 show that $\mathbb{C}^m$ may be regarded as a Banach space Y whose unit ball is Ω_2. The corresponding norm is

(1) $$\|w\| = \inf\{c > 0: c^{-1}w \in \Omega_2\}.$$

Fix $z \in r\Omega_1$, where $0 < r \le 1$. Since Ω_1 is open, $z \in t\Omega_1$ for some $t < r$. Let L be a linear functional on Y, of norm 1. Then

$$(2) \qquad\qquad g(\lambda) = LF(\lambda t^{-1}z)$$

defines a holomorphic map g of U into U. By the chain rule,

$$(3) \qquad\qquad g'(0) = LF'(0)t^{-1}z.$$

Since $|g'(0)| \le 1$, by 8.1.1(a), and since this holds for every L of norm 1, the Hahn–Banach theorem implies that

$$(4) \qquad\qquad \|F'(0)t^{-1}z\| \le 1.$$

Thus $F'(0)z \in t\bar{\Omega}_2 \subset r\Omega_2$. This proves (a).

If also $F(0) = 0$ and g is given by (2), then $g(0) = 0$, hence $|g(\lambda)| \le |\lambda|$, and (b) follows by the same argument that gave (a).

Remark. If Ω_1 is also convex and bounded, then $\mathbb{C}^n$ is a Banach space X with unit ball Ω_1, and (a) asserts that $F'(0): X \to Y$ is a linear operator of norm at most 1. By analogy with the classical Schwarz lemma, one may ask whether F must then be linear whenever $\|F'(0)\| = 1$. This is not so when $n > 1$, even in the case $\Omega_1 = B_n$, $\Omega_2 = B_m$; we shall see this in §8.1.5. But the linearity of F does follow if $F'(0)$ is assumed to be an isometry:

8.1.3. Theorem. *If $F: B_n \to B_m$ is holomorphic and $F'(0)$ is an isometry of $\mathbb{C}^n$ into $\mathbb{C}^m$, then $F(z) = F'(0)z$ for all $z \in B_n$.*

Proof. Put $F'(0) = A$, $F(0) = a$, $G = \varphi_a \circ F$, where $\varphi_a \in \text{Aut}(B_m)$ is as in §2.2.1. We claim that $a = 0$.

If $z \in B_n$ and $w = Az$, the chain rule gives

$$(1) \qquad\qquad G'(0)z = \varphi_a'(a)w.$$

By hypothesis, $|w| = |Az| = |z|$. By Theorem 8.1.2, $|G'(0)z| \le |z|$. Hence Theorem 2.2.2 shows that

$$(2) \qquad\qquad |s^{-2}Pw + s^{-1}Qw| \le |w| = |Pw + Qw|,$$

where $s = (1 - |a|^2)^{1/2}$ and $Pw \perp Qw$. This can only happen when $s = 1$, i.e., $a = 0$.

Thus $F(0) = 0$, hence $|F(z)| \le |z|$, by Theorem 8.1.2.

Pick $\zeta \in \mathbb{C}^n$, $|\zeta| = 1$, and define

$$(3) \qquad\qquad h(\lambda) = \langle F(\lambda\zeta), A\zeta \rangle \qquad (\lambda \in U).$$

Then h is a holomorphic map of U into U with $h'(0) = |A\zeta|^2 = 1$, so that $h(\lambda) = \lambda$, or

(4) $$\langle \lambda^{-1} F(\lambda\zeta), A\zeta \rangle = 1 \qquad (0 < |\lambda| < 1).$$

Since $|F(\lambda\zeta)| \leq |\lambda|$, the left side of (4) is the inner product of two vectors in B_m. This can only be 1 when the two vectors are equal (and have norm 1). Hence $F(\lambda\zeta) = \lambda A\zeta$, which gives the desired conclusion, since A is linear.

As in the case in one variable, part (b) of the Schwarz lemma can be generalized by applying automorphisms to both the domain and the range of F:

8.1.4. Theorem. *If $F: B_n \to B_m$ is holomorphic, $a \in B_n$, and $F(a) = b$, then*

(1) $$|\varphi_b(F(z))| \leq |\varphi_a(z)| \qquad (z \in B_n).$$

Equivalently,

(2) $$\frac{|1 - \langle F(z), F(a) \rangle|^2}{(1 - |F(z)|^2)(1 - |F(a)|^2)} \leq \frac{|1 - \langle z, a \rangle|^2}{(1 - |z|^2)(1 - |a|^2)}.$$

It is of course understood that $\varphi_a \in \mathrm{Aut}(B_n)$ and $\varphi_b \in \mathrm{Aut}(B_m)$; see §2.2.1. Assertion (1) can be stated in geometric terms: F maps each ellipsoid $E(a, \varepsilon)$ (see §2.2.7) into the ellipsoid $E(F(a), \varepsilon)$.

Proof. Since $\varphi_b \circ F \circ \varphi_a$ maps B_n into B_m and takes 0 to 0, Theorem 8.1.2 shows that

$$|\varphi_b(F(\varphi_a(z)))| \leq |z|,$$

which gives (1) if z is replaced by $\varphi_a(z)$. If we square (1), subtract from 1, and apply the identity 2.2.2(iv), we obtain (2).

Note: If $m = n$ and $F \in \mathrm{Aut}(B_n)$, then equality holds in (2). To see this, apply (2) to F^{-1} as well as to F.

8.1.5. Examples. (i) Suppose $f: B_n \to U$ is holomorphic. Then $f'(0)$ is the linear functional that takes $z \in B_n$ to

(1) $$\sum_{k=1}^{n} (D_k f)(0) z_k$$

which lies in U, by Theorem 8.1.2 with $m = 1$. It follows that

(2)
$$\sum_{k=1}^{n} |(D_k f)(0)|^2 \le 1.$$

(ii)   Suppose $F: U \to B_m$ is holomorphic, $F = (f_1, \ldots, f_m)$. Then $F'(0)$ is the linear map that takes $\lambda \in U$ to the vector

(3)
$$(f'_1(0)\lambda, \ldots, f'_m(0)\lambda)$$

in B_m, by Theorem 8.1.2 with $n = 1$. Hence

(4)
$$\sum_{i=1}^{m} |f'_i(0)|^2 \le 1.$$

(iii)   As regards the remark that precedes Theorem 8.1.3, we shall now see that the extremal functions related to the Schwarz lemma need not be unique, even in the simplest case $\Omega_1 = B_2, \Omega_2 = U$.
The power series

(5)
$$1 - \sqrt{1 - t} = \sum_{k=1}^{\infty} c_k t^k \qquad (|t| < 1)$$

has $c_k > 0$ for all k. Let the functions g_k be arbitrary members of $H^\infty(B_2)$, subject only to the inequality $\|g_k\|_\infty \le c_k$, and define

(6) $f(z, w) = z + w^2 g_1(z, w) + w^4 g_2(z, w) + w^6 g_3(z, w) + \cdots.$

If $|z|^2 + |w|^2 < 1$, it follows that

(7)
$$|f(z, w)| \le |z| + 1 - \sqrt{1 - |w|^2} < 1.$$

Every f given by (6) is thus a holomorphic map of B_2 into U.
If $h \in H^\infty(B_2)$ and $\|h\|_\infty \le 1$, Theorem 8.1.2(a) implies that $\|h'(0)\| \le 1$. Equality holds for *every f* of the form (6), since $f'(0)e_1 = 1$ and $f'(0)e_2 = 0$.
If $h \in H^\infty(B_2)$, $\|h\|_\infty \le 1$, and $h(0, 0) = 0$, Theorem 8.1.2(b) implies that $|h(z, 0)| \le |z|$; again, equality holds for *every f* given by (6)
Simple examples of (6) are

(8)
$$z + \tfrac{1}{2}w^2 \quad \text{or} \quad z + 1 - \sqrt{1 - w^2}.$$

8.2. Fixed-Point Sets in B

In Section 2.4 we saw that the fixed-point sets of automorphisms of B are affine. Theorem 8.2.3 will show that this property is shared by all holomorphic maps of B into B. But we first consider a somewhat more general situation.

8.2.1. Definition. Let Ω be a balanced, convex, bounded region in $\mathbb{C}^n$. As pointed out in the proof of Theorem 8.1.2, $\mathbb{C}^n$ may then be regarded as a Banach space X whose unit ball is Ω. We say that Ω is *strictly convex* if to every linear functional L on X, with $\|L\| = 1$, corresponds just one $z \in \overline{\Omega}$ (the closure of Ω) such that $Lz = 1$.

Evidently, B is strictly convex.

8.2.2. Theorem (Rudin [13]). *Let Ω be a balanced, bounded, strictly convex region in $\mathbb{C}^n$. If $F: \Omega \to \Omega$ is holomorphic and $F(0) = 0$, then F and the linear operator $F'(0)$ fix the same points of Ω.*

Proof. Let X be the Banach space whose unit ball is Ω. We shall use $\|\cdot\|$ for the norm in X, for the corresponding norms of linear functionals on X, and for the norms of linear operators on X.

Put $F'(0) = A$. By Theorem 8.1.2,

$$(1) \qquad \|A\| \leq 1 \quad \text{and} \quad \|F(z)\| \leq \|z\| \qquad (z \in \Omega).$$

Fix $z \in \Omega$, $z = ru$, where $0 < r < 1$, $\|u\| = 1$. By the Hahn–Banach theorem, there is a linear functional L on X with

$$(2) \qquad \|L\| = 1, \qquad Lu = 1.$$

Put

$$(3) \qquad g(\lambda) = LF(\lambda u) \qquad (\lambda \in U).$$

Then $g: U \to U$ is holomorphic, $g(0) = 0$, and $g'(0) = LAu$.

If $F(z) = z$, then $g(r) = L(ru) = r$, hence $g(\lambda) = \lambda$ for all λ, hence $g'(0) = 1$. Thus $LAu = 1$. The strict convexity of Ω, combined with (2), implies now that $Au = u$. Hence $Az = z$.

Conversely, assume $Az = z$. Then $g'(0) = Lu = 1$, hence $g(r) = r$, or

$$(4) \qquad L(r^{-1}F(ru)) = 1.$$

By (1), $\|r^{-1}F(ru)\| \leq 1$. The strict convexity of Ω, combined with (2) and (4), implies now that $r^{-1}F(ru) = u$, hence $F(z) = z$.

8.2.3. Theorem (Rudin [13]). *If $F: B \to B$ is holomorphic, then the fixed-point set E of F is affine.*

Proof. Suppose $a \in E$, and let E_a denote the fixed point set of $\varphi_a \circ F \circ \varphi_a$. Then $0 \in E_a$, and Theorem 8.2.2 implies that E_a is affine. Since $E = \varphi_a(E_a)$, it follows from Proposition 2.4.2 that E is affine.

8.2.4. Holomorphic Retracts. A map $F: B \to B$ is said to be a *retraction* of B if $F(F(z)) = F(z)$ for every $z \in B$. The range of F is then exactly its fixed-point set. A *holomorphic retract* of B is, by definition, the range of some holomorphic retraction of B.

Theorem 8.2.3 thus has the following corollary.

Corollary (Suffridge [1]). *The holomorphic retracts of B are exactly the affine subsets of B.*

Indeed, if $E \subset B$ is affine and $a \in E$, then $\varphi_a(E) = B \cap Y$, where Y is a subspace of $\mathbb{C}^n$. Let P be the orthogonal projection of $\mathbb{C}^n$ onto Y. Then $\varphi_a P \varphi_a$ is a holomorphic retraction of B onto E. The converse follows from Theorem 8.2.3.

Although the holomorphic retracts of B are thus very simple, there exist very complicated holomorphic retractions. For example, let $f \in H^\infty(B_2)$ be any one of the functions described by 8.1.5(6), and put $F(z, w) = (f(z, w), 0)$. Since $f(z, 0) = z$, F retracts B onto the set $\{(z, 0): |z| < 1\}$.

8.3. An Extension Problem

8.3.1. Statement of the Problem. Suppose $1 \le n < m$, and let $\Phi: B_n \to B_m$ be holomorphic. Let us say that Φ has the *norm-preserving H^∞ extension property* (or property (∗), for brevity) if the following is true:

(∗)   To every $f \in H^\infty(B_n)$ corresponds a $g \in H^\infty(B_m)$ such that
(a) $g \circ \Phi = f$, and
(b) $\|g\|_\infty = \|f\|_\infty$.

The problem is: *Which Φ have property (∗)?*

The reason for calling this an extension problem is quite simple. Clearly, (∗) implies that Φ is one-to-one. Every $f \in H^\infty(B_n)$ corresponds therefore to a function $\tilde{f}$ on $\Phi(B_n)$ such that $\tilde{f} \circ \Phi = f$, and any g that satisfies (a) is an extension of $\tilde{f}$. The requirement (b) is of course extremely strong, and one should expect that only very special Φ's can satisfy it. Theorem 8.3.2 confirms this expectation.

If Φ has property (∗) one sees very easily that $\psi \circ \Phi$ has property (∗) for every $\psi \in \text{Aut}(B_m)$. Theorem 8.3.2 implies therefore that every Φ with property (∗) has affine range.

8.3.2. Theorem. *For a holomorphic map* $\Phi: B_n \to B_m$ *with* $\Phi(0) = 0$, *the follow-ing are equivalent*:

(i) Φ *has property* (∗).

(ii) Φ *is a linear isometry.*

(iii) *There is a multiplicative linear operator*

$$E: H^\infty(B_n) \to H^\infty(B_m)$$

such that $(Ef) \circ \Phi = f$ *for every* $f \in H^\infty(B_n)$.

Proof. Assume (i). Pick $\zeta \in \mathbb{C}^n$, $|\zeta| = 1$, and put $f(z) = \langle z, \zeta \rangle$. Then $f \in H^\infty(B_n)$, $\|f\|_\infty = 1$. Hence there is a $g \in H^\infty(B_m)$, with $\|g\|_\infty = 1$, such that $g(\Phi(z)) = \langle z, \zeta \rangle$. With $z = \lambda\zeta$, this becomes

$$(1) \qquad\qquad g(\Phi(\lambda\zeta)) = \lambda \qquad (\lambda \in U).$$

Since $\Phi(0) = 0$, differentiation of (1) gives

$$(2) \qquad\qquad g'(0)\Phi'(0)\zeta = 1.$$

By Theorem 8.1.2, $\Phi'(0)\zeta \in \bar{B}_m$ and $g'(0)$ is a linear functional on $\mathbb{C}^m$, of norm at most 1. Hence (2) implies that $\Phi'(0)\zeta$ is a unit vector in $\mathbb{C}^m$ for every unit vector ζ in $\mathbb{C}^n$. This says that $\Phi'(0)$ is an isometry, hence $\Phi(z) = \Phi'(0)z$, by Theorem 8.1.3. Thus (i) implies (ii).

If (ii) holds, then $\Phi(z) = Az$, where A is a linear isometry of $\mathbb{C}^n$ onto a subspace Y of $\mathbb{C}^m$. Let P be the orthogonal projection of $\mathbb{C}^m$ onto Y, and define

$$(3) \qquad\qquad (Ef)(w) = f(A^{-1}Pw) \qquad (w \in B_m)$$

for all $f \in H^\infty(B_n)$. (Note that A^{-1} is linear and well-defined on the range of P, and that P maps B_m onto $Y \cap B_m$.) It is clear that E is linear and multi-plicative; also, $(Ef) \circ \Phi = f$, because $A^{-1}P\Phi(z) = z$. Thus (ii) implies (iii).

Finally, assume (iii). Since E is multiplicative, $Ef = E(f \cdot 1) = (Ef) \cdot (E1)$, hence $E1 = 1$. (Note that $Ef \equiv 0$ implies $f \equiv 0$.) If $fg = 1$, it follows that $(Ef) \cdot (Eg) = E(fg) = 1$. Thus Ef is invertible in $H^\infty(B_m)$ whenever f is invertible in $H^\infty(B_n)$. It follows that the sets $f(B_n)$ and $(Ef)(B_m)$ have the same closures in $\mathbb{C}$. In particular, $\|Ef\|_\infty = \|f\|_\infty$. Thus (iii) implies (i).

Note: The only $f \in H^\infty(B_n)$ that were needed to prove the implication (i) → (ii) were the linear functions $\langle z, \zeta \rangle$.

8.3.3. This problem has been treated by Stanton [1] with finite Riemann surfaces in place of B_n.

If one drops condition (b) in §8.3.1, one obtains what is usually called the H^∞ extension problem. Henkin [5] and Adachi [1] have studied this in strictly pseudoconvex domains. With polydiscs in place of balls, extension problems of this type occur in Chapter 7 of Rudin [1].

For extension theorems in the context of Bergman spaces and Hardy spaces, we refer to Amar [1], [3], and to Cumenge [1]. Extension theorems with C^∞-data were investigated by Elgueta [1].

8.4. The Lindelöf–Čirka Theorem

The classical theorem of Lindelöf which Čirka extended to several variables concerns the limit of a function $f \in H^\infty(U)$ at a single boundary point. It is thus not a theorem of Fatou type. Although Lindelöf's theorem is an elementary consequence of the maximum modulus principle, it does not seem to appear in the standard elementary texts. For this reason, a proof is included here.

8.4.1. Theorem (Lindelöf [1]). *Suppose $f \in H^\infty(U)$ and $\gamma: [0, 1) \to U$ is a continuous curve such that $\gamma(t) \to 1$ as $t \to 1$. If*

$$\lim_{t \to 1} f(\gamma(t)) = L \tag{1}$$

exists, then f has nontangential limit L at the point 1.

Note that there is no restriction on the manner in which $\gamma(t)$ tends to 1, except that $\gamma(t)$ must lie in U for all $t < 1$.

Proof. Without loss of generality, assume $\|f\|_\infty = 1$ and $L = 0$. Let Σ be the strip defined by $|\operatorname{Re} z| < 1$. Let φ be a conformal map of U onto Σ, with $\varphi(0) = 0$, such that, setting $\Gamma = \varphi \circ \gamma$, we have $\operatorname{Im} \Gamma(t) \to +\infty$ as $t \to 1$. Replace f by $F = f \circ \varphi^{-1}$. Then $F \in H^\infty(\Sigma)$, $|F| \le 1$, $F(\Gamma(t)) \to 0$ as $t \to 1$. Given $\delta \in (0, 1)$, we have to prove that $F(x + iy) \to 0$ as $y \to +\infty$, uniformly in $|x| \le 1 - \delta$.

Fix ε, $0 < \varepsilon < 1$. Choose any $y > \operatorname{Im} \Gamma(0)$, so large that $|F(\Gamma(t))| < \varepsilon$ whenever $\operatorname{Im} \Gamma(t) \ge y$. We claim that then

$$|F(x + iy)| \le \varepsilon^{\delta/4} \quad \text{if} \quad |x| \le 1 - \delta. \tag{2}$$

The theorem follows obviously from (2).

To prove (2), assume $y = 0$, without loss of generality (by a vertical translation of Σ), choose t_0 so that $\operatorname{Im} \Gamma(t_0) = 0$ but $\operatorname{Im} \Gamma(t) > 0$ if $t_0 < t < 1$, let $E = \{\Gamma(t): t_0 \le t < 1\}$, and let $\bar{E}$ be the reflection of E in the real axis. Then $E \cup \bar{E}$ intersects the real axis in a unique point x_0.

Assume $x_0 < x \leq 1 - \delta$. Define

$$(3) \qquad\qquad G_\eta(z) = \frac{F(z)\overline{F(\bar{z})}\varepsilon^{(1+z)/2}}{1 + \eta(1 + z)} \qquad (z \in \Sigma)$$

where η is a positive parameter. Then $G_\eta \in H^\infty(\Sigma)$. On E, $|F(z)| < \varepsilon$; on $\bar{E}$, $|\overline{F(\bar{z})}| < \varepsilon$; hence $|G_\eta| < \varepsilon$ on $E \cup \bar{E}$. On the right edge of Σ, the boundary values of $|G_\eta|$ are $<\varepsilon$. When $|\operatorname{Im} z|$ is sufficiently large, then $|G_\eta(z)| < \varepsilon$ because of the denominator in (3). These facts imply that $|G_\eta(x)| < \varepsilon$, by the maximum modulus principle, applied to G_η in the component of $\Sigma \backslash (E \cup \bar{E})$ that contains x. Letting $\eta \to 0$, we obtain therefore

$$|F(x)|^2 \leq \varepsilon \cdot \varepsilon^{-(1+x)/2} = \varepsilon^{(1-x)/2} \leq \varepsilon^{\delta/2}$$

since $1 - x \geq \delta$.

If $-1 + \delta \leq x \leq x_0$, replace $1 + z$ by $1 - z$ in (3); this leads to the same conclusion.

Thus (2) holds, and the proof is complete.

8.4.2. Remark. We stated in Lindelöf's theorem in the disc U but proved it in the strip Σ. Other conformal maps will of course transfer the theorem to other regions in $\mathbb{C}$.

For example, let $\Pi_\alpha = \{z = re^{i\theta} : r > 0, |\theta| < \alpha\}$. If $f \in H^\infty(\Pi_\alpha)$, $f \to L$ along *some* curve γ_0 in Π_α that approaches 0, and $\beta < \alpha$, then f tends to L along *every* curve γ that approaches 0 within Π_β.

8.4.3. Approach Curves in B. A curve in B that approaches a point $\zeta \in S$ will be called a ζ-*curve*. More precisely, a ζ-curve is a continuous map $\Gamma \colon [0, 1) \to B$ such that $\Gamma(t) \to \zeta$ as $t \to 1$. Usually, however, it will not be necessary to refer to any parametrization.

With each ζ-curve Γ we associate its orthogonal projection

$$(1) \qquad\qquad \gamma = \langle \Gamma, \zeta \rangle \zeta$$

into the complex line through 0 and ζ. Then $(\Gamma - \gamma) \perp \gamma$, so that

$$(2) \qquad\qquad |\Gamma - \gamma|^2 + |\gamma|^2 = |\Gamma|^2.$$

Since $|\Gamma| < 1$, (2) implies

$$(3) \qquad\qquad \frac{|\Gamma - \gamma|^2}{1 - |\gamma|^2} < 1.$$

As ζ-curve Γ is said to be *special* if

(4)
$$\lim_{t \to 1} \frac{|\Gamma(t) - \gamma(t)|^2}{1 - |\gamma(t)|^2} = 0$$

and is said to be *restricted* if it satisfies both (4) and

(5)
$$\frac{|\gamma(t) - \zeta|}{1 - |\gamma(t)|} \leq A \qquad (0 \leq t < 1)$$

for some $A < \infty$.

The restricted ζ-curves Γ are thus the special ones whose projection γ is nontangential.

There is a simple relation between restricted ζ-curves and the Korányi regions $D_\alpha(\zeta)$. Recall that $z \in D_\alpha(\zeta)$ precisely when

(6)
$$|1 - \langle z, \zeta \rangle| < \frac{\alpha}{2}(1 - |z|^2).$$

Assume that Γ satisfies (5), and also (3), but with some $c < 1$ in place of 1. Then (2) leads to

$$1 - |\Gamma|^2 = 1 - |\gamma|^2 - |\Gamma - \gamma|^2 > (1 - c)(1 - |\gamma|^2)$$

and (5) shows that

$$|1 - \langle \Gamma, \zeta \rangle| = |\langle \zeta - \gamma, \zeta \rangle| \leq |\zeta - \gamma| \leq A(1 - |\gamma|).$$

Thus

(7)
$$\frac{|1 - \langle \Gamma, \zeta \rangle|}{1 - |\Gamma|^2} < \frac{A}{(1 - c)(1 + |\gamma|)}$$

which tends to $A/2(1 - c)$ as $t \to 1$.

We conclude: If $\alpha > A/(1 - c)$, then Γ lies in $D_\alpha(\zeta)$ eventually; that is to say, $\Gamma(t) \in D_\alpha(\zeta)$ for all t that are sufficiently close to 1.

If (3) is replaced by (4), the above holds for arbitrarily small c. Thus:

Every restricted ζ-curve Γ satisfying (5) lies eventually in $D_\alpha(\zeta)$, for all $\alpha > A$.

Conversely, every ζ-curve Γ that lies in $D_\alpha(\zeta)$ satisfies (5) with $A = \alpha$.

We shall say that a function $f: B \to \mathbb{C}$ has *restricted K-limit L* at ζ if $\lim f(\Gamma(t)) = L$ as $t \to 1$, for every restricted ζ-curve Γ.

The preceding discussion shows that this happens whenever f has a K-limit at ζ. However, an $f \in H^\infty(B)$ may have a restricted K-limit at a point ζ, without having a K-limit at ζ. The simplest example of this is probably given by the function

$$(8) \qquad\qquad f(z, w) = \frac{w^2}{1 - z^2}$$

which is in $H^\infty(B_2)$, has restricted K-limit 0 at $(1, 0)$, but fails to have a K-limit there, since

$$(9) \qquad\qquad f(t, c\sqrt{1 - t^2}) = c^2 \qquad (0 \le t < 1)$$

for every $c \in U$. The expansion of (8), namely

$$(10) \qquad\qquad f(z, w) = \sum_{k=0}^{\infty} z^{2k} w^2$$

is a very simple example of a power series that converges absolutely at every point of S although the convergence is not uniform.

8.4.4. Theorem (Čirka [1]). *Suppose $f \in H^\infty(B)$, $\zeta \in S$, Γ_0 is a special ζ-curve, and*

$$(1) \qquad\qquad \lim_{t \to 1} f(\Gamma_0(t)) = L.$$

Then f has restricted K-limit L at ζ.

Proof. Let Γ be any special ζ-curve. Fix $t \in [0, 1)$ for the moment. Since $(\Gamma - \gamma) \perp \gamma$, the point $(1 - \lambda)\gamma(t) + \lambda\Gamma(t)$ lies in B whenever $|\gamma|^2 + |\lambda|^2|\Gamma - \gamma|^2 < 1$, i.e., whenever $|\lambda| < R = R(t)$, where

$$(2) \qquad\qquad R^2 = \frac{1 - |\gamma|^2}{|\Gamma - \gamma|^2}.$$

By 8.4.3(3), $R > 1$. Since Γ is special, $R(t) \to \infty$ as $t \to 1$.

If $|\lambda| < R$ we can define

$$(3) \qquad\qquad g(\lambda) = f((1 - \lambda)\gamma(t) + \lambda\Gamma(t)).$$

The Schwarz lemma, applied to $g(\lambda) - g(0)$ in the disc $\{|\lambda| < R\}$, shows that

$$(4) \qquad\qquad |g(1) - g(0)| \le \frac{2\|f\|_\infty}{R(t)}.$$

Since $R(t) \to \infty$, we conclude from (3) and (4) that

(5) $$\lim_{t \to 1} \{ f(\Gamma(t)) - f(\gamma(t)) \} = 0.$$

We now apply (5) to the given curve Γ_0 and to an arbitrary restricted ζ-curve Γ. By (1) and (5), $f(\gamma_0(t)) \to L$. Since γ is nontangential, Lindelöf's theorem (applied in the disc $\{ \lambda\zeta : \lambda \in U \}$) shows that $f(\gamma(t)) \to L$. Hence $f(\Gamma(t)) \to L$, by (5), and the proof is complete.

8.4.5. Asymptotic Values. If f is a function in B, Γ is a ζ-curve, and $f(z)$ tends to L as z tends to ζ along Γ, then L is said to be an *asymptotic value* of f at ζ.

Lindelöf's theorem implies that no $f \in H^\infty(U)$ can have more than one asymptotic value at any boundary point. This is false if U is replaced by B; the function f mentioned at the end of §8.4.3 has every c with $|c| \le 1$ as an asymptotic value at $(1, 0)$, even though $|f| \le 1$. But Čirka's theorem shows that we still have uniqueness if we restrict ourselves to special ζ-curves:

If $f \in H^\infty(B)$, $\zeta \in S$, and f tends to L_1 and L_2 along special ζ-curves Γ_1 and Γ_2, then $L_1 = L_2$.

8.4.6. Example (Nagel–Rudin [2]). Here is an example that is a bit more ambitious than the one given at the end of §8.4.3. It exhibits a function $f \in H^\infty(B_2)$ whose restricted K-limit is 0 at every point on the circle $\{ (e^{i\theta}, 0) : -\pi \le \theta \le \pi \}$ but which has no K-limit at any of these points.

To do this, pick positive integers n_j and corresponding radii $r_j = 1 - 1/n_j$, so that $n_1 = 2$,

(1) $$n_j > 10(n_1 + \cdots + n_{j-1}) \qquad (j = 2, 3, 4, \ldots)$$

and

(2) $$n_j \exp \left\{ -\frac{n_j}{n_k} \right\} < j^{-2} \qquad (1 \le k < j).$$

Define

(3) $$f(z, w) = w^2 g(z) = w^2 \sum_{j=1}^\infty n_j z^{n_j}.$$

Since $n_j |z|^{n_j} < 2(n_j - n_{j-1})|z|^{n_j} < 2 \sum |z|^m$, where the sum extends over all m with $n_{j-1} < m \le n_j$, it follows that $|g(z)| < 2/(1 - |z|)$, hence $|f(z, w)| < 4$ in B_2. Thus $f \in H^\infty(B_2)$.

Since $f(z, 0) = 0$, Čirka's theorem implies that f has restricted K-limit 0 at all points $(z, 0)$ with $|z| = 1$.

For $k \geq 2$, $\frac{1}{4} \leq (1 - 1/k)^k < 1/e$. If $|z| = r_p$, it follows from (1), (2), (3) that

(4) $$|g(z)| > \frac{n_p}{4} - \frac{n_p}{10} - \sum_{p+1}^{\infty} j^{-2}.$$

Thus $|g(z)| > n_p/20 = 1/(20(1 - r_p))$ as soon as n_p is large enough, and therefore

(5) $$|f(r_p e^{i\theta}, c\sqrt{1 - r_p^2})| > \frac{|c|^2}{20}$$

if $|c| < 1$. The points at which f is evaluated in (5) lie in $D_\alpha(e^{i\theta}, 0)$ when $\alpha > 2/(1 - |c|^2)$. Hence f has no K-limit at $(e^{i\theta}, 0)$.

8.4.7. Example. Fix a constant $c > \frac{1}{2}$ and define f in B_2 by

(1) $$f(z, w) = (1 - z)^{-c} w.$$

Then $f \notin H^\infty(B_2)$, but $f \in H^p(B_2)$ for all $p < 4/(2c - 1)$. If $\frac{1}{2} < \delta < c$ and

(2) $$\Gamma(t) = (t, (1 - t)^\delta) \qquad (0 \leq t < 1)$$

then Γ tends to $(1, 0)$ restrictedly, and

(3) $$f(\Gamma(t)) = (1 - t)^{\delta - c} \to \infty.$$

Since $f(z, 0) = 0$, we see that f has no restricted K-limit at $(1, 0)$.

Take a point $(a, b) \in S$, $a \neq 1$, and consider the rectilinear path

(4) $$\Gamma(t) = (t + (1 - t)a, (1 - t)b)$$

from (a, b) to $(1, 0)$. On this path,

(5) $$f(\Gamma(t)) = \frac{b}{(1 - a)^c} \cdot (1 - t)^{1 - c}.$$

When $c < 1$, this tends to 0 as $t \to 1$. Thus all "rectilinear limits" of f at $(1, 0)$ are 0. In fact, $f(z, w) \to 0$ as $(z, w) \to (1, 0)$ within any cone in B whose vertex is at $(1, 0)$, although (as we saw above) the restricted K-limit of f does not exist there.

When $c = 1$, then $f(z, w) = w/(1 - z)$, and (5) shows that f is constant on each of the lines (4). All rectilinear limits of f exist therefore at $(1, 0)$, but they are not equal. In fact, they cover $\mathbb{C}$.

By Čirka's theorem, no $f \in H^\infty(B)$ can behave in this way.

The following version of Čirka's theorem will be used in Section 8.5. It will be clear from the proof that the hypotheses could be varied considerably, but it seems best to stick to a simple statement.

8.4.8. Theorem. *Suppose* $f \in H(B)$, $\zeta \in S$, f *is bounded in every region* $D_\alpha(\zeta)$, *and the radial limit of* f *exists at* ζ. *Then the restricted K-limit of* f *exists at* ζ.

Proof. Let $\gamma_0(t) = t\zeta$, $0 \leq t < 1$, and let Γ be any restricted ζ-curve, with projection γ as in §8.4.3. Then γ is nontangential. Lindelöf's theorem shows therefore that f has the same limit along γ and γ_0. It is thus enough to prove that

(1) $$\lim_{t \to 1}\{f(\Gamma(t)) - f(\gamma(t))\} = 0.$$

We saw in §8.4.3 that $\Gamma(t) \in D_\alpha(\zeta)$ eventually, for some α. Choose $\beta > \alpha$. A slight modification of the proof of Theorem 8.4.4 shows that $(1 - \lambda)\gamma + \lambda\Gamma \in D_\beta$ whenever $|\lambda| < R = R(t)$, where

(2) $$R^2 = \frac{1 - |\gamma|^2 - (2/\beta)|1 - \gamma|}{|\Gamma - \gamma|^2}.$$

If $\Gamma(t) \in D_\alpha$, then $|1 - \gamma| < (\alpha/2)(1 - |\gamma|^2)$, so that

(3) $$R^2 > \frac{\beta - \alpha}{\beta} \cdot \frac{1 - |\gamma|^2}{|\Gamma - \gamma|^2}$$

which tends to ∞ as $t \to 1$, since Γ is special.

Now define $g(\lambda) = f((1 - \lambda)\gamma + \lambda\Gamma)$ for $|\lambda| < R$, use the fact that f is bounded in D_β, and estimate $g(1) - g(0)$ by the Schwarz lemma, as in the proof of Theorem 8.4.4. This leads to (1).

8.5. The Julia–Carathéodory Theorem

8.5.1. In the present section, the following one-variable facts will be generalized to holomorphic maps from one ball into another:

Suppose $f: U \to U$ is holomorphic. If there is some sequence $\{z_i\}$ in U, with $z_i \to 1$ and $f(z_i) \to 1$, along which

$$\frac{1 - |f(z_i)|}{1 - |z_i|}$$

is bounded, then f maps each circular disc in U that has 1 in its boundary into a disc of the same sort. This (in a more quantitative form) is Julia's theorem. Carathéodory added that $f'(z)$ then has a nontangential positive finite limit at $z = 1$.

Full details of this may be found in vol. 2 of Carathéodory's book [3].

The generalizations to several variables will be proved directly, without any reference to the theorems just mentioned. In fact, if one takes $n = 1$ and $m = 1$, the proofs that follow are the classical ones.

8.5.2. The Setting. Throughout this section, m and n will be fixed, F will be a holomorphic map of B_n into B_m, ζ will be a fixed boundary point of B_n, and we define

$$(1) \qquad L = \liminf_{z \to \zeta} \frac{1 - |F(z)|^2}{1 - |z|^2}.$$

The basic assumption we make is that $L < \infty$.

There is then a sequence $\{a_i\}$ in B_n that converges to ζ, such that

$$(2) \qquad \lim_{i \to \infty} \frac{1 - |F(a_i)|^2}{1 - |a_i|^2} = L,$$

and such that $F(a_i)$ converges to some boundary point of B_m. By unitary transformations we may choose coordinates so that $\zeta = e_1$ and $F(a_i)$ converges to e_1. (The symbol e_1 is here used with two meanings; it designates the first element in the standard basis of $\mathbb{C}^n$ as well as $\mathbb{C}^m$. It is unlikely that this will cause any confusion.)

Let $f_1, \ldots, f_m$ be the components of F.

The Schwarz lemma (Theorem 8.1.4) states that

$$(3) \qquad \frac{|1 - \langle F(z), F(a_i) \rangle|^2}{1 - |F(z)|^2} \leq \frac{1 - |F(a_i)|^2}{1 - |a_i|^2} \cdot \frac{|1 - \langle z, a_i \rangle|^2}{1 - |z|^2}$$

for all $z \in B_n$. As $i \to \infty$, $\langle z, a_i \rangle \to z_1$ and $\langle F(z), F(a_i) \rangle \to f_1(z)$. Hence (2) and (3) yield

8.5.3. Julia's Theorem. *Under the above hypotheses*

$$(1) \qquad \frac{|1 - f_1(z)|^2}{1 - |F(z)|^2} \leq L \frac{|1 - z_1|^2}{1 - |z|^2} \qquad (z \in B_n).$$

One incidental consequence of (1) is that $L > 0$.

The inequality (1) has an appealing geometric interpretation that involves ellipsoids: For $0 < c < 1$, let E_c be the set of all $z \in B_n$ that satisfy

$$(2) \qquad \frac{|1 - z_1|^2}{1 - |z|^2} < \frac{c}{1 - c}.$$

Writing $z = (z_1, z')$ in the usual way, a little computing shows that (2) is the same as

$$(3) \qquad \frac{|z_1 - (1 - c)|^2}{c^2} + \frac{|z'|^2}{c} < 1.$$

Thus E_c is an ellipsoid in B_n that has e_1 as a boundary point, has its center at $(1 - c)e_1$, has radius c in the e_1-plane, and has radius $\sqrt{c}$ in the directions orthogonal to e_1.

If $\gamma/(1 - \gamma) = Lc/(1 - c)$ and if E_γ denotes the corresponding ellipsoid in B_m, then it follows from (1) and (2) *that F maps E_c into E_γ, where*

$$(4) \qquad \gamma = \frac{Lc}{1 + Lc - c}.$$

Let us now add an inessential assumption that will simplify the statements of some inequalities, namely: $F(0) = 0$. Then $|F(z)| \le |z|$ (Theorem 8.1.2), hence $L \ge 1$, and thus (4) implies the simpler statement $\gamma \le Lc$.

This proves the first part of the following geometric version of Julia's theorem:

8.5.4. Theorem. *If F is as in §8.5.2 and if also $F(0) = 0$, then*

    (i) $F(E_c) \subset E_{Lc}$ *when* $0 < c < 1/L$, *and*
    (ii) $F(D_\alpha) \subset D_{\alpha\sqrt{L}}$ *for all* $\alpha > 1$.

To prove the assertion about the Korányi regions $D_\alpha = D_\alpha(e_1)$, simply multiply the inequalities 8.5.3(1) and

$$\frac{1}{1 - |F(z)|^2} \le \frac{1}{1 - |z|^2}.$$

Then take square roots, to obtain

$$\frac{|1 - f_1(z)|}{1 - |F(z)|^2} \le \sqrt{L} \cdot \frac{|1 - z_1|}{1 - |z|^2} < \alpha\sqrt{L}$$

if $z \in D_\alpha$.

We shall need the following relation between D_α and D_β.

8.5.5. Lemma. *Suppose $1 < \alpha < \beta$, $\delta = \frac{1}{3}(1/\alpha - 1/\beta)$, and $z = (z_1, z') \in D_\alpha$.*

    (i)   *If $|\lambda| \le \delta|1 - z_1|$ then $(z_1 + \lambda, z') \in D_\beta$.*
    (ii) *If $|w'| \le \delta|1 - z_1|^{1/2}$ then $(z_1, z' + w') \in D_\beta$.*

Proof. The condition that $z \in D_\alpha$ can be written in the form

(1) $$|z'|^2 < 1 - |z_1|^2 - \frac{2}{\alpha}|1 - z_1|$$

in which z_1 and z' are separated.

Since $|z_1| < 1, |\lambda| < 1, \beta > 1$, and $5\delta + 2/\beta < 2/\alpha$, we have

$$|z_1 + \lambda|^2 + \frac{2}{\beta}|1 - z_1 - \lambda| < |z_1|^2 + 5|\lambda| + \frac{2}{\beta}|1 - z_1|$$

$$< |z_1|^2 + \frac{2}{\alpha}|1 - z_1| < 1 - |z'|^2,$$

which proves (i). Since $2|z'| < 3|1 - z_1|^{1/2}$ for all $z \in B$, we have

$$|z' + w'|^2 \le |z'|^2 + (3\delta + \delta^2)|1 - z_1|$$

$$< 1 - |z_1|^2 + \left(4\delta - \frac{2}{\alpha}\right)|1 - z_1|$$

$$< 1 - |z_1|^2 - \frac{2}{\beta}|1 - z_1|,$$

which proves (ii).

We are now ready for the generalization of Carathéodory's theorem. Recall that $D_1, \ldots, D_n$ denote the partial derivatives with respect to $z_1, \ldots, z_n$.

8.5.6. Theorem. *Suppose* $F = (f_1, \ldots, f_m)$ *is a holomorphic map of* B_n *into* B_m, $F(0) = 0$,

(1) $$L = \liminf_{z \to e_1} \frac{1 - |F(z)|^2}{1 - |z|^2} < \infty,$$

and $F(a_i) \to e_1$ *for some sequence* $\{a_i\}$ *in* B_n *such that* $a_i \to e_1$ *and*

(2) $$\lim_{i \to \infty} \frac{1 - |F(a_i)|^2}{1 - |a_i|^2} = L.$$

Suppose $2 \le j \le m$ *and* $2 \le k \le n$.

The following functions are then bounded in every region $D_\alpha(e_1)$:

(i) $(1 - f_1(z))/(1 - z_1)$
(ii) $(D_1 f_1)(z)$
(iii) $f_j(z)/(1 - z_1)^{1/2}$
(iv) $(1 - z_1)^{1/2}(D_1 f_j)(z)$
(v) $(D_k f_1)(z)/(1 - z_1)^{1/2}$
(vi) $(D_k f_j)(z)$.

Moreover, the functions (i), (ii) *have restricted K-limit L at e_1 and the functions* (iii), (iv), (v) *have restricted K-limit 0 at e_1.*

Corollary. *In the case $m = n$, the Jacobian JF of F is bounded in every region $D_\alpha(e_1)$.*

Because of its length, the proof will be divided into several steps.

Step 1. Radial Behavior. We shall first prove that

$$(3) \qquad\qquad \lim_{x \to 1} \frac{1 - f_1(xe_1)}{1 - x} = L$$

and

$$(4) \qquad\qquad \lim_{x \to 1} \frac{f_j(x)}{(1 - x)^{1/2}} = 0 \qquad (2 \le j \le m),$$

where it is understood that $0 < x < 1$.

Suppose that actually $1 - x < 1/L$. Put $1 - x = 2c$. Then xe_1 is a boundary point of E_c. By Theorem 8.5.4, $F(xe_1)$ lies in the closure of E_{Lc}. Since $2Lc < 1$, it follows that $|F(xe_1)| \ge 1 - 2Lc$, which is the same as

$$(5) \qquad\qquad 1 - |F(xe_1)| \le L(1 - x).$$

Since $F(0) = 0$, $1 + |F(x)| \le 1 + x$. Hence (5) implies

$$(6) \qquad\qquad \frac{1 - |F(xe_1)|^2}{1 - x^2} \le \frac{1 - |F(xe_1)|}{1 - x} \le L.$$

By the definition of L as the lower limit (1), it follows from (6) that

$$(7) \qquad\qquad \lim_{x \to 1} \frac{1 - |F(xe_1)|^2}{1 - x^2} = L.$$

To simplify the notation, we now write $w = w(x)$ for $f_1(xe_1)$. By (6) and Theorem 8.5.3,

$$(8) \qquad \frac{|1 - w|^2}{(1 - x)^2} \le L \cdot \frac{1 - |F(xe_1)|^2}{1 - x^2} \le L^2.$$

Since $1 - |F(xe_1)| \le 1 - |w| \le |1 - w|$, we conclude from (6), (7), and (8) that

$$(9) \qquad \lim_{x \to 1} \frac{1 - |w(x)|}{1 - x} = \lim_{x \to 1} \frac{|1 - w(x)|}{1 - x} = L.$$

The ratio of the two numerators in (9) converges therefore to 1 as $x \to 1$. This implies that also

$$(10) \qquad \lim_{x \to 1} \frac{1 - w(x)}{1 - |w(x)|} = 1,$$

and (3) is thus a consequence of (9).

Since $w(x) \to 1$ as $x \to 1$, (3) is the same as

$$(11) \qquad \lim_{x \to 1} \frac{1 - |f_1(xe_1)|^2}{1 - x^2} = L.$$

Now (4) follows from (7) and (11), because

$$(12) \qquad |F|^2 = |f_1|^2 + \cdots + |f_m|^2.$$

Step 2. The Functions (i) *and* (iii). Fix $\alpha > 1$, and assume $z \in D_\alpha(e_1)$ is so close to e_1 that $Lc < 1$ if $c = (\alpha/2)|1 - z_1|$.

Then $|1 - z_1|^2 = (2c/\alpha)|1 - z_1| < c(1 - |z|^2)$. Since $c < c/(1 - c)$, it follows that $z \in E_c$ (see 8.5.3(2)), hence $F(z) \in E_{Lc}$, and therefore

$$(13) \qquad |1 - f_1(z)| < 2Lc = \alpha L|1 - z_1|.$$

Since (13) holds for every $z \in D_\alpha(e_1)$ that is sufficiently close to e_1, we conclude that the function $(1 - f_1)/(1 - z_1)$ is bounded in every $D_\alpha(e_1)$; by (3) and Theorem 8.4.8, its restricted K-limit at e_1 is L.

If $2 \le j \le m$, the inclusion $F(z) \in E_{Lc}$ shows that

$$(14) \qquad |f_j(z)|^2 < Lc = \tfrac{1}{2}\alpha L|1 - z_1|.$$

Hence $f_j(z)/(1 - z_1)^{1/2}$ is bounded in every $D_\alpha(e_1)$, and its restricted K-limit at e_1 is 0, because of (4) and Theorem 8.4.8.

Step 3. The Functions (ii) *and* (iv). These involve differentiation with respect to z_1. Suppose $1 < \alpha < \beta$, choose δ as in Lemma 8.5.5, let $z \in D_\alpha$, and put

$$(15) \qquad r = r(z) = \delta|1 - z_1|.$$

Then $(z_1 + \lambda, z') \in D_\beta$ for all λ with $|\lambda| \le r$. By the Cauchy formula,

$$(16) \qquad (D_1 f_1)(z) = \frac{1}{2\pi i} \int_{|\lambda|=r} f_1(z_1 + \lambda, z')\lambda^{-2}\, d\lambda.$$

The integral is unchanged if f_1 is replaced by $f_1 - 1$. Do this, then multiply and divide the integrand by $z_1 + \lambda - 1$, and put $\lambda = re^{i\theta}$, to obtain

$$(17) \qquad (D_1 f_1)(z) = \frac{1}{2\pi} \int_{-\pi}^{\pi} \frac{1 - f_1(z_1 + re^{i\theta}, z')}{1 - (z_1 + re^{i\theta})} \cdot \left\{1 - \frac{1 - z_1}{re^{i\theta}}\right\} d\theta.$$

The first factor in the integrand is bounded, by Step 2, since $(z_1 + re^{i\theta}, z') \in D_\beta(e_1)$. The second factor is at most $1 + 1/\delta$, by (15). We conclude that $D_1 f_1$ is bounded in $D_\alpha(e_1)$.

When $z = xe_1$ in (17), then the second factor in the integrand is $1 - \delta^{-1}e^{-i\theta}$, and the first factor converges boundedly to L as $x \to 1$, since $x + r(x)e^{i\theta} \to 1$ nontangentially, for every θ, by (15). Hence $(D_1 f_1)(xe_1) \to L$ as $x \to 1$, by the dominated convergence theorem. Another application of Theorem 8.4.8 shows now that $D_1 f_1$ has restricted K-limit L at e_1.

If $2 \le j \le m$, a similar application of the Cauchy formula gives

$$(18) \qquad (D_1 f_1)(z) = \frac{1}{2\pi} \int_{-\pi}^{\pi} \frac{f_j(z_1 + re^{i\theta}, z')}{(1 - z_1 - re^{i\theta})^{1/2}} \cdot \frac{(1 - z_1 - re^{i\theta})^{1/2}}{re^{i\theta}}\, d\theta,$$

from which it follows exactly as above (using Step 2 and Theorem 8.4.8) that $(1 - z_1)^{1/2}(D_1 f_j)(z)$ is bounded in $D_\alpha(e_1)$ and that its restricted K-limit at e_1 is 0.

Step 4. The Functions (v) *and* (vi). These involve differentiation with respect to z_k for $2 \le k \le n$. Without loss of generality, take $k = 2$.

Suppose $1 < \alpha < \beta$, choose δ as in Lemma 8.5.5, let $z \in D_\alpha(e_1)$, and put

$$(19) \qquad \rho = \rho(z) = \delta|1 - z_1|^{1/2}.$$

Then $(z_1, z' + w') \in D_\beta(e_1)$ for all w' with $|w'| \le \rho$. If we apply the Cauchy formula as in Step 3, we obtain

$$(20) \qquad \frac{(D_2 f_1)(z)}{(1 - z_1)^{1/2}} = -\frac{(1 - z_1)^{1/2}}{\rho(z)} \cdot \frac{1}{2\pi} \int_{-\pi}^{\pi} \frac{1 - f_1(z_1, z_2 + \rho e^{i\theta}, \ldots)}{1 - z_1} e^{-i\theta}\, d\theta$$

and, for $j \geq 2$,

$$(21) \quad (D_2 f_j)(z) = \frac{(1 - z_1)^{1/2}}{\rho(z)} \cdot \frac{1}{2\pi} \int_{-\pi}^{\pi} \frac{f_j(z_1, z_2 + \rho e^{i\theta}, \ldots)}{(1 - z_1)^{1/2}} e^{-i\theta} \, d\theta.$$

The integrands are bounded, by the bounds of (i) and (iii) in $D_\beta(e_1)$. In view of (19), the left sides of (20) and (21) are therefore bounded in $D_\alpha(e_1)$.

To finish, we have to prove that the left side of (20) has restricted K-limit 0 at e_1. By Theorem 8.4.8 it is enough to prove this for the radial limit. Moreover, it involves now no loss of generality to assume $n = 2$, $m = 1$, in which case $f_1 = F$. Writing (z, w) in place of (z_1, z_2), we can expand F in the form

$$(22) \qquad F(z, w) = f(z) + 2w(1 - z)^{1/2} g(z) + \sum_{j=2}^{\infty} g_j(z) w^j.$$

Then $(D_2 F)(z, 0)/(1 - z)^{1/2} = 2g(z)$. It is therefore enough to show that

$$(23) \qquad\qquad g(x) \to 0 \quad \text{as} \quad x \nearrow 1.$$

We know that $(1 - f(z))/(1 - z) \to L$ as $z \to 1$ nontangentially, and that g is nontangentially bounded at 1, and we make one further reduction:

If $|\sum_0^{\infty} c_k w^k| < 1$ in a certain disc with center at 0, then also $|c_0 + \frac{1}{2} c_1 w| < 1$ in this same disc.

This is so because $c_0 + \frac{1}{2} c_1 w$ is the arithmetic mean of the first two partial sums of the power series. If we apply this to (22), we see that (23) is a consequence of the following proposition (in which there is some redundancy in the hypotheses):

8.5.7. Proposition. *Suppose $h: B_2 \to U$ has the form*

$$(1) \qquad\qquad h(z, w) = f(z) + w(1 - z)^{1/2} g(z)$$

where f, $g \in H(U)$, $(1 - f(z))/(1 - z)$ has finite nontangential limit L at $z = 1$, and g is nontangentially bounded at 1. Then

$$(2) \qquad\qquad\qquad g(x) \to 0 \quad as \quad x \nearrow 1.$$

Proof. Choose $\varepsilon > 0$, put $c = L^2/\varepsilon^2$, let z tend to 1 along the line $z = x + ic(1 - x)$. Then $1 - z = (1 - ic)(1 - x)$, hence

$$(3) \qquad\qquad\qquad |1 - z| \geq c(1 - x),$$

and also $1 - |z|^2 > 1 - x$ if $c^2/(1 + c^2) < x < 1$, an assumption that will be made in the rest of this proof. Note that

(4) $$f(z) = 1 - (L + o(1))(1 - ic)(1 - x)$$

so that

(5) $$\text{Re } f(z) = 1 - (L + o(1))(1 - x).$$

Associate with every z under consideration a $w \in \mathbb{C}$ with $|w|^2 = 1 - |z|^2 > 1 - x$, whose argument is so chosen that

(6) $$w(1 - z)^{1/2}g(z) = |w(1 - z)^{1/2}g(z)| \geq c^{1/2}(1 - x)|g(z)|,$$

by (3). Hence, by (5) and (6),

(7) $$1 \geq \text{Re } h(z, w) \geq 1 + \{c^{1/2}|g(z)| - L - o(1)\}(1 - x).$$

Consequently,

(8) $$\limsup_{x \to 1} |g(x + ic(1 - x))| \leq Lc^{-1/2} = \varepsilon.$$

The same estimate holds on the line $z = x - ic(1 - x)$. Since $g(z)$ is bounded as $z \to 1$ between these two lines, it follows that

(9) $$\limsup_{x \to 1} |g(x)| \leq \varepsilon,$$

which proves (2), since ε was arbitrary.

8.5.8. Examples. We shall now show that the conclusions of Theorem 8.5.6 are optimal. The numbers (i) through (vi) will refer to Theorem 8.5.6.
    The first two examples will use the function

(1) $$g(z) = \exp\left\{-\frac{\pi}{2} - i \log(1 - z)\right\} \qquad (z \in U).$$

Note that $|g| < 1$ in U, and that

(2) $$g'(z) = \frac{ig(z)}{1 - z}.$$

As $z \to 1$, $g(z)$ spirals around the origin without approaching it.

First Example. Take $n = m = 2$, define $F : B_2 \to B_2$ by

$$(3) \qquad\qquad F(z, w) = (z, wg(z)).$$

The hypotheses of Theorem 8.5.6 hold, with $L = 1$. Since $D_1 f_1 = 1$ and $D_2 f_1 = 0$, we have

$$(4) \qquad\qquad (JF)(z, w) = (D_2 f_2)(z, w) = g(z).$$

Therefore the radial limit of $D_2 f_2$ and of JF does not exist at e_1.
This dealt with (vi). As regards (iv),

$$(5) \qquad\qquad (1 - z)^{1/2}(D_1 f_2)(z, w) = \frac{iw}{(1 - z)^{1/2}} \cdot g(z).$$

This has no K-limit at e_1, although its restricted K-limit is 0. We see also that the boundedness assertion made about (iv) becomes false if the exponent $\frac{1}{2}$ is replaced by any smaller one.
   Second Example. Take $n = 2$, $m = 1$, put

$$(6) \qquad\qquad F(z, w) = z + \tfrac{1}{2}w^2 g(z).$$

Example 8.1.5 shows that F maps B_2 into U. The hypotheses of Theorem 8.5.6 hold again with $L = 1$. Since $F = f_1$,we now have

$$(7) \qquad\qquad \frac{1 - f_1(z, w)}{1 - z} = 1 - \frac{w^2}{2(1 - z)} \cdot g(z),$$

$$(8) \qquad\qquad (D_1 f_1)(z, w) = 1 + \frac{iw^2}{2(1 - z)} \cdot g(z),$$

$$(9) \qquad\qquad \frac{(D_2 f_1)(z, w)}{(1 - z)^{1/2}} = \frac{w}{(1 - z)^{1/2}} \cdot g(z).$$

Hence (i), (ii), *and* (v) *need have no K-limit at e_1, and the boundedness assertion made about* (v) *becomes false if $\frac{1}{2}$ is replaced by any larger exponent.*
   Third Example. This will show that the exponent $\frac{1}{2}$ is best possible in (iii).
   Take $n = 1$, $m = 2$. Pick $\varepsilon > 0$, put

$$(10) \qquad\qquad h(z) = \frac{1}{2\pi} \int_{-\pi}^{\pi} \frac{e^{i\theta} + z}{e^{i\theta} - z} |\theta|^{1 + \varepsilon} \, d\theta,$$

and note that h lies in the disc algebra, that $h(1) = 0$, and that $\operatorname{Re} h(z) > 0$ for all other $z \in \overline{U}$.

Put $c = 1/2\pi^{1+\varepsilon}$ and define $F = (f_1, f_2)$ by

(11) $$f_1(z) = ze^{-\mathrm{ch}(z)}, \qquad f_2(z) = c^{1/2}(1 - z)^{(1+\varepsilon)/2}z.$$

Let $u = \mathrm{Re}[\mathrm{ch}]$. Since $|1 - e^{i\theta}| \le |\theta|$ if $|\theta| \le \pi$, we have $|f_2|^2 \le u$ on the unit circle, hence $|f_2|^2 \le u$ in U, because $|f_2|^2$ is subharmonic. Therefore

(12) $$|f_1|^2 + |f_2|^2 \le u + e^{-2u} < 1$$

in U; the last inequality holds because $0 < u < \frac{1}{2}$ by our choice of c.

Thus F maps U into B_2 and $F(0) = 0$.

To show that F satisfies the other hypothesis of Theorem 8.5.6, it is enough to show that $f_1'(x)$ has a finite limit as $x \nearrow 1$, since then $(1 - |F(x)|^2)/(1 - x^2)$ is bounded. By (10),

(13) $$h'(x) = \frac{1}{\pi} \int_{-\pi}^{\pi} \frac{e^{i\theta}}{(e^{i\theta} - x)^2} |\theta|^{1+\varepsilon} \, d\theta.$$

Since $|e^{i\theta} - x| \ge \sin|\theta/2|$ if $0 < x < 1$, $|\theta| \le \pi$, the dominated convergence theorem leads from (13) to

(14) $$\lim_{x \to 1} h'(x) = \frac{-1}{2\pi} \int_{-\pi}^{\pi} \frac{|\theta|^{1+\varepsilon}}{1 - \cos\theta} \, d\theta$$

which is finite because $\varepsilon > 0$. By (14) and (11), $\lim f_1'(x)$ is also finite. (Ahern and Clark [1] have proved much more general theorems about derivatives of functions of the form (10).)

By (11), $f_2(x)/(1 - x)^{(1/2)+\varepsilon}$ is unbounded as $x \nearrow 1$. *The boundedness assertion concerning* (iii) *becomes therefore false if* $\frac{1}{2}$ *is replaced by any larger exponent.*

Finally, we note that the map F defined by (3) furnishes an example in which the function (iii) has no K-limit at e_1.

Chapter 9

Measures Related to the Ball Algebra

9.1. Introduction

This chapter deals with two types of topics. The material of Sections 9.2 and 9.3 is function-theoretic. The measures that are discussed there are intimately related to the holomorphic functions in B. On the other hand, Sections 9.4 and 9.5 describe some measure-theoretic aspects of the theory of function algebras in general. These would not become any simpler by specializing to the ball algebra. Both aspects are used in the proof of the Cole–Range theorem (Section 9.6), which is one of several modern generalizations of the classical theorem of F. and M. Riesz.

The results of this chapter will be used in Chapters 10 and 11.

Here are some of the relevant definitions.

9.1.1. Function Algebras. Let X be a compact Hausdorff space. A *function algebra on X* is a subalgebra A of $C(X)$ which is closed in the sup-norm topology, which contains the constants, and which separates points on X.

In the case of greatest interest to us, X will be S, and A will be $A(S)$, the restriction of the ball algebra $A(B)$ to the sphere S. Let us note that $A(S)$ and $A(B)$ are isometrically isomorphic Banach algebras. This is an immediate consequence of the maximum modulus theorem.

9.1.2. Representing Measures. If X is a compact Hausdorff space, the set of all regular complex Borel measures on X will be denoted by $M(X)$. With respect to the total variation norm, $M(X)$ is a Banach space which can be identified with $C(X)^*$, the dual space of $C(X)$. Hence $M(X)$ also has the corresponding weak*-topology.

Suppose A is a function algebra on X, and h is a multiplicative linear functional on A, $h \neq 0$. (I.e., h is a homomorphism of A onto $\mathbb{C}$.) Since $h(1) = 1$ and $\|h\| = 1$, there is at least one probability measure $\rho \in M(X)$ which *represents h*, in the sense that

$$(1) \qquad h(f) = \int_X f \, d\rho \qquad (f \in A).$$

185

The set M_h of all such ρ is clearly a convex subset of $M(X)$ which is also weak*-compact.

In the special case $A = A(S)$, we associate to every $z \in B$ the set M_z of all probability measures $\rho \in M(S)$ that "represent z" in the sense that

$$(2) \qquad\qquad f(z) = \int_S f \, d\rho$$

for every $f \in A(B) = A(S)$.

For example, if $P_z(\zeta) = P(z, \zeta)$ (the invariant Poisson kernel), then $P_z \sigma \in M_z$. In particular, $\sigma \in M_0$.

There is an interesting difference here between the cases $n = 1$ and $n > 1$. When $n = 1$, every point of the open unit disc U has a *unique* representing measure on the circle T. But when $n > 1$, every $z \in B$ has many representing measures on S. For example, for every $\zeta \in S$ we have

$$(3) \qquad\qquad f(0) = \frac{1}{2\pi} \int_{-\pi}^{\pi} f(e^{i\theta}\zeta) d\theta \qquad (f \in A(B)).$$

Thus there is a $\rho \in M_0$ that is concentrated on the circle $\{e^{i\theta}\zeta : -\pi \le \theta \le \pi\}$.

9.1.3. Totally Singular Measures. We say that a measure $\mu \in M(S)$ is *totally singular* if $\mu \perp \rho$ for every $\rho \in M_0$, and we say that a Borel set $E \subset S$ is *totally null* if $\rho(E) = 0$ for every $\rho \in M_0$.

It is important to realize that these concepts are not changed if M_0 is replaced by M_z, for any $z \in B$. The following observation shows this.

Suppose $z \in B$, $w \in B$, $c > 0$ is chosen so that $P_z > cP_w$ on S, and $\rho_w \in M_w$. Define

$$\rho_z = (P_z - cP_w)\sigma + c\rho_w.$$

Then $\rho_z \in M_z$ and $\rho_w \ll \rho_z$. We conclude:

(a)   If $\rho(E) = 0$ for every $\rho \in M_z$ then also $\rho(E) = 0$ for every $\rho \in M_w$.

(b)   If $\mu \perp \rho$ for every $\rho \in M_z$ then also $\mu \perp \rho$ for every $\rho \in M_w$.

9.1.4. Annihilating Measures. If $A \subset C(X)$, and if $v \in M(X)$ satisfies $\int f \, dv = 0$ for every $f \in A$, we write $v \perp A$ or $v \in A^\perp$. The members of $A^\perp$ are the *annihilating measures* for A.

As regards notation, the letter v will denote annihilating measures, although we used it earlier for Lebesgue measure on $\mathbb{C}^n$. However, the letter σ will continue to stand for the usual rotation-invariant probability measure on S.

9.1.5. Henkin Measures. Suppose $f_i \in A(B)$ for $i = 1, 2, 3, \ldots$, the sequence $\{f_i\}$ is uniformly bounded on $\bar{B}$, and $f_i(z) \to 0$ as $i \to \infty$, for every $z \in B$. (No convergence is assumed at points of S.) Under these circumstances, $\{f_i\}$ is said to be a *Montel sequence*, following Pelczynski [1]. By Montel's theorem on normal families, every Montel sequence $\{f_i\}$ converges uniformly to 0 on every compact subset of B, and the same is true of the derivatives $D^\alpha f_i$, for every multi-index α.

A measure $\mu \in M(S)$ is a *Henkin measure* if

$$\lim_{i \to \infty} \int_S f_i \, d\mu = 0$$

for every Montel sequence $\{f_i\}$. These measures were introduced in Henkin's paper [1]. They have also been called L-measures, A-measures, and analytic measures.

Examples of Henkin measures are

   (i)   every $v \in A(S)^\perp$,
  (ii)   every ρ that represents a point of B,
 (iii)   every $\mu \ll \sigma$.

Of these, the first two are quite obvious. To prove (iii), we have to show that $g\sigma$ is a Henkin measure for every $g \in L^1(\sigma)$. Since the class of all Henkin measures is a norm-closed subspace of $M(S)$, it suffices to prove (iii) for $\mu = g\sigma$, where g is a monomial, say $g(z) = z^\alpha \bar{z}^\beta$. If $\{f_i\}$ is a Montel sequence, then so is $\{z^\alpha f_i(z)\}$, hence $D^\beta(z^\alpha f_i) \to 0$ as $i \to \infty$. Since

$$\int_S f_i g \, d\sigma = \frac{(n-1)!}{(n-1+|\beta|)!} \, D^\beta(z^\alpha f_i)(0),$$

(iii) follows.

9.2. Valskii's Decomposition

We shall now prove that every Henkin measure is a sum of the types that we just looked at.

9.2.1. Theorem (Valskii [1]). *If μ is a Henkin measure than there exist $v \in A(S)^\perp$ and $g \in L^1(\sigma)$ such that $\mu = v + g\sigma$.*

Proof. Let us write A for $A(S)$, and A^* for the dual of A. Thus A^* is (iso-metrically isomorphic to) the quotient space $M(S)/A^\perp$. For $\mu \in M(S)$, let $\|\mu\|_{A^*}$ denote its norm as a linear functional on A, and let $\|\mu\|$ be its total variation, as usual.

We break the proof into two steps.

Step 1. If λ is a Henkin measure and $\varepsilon > 0$, then there exists $h \in L^1(\sigma)$ such that $\|h\|_1 \leq \|\lambda\|$ and $\|\lambda - h\sigma\|_{A*} < \varepsilon$.

To prove this, put $u = P[\lambda]$, the Poisson integral of λ, and let $u_r(\zeta) = u(r\zeta)$, as usual, where $\zeta \in S, 0 < r < 1$. We claim that

$$\text{(1)} \qquad\qquad \lim_{r \to 1} \|\lambda - u_r\sigma\|_{A*} = 0.$$

Since $\|u_r\|_1 \leq \|\lambda\|$, (1) implies that $h = u_r$ has the desired properties if r is close enough to 1.

Assume, to reach a contradiction, that (1) fails. Then there exist $\delta > 0$, $r_i \nearrow 1$, and $f_i \in A$ with $\|f_i\|_\infty \leq 1$, such that

$$\text{(2)} \qquad\qquad \left| \int_S f_i \, d\lambda - \int_S f_i u_{r_i} \, d\sigma \right| \geq \delta \qquad (i = 1, 2, 3 \ldots).$$

Since $u_r = P[\lambda]_r$, Fubini's theorem gives

$$\text{(3)} \qquad\qquad \int_S f u_r \, d\sigma = \int_S f_r \, d\lambda$$

for $f \in A, 0 < r < 1$. Thus (2) becomes

$$\text{(4)} \qquad\qquad \left| \int_S [f_i(\zeta) - f_i(r_i\zeta)] d\lambda(\zeta) \right| \geq \delta.$$

But if $g_i(z) = f_i(z) - f_i(r_i z)$ for $z \in \bar{B}$, then $\{g_i\}$ is a Montel sequence (since $\{f_i\}$ is equicontinuous on every compact subset of B), and the assumption that λ is a Henkin measure shows therefore that $\int_S g_i \, d\lambda \to 0$ as $i \to \infty$, contradicting (4).

Step 2. We now complete the proof of the theorem. Choose $\varepsilon_i > 0$, so that $\varepsilon_1 > \|\mu\|_{A*}$ and $\sum_1^\infty \varepsilon_i < \infty$. Put $\mu_1 = \mu$ and make the induction hypothesis that $k \geq 1$ and that μ_k is a Henkin measure with $\|u_k\|_{A*} < \varepsilon_k$. By the Hahn–Banach theorem, this means that $\|\mu_k - v_k\| < \varepsilon_k$ for some $v_k \in A^\perp$. Apply Step 1 to $\mu_k - v_k$: there is an $h_k \in L^1(\sigma)$, such that $\|h_k\|_1 < \varepsilon_k$ and

$$\text{(5)} \qquad\qquad \|\mu_k - v_k - h_k\sigma\|_{A*} < \varepsilon_{k+1}.$$

Define $\mu_{k+1} = \mu_k - v_k - h_k\sigma$, and proceed.

It follows, for $k = 1, 2, 3, \ldots$, that

$$\text{(6)} \qquad\qquad \mu = \mu_{k+1} + \sum_{i=1}^k v_i + \sum_{i=1}^k h_i\sigma.$$

Put $g = \sum_1^\infty h_i$. Then $g \in L^1(\sigma)$, $\|g\|_1 \leq \sum_1^\infty \varepsilon_i$, and

$$(7) \qquad \mu - g\sigma = \mu_{k+1} + \sum_{i=1}^k v_i - \sum_{i=k+1}^\infty h_i \sigma.$$

Since $v_i \in A^\perp$,

$$(8) \qquad \|\mu - g\sigma\|_{A*} \leq \|\mu_{k+1}\|_{A*} + \sum_{k+1}^\infty \|h_i\|_1.$$

The right side of (8) tends to 0 as $k \to \infty$. Thus $\mu - g\sigma \in A^\perp$, and the proof is complete.

9.2.2. Remarks. (a) if $\varepsilon > 0$, the ε_i can be so chosen in the preceding proof that $\sum_1^\infty \varepsilon_i < \|\mu\|_{A*} + \varepsilon$. The conclusion of the theorem can therefore be strengthened: $g \in L^1(\sigma)$ can be so chosen that $\|g\|_1 < \|\mu\|_{A*} + \varepsilon$ and $\mu - g\sigma \in A^\perp$.

(b) The Valskii decomposition is far from unique, since $A^\perp \cap L^1(\sigma) \neq \{0\}$. In fact, if $f \in H^1(B)$ and $f(0) = 0$, then $f * \sigma \in A^\perp \cap L^1(\sigma)$.

(c) When $n = 1$, the F. and M. Riesz theorem asserts that $A^\perp \subset L^1(\sigma)$. In that case, the Henkin measures are thus exactly those that are absolutely continuous with respect to σ, the Lebesgue measure on the unit circle.

9.3. Henkin's Theorem

At the end of §9.1.5 we saw that the relation $\mu \ll \sigma$ implies that μ is a Henkin measure. This can be significantly generalized:

9.3.1. Theorem (Henkin [1]). *If λ is a Henkin measure and $\mu \ll \lambda$, then μ is a Henkin measure.*

Recall that $\mu \ll \lambda$ is the same, by definition, as $\mu \ll |\lambda|$. By the Radon–Nikodym theorem, the hypothesis implies therefore that $\mu = \varphi\lambda$, for some $\varphi \in L^1(|\lambda|)$.

Proof. Since the set of Henkin measures is norm-closed in $M(S)$, it suffices to prove the theorem under the assumption $\mu = \varphi\lambda$, $\varphi \in C^1$. Valskii's decomposition shows therefore that it is enough to prove the following proposition:

If $\mu \in A^\perp$ and $\varphi \in C^1(\bar{B})$, then

$$(1) \qquad \lim_{i \to \infty} \int_S f_i \varphi \, d\mu = 0$$

for every Montel sequence $\{f_i\}$.

To do this, we use the Toeplitz operator T_φ and the related operator V_φ, defined as in Theorem 6.5.4 by

(2) $$(T_\varphi f)(z) = \int_S f(\zeta)\varphi(\zeta)C(z, \zeta)d\sigma(\zeta)$$

and

(3) $$(V_\varphi f)(z) = \int_S f(\zeta)[\varphi(z) - \varphi(\zeta)]C(z, \zeta)d\sigma(\zeta).$$

By Theorem 6.5.4, V_φ maps the unit ball of $H^\infty(S)$ into a uniformly bounded equicontinuous subset of $C(\bar{B})$. Since (2) and (3) imply that

(4) $$f_i(z)\varphi(z) = (T_\varphi f_i)(z) + (V_\varphi f_i)(z) \qquad (z \in B)$$

it follows that $T_\varphi f_i$ extends continuously to $\bar{B}$. Thus $T_\varphi f_i \in A(B)$. Since $\mu \in A^\perp$, (4) gives

(5) $$\int_S f_i \varphi \, d\mu = \int_S (V_\varphi f_i)d\mu.$$

Return to (3) and appeal once again to the fact that every measure absolutely continuous with respect to σ is a Henkin measure, to conclude that $(V_\varphi f_i)(z) \to 0$ as $i \to \infty$, for every $z \in B$. But $\{V_\varphi f_i\}$ is uniformly bounded and equicontinuous on $\bar{B}$. Hence $V_\varphi f_i \to 0$ uniformly on $\bar{B}$. The integrals on the right of (5) converge therefore to 0 as $i \to \infty$. This proves (1), and completes the theorem.

Here is an application:

9.3.2. Theorem. *Suppose* $f \in H^\infty(B)$, $|f(z)| < 1$ *if* $z \in B$, *and* E *is the set of all* $\zeta \in S$ *for which* $\lim_{r \to 1} f(r\zeta) = 1$. *Then* E *is totally null.*

Note that $1 - f \not\equiv 0$, so that $\int_S \log|1 - f^*|d\sigma > -\infty$. (Theorem 5.6.4.) Hence $\sigma(E) = 0$. The theorem asserts much more, namely that $\rho(E) = 0$ for every $\rho \in M_0$. (See §9.1.3.)

Proof. Put $g = \frac{1}{2}(1 + f)$. Then $|g^*| = 1$ if and only if $g^* = 1$, which happens if and only if $f^* = 1$.

Pick a representing measure $\rho \in M_0$. Let $\lambda = \rho|_E$, the restriction of ρ to E; i.e., $\lambda(X) = \rho(E \cap X)$. Then $\lambda \ll \rho$, hence λ is a Henkin measure, by Theorem 9.3.1, and therefore λ has a Valskii decomposition

(1) $$\lambda = v + h\sigma \qquad (v \in A^\perp, h \in L^1(\sigma)).$$

If $r < 1$ then $g_r \in A$, so that

(2) $$\int_S g_r^k \, d\lambda = \int_S g_r^k h \, d\sigma \qquad (k = 1, 2, 3, \ldots).$$

The integral on the left of (2) extends only over E, and $g^*(\zeta) = 1$ for $\zeta \in E$. Letting $r \nearrow 1$, it follows that

(3) $$\rho(E) = \lambda(E) = \int_S (g^*)^k h \, d\sigma \qquad (k = 1, 2, 3, \ldots).$$

Since $|g^*| < 1$ a.e. $[\sigma]$, the last integral converges to 0 as $k \to \infty$. Thus $\rho(E) = 0$.

Corollary. *Suppose $f \in H(B)$, Re $f > 0$, and E is the set of all $\zeta \in S$ at which Re $f(r\zeta) \to +\infty$ as $r \nearrow 1$. Then E is totally null.*

Proof. Apply the theorem to $f/(1 + f)$.

9.4. A General Lebesgue Decomposition

9.4.1. In the present Section, X is a compact Hausdorff space, $M(X) = C(X)^*$ is the space of all regular Borel measures on X, and K is a nonempty convex weak*-compact subset of $M(X)$, consisting of probability measures. The object is to obtain a "Lebesgue decomposition" of an arbitrary $\mu \in M(X)$ relative to K.

When K is a singleton, say $K = \{\rho\}$, then every $\mu \in M(X)$ decomposes into $\mu = \mu_a + \mu_s$, where $\mu_a \ll \rho$, and μ_s is concentrated on a set E with $\rho(E) = 0$. This is the ordinary Lebesgue decomposition of μ relative to ρ. There are at least two ways in which one can try to extend this to larger sets K. We shall describe these before proving anything.

Glicksberg [1] used the following approach: Among all Borel sets $E \subset X$ that are *K-null* (this means, by definition, that $\rho(E) = 0$ for every $\rho \in K$) find one, say H, that maximizes $|\mu|(E)$, put $\mu_s = \mu|_H$ and $\mu_a = \mu - \mu_s$. Then μ_s is concentrated on a set that is K-null, and $|\mu_a|(E) = 0$ for every E that is K-null. But it is not at all clear whether μ_a is absolutely continuous with respect to any $\rho \in K$.

König and Seever [1] attacked the problem from the opposite direction: For every $\rho \in K$, let $\mu = \mu_\rho + \mu_\rho'$ be the Lebesgue decomposition of μ relative to ρ, with $\mu_\rho \ll \rho$, $\mu_\rho' \perp \rho$. Find $\rho_0 \in K$ for which $\|\mu_\rho\|$ is maximal. Let μ_a be this μ_{ρ_0} and put $\mu_s = \mu - \mu_a$. Then $\mu_a \ll \mu_{\rho_0}$ and $\mu_s \perp \rho$ for every $\rho \in K$, but it is not clear whether μ_s is *K-singular* in the strong sense of being concentrated on a set that is K-null.

Rainwater [1] proved that this is in fact the case, thereby showing that the two decompositions are actually the same. Theorem 9.4.4 summarizes this line of development.

Rainwater's lemma 9.4.3 depends on Glickberg's version of von Neumann's minimax theorem. The main point of the simple proof given by Glicksberg [1] is that it requires only one of the two convex sets to be compact. This is Theorem 9.4.2.

When F is a function of two variables, we define F_x and F^y in the customary way by

$$F_x(y) = F(x, y) = F^y(x).$$

The vector spaces that occur in the minimax theorem are understood to have real scalars.

9.4.2. The Minimax Theorem. *Suppose*

   (i)   *G is a convex subset of some vector space,*
   (ii) *K is a compact convex subset of some topological vector space, and*
  (iii) *$F: G \times K \to R$ satisfies*

        (a) *F^y is convex on G for every $y \in K$,*
        (b) *F_x is concave and continuous on K for every $x \in G$.*

Then

$$\sup_{y \in K} \inf_{x \in G} F(x, y) = \inf_{x \in G} \sup_{y \in K} F(x, y).$$

Proof. Let $\alpha = \sup \inf$, $\beta = \inf \sup$. Since

$$\inf_x f(x, y_0) \le f(x_0, y_0) \le \sup_y f(x_0, y)$$

for all $(x_0, y_0) \in G \times K$, we have $\alpha \le \beta$. If $\beta = -\infty$ there is nothing left to prove. Since K is compact, $\beta < \infty$. So we have to consider the case $-\infty < \beta < \infty$. Replacement of F by $F - \beta$ shows that the assumption $\beta = 0$ involves no loss of generality.

Let H be the convex hull of the set $\{F_x : x \in G\} \subset C_R(K)$, the space of all real-valued continuous functions on K. We claim that $\sup\{h(y): y \in K\} \ge 0$ for every $h \in H$.

Let $h = \sum_1^N t_i F_{x_i}$, where $x_i \in G$, $t_i \ge 0$, $\sum t_i = 1$. Put $x' = \sum t_i x_i$. Then $x' \in G$ and $F(x', y') \ge \beta = 0$ for some $y' \in K$. The convexity of $F^{y'}$ shows therefore that

$$0 \le F(x', y') \le \sum t_i F(x_i, y') = h(y').$$

We conclude from this that the convex set H is disjoint from the (open) negative cone in $C_R(K)$. By the Hahn–Banach separation theorem and the

Riesz representation theorem there is a probability measure μ on K such that $\int_K h \, d\mu \geq 0$ for every $h \in H$.

Put $y_0 = \int_K y \, d\mu(y)$. Then $y_0 \in K$, since K is compact and convex. (The simple facts about vector-valued integrals that are needed here may be found in Theorem 3.28 of Rudin [2]; the point y_0 is often called the *barycenter* of μ.) For every $x \in G$, F_x is concave and $F_x \in H$. It follows that

$$F(x, y_0) \geq \int_K F(x, y) d\mu(y) \geq 0.$$

Hence $\alpha \geq 0 = \beta$, and the proof is complete.

Note: The function $F(x, y) = y/(x + y)$ on $(0, \infty) \times (0, \infty)$ shows that the compactness of K cannot be omitted from the hypotheses.

9.4.3. Rainwater's Lemma. *Let X and K be as in §9.4.1. Suppose $v \in M(X)$, and $v \perp \rho$ for every $\rho \in K$. Then v is concentrated on a set $E \subset X$, of type F_σ, such that $\rho(E) = 0$ for every $\rho \in K$.*

(Recall that a set is of type F_σ if it is a union of countably many closed sets.)

Proof. Replacing v by $|v|$, we see that we may assume $v \geq 0$, without loss of generality. Let G be the set of all continuous functions taking X into $[0, 1]$. Define F on $G \times K$ by

$$(1) \qquad\qquad F(h, \rho) = \int_X h \, dv + \int_X (1 - h) d\rho.$$

By the definition of the weak*-topology in $M(X)$, F is a continuous function of ρ, for every fixed h. Since F is affine in each variable, F satisfies the hypotheses of the minimax theorem.

Now fix $\rho \in K$ and recall that $v \perp \rho$. This implies that there are disjoint compact sets in X, one of which carries most of the mass of ρ, whereas the other carries most of the mass of v. Urysohn's lemma provides therefore an $h \in G$ for which both integrals in (1) are very small. In other words,

$$(2) \qquad\qquad \inf_{h \in G} F(h, \rho) = 0$$

for every $\rho \in K$, so that

$$(3) \qquad\qquad \inf_{h \in G} \sup_{\rho \in G} F(h, \rho) = 0$$

by the minimax theorem. By (3), there are functions $h_i \in G$ ($i = 1, 2, 3, \ldots$) such that

(4)
$$\int_X h_i \, dv + \int_X (1 - h_i) d\rho < 2^{-i}$$

for *every* $\rho \in K$.

Put $g_0 = \sum h_i$, $g_1 = \sum(1 - h_i)$, and let E be the set where $g_0 < \infty$. Since g_0 is lower semi-continuous, E is of type F_σ. By (4) $g_0 \in L^1(v)$, so that v is concentrated on E. But (4) shows also that $g_1 \in L^1(\rho)$ for every $\rho \in K$; since $g_1 = \infty$ on E, $\rho(E) = 0$. This proves the lemma.

The initials that name the following Theorem refer to Glicksberg, König, and Seever; see §9.4.1.

9.4.4. The GKS Decomposition Theorem. *If K is a weak*-compact convex non-empty set of regular Borel probability measures on a compact Hausdorff space X, then every $\mu \in M(X)$ has a unique decomposition*

$$\mu = \mu_a + \mu_s$$

in which $\mu_a \ll \rho_0$ for some $\rho_0 \in K$, and μ_s is concentrated on a set of type F_σ that is K-null.

Proof. The uniqueness of such a decomposition is trivial. To prove existence, we begin by associating to every $\rho \in K$ the Lebesgue decomposition of μ with respect to ρ. Thus $\mu = \mu_\rho + \mu'_\rho$, where $\mu_\rho \ll \rho$, $\mu'_\rho \perp \rho$. Put

(1)
$$t = \sup\{\|\mu_\rho\| : \rho \in K\}.$$

Choose $t_1 < t_2 < \cdots$, so that $t_i \to t$ as $i \to \infty$. There exist $\rho_i \in K$ ($i = 1, 2, 3, \ldots$) with $\|\mu_{\rho_i}\| > t_i$, and there are Borel sets $A_i \subset X$ such that $\mu_{\rho_i} = \mu|_{A_i}$. Put $A = \bigcup_i A_i$, and define $\mu_a = \mu|_A$.

Setting $\rho_0 = \sum_1^\infty 2^{-i} \rho_i$, the convexity and compactness of K imply that $\rho_0 \in K$. Suppose $E \subset X$ satisfies $\rho_0(E) = 0$. Then $\rho_i(E) = 0$ for all i, hence also $|\mu_{\rho_i}|(E) = 0$, or $|\mu|(E \cap A_i) = 0$. Taking the union over all i, it follows that $|\mu|(E \cap A) = 0$, or $|\mu_a|(E) = 0$. This proves that $\mu_a \ll \rho_0$.

Note also that $A \supset A_i$, so that $\|\mu_a\| \geq \|\mu_{\rho_i}\| > t_i$ for all i. Consequently, $\|\mu_a\| = t$.

Finally, put $\mu_s = \mu - \mu_a = \mu|_{X \backslash A}$. Pick $\rho \in K$. Then

(2)
$$\mu_a + (\mu_s)_\rho \ll \tfrac{1}{2}(\rho_0 + \rho) \in K,$$

so that $\|\mu_a + (\mu_s)_\rho\| \leq t$. Since μ_a and $(\mu_s)_\rho$ are concentrated on A and $X \backslash A$, respectively, the norm of their sum is the sum of their norms, and since

$\|\mu_a\| = t$, we conclude that $\|(\mu_s)_\rho\| = 0$. Thus $\mu_s \perp \rho$, for every $\rho \in K$, and an appeal to Rainwater's lemma shows that μ_s is concentrated on a set of type F_σ that is K-null.

9.5. A General F. and M. Riesz Theorem

9.5.1. In this Section, X is a compact Hausdorff space, as before, A is a function algebra on X, Φ is a multiplicative linear functional on A, and M_Φ is the set of all probability measures $\rho \in M(X)$ that represent Φ. As was pointed out in §9.1.2, M_Φ is nonempty, convex, and weak*-compact. Hence M_Φ can play the role of K in Theorem 9.4.4.

The main result (Theorem 9.5.6) states that if $v \in A^\perp$ and if $v = v_a + v_s$ is its GKS decomposition relative to M_Φ, then both v_a and v_s lie in $A^\perp$. (In the original version of the F. and M. Riesz theorem, A was the disc algebra on the unit circle, and the conclusion was that $v_s = 0$, i.e. that $v \ll \sigma$.)

9.5.2. Definition. If $f \in A$ and $u = \operatorname{Re} f$, we define $\Phi u = \operatorname{Re} \Phi f$. Then

$$(1) \qquad\qquad \Phi u = \int_X u \, d\rho$$

for every $\rho \in M_\Phi$, simply because every such ρ is a real measure.

The set of all real parts of members of A will be denoted by $\operatorname{Re} A$. Clearly, $\operatorname{Re} A$ is a subspace of $C_R(X)$.

Observe that if $\rho \in M(X)$ is a probability measure that satisfies (1) for every $u \in \operatorname{Re} A$, then $\rho \in M_\Phi$. For if $f \in A$, then (1) holds for the real parts of f and of if.

9.5.3. If $g \in C_R(X)$, $u \in \operatorname{Re} A$, $u \geq g$, and $\rho \in M_\Phi$, then obviously

$$(1) \qquad\qquad \int_X d\rho \leq \Phi u.$$

Hence the supremum of the left side, over all ρ, is at most equal to the infimum of the right side, over all $u \in \operatorname{Re} A$ such that $u \geq g$. The following lemma asserts that equality actually holds.

9.5.4. Lemma. *If* $g \in C_R(X)$ *then there exists* $\rho_0 \in M_\Phi$ *such that*

$$(1) \qquad\qquad \int_X g \, d\rho_0 = \inf\{\Phi u : u \geq g, u \in \operatorname{Re} A\}.$$

Proof. For $h \in C_R(X)$, define

(2) $$p(h) = \inf\{\Phi u : u \geq h, u \in \text{Re } A\}.$$

Then $p(h_1 + h_2) \leq p(h_1) + p(h_2)$, and $p(th) = tp(h)$ for scalars $t \geq 0$. In particular, $0 = p(0) \leq p(h) + p(-h)$, hence $-p(h) \leq p(-h)$, so that $tp(h) \leq p(th)$ for all real scalars t and all $h \in C_R(X)$. If we define $\Lambda(tg) = tp(g)$, it follows that Λ is a linear functional on the one-dimensional space spanned by g, and that $\Lambda \leq p$ there. One of the standard versions of the Hahn–Banach theorem (see, for example, Theorem 3.2 in Rudin [2]) asserts now that Λ extends to a linear functional on $C_R(X)$ that satisfies

(3) $$-p(-h) \leq \Lambda h \leq p(h)$$

for every $h \in C_R(X)$.

Since $|p(h)| \leq \|h\|_\infty$, Λ is continuous, $\|\Lambda\| \leq 1$.

When $h \in \text{Re } A$, then (2) gives $p(h) = \Phi h$, so that (3) gives $\Lambda h = \Phi h$. Thus Λ is a norm-preserving linear extension of Φ.

Consequently, there is a $\rho_0 \in M_\Phi$ such that

(4) $$\Lambda h = \int_X h \, d\rho_0 \qquad (h \in C_R(X)).$$

Since $\Lambda g = p(g)$, the proof is complete.

9.5.5. Lemma (Forelli [1]). *Suppose $E \subset X$ is a set of type F_σ, and $\rho(E) = 0$ for every $\rho \in M_\Phi$. Then there is a sequence $\{f_m\}$ in A, with $\|f_m\| \leq 1$ for all m, such that*

(i) $\lim_{m \to \infty} f_m(x) = 0$ *for every $x \in E$, but*

(ii) $\lim_{m \to \infty} f_m(x) = 1$ *a.e. $[\rho]$ for every $\rho \in M_\Phi$.*

Proof. By assumption, $E = \bigcup E_m$, where each E_m is compact, and $E_m \subset E_{m+1}$. Fix m, for the moment. By the minimax theorem,

(1) $$\inf_h \sup_\rho \int_X h \, d\rho = \sup_\rho \inf_h \int_X h \, d\rho,$$

where ρ ranges over M_Φ and h ranges over all $h \in C_R(X)$ that satisfy $h \geq m$ on E_m, $h \geq 0$ on X. Since $\rho(E_m) = 0$ for every ρ, the right side of (1) is 0. Hence (1) shows that for some $h = h_m$ in our class,

(2) $$\int_X h_m \, d\rho < \frac{1}{m^2}.$$

for every $\rho \in M_\Phi$. Lemma 9.5.4 (with h_m in place of g) implies therefore that *there exists* $u_m \in \mathrm{Re}\, A$ *such that* $u_m \geq m$ *on* E_m, $u_m \geq 0$ *on* X, *and* $\Phi u_m \leq 1/m^2$.

Now choose $g_m \in A$ so that $u_m = \mathrm{Re}\, g_m$ and Φg_m is real, and put $f_m = \exp(-g_m)$.

Then $|f_m| = \exp(-u_m) \leq 1$ on X.

On E_m, $|f_m| \leq e^{-m}$, which proves (i).

If $\rho \in M_\Phi$ then

$$(3) \qquad \int_X f_m \, d\rho = \Phi f_m = \exp(-\Phi g_m) = \exp(-\Phi u_m),$$

and since $\Phi u_m \leq 1/m^2$,

$$(4) \qquad \int_X (1 - f_m)d\rho \leq 1 - \exp\left(-\frac{1}{m^2}\right) < \frac{1}{m^2}.$$

Consequently, $\sum_m \mathrm{Re}(1 - f_m) < \infty$ a.e. $[\rho]$. Thus $\mathrm{Re}\, f_m \to 1$ a.e. $[\rho]$, and since $|f_m| \leq 1$, it follows that $f_m \to 1$ a.e. $[\rho]$. This proves (ii).

We are now ready for the Glicksberg–König–Seever generalization of the F. and M. Riesz theorem. (Part 3 of Glicksberg [2], Chapter II.7 of Gamelin [1], and Section 23 of Stout [1] contain more information on this topic.)

9.5.6. Theorem. *Let* Φ *be a multiplicative linear functional on a function algebra* A *on* X, *let* M_Φ *be the set of representing measures for* Φ, *let* $v \in A^\perp$, *and let*

$$v = v_a + v_s$$

be the GKS-decomposition of v *relative to* M_Φ.

Then $v_a \in A^\perp$ *and* $v_s \in A^\perp$.

Proof. By Theorem 9.4.4, v_s is concentrated on a set $E \subset X$, of type F_σ, such that $\rho(E) = 0$ for every $\rho \in M_\Phi$, and v_a is the restriction of v to $X \backslash E$. Associate $\{f_m\}$ to E, as in Lemma 9.5.5. For any $f \in A$, we have $ff_m \in A$, hence $\int_X ff_m \, dv = 0$, or

$$\int_X ff_m dv_s + \int_X ff_m \, dv_a = 0.$$

The first of these integrals tends to 0 as $m \to \infty$, since $f_m \to 0$ at every point of E. By Theorem 9.4.4, $v_a \ll \rho_0$ for some $\rho_0 \in M_\Phi$. Since $f_m \to 1$ a.e. $[\rho_0]$, the second integral converges to $\int_X f \, dv_a$ as $m \to \infty$. Thus $\int f \, dv_a = 0$, $v_a \in A^\perp$, and the same is then true of $v_s = v - v_a$.

9.6. The Cole–Range Theorem

We return now to the ball algebra $A = A(B)$ which, as noted in §9.1.1, can be regarded as a function algebra on S. As in §9.1.2, we let M_0 denote the set of all probability measures $\rho \in M(S)$ that represent the evaluation functional at the origin of B, in the sense that

$$f(0) = \int_S f \, d\rho$$

for every $f \in A(B)$.

9.6.1. Theorem (Cole–Range [1]). *Every Henkin measure $\mu \in M(S)$ is absolutely continuous with respect to some $\rho \in M_0$.*

Proof. Let $\mu = \mu_a + \mu_s$ be the GKS-decomposition of μ with respect to M_0 (Theorem 9.4.4): μ_s is concentrated on a Borel set that is totally null (see §9.1.3) and $\mu_a \ll \rho$ for some $\rho \in M_0$. We shall prove that $\mu_s = 0$.

Let h be any bounded Borel function on S. Since $h\mu_s \ll \mu$, Theorem 9.3.1 asserts that $h\mu_s$ is a Henkin measure, so that $h\mu_s$ has a Valskii decomposition

$$h\mu_s = v + g\sigma$$

where $v \in A^\perp$ and $g \in L^1(\sigma)$. Since $h\mu_s$ is concentrated on a set that is totally null, the uniqueness part of Theorem 9.4.4 shows that

$$v = -g\sigma + h\mu_s$$

is the GKS decomposition of v with respect to M_0. Thus $h\mu_s = v_s \in A^\perp$, by Theorem 9.5.6. In particular, $\int h \, d\mu_s = 0$. The arbitrariness of h implies now that $\mu_s = 0$, and the proof is complete.

9.6.2. Remark. It is a corollary of Theorem 9.6.1 that every $\mu \in A^\perp$ satisfies $\mu \ll \rho$ for some $\rho \in M_0$.

This makes it clear that the Cole–Range theorem contains the original F. and M. Riesz theorem, since in the original setting M_0 had only one member, namely Lebesgue measure on the unit circle.

9.7. Pluriharmonic Majorants

9.7.1. In this section we return to Lumer's Hardy spaces $(LH)^p(B)$ that were discussed in Section 7.4. The reason is that Lemma 9.5.4 makes it possible to express the norms defined in §7.4.3 in terms of representing measures (Theorem 9.7.4), and this leads to further information about these norms.

As regards notation, recall that the real pluriharmonic functions in B are exactly the real parts of holomorphic functions (Theorem 4.4.9). We therefore denote them by $RP(B)$, and we define

(1) $$\alpha(g) = \inf\{u(0): u \geq g, u \in RP(B)\}$$

if g is a real function with domain B, and

(2) $$\beta(g) = \inf(u(0): u \geq g, u \in \text{Re } A(B)\}$$

if g is bounded above on $\bar{B}$.

In terms of this notation, the norm $\|\|f\|\|_p$ of a function $f \in (LH)^p(B)$, as defined in §7.4.3, satisfies

(3) $$\|\|f\|\|_p^p = \alpha(|f|^p).$$

Here $0 < p < \infty$.

When $f \in A(B)$, then α can be replaced by β in (3):

9.7.2. Lemma. *If $f \in A(B)$ then $\|\|f\|\|_p^p = \beta(|f|^p)$.*

Proof. Pick $\varepsilon > 0$. Since $|f|^p$ is uniformly continuous on $\bar{B}$, there exists $r < 1$ such that

$$||f|^p - |f_r|^p| < \varepsilon$$

on $\bar{B}$. The definition of $\|\|f\|\|_p$ shows that there is a $w \in RP(B)$, $w \geq |f|^p$, such that

$$w(0) \leq \|\|f\|\|_p^p + \varepsilon.$$

Put $u = \varepsilon + w_r$. Then $u \in \text{Re } A(B)$,

$$u \geq \varepsilon + |f_r|^p \geq |f|^p$$

on $\bar{B}$, and $u(0) = \varepsilon + w(0) \leq \|\|f\|\|_p^p + 2\varepsilon$. Hence

$$\beta(|f|^p) \leq \|\|f\|\|_p^p.$$

Since $\alpha \leq \beta$ is trivial, the lemma follows from 9.7.1(3).

9.7.3. Lemma. *If $f \in (LH)^p(B)$ then*

(1) $$\lim_{r \to 1} \|\|f_r\|\|_p = \|\|f\|\|_p = \sup_{r < 1} \|\|f_r\|\|_p.$$

Note that the first of these equalities holds although $\|\|f - f_r\|\|_p$ need not tend to 0 as $r \to 1$ (Theorem 7.4.6). The second equality in (1) implies that $\|\|f_r\|\|_p$ is a nondecreasing function of r.

Proof. If $u \geq |f|^p$ then $u_r \geq |f_r|^p$, hence

$$(2) \qquad \|\|f_r\|\|_p^p \leq u_r(0) = u(0).$$

Taking the infimum over all such $u \in RP(B)$ gives

$$(3) \qquad \|\|f_r\|\|_p \leq \|\|f\|\|_p \qquad (0 < r < 1).$$

Next, associate to each $r < 1$ a function $u^{(r)} \in RP(B)$ such that

$$(4) \qquad u^{(r)} \geq |f_r|^p \quad \text{and} \quad u^{(r)}(0) \leq \|\|f_r\|\|_p^p + \varepsilon_r,$$

where $\varepsilon_r \searrow 0$ as $r \nearrow 1$. Since $u^{(r)} \geq 0$ and $\{u^{(r)}(0)\}$ is bounded, $\{u^{(r)}\}$ is equicontinuous on every compact subset of B. Some subsequence of $\{u^{(r)}\}$ converges therefore, uniformly on compact subsets of B, to a function $u \in RP(B)$ which satisfies $u \geq |f|^p$ and

$$(5) \qquad \|\|f\|\|_p \leq u(0)^{1/p} \leq \liminf_{r \to 1} \|\|f_r\|\|_p,$$

by (4). Now (1) follows from (3) and (5).

9.7.4. Theorem. *If $0 < p < \infty$ and $f \in (LH)^p(B)$, then*

$$(1) \qquad \|\|f\|\|_p^p = \sup\left\{\int_S |f_r|^p \, d\rho : 0 < r < 1, \rho \in M_0\right\}.$$

Here M_0 is the set of all representing measure corresponding to the origin of B, as in Section 9.6.

The fact that $\|\|f\|\|_p$ is the supremum of a large family of L^p-norms may account for some of the pathology of the spaces $(LH)^p(B)$.

Proof. For $g \in C_R(S)$, a special case of Lemma 9.5.4 asserts that

$$(2) \qquad \sup\left\{\int_S g \, d\rho : \rho \in M_0\right\} = \inf\{u(0) : u \geq g \text{ on } S, u \in \text{Re } A(B)\}.$$

If $r < 1$ then $f_r \in A(B)$, so that

$$(3) \qquad \|\|f_r\|\|_p^p = \beta(|f_r|^p)$$

by Lemma 9.7.2. If $u \in \operatorname{Re} A(B)$ satisfies $u \geq |f_r|^p$ on S, then the same inequality holds in B, since $|f_r|^p$ is subharmonic. Hence it follows from (2), (3), and the definition of β in 9.7.1(2) that

(4)
$$\||f_r|\|_p^p = \sup\left\{\int_S |f_r|^p \, d\rho : \rho \in M_0\right\}.$$

If we now take the supremum over r in (4), the theorem follows from Lemma 9.7.3.

9.7.5. The "obvious" members of M_0 are the *circular* probability measures $\rho \in M(S)$. By definition, these satisfy

(1)
$$\int_S v(e^{i\theta}\zeta)d\rho(\zeta) = \int_S v \, d\rho$$

for every $v \in C(S)$ and for every real θ. By Fubini's theorem, (1) implies that

(2)
$$\int_S v \, d\rho = \int_S d\rho(\zeta) \cdot \frac{1}{2\pi} \int_{-\pi}^{\pi} v(e^{i\theta}\zeta)d\theta.$$

With $f \in A(B)$ in place of v, the fact that the slice functions f_ζ are in $A(U)$ shows therefore that $\int f \, d\rho = f(0)$. Thus $\rho \in M_0$, as asserted.

If $f \in H(B)$ and v is replaced by $|f_r|^p$ in (2), consideration of the slice functions shows that

(3)
$$\int_S |f_r|^p \, d\rho$$

is a *nondecreasing* function of r.

But if $\rho \in M_0$ is not circular, then (3) may fail to be a monotonic function of r, for certain f, in spite of the monotonicity of $\||f_r|\|_p$ (Lemma 9.7.3) and Theorem 9.7.4. Theorem 9.7.6 will show this.

To construct some noncircular $\rho \in M_0$, take $n = 2$, for simplicity, and let τ be *any* probability measure on $\bar{U} \subset \mathbb{C}$ that satisfies

(4)
$$\int_{\bar{U}} g \, d\tau = g(0)$$

for every $g \in A(U)$. For example, τ might be concentrated on a simple closed curve Γ in U that surrounds the origin, in such a way that τ solves the Dirichlet problem at 0 relative to the domain bounded by Γ. The measure ρ that satisfies

(5)
$$\int_S v \, d\rho = \int_{\bar{U}} d\tau(z) \cdot \frac{1}{2\pi} \int_{-\pi}^{\pi} v(z, e^{i\theta}\sqrt{1 - |z|^2})d\theta$$

for every $v \in C(S)$ belongs then to M_0. To see this, simply note that the inner integral on the right side of (5), with v replaced by $f \in A(B)$, equals $f(z, 0)$.

The support of this ρ is the set of all $(z, w) \in S$ for which z lies in the support of τ.

9.7.6. Theorem. *Put* $f(z, w) = (1 - z)w$ *in* $\mathbb{C}^2$. *Then there exists* $\rho \in M_0$ *such that* $\int_S |f_r|^p \, d\rho$ *is not a monotonic function of* r *in* $[0, 1]$, *for any* $p \in (0, \infty)$.

Proof. Let I be the interval $[\frac{3}{4}, \frac{7}{8}]$ on the real axis in $\mathbb{C}$, let $\Omega = U \backslash I$, and let τ be the probability measure on $\partial\Omega = I \cup T$ that satisfies

$$(1) \qquad\qquad \int_{\partial\Omega} h \, d\tau = h(0)$$

for every $h \in C(\overline{\Omega})$ that is harmonic in Ω. Use this τ to define ρ as in 9.7.5(5). If we replace v by $|f_r|^p$ in that formula, the inner integral is

$$(2) \qquad\qquad |r - r^2 z|^p (1 - |z|^2)^{p/2}.$$

This vanishes on T, where $|z| = 1$. Hence

$$(3) \qquad\qquad \int_S |f_r|^p \, d\rho = \int_I (r - r^2 x)^p (1 - x^2)^{p/2} \, d\tau(x).$$

The integral on the right is 0 when $r = 0$. It is positive for every $r \in (0, 1]$. It decreases as r increases from $\frac{2}{3}$ to 1, since the integrand is then a decreasing function of r, for every $x \in I$.

9.8. The Dual Space of $A(B)$

9.8.1. The dual X^* of a given Banach space X can often be described in several ways. For instance, $A(B)$ is isometrically isomorphic to $A(S)$, a closed subspace of $C(S)$. Since $C(S)^* = M(S)$, the space of complex Borel measures on S, standard duality theory gives one description of $A(B)^*$ as a quotient space, namely

$$(1) \qquad\qquad A(B)^* = M(S)/A^\perp.$$

(In this context, equality is understood to mean isometric isomorphism.)

Here is another description:

Let HM be the space of all Henkin measures on S, and let TS be the space of all totally singular ones. If we combine the theorems of Henkin and Cole–Range, we see that HM consists of precisely those measures that are absolutely continuous with respect to some representing measure $\rho \in M_0$. The GKS decomposition theorem (with respect to M_0) says therefore that there is a direct sum decomposition

$$(2) \qquad M(S) = HM \oplus TS.$$

Observe next that $A^\perp \subset HM$. Thus (1) can be replaced by

$$(3) \qquad A(B)^* = (HM/A^\perp) \oplus TS.$$

By Valskii's theorem, $HM = L^1(\sigma) + A^\perp$. (Note that this sum is not direct, since $L^1(\sigma) \cap A^\perp$ is far from being $\{0\}$.) The first summand on the right of (3) is thus the image of $L^1(\sigma)$ under the quotient map with null space $A^\perp$. In particular, $HM/A^\perp$ is separable.

If $\mu \in TS$ and $\lambda \ll \mu$ then (trivially) $\lambda \in TS$. Thus TS is what Kakutani [1] has called an (L)-space. (Further bibliographic information on this topic may be found in Dunford–Schwartz [1], pp. 394–395.)

Let us summarize these observations:

9.8.2. Theorem. $A(B)^*$ *is the direct sum of an* (L)*-space and a separable Banach space.*

Pelczynski [1; Theorem 11.5] used this information about $A(B)^*$ to prove the following: *If* $k \geq 2$ *and* $A(U^k)$ *is the polydisc algebra in k variables, then* $A(U^k)^*$ *is not isomorphic to any closed subspace of* $A(B_n)^*$.

Here "isomorphic" means: linearly homeomorphic. The proof uses more Banach space theory than can be presented here. We therefore omit it. But let us state the dual formulation of the result:

9.8.3. Theorem. *If* $n \geq 1$ *and* $k \geq 2$, *then* $A(U^k)$ *is not isomorphic to any quotient space of* $A(B_n)$.

In other words, there is no continuous linear mapping of any $A(B_n)$ onto $A(U^k)$ if $k \geq 2$.

Henkin [1] proved earlier that $A(B_n)$ and $A(U^k)$ are not isomorphic as Banach spaces. In fact, it was apparently for this purpose that he introduced his class of measures.

Chapter 10

Interpolation Sets for the Ball Algebra

10.1. Some Equivalences

10.1.1. Definitions. We shall be concerned with compact sets $K \subset S$. For convenience, we introduce six labels, (Z), (P), (I), $PI)$, (N), and (TN), to denote certain properties that K may or may not have in relation to the ball algebra $A(B)$.

K is a (Z)-set (*zero set*) if there is an $f \in A(B)$ such that $f(\zeta) = 0$ for every $\zeta \in K$ and $f(z) \neq 0$ for every $z \in \bar{B} \backslash K$.

K is a (P)-set (*peak set*) if there is an $f \in A(B)$ such that $f(\zeta) = 1$ for every $\zeta \in K$ and $|f(z)| < 1$ for every $z \in \bar{B} \backslash K$. (Such an f is said to *peak on* K.)

K is an (I)-set (*interpolation set*) if every complex continuous function on K extends to a member of $A(B)$.

K is a (PI)-set (*peak-interpolation set*) if the following is true: to every $g \in C(K)$ ($g \not\equiv 0$) corresponds an $f \in A(B)$ such that $f(\zeta) = g(\zeta)$ for every $\zeta \in K$ and $|f(z)| < \|g\|_K$ for every $z \in \bar{B} \backslash K$. (Here $\|g\|_K$ denotes the maximum of $|g(\zeta)|$ on K.)

K is an (N)-set if K is a *null set* for every $v \in A^{\perp}$. More explicitly, the requirement is that $|v|(K) = 0$ whenever $v \in M(S)$ satisfies $\int f \, dv = 0$ for every $f \in A(B)$.

K is a (TN)-set if K is *totally null*, i.e., if $\rho(K) = 0$ for every representing measure $\rho \in M_0$. (See §9.1.2.)

These six properties turn out to be equivalent:

10.1.2. Theorem. *If a compact set $K \subset S$ has one of the six properties (Z), (P), (I), (PI), (N), (TN), then it has the other five.*

This will be proved in accordance with the following diagram, in which single arrows indicate easy implications, whereas double arrows indicate substantial theorems.

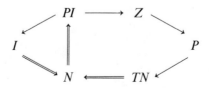

It is trivial that (PI) implies (I) and (Z).

Suppose $f \in A(B)$ has K as its zero set in $\bar{B}$, and $|f| < 1$. Since $\bar{B} \setminus K$ is simply connected, $f = \exp g$ for some g that is continuous on $\bar{B} \setminus K$ and holomorphic in B. Clearly, $\operatorname{Re} g < 0$ and $\operatorname{Re} g(z) \to -\infty$ as z approaches any point of K. Hence $h = g/(g - 1)$ is in $A(B)$ and peaks on K. This shows that $(Z) \to (P)$.

If $f \in A(B)$ peaks on K, and $\rho \in M_0$, then $f^m(0) = \int f^m \, d\rho$ for $m = 1, 2, 3, \ldots$. The integral converges to $\rho(K)$ as $m \to \infty$, by the dominated convergence theorem, and $f^m(0) \to 0$, since $|f(0)| < 1$. Thus $\rho(K) = 0$. This shows that $(P) \to (TN)$.

That $(TN) \Rightarrow (N)$ follows from the Cole–Range theorem: Suppose K is totally null, and $v \in A^\perp$. Then v is a Henkin measure, hence $v \ll \rho$ for some $\rho \in M_0$. Since $\rho(K) = 0$, it follows that $|v|(K) = 0$.

Finally, the implications $(I) \Rightarrow (N) \Rightarrow (PI)$ are special cases of the theorems of Varopoulos and Bishop which will be proved in Sections 10.2 and 10.3.

10.1.3. Some Background. The case $n = 1$ of Theorem 10.1.2 (dealing with the *disc algebra* and with compact sets K on the unit circle) represents theorems that were proved over a rather long span of time. [When $n = 1$, (TN) is simply the property of having Lebesgue measure 0. We shall nevertheless continue to write (TN), for the sake of uniformity and brevity.]

In his thesis, Fatou [1] showed that (TN) implies (Z). The brothers F. and M. Riesz [1], in their only joint paper, proved that (TN) implies (N). That (TN) also implies (P) is implicit in their proof. About 40 years later, Carleson [2] and Rudin [4] showed, independently, that (TN) implies (I). Since it is easy to see that no set of positive Lebesgue measure has any of the properties (Z), (N), (I), these three were thus known to be equivalent in the disc algebra context, but only because each of them was known to be equivalent to (TN).

Bishop [1] threw an entirely new light on this subject by proving that (N) implies (PI) in very general situations that have nothing to do with holomorphic functions or with (TN). This theorem occupies Section 10.3.

In the polydisc setting (where $K \subset T^n$) it is known that (Z), (P), (I), (PI), and (N) are equivalent. Proofs may be found in Chap. 6 of Rudin (1) and in §21 of Stout [1].

Valskii [1] proved part of Theorem 10.1.2, namely the equivalence of (PI), (Z), (P), and (N), for smoothly bounded, star-shaped, strictly pseudoconvex domains in $\mathbb{C}^n$. Chollet [1] showed that Theorem 10.1.2 holds in arbitrary strictly pseudoconvex domains (not necessarily simply connected) by giving a different proof of $(Z) \to (P)$.

10.1.4. Among the implications marked in the diagram that follows Theorem 10.1.2 there is only one, namely $(TN) \Rightarrow (N)$, whose proof involves holomorphic functions in a significant way. The others (except for $(Z) \to (P)$, where holomorphic functions do play a small role) are function algebra

theorems, and it is reasonable to ask questions such as the following: Does
(P) imply (N) in every function algebra? Do (P) and (I) always imply each
other? Here is a simple example that answers these negatively.

Let $X = K_1 \cup K_2$ be the following compact subset of $\mathbb{C} \times R$: K_1 consists
of all points $(z, 0)$ with $|z| \leq 1$, K_2 consists of all points $(0, t)$ with $0 \leq t \leq 1$.
Thus X looks like a thumbtack, and K_1 can be identified with the closed unit
disc in $\mathbb{C}$. Let A consist of all $f \in C(X)$ whose restriction to K_1 is holomorphic
in the interior of K_1. Then A is a function algebra on X (and, in fact, X is the
maximal ideal space of A).

Clearly, K_1 is a (P)-set of A which is neither (I) nor (N). (Note that K_1 is
a (P)-set which contains compact sets that are *not* (P)-sets!)

On the other hand, K_2 is an (I)-set for A which is neither (P) nor (N). In
view of Varopoulos' theorem (Section 10.2), it should be mentioned that
only one point of K_2, namely $(0, 0)$, fails to be a peak point for A. That K_2
is not (P) is due to the maximum modulus theorem. To see that K_2 is not
(N), let v be the sum of normalized Lebesgue measure on the boundary of
K_1 and the point measure of mass -1 at $(0, 0) \in K_1 \cap K_2$.

Both K_1 and K_2 are (Z)-sets for A.

10.1.5. Consequences of Theorem 10.1.2. Properties (N) and (TN) are obviously hereditary (if K has one of them, so does every compact $K_1 \subset K$), and
they are preserved under the formation of countable unions. Since this is
not so obvious for the other properties under consideration, Theorem 10.1.2
leads to some nontrivial conclusions, such as the following.

(a) *If a compact $K \subset S$ is the union of countably many (PI)-sets, then
     K is a (PI)-set.*

(b) *If $K \subset S$ is a (Z)-set, then every compact $K_1 \subset K$ is a (Z)-set.*

(c) For explicitly described sets K, properties (P) or (Z) are often the
     ones that are most easily verified. For example, let K be the set of
     all $z \in S = \partial B_n$ all of whose coordinates are real. Thus K is an
     $(n - 1)$-sphere. Define

     $$g(z) = z_1^2 + \cdots + z_n^2.$$

     Then $\frac{1}{2}(1 + g)$ peaks on K. *Conclusion:* K is totally null, K is a
     (PI)-set, etc.

(d) Here is an elaboration of (c): Let E be a compact subset of the unit
     circle, of Lebesgue measure 0, let g be as in (c), and put $K = S \cap g^{-1}(E)$.

     The case $E = \{1\}$ occurred in (c).

     Since E has measure 0, E is a peak-set for $A(U)$. Let $h \in A(U)$
     peak on E. Then $h \circ g \in A(B)$ and $h \circ g$ peaks on K.

     If E is a Cantor set of measure 0, we conclude that there exist
     "Cantor sets of $(n - 1)$-spheres" in S that are (PI)-sets.

(e)   Let E be as in (d), but this time define

$$g(z) = n^{n/2} z_1 \cdots z_n.$$

The set of all $z \in \bar{B}$ where $|g(z)| = 1$ consists of the n-torus defined by

$$|z_1| = \cdots = |z_n| = n^{-1/2}.$$

For any $\alpha \in E$, $g^{-1}(\alpha) \cap \bar{B}$ is thus an $(n - 1)$-torus in S. Setting $K = S \cap g^{-1}(E)$, it follows, as in (d), that there exist "Cantor sets of $(n - 1)$-tori" in S that are (PI)-sets.

10.2. A Theorem of Varopoulos

10.2.1. The original proof of this theorem, as given by Varopoulos [1], relied on the fact that every compact Hausdorff space can be homeomorphically embedded in some (large) compact abelian group, and used some harmonic analysis on such groups. Glicksberg [2; p. 9] simplified the proof a great deal: the only group that appears is a finite cyclic one, and no harmonic analysis is needed beyond the ability to sum a finite geometric progression.

10.2.2. Theorem. *Suppose*

(i)   *A is a function algebra on a compact Hausdorff space X,*
(ii)   *$K \subset X$ is a compact interpolation set for A,*
(iii) *every point of K is a peak-point for A,*
(iv)   *$v \in M(X)$ and $v \in A^{\perp}$.*

Then $|v|(K) = 0$.

Proof. Fix $\varepsilon > 0$. There is an open set $\Omega \subset X$ such that $K \subset \Omega$ and $|v|(\Omega \backslash K) < \varepsilon$. By (iii) there corresponds to every point $x \in K$ a function $f_x \in A$ such that $f_x(x) = 1$ but $|f_x(y)| < 1$ for every $y \neq x$, $y \in X$. We can replace f_x by f_x^m, for some large integer m, so that $|f_x| < \varepsilon$ outside Ω. The sets where $|1 - f_x| < \varepsilon$ form an open cover of K. Hence (writing f_i in place of f_{x_i}) there are finitely many functions $f_1, \ldots, f_N \in A$, with $\|f_i\| = 1$, such that $|f_i| < \varepsilon$ outside Ω, and such that the sets

(1) $E_i = \{|1 - f_i| < \varepsilon\}$ $(i = 1, \ldots, N)$

cover K.

Next, there are *pairwise disjoint* compact sets $K_i \subset E_i \cap K$ such that

$$(2) \qquad \left| v \right| \left(\Omega \setminus \bigcup_1^N K_i \right) < 2\varepsilon.$$

To say that K is an interpolation set for A is the same as to say that the bounded linear operator that assigns to every $f \in A$ its restriction to K has all of $C(K)$ as its range. The open mapping theorem for Banach spaces implies therefore that there is a constant $\alpha < \infty$ with the following property: *every $g \in C(K)$ has an extension $f \in A$ such that $\|f\|_X \leq \alpha \|g\|_K$.*

Briefly: K is an (I)-set with constant α.

Since $K_1 \cup \cdots \cup K_N \subset K$, it follows from Tietze's extension theorem that $K_1 \cup \cdots \cup K_N$ is also an (I)-set with constant α.

Now put $\omega = \exp(2\pi i/N)$. Since the sets K_s are pairwise disjoint, the preceding remarks show that there are functions $g_r \in A$, $1 \leq r \leq N$, with $\|g_r\|_X \leq \alpha$, such that

$$(3) \qquad g_r(x) = \omega^{rs} \quad \text{if} \quad x \in K_s.$$

Put

$$(4) \qquad h_p(x) = N^{-1} \sum_{r=1}^N \omega^{-rp} g_r(x) \qquad (x \in X, 1 \leq p \leq N)$$

and

$$(5) \qquad H = \sum_{p=1}^N h_p^2 f_p.$$

Then, obviously, $H \in A$. If $x \in K_s$, (3) and (4) give

$$(6) \qquad h_p(x) = N^{-1} \sum_{r=1}^N \omega^{r(s-p)} = \delta_{ps}$$

where $\delta_{ps} = 1$ if $p = s$, $\delta_{ps} = 0$ if $p \neq s$. Hence $H(x) = f_s(x)$ on $K_s \subset E_s$, so that

$$(7) \qquad |1 - H(x)| < \varepsilon \quad \text{on} \quad K_1 \cup \cdots \cup K_N.$$

Next, for any $x \in X$,

$$\sum_p |h_p(x)|^2 = N^{-2} \sum_{r,s} g_r(x)\overline{g_s(x)} \sum_p \omega^{-rp}\omega^{sp} = N^{-1} \sum_r |g_r(x)|^2 \leq \alpha^2.$$

(The indices p, r, s run from 1 to N.) Hence

$$(8) \qquad \|H\|_X \leq \alpha^2$$

and

(9) $$|H(x)| \leq \alpha^2 \varepsilon \quad \text{if} \quad x \in X \backslash \Omega.$$

Let us write H_ε for H, and let $\varepsilon \to 0$ through the sequence $1/n^2$, for example. By (7), (8), (9), we obtain a uniformly bounded sequence $\{H_\varepsilon\}$ in A which converges pointwise, a.e. $[|v|]$, to the characteristic function of K, by (2). Since $v \in A^\perp$ and $H_\varepsilon \in A$, it follows, by the dominated convergence theorem, that $v(K) = 0$.

Finally, the hypotheses (ii) and (iii) hold for every compact $K_1 \subset K$. Thus $v(K_1) = 0$ for all such K_1, and therefore $|v|(K) = 0$.

10.2.3. Every $\zeta \in S$ is a peak point for $A(B)$. The function

$$\tfrac{1}{2}(1 + \langle z, \zeta \rangle)$$

shows this. Consequently, Theorem 10.2.2 gives the implication $(I) \Rightarrow (N)$ in Theorem 10.1.2.

10.3. A Theorem of Bishop

This will complete the proof of Theorem 10.2.1, by establishing the implication $(N) \Rightarrow (PI)$.

10.3.1. Theorem (Bishop [1]). *Suppose*

(a) *X is a compact Hausdorff space,*
(b) *A is a closed linear subspace of $C(X)$,*
(c) *K is a compact subset of X, of type G_δ, such that $|v|(K) = 0$ for every $v \in M(X)$ that annihilates A.*

Then K is a peak-interpolation set for A.

Note that A does not have to be an algebra. The assumption that K is of type G_δ is of course satisfied by every compact K if X is metric; it will only play a role in the last of the three steps into which we split the proof.

Step 1. If $g \in C(K)$, $|g| < 1$, then there exists $f \in A$ such that $f = g$ on K and $|f| < 1$ on X.

Proof. Let U_A and U_C be the open unit balls in A and $C(K)$, respectively. Let R be the restriction map that takes A into $C(K)$. Then $R(U_A)$ is a convex balanced subset of U_C.

Assume, to reach a contradiction, that $R(U_A)$ is not dense in U_C. By the Hahn–Banach separation theorem, there is then a measure $\mu \in M(X)$, concentrated on K, and a constant $\eta < 1$, such that $|\int_K G \, d\mu| < \eta$ for every $G \in U_A$, whereas $\int_K F \, d\mu = 1$ for some $F \in U_C$.

Another application of the Hahn–Banach theorem shows that the linear functional $G \to \int_K G \, d\mu$, whose norm is $\leq \eta$, has a norm-preserving extension to $C(X)$. Hence there is a $\tau \in M(X)$, $\|\tau\| \leq \eta$, such that $\int G \, d\tau = \int G \, d\mu$ for all $G \in A$. In other words, $\mu - \tau \in A^\perp$. Hypothesis (c) implies therefore that $\tau(E) = \mu(E)$ for every $E \subset K$. Hence

$$1 = \int_K F \, d\mu = \int_K F \, d\tau \leq \|F\|_K \|\tau\| \leq \eta < 1.$$

This contraction shows that $R(U_A)$ is dense in U_C.

If how $g \in C(K)$ and $\|g\|_K < r < 1$, it follows that there is an $f_1 \in A$ with $\|f_1\|_X < r$ and $\|g - f_1\|_K < \frac{1}{2}(1 - r)$. Next, there is an $f_2 \in A$, $\|f_2\|_X < \frac{1}{2}(1 - r)$, and $\|g - f_1 - f_2\|_K < \frac{1}{4}(1 - r)$, etc. If $f = \Sigma f_i$, then $f = g$ on K, $f \in A$, and

$$\|f\|_X < r + (1 - r)(\tfrac{1}{2} + \tfrac{1}{4} + \tfrac{1}{8} + \cdots) = 1.$$

Step 2. Suppose $g \in C(K)$, $b \in C(X)$, $b > 0$ on X, and $|g| < b$ on K. Then there exists $f \in A$ such that $f = g$ on K and $|f| < b$ on X.

Proof. Let $A_1 = \{f/b : f \in A\}$, and apply Step 1 to g/b, with A_1 in place of A. (Note that $\mu \perp A$ if and only if $b\mu \perp A_1$, and that $|\mu|$ and $|b\mu|$ vanish on the same sets.)

Step 3. We now prove the theorem. There are open sets $V_1 \supset V_2 \supset V_3 \supset \cdots$ so that $K = \cap V_i$. Put $r_i = 2^{-i-2}$. Choose $g \in C(K)$, $\|g\|_K = 1$. We will construct functions $f_i \in A$ such that

(1) $f_i = (1 - 8r_i)g$ on K,
(2) $|f_i| < 1 - 2r_i$ on X,
(3) $|f_{i+1} - f_i| < 6r_i$ on X,
(4) $|f_{i+1} - f_i| < r_i$ on $X \backslash V_i$.

Put $f_1 = 0$. Then (1) and (2) hold when $i = 1$.

Assume $i \geq 1$ and $f_i \in A$ satisfies (1) and (2). By (1), there is an open set W, $K \subset W \subset V_i$, such that $|f_i| < 1 - 7r_i$ in W. Choose $b \in C(X)$ so that $b = 5r_i$ on K, $b = r_i$ off W, and $0 < b < 6r_i$ on X. By Step 2 there is an $h_i \in A$ such that

(α) $h_i = 4r_i g = (8r_i - 8r_{i+1})g$ on K,
(β) $|h_i| < 6r_i$ on X,
(γ) $|h_i| < r_i$ on $X \backslash W$, hence on $X \backslash V_i$.

Put $f_{i+1} = f_i + h_i$. Then (β) and (γ) give (3) and (4); (α) shows that (1) holds with $i + 1$ in place of i. In W, $|f_{i+1}| < (1 - 7r_i) + 6r_i$. Outside W, $|f_{i+1}| < (1 - 2r_i) + r_i$. Since $r_i = 2r_{i+1}$ it follows that (2) holds with $i + 1$ in place of i. Thus $\{f_i\}$ is constructed, by induction.

Put $f = \lim f_i$. By (3), $f \in A$. By (1), $f = g$ on K. If $x \notin K$ there exists m such that $x \notin V_m$, so that

$$|f(x)| \leq |f_m(x)| + \sum_m^\infty |h_i(x)| < 1 - 2r_m + \sum_m^\infty r_i = 1,$$

since $|h_i| < r_i$ outside $V_i \subset V_m$. Thus $|f(x)| < 1$ outside K, and peak-interpolation is established.

10.3.2. To prove the implication $(N) \Rightarrow (PI)$ in Theorem 10.1.2 we apply Theorem 10.3.1 with $X = S$. If K satisfies (N) and $g \in C(K)$, $g \not\equiv 0$, Theorem 10.3.1 yields an $f \in A(B)$ such that $f = g$ on K and $|f(\zeta)| < \|g\|_K$ for every $\zeta \in S \backslash K$. Thus f is not constant, and now $|f(z)| < \|g\|_K$ follows for every $z \in B$, by the maximum modulus theorem.

10.4. The Davie–Øksendal Theorem

10.4.1. The shortest as well as the most elementary proof of the F. and M. Riesz theorem is undoubtedly the one found by Øksendal [1]. Its idea was used by Davie and Øksendal [1] to formulate a covering condition that implies peak-interpolation, in the context of strictly pseudoconvex domains. Specialized to the ball, this is Theorem 10.4.3 below. Øksendal's proof is the case $n = 1$ of the proof of Theorem 10.4.3.

10.4.2. Definition. For $\zeta \in S$ and $\delta > 0$, put

$$V(\zeta, \delta) = \{z \in S : |1 - \langle z, \zeta \rangle| < \delta\}.$$

Each $V(\zeta, \delta)$ is a "nonisotropic ball" Q defined in §5.1.1; however, $Q(\zeta, \delta) = V(\zeta, \delta^2)$. For our present purpose, it seems desirable to make this change of notation.

When $n = 1$, then $V(\zeta, \delta)$ is an arc on the unit circle, of length 2δ, having ζ as midpoint.

10.4.3. Theorem. *Suppose K is a compact subset of S, with the following property:*

To every $\varepsilon > 0$ correspond finitely many sets $V(\zeta_i, \delta_i)$, $1 \leq i \leq m$, with $\zeta_i \in K$ and $\Sigma \delta_i < \varepsilon$, such that

$$K \subset V(\zeta_1, \delta_1) \cup \cdots \cup V(\zeta_m, \delta_m).$$

Then K is an (N)-set.

Note: When $n = 1$, the hypothesis says simply that K has Lebesgue measure 0.

Proof. For each $\varepsilon > 0$, choose ζ_i and δ_i as in the hypothesis, put $c_i = \delta_i/\sqrt{\varepsilon}$, and define $f_\varepsilon \in A(B)$ by

(1) $$f_\varepsilon(z) = 1 - \prod_{i=1}^{m} \frac{1 - \langle z, \zeta_i \rangle}{1 + c_i - \langle z, \zeta_i \rangle} \qquad (z \in \bar{B}).$$

We claim that

    (a) $|f_\varepsilon(z)| < 2$ for all $z \in \bar{B}$,

    (b) $|1 - f_\varepsilon(z)| < \sqrt{\varepsilon}$ for every $z \in K$, and

    (c) $\lim_{\varepsilon \to 0} f_\varepsilon(z) = 0$ for every $z \in \bar{B} \setminus K$.

Once these are proved, pick $v \in A(B)^\perp$. Then $\int f_\varepsilon \, dv = 0$. By the dominated convergence theorem, $v(K) = 0$. The same applies to every compact $K_1 \subset K$. Hence $|v|(K) = 0$, which is what the theorem asserts.

Since $\mathrm{Re}(1 - \langle z, \zeta_i \rangle) \geq 0$, each factor in (1) has absolute value < 1, for every $z \in \bar{B}$. This proves (a).

If $z \in K$, then at least one factor in (1) has absolute value $< \delta_i/c_i = \sqrt{\varepsilon}$. This proves (b).

Fix $z \in \bar{B} \setminus K$. Then there exists $t > 0$ such that $|1 - \langle z, \zeta \rangle| > t$ for every $\zeta \in K$. Put

(2) $$u_i = \frac{c_i}{1 + c_i - \langle z, \zeta_i \rangle}.$$

Then $|u_i| < c_i/t$, $\sum |u_i| < t^{-1} \sum c_i < t^{-1}\sqrt{\varepsilon}$, and

(3) $$f_\varepsilon(z) = 1 - \prod_{1}^{m} (1 - u_i),$$

so that

$$|f_\varepsilon(z)| = |-1 + \prod(1 - u_i)| \leq -1 + \prod(1 + |u_i|)$$
$$\leq -1 + \exp \sum |u_i| < -1 + \exp(t^{-1}\sqrt{\varepsilon}).$$

This proves (c).

10.4.4. An Application. Let γ be a complex-tangential curve, as defined in §6.4.1. Then γ (or, more precisely, the range of γ) is a (*PI*)-set.

This follows from Theorems 10.4.3 and 10.1.2, since it can be verified that the hypothesis of 10.4.3 holds. Since we shall find a different proof in Section 10.5, we omit the details.

10.4.5. There are variations of the Davie–Øksendal technique that allow one to prove that certain *uncountable* unions of (*PI*)-sets are (*PI*).

For instance, let $H \subset S$ be the set where a certain $h \in A(B)$ peaks, and assume also that $h \in \text{Lip } 1$, i.e. that $|h(z) - h(w)| \leq c|z - w|$ for some fixed c and for all $z, w \in \bar{B}$.

Let E be a compact set of real numbers, of Lebesgue measure 0, and associate with each $t \in E$ an automorphism $\psi_t \in \text{Aut}(B)$, in such a way that $|\psi_t(z) - \psi_s(z)| \leq |t - s|$ for all $z \in \bar{B}$, $t \in E$, $s \in E$.

Put $h_t = h \circ \psi_t$. Then h_t peaks on $H_t = \psi_t^{-1}(H)$.

Define $K = \bigcup_{t \in E} H_t$. Then K is compact.

10.4.6. Proposition. *The set K described above is a (PI)-set.*

Proof. Since $m(E) = 0$, there are intervals L_i ($i = 1, 2, 3, \ldots$) with centers $t_i \in E$, of length δ_i, so that $\Sigma \delta_i < \infty$ and so that every point of E lies in infinitely many L_i. Define

$$g_i = \frac{1 - h_{t_i}}{c\delta_i + 1 - h_{t_i}}, \qquad f_N = \prod_{i=1}^{N} g_i,$$

where c is the Lipschitz constant of h.

Note that $\text{Re}(1 - h) \geq 0$ on $\bar{B}$. Hence $|g_i| < 1$.

Fix $z \in K$. Then $z \in H_t$ for some $t \in E$. Hence there are infinitely many i with $t \in L_i$, so that $|t - t_i| \leq \delta_i/2$. For any such i,

$$|1 - h_{t_i}(z)| = |h_t(z) - h_{t_i}(z)| \leq c|t - t_i| \leq \tfrac{1}{2}c\delta_i,$$

which implies that $|g_i(z)| \leq \tfrac{1}{2}$. This happens for infinitely many i. *Thus $f_N(z) \to 0$, as $N \to \infty$, for every $z \in K$.*

Next, let K_0 be a compact subset of $\bar{B} \backslash K$. Note that $(z, t) \to 1 - h_t(z)$ is a continuous zero-free function on $K_0 \times E$. Hence there is an $\eta > 0$ (depending on K_0) such that $|1 - h_t(z)| \geq \eta$ for all $z \in K_0, t \in E$. This implies

$$|1 - g_i(z)| = \frac{c\delta_i}{|c\delta_i + 1 - h_{t_i}(z)|} \leq \frac{c\delta_i}{\eta}$$

so that $\sum |1 - g_i(z)|$ converges uniformly on K_0. It follows that $\{f_N\}$ converges, uniformly on K_0, to a continuous limit which is $\neq 0$ at every point of K_0.

Thus $f(z) = \lim_{N \to \infty} f_N(z)$ exists for all $z \in \bar{B}$, $f(z) = 0$ if and only if $z \in K$, and f is continuous on $\bar{B} \backslash K$. To prove that $f \in A(B)$, it is now enough to show that f is continuous at every point of K. Since $K = \{f = 0\}$, this amounts to showing that $|f|$ is continuous at every point where it vanishes. But $|f|$ is upper-semicontinuous (since $|f_{N+1}| \leq |f_N|$) and ≥ 0, and all such functions are continuous wherever they are 0.

We have now proved that K is a (Z)-set, hence it is a (PI)-set, by Theorem 10.1.2.

10.4.7. Example. Take $n = 2$, and let K be the set of all points

$$(\cos\theta, e^{it}\sin\theta) \qquad (-\pi \le \theta \le \pi, t \in E)$$

where E is compact, $m(E) = 0$. Since $\{(\cos\theta, \sin\theta)\}$ is the peak set of $(1 + z^2 + w^2)/2$, which is in Lip 1, Proposition 10.4.6 shows that K is a (PI)-set.

   K is a union of circles, one for each $t \in E$; they all pass through the points $(1, 0)$ and $(-1, 0)$.

10.5. Smooth Interpolation Sets

10.5.1. The following simple example illustrates the theme of this section. Fix $\zeta \in S$, and put

$$\Gamma(t) = e^{it}\zeta, \qquad \gamma(t) = (\cos t, \sin t, 0, \ldots, 0),$$

where $-\pi \le t \le \pi$; then Γ and γ are circles in S.

   It is easy to see that the range of γ is a (PI)-set whereas the range of Γ is not: The function $\frac{1}{2}(1 + z_1^2 + z_2^2)$ peaks on the range of γ, but if $f \in A(B)$ and $f(\Gamma(t)) = 0$ for all t (or even only on some set of positive measure on $(-\pi, \pi)$), then the slice function $f_\zeta \in A(U)$ vanishes on all of U, so that f has zeros in B; the range of Γ is thus not a (Z)-set.

   Observe now that $\langle \gamma', \gamma \rangle = 0$, but $\langle \Gamma', \Gamma \rangle = i$. Thus γ is complex-tangential (see §6.4.1, §6.4.2) whereas Γ is not. This is not an accident. We shall see that *a C^1-curve γ in S is a (PI)-set if and only if γ is complex-tangential*. The sufficiency of this condition is a special case of Theorem 10.5.4. The necessity will follow from Theorem 11.2.5.

   Instead of curves, we shall consider higher-dimensional smooth sets in S.

10.5.2. Definitions. Suppose that m and n are positive integers, $S = \partial B_n \subset \mathbb{C}^n$, as usual, Ω is an open set in the euclidean space R^m, and $\Phi : \Omega \to S$ is a map of class C^1. The first-order partial derivatives of the components of Φ are thus assumed to be continuous functions in Ω.

   For each $x \in \Omega$, Φ has then a Fréchet derivative $\Phi'(x)$; this is the real-linear map of R^m into $\mathbb{C}^n$ that satisfies

$$(1) \qquad \lim_{h \to 0} \frac{|\Phi(x + h) - \Phi(x) - \Phi'(x)h|}{|h|} = 0.$$

   We say that Φ is *complex-tangential* if the orthogonality relation

$$(2) \qquad \langle \Phi'(x)h, \Phi(x) \rangle = 0$$

holds for every $x \in \Omega$ and for every $h \in R^m$.

Let $M = \Phi(\Omega)$ and assume (to simplify this discussion) that Φ is one-to-one. Put $\zeta = \Phi(x)$. The tangent space $T_\zeta(M)$ consists then of all vectors $\Phi'(x)h$, as h ranges over R^m, and (2) is equivalent to the inclusion

$$(3) \qquad\qquad T_\zeta(M) \subset T_\zeta^{\mathbb{C}}(S).$$

The latter is the complex tangent space of S at ζ. (See §5.4.2.)

Here is yet another interpretation of (2): Let $\gamma: [0, 1] \to \Omega$ be any C^1-curve, and define $\Gamma: [0, 1] \to S$ by $\Gamma = \Phi \circ \gamma$. The chain rule shows then that Φ is complex-tangential if and only if every Γ obtained in this way is complex-tangential.

We shall say that Φ is *nonsingular* if the rank of $\Phi'(x)$ is m for every $x \in \Omega$. In that case, Φ is locally one-to-one, and there is a continuous *positive* function c on Ω such that

$$(4) \qquad\qquad |\Phi'(x)h| \geq c(x)|h| \qquad (x \in \Omega, h \in R^m).$$

The precise manner in which the crucial hypothesis (2) enters the proof of Theorem 10.5.4 is contained in the following lemma.

10.5.3. Lemma. *If* $\Phi: \Omega \to S$ *is complex-tangential, then the inner products*

$$(1) \qquad\qquad \left\langle \frac{\Phi(x + \delta v) - \Phi(x + \delta u)}{\delta^2}, \Phi(x + \delta v) \right\rangle$$

converge, as $\delta \searrow 0$, *to*

$$(2) \qquad\qquad \tfrac{1}{2}|\Phi'(x)(v - u)|^2$$

for all $x \in \Omega$ *and all* $u, v \in R^m$.

Note that the denominator in (1) is δ^2, not δ.

Proof. Fix x, u, v. If δ is small enough then (since Ω is open) all points

$$(3) \qquad\qquad y(t) = x + (1 - t)\delta u + t\delta v \qquad (0 \leq t \leq 1)$$

lie in Ω. Fix such a δ, for the moment, and define $\gamma: [0, 1] \to S$ by $\gamma(t) = \Phi(y(t))$. Then

$$(4) \qquad\qquad \gamma'(t) = \delta\Phi'(y(t))(v - u).$$

Since Φ is complex tangential, it follows that

(5) $$\langle \gamma'(t), \gamma(t) \rangle = 0 \qquad (0 \le t \le 1).$$

Note that (1) equals $\delta^{-2} \langle \gamma(1) - \gamma(0), \gamma(1) \rangle$. Using (5), one obtains

$$\langle \gamma(1) - \gamma(0), \gamma(1) \rangle = \left\langle \int_0^1 \gamma'(t)dt, \gamma(1) \right\rangle$$

$$= \int_0^1 \langle \gamma'(t), \gamma(1) - \gamma(t) \rangle dt$$

$$= \int_0^1 dt \int_t^1 \langle \gamma'(t), \gamma'(s) \rangle ds.$$

Hence (4) shows that (1) equals

(6) $$\int_0^1 dt \int_t^1 \langle \Phi'(y(t))(v - u), \Phi'(y(s))(v - u) \rangle ds.$$

As $\delta \searrow 0$, $y(t) \to x$, uniformly for $0 \le t \le 1$. Since Φ' is continuous, the integrand in (6) converges uniformly to $|\Phi'(x)(v - u)|^2$. Since the double integral extends over one half of the unit square in the (s, t)-plane, (6) converges to (2) as $\delta \searrow 0$.

10.5.4. Theorem. *If Ω is open in R^m and*

$$\Phi : \Omega \to S = \partial B_n$$

is a C^1-map that is nonsingular and complex-tangential, then $\Phi(K)$ is a (PI)-set for every compact $K \subset \Omega$.

Under stronger smoothness assumptions on Φ this was proved (with strictly pseudoconvex domains in place of balls) by Henkin (see Čirka–Henkin [1]), by Nagel [1], and by Burns and Stout [1]. The proof that follows is basically the one in Rudin [14], but the details are considerably simpler since we now deal only with the unit ball.

Proof. By Bishop's theorem (Section 10.3) it suffices to show that every $p \in \Omega$ is the center of a compact ball K such that $|v|(\Phi(K)) = 0$ for every $v \in A^\perp$.

Since Φ is nonsingular, we can choose K so that there is an $a > 0$ such that

(1) $$|\Phi(x) - \Phi(y)| \ge a|x - y| \qquad (x, y \in K).$$

For $y \in \Omega$, define

$$(2) \qquad g(y) = \int_{R^m} \{1 + \tfrac{1}{2}|\Phi'(y)v|^2\}^{-m} \, dv.$$

Since $|\Phi'(y)v| \geq c(y)|v|$, by 10.5.2(4), g is continuous, and of course >0.

Now pick a continuous $f: R^m \to \mathbb{C}$, with support in K, and define

$$(3) \qquad h_\delta(z) = \int_{R^m} \left\{\frac{\delta}{\delta^2 + 1 - \langle z, \Phi(x)\rangle}\right\}^m \frac{f(x)}{g(x)} \, dx$$

for $\delta > 0$, $z \in \bar{B}$. (The integrand is understood to be 0 outside K.) Since $|\langle z, \Phi(x)\rangle| \leq 1$, it is clear that $h_\delta \in A(B)$.

We claim that $\{h_\delta\}$ has the following properties:

$$(4) \qquad \sup\{|h_\delta(z)|: z \in \bar{B}, \delta > 0\} < \infty.$$

$$(5) \qquad \lim_{\delta \to 0} h_\delta(z) = 0 \quad \text{if} \quad z \in \bar{B} \backslash \Phi(K).$$

$$(6) \qquad \lim_{\delta \to 0} h_\delta(\Phi(y)) = f(y). \quad \text{if} \quad y \in K.$$

Of these, (5) is obvious, since $1 - \text{Re}\langle z, \Phi(x)\rangle$ has a positive lower bound as x ranges over K.

To prove (4) and (6), associate to each $z \in \bar{B}$ a point $y \in K$ such that

$$(7) \qquad |\Phi(y) - z| \leq |\Phi(x) - z|$$

for all $x \in K$. By the triangle inequality, (7) gives

$$(8) \qquad |\Phi(y) - \Phi(x)| \leq 2|z - \Phi(x)|.$$

The change of variables $x = y + \delta v$ converts (3) to

$$(9) \qquad h_\delta(z) = \int_{R^m} \frac{(f/g)(y + \delta v)dv}{\{1 + \delta^{-2}[1 - \langle z, \Phi(y + \delta v)\rangle]\}^m}.$$

By (8) and (1),

$$8 \, \text{Re}(1 - \langle z, \Phi(x)\rangle) \geq 4|z - \Phi(x)|^2$$
$$\geq |\Phi(y) - \Phi(x)|^2 \geq a^2|x - y|^2 = a^2\delta^2|v|^2.$$

The integrand in (9) is thus dominated by the L^1-function

(10)
$$\left\|\frac{f}{g}\right\|_\infty \left\{1 + \frac{a^2}{8}|v|^2\right\}^{-m}.$$

This establishes (4), and shows that (6) follows from (9), (2), and the dominated convergence theorem, since

(11)
$$\lim_{\delta \to 0} \frac{1 - \langle \Phi(y), \Phi(y + \delta v) \rangle}{\delta^2} = \frac{1}{2}|\Phi'(y)v|^2,$$

by Lemma 10.5.3 (with $u = 0$).

To complete the proof, note that Φ is one-to-one on K, so that Φ^{-1}: $\Phi(K) \to K$ is well defined and continuous. Since each $h_\delta \in A(B)$, it follows from (4), (5) and (6) that

(12)
$$\int_{\Phi(K)} (f \circ \Phi^{-1}) dv = 0$$

for every $v \in A^\perp$, and for every $f \in C(K)$ that vanishes on the boundary of the ball K. Hence $|v|(\Phi(K_1)) = 0$ for every compact K_1 in the interior of K. As noted at the start of the proof, this gives the desired conclusion.

10.5.5. Totally Real Vector Spaces. A real vector space $V \subset \mathbb{C}^n$ is said to be *totally real* if $V \cap (iV) = \{0\}$, i.e., if V contains no complex subspace of positive dimension.

Suppose V is totally real, and $V \subset T_\zeta^\mathbb{C}$, the complex tangent space of S at ζ, whose real dimension is $2n - 2$. Then iV also lies in $T_\zeta^\mathbb{C}$. Consequently

$$\dim_R V \le n - 1.$$

This leads to the following upper bound on m in Theorem 10.5.4.

10.5.6. Theorem. *Suppose that the hypotheses of Theorem 10.5.4 hold. Associate to each $x \in \Omega$ the real vector space*

$$V_x = \{\Phi'(x)u : u \in R^m\}.$$

Then

(a) $V_x \cdot (iV_x) = 0,$
(b) V_x *is totally real, and*
(c) $m \le n - 1.$

Assertion (a) means, more explicitly, that z and iw are perpendicular to each other, with respect to the real dot-product on $R^{2n} = C^n$, for all $z \in V_x$, $w \in V_x$. (See §5.4.2.) This amounts to proving that

(a') $\langle z, w \rangle$ *is real for all* $z, w \in V_x$.

Proof. Take $x = 0$, for simplicity, and write V in place of V_x. Obviously, $V \cap (iV) = \{0\}$ of $V \cdot (iV) = 0$. Thus (a) implies (b). Since Φ is assumed to be nonsingular, $\dim_R V = m$. Thus (b) implies (c). So we have to prove (a').

Let $z = \Phi'(0)u$, $w = \Phi'(0)v$, where $u, v \in R^m$. Define

$$Q_\delta(u, v) = \delta^{-2} \langle \Phi(\delta v) - \Phi(\delta u), \Phi(\delta v) \rangle.$$

Since

$$\Phi'(0)u = \lim_{\delta \to 0} \frac{\Phi(\delta u) - \Phi(0)}{\delta}$$

it is easy to check that

$$\langle z, w \rangle = \lim_{\delta \to 0} \{Q_\delta(u, 0) + Q_\delta(0, v) - Q_\delta(u, v)\},$$

which is real, by Lemma 10.5.3.

Note. If Φ is assumed to be of class C^2 rather than just C^1, there is a more appealing proof of Theorem 10.5.6:

The hypothesis $\langle \Phi'(x)v, \Phi(x) \rangle = 0$, when applied to basis vectors v, shows, for $1 \le j \le m$, that $\langle D_j \Phi, \Phi \rangle = 0$, where $D_j = \partial/\partial x_j$. Hence

(1) $$\langle D_k D_j \Phi, \Phi \rangle + \langle D_j \Phi, D_k \Phi \rangle = 0$$

for all j, k. Since $\Phi \in C^2$, the first inner product in (1) is unchanged if j and k are switched. Hence so is the second. But switching j and k replaces $\langle D_j \Phi, D_k \Phi \rangle$ by its complex conjugate. It follows that $\langle D_j \Phi, D_k \Phi \rangle$ is *real*, for all j, k.

This implies the theorem, as before.

10.5.7. We shall see later (as a Corollary to Theorem 11.2.5) that a C^1-curve in $S = \partial B_n$ cannot be a (*PI*)-set unless it is complex-tangential. All smooth (*PI*)-sets in S are thus described (at least locally) by Theorem 10.5.4, and we conclude from Theorem 10.5.6 that their dimension is at most $n - 1$.

This top dimension can be attained: in §10.1.5 we met an $(n - 1)$-sphere and an $(n - 1)$-torus in S that are (*PI*)-sets. Stout [7] noticed that these are the only known examples of (*PI*)-sets of dimension $n - 1$ that are compact connected manifolds, and asked whether there were any others. The case

$n = 3$ is the first one that is of interest in this connection. Here the question may be asked as follows:

Which compact connected 2-manifolds admit complex-tangential embeddings in the 5-sphere ∂B_3?

Changing the subject slightly, it seems reasonable to conjecture that the topological dimension of no (PI)-set K in ∂B_n exceeds $n - 1$, even if no smoothness condition is imposed on K.

In this connection, Tumanov [1] has constructed a (PI)-set in ∂B_2 which does not directly deal with this conjecture (it is totally disconnected, thus has topological dimension 0) but whose Hausdorff dimension is surprisingly large, namely $\frac{5}{2}$. The construction is quite intricate, and we shall not include it here.

10.5.8. Although it is difficult to find complex-tangential embeddings of *compact* manifolds of dimension $n - 1$, it is quite easy to do this with R^{n-1}:

Let $\alpha = (\alpha_1, \ldots, \alpha_n)$ be a nonsingular C^2-map of R^{n-1} onto a hypersurface in R^n whose normal has all components positive. This implies that there are positive functions F_j on R^{n-1}, of class C^1, such that

$$(1) \qquad \sum_{j=1}^{n} F_j^2(x) \frac{\partial \alpha_j}{\partial x_k}(x) = 0 \qquad (1 \le k \le n - 1).$$

Moreover, one can adjust them so that

$$(2) \qquad \sum_{j=1}^{n} F_j^2(x) = 1.$$

Now put $\Phi = (\varphi_1, \ldots, \varphi_n)$, where

$$(3) \qquad \varphi_j(x) = F_j(x)\exp\{i\alpha_j(x)\} \qquad (1 \le j \le n).$$

Then Φ is a nonsingular C^1-map of R^{n-1} into ∂B_n that satisfies

$$(4) \qquad \sum_{j=1}^{n} \frac{\partial \varphi_j}{\partial x_k}(x)\bar{\varphi}_j(x) = 0 \qquad (1 \le k \le n - 1),$$

by (1) and (2). If e_k denotes the kth unit vector in the standard basis of R^{n-1}, (4) can be rewritten as

$$(5) \qquad \langle \Phi'(x)e_k, \Phi(x) \rangle = 0 \qquad (1 \le k \le n - 1).$$

Thus Φ is complex-tangential.

10.5.9. Curves in T^2. Take $n = 2$, take $r > 0$, $s > 0$ so that $r^2 + s^2 = 1$. The boundary of B_2 contains the torus consisting of the points $(re^{i\theta}, se^{i\varphi})$, $-\pi \le \theta$, $\varphi \le \pi$. Let γ be a C^1-curve in this torus; thus

$$(1) \qquad\qquad \gamma(t) = (re^{i\theta(t)}, se^{i\varphi(t)}) \qquad (a \le t \le b).$$

Then $\langle \gamma', \gamma \rangle = 0$ if and only if $r^2\theta' + s^2\varphi' = 0$. In other words, the range of γ is a (PI)-set for $A(B_2)$ if and only if

$$(2) \qquad\qquad r^2\theta(t) + s^2\varphi(t) = \text{const.}$$

In the (θ, φ)-plane, this represents a line with slope $-r^2/s^2$. Consequently, γ cannot be a circle unless r^2/s^2 is rational.

The torus contains many other smooth curves that are (PI)-sets for the polydisc algebra $A(rU \times sU)$ (see Rudin [1] or Stout [1]), but only those that satisfy (2) work in B_2 which is of course much larger than $rU \times sU$.

10.5.10. An Annulus in B_2. Fix a, $0 < a < 1$, and consider the map Φ that takes $\lambda \in \mathbb{C}\setminus\{0\}$ to $(z, w) \in \mathbb{C}^2$, by

$$z = \frac{a}{1 + a^2}\left(\frac{\lambda}{a} + \frac{a}{\lambda}\right), \qquad w = \frac{a}{1 + a^2}\left(\lambda - \frac{1}{\lambda}\right).$$

Assume $\lambda \ne \mu$. If $\lambda\mu \ne a^2$ then $z(\lambda) \ne z(\mu)$. If $\lambda\mu \ne -1$ then $w(\lambda) \ne w(\mu)$. Thus Φ is $1 - 1$. A simple computation gives

$$|z|^2 + |w|^2 = \frac{a}{1 + a^2}\left(\frac{|\lambda|^2}{a} + \frac{a}{|\lambda|^2}\right).$$

This is 1 if and only if $|\lambda| = a$ or $|\lambda| = 1$, and it is <1 if $a < |\lambda| < 1$.

The annulus $\Omega = \{\lambda : a < |\lambda| < 1\}$ is thus mapped into B_2 by Φ, and Φ carries the two bounding circles to curves Γ_1, Γ_a in S.

Neither of these curves is a (Z)-set. For if $f \in A(B)$ vanishes on Γ_1, say, then $f \circ \Phi$ vanishes on one of the circles that bound Ω, hence $f \circ \Phi \equiv 0$ on $\bar{\Omega}$. In other words, f vanishes on $\Phi(\bar{\Omega})$, and, in particular, on Γ_a.

This suggests a problem that seems quite hard:

Suppose K is a compact set in S, and K is *not* a (Z)-set. Every $f \in A(B)$ with $f|_K = 0$ must then have some zeros somewhere on $\bar{B}\setminus K$, by definition of (Z)-set. But let E_K be the set of all $z \in \bar{B}\setminus K$ such that $f(z) = 0$ for *every* $f \in A(B)$ that vanishes at every point of K. The question is: *Must E_K be nonempty?*

In the above example of the annulus, the answer was obvious. It is equally obvious whenever K contains any arc that forms part of the boundary of (say) an analytic disc embedded in B. But when there is no such analytic structure in evidence, the question remains open.

Here is a related (probably easier) question:

If E_K intersects S, must E_K intersect B?

10.5.11. Remark. Suppose $0 < \alpha < \frac{1}{2}$, γ is a complex-tangential curve in S, and $f \in A(B) \cap \text{Lip } \alpha$. Theorems 6.4.9 and 6.4.10 imply then that the restriction of f to γ lies in $\text{Lip}(2\alpha)$. (For the moment, we ignore the distinction between the mapping γ and its range.) One might therefore expect (replacing α by 0) that all restrictions of members of $A(B)$ to γ ought to be just a little smoother than merely continuous. However, γ is an interpolation set, and thus exactly the opposite is true: *every* continuous function on γ has an $A(B)$-extension.

Since restriction to γ improves the Lipschitz behavior, some smoothness is lost in the extension process. For example, if some function on γ lies in $\text{Lip}(\frac{1}{2} - \varepsilon)$ for every $\varepsilon > 0$, but is not in $\text{Lip}(\frac{1}{2})$, then none of its $A(B)$-extensions can lie in $\text{Lip}(\frac{1}{4})$.

10.6. Determining Sets

10.6.1. Definition. A set $K \subset S$ is said to be a *determining set*, or simply a *(D)-set*, if only one $f \in A(B)$ (namely $f \equiv 0$) has $f(\zeta) = 0$ for all $\zeta \in K$.

In other words, every $f \in A(B)$ is determined by its restriction to K. To be a *(D)-set*, K thus has to be "sufficiently large" in some sense. However, as is the case for *(Z)-sets*, the important thing in this context is not just the size of a set, but its positioning.

Before we develop the machinery that is needed for the main result of this section (Theorem 10.6.8), let us look at some simple examples.

10.6.2. Examples. (a) If $\sigma(K) > 0$ then K is a *(D)-set*, by Theorem 5.6.4.

(b) If K is the set of all $z \in S$ with $z_n = 0$, then K is obviously not a *(D)-set*. Note that K is a sphere of dimension $2n - 3$. This should be compared to Theorem 10.6.9.

(c) When $n \geq 2$ then $1 < 2n - 2$. Nevertheless, there exist smooth arcs in S that are *(D)-sets*:

Let $\alpha_1, \ldots, \alpha_n$ be positive numbers that are linearly independent over the rationals. Define $\Phi: \mathbb{C} \to \mathbb{C}^n$ by

$$\Phi(\lambda) = n^{-1/2}(e^{i\alpha_1 \lambda}, \ldots, e^{i\alpha_n \lambda}).$$

Let I be an interval (or any set of positive measure) on the real axis R, and put $K = \Phi(I)$. We claim that K is a *(D)-set*.

Suppose $f \in A(B), f|_K = 0$. Then $f \circ \Phi$ is continuous on the closed upper half-plane, holomorphic in its interior, 0 on I. Hence $f \circ \Phi \equiv 0$. Thus $f(\zeta) = 0$ for all $\zeta \in \Phi(R)$. The arithmetic assumptions about $\{\alpha_1, \ldots, \alpha_n\}$ imply that $\Phi(R)$ is dense in the torus defined by $|\zeta_1| = \cdots = |\zeta_n| = n^{-1/2}$

(Kronecker's theorem). By continuity, f vanishes at every point of this torus, hence (see §1.2.1) at every point of the polydisc defined by $|\zeta_i| < n^{-1/2}$, $1 \le i \le n$. It follows that $f \equiv 0$.

10.6.3. Totally Real Manifolds. A smooth manifold M in $\mathbb{C}^n$ is said to be *totally real* if the tangent spaces $T_p(M)$ are totally real for every $p \in M$, as in Definition 10.5.5. We saw in Section 10.5 that there exist smooth (PI)-sets of dimension $n - 1$ in S, and that these are all totally real. By way of contrast, Theorem 10.6.8 will show that every totally real $M \subset S$ (of class C^2), whose dimension is n, is a (D)-set. The torus that occurred in 10.6.2(c) is an example of such a set.

The proof of Theorem 10.6.8 uses carefully controlled holomorphic maps of U into $\mathbb{C}^n$ that carry part of the unit circle T into M. The existence of such maps will be proved by applying the Banach contraction theorem in an appropriate function space H:

10.6.4. Definition. For a fixed $n \ge 1$, let H be the space of all absolutely continuous maps $u: T \to R^n$ whose derivative is in L^2, with the norm

(1) $$\|u\| = \|u\|_2 + \|u'\|_2.$$

Here $u' = du/d\theta$ and

(2) $$\|u\|_2 = \left\{ \frac{1}{2\pi} \int_{-\pi}^{\pi} |u(e^{i\theta})|^2 \, d\theta \right\}^{1/2},$$

where $|u(e^{i\theta})|$ is the euclidean norm of the vector $u(e^{i\theta}) \in R^n$.

With every $u \in H$ one can associate its "harmonic conjugate" u^*. One way to do this is to use the Fourier series

(3) $$u(e^{i\theta}) = \sum_{-\infty}^{\infty} a_k e^{ik\theta}$$

(whose coefficients lie in $\mathbb{C}^n = R^n + iR^n$). Put $\varepsilon_0 = 0$, $\varepsilon_k = 1$ if $k > 0$, $\varepsilon_k = -1$ if $k < 0$, and define

(4) $$u^*(e^{i\theta}) = -i \sum_{-\infty}^{\infty} \varepsilon_k a_k e^{ik\theta},$$

so that

(5) $$(u + iu^*)(e^{i\theta}) = a_0 + 2\sum_{1}^{\infty} a_k e^{ik\theta}.$$

Setting $f = u + iu^*$, we see that f can be extended to a continuous map of $\bar{U}$ into $\mathbb{C}^n$ which is holomorphic in U, such that $f(0) = a_0$.

By Parseval's theorem and (4), $u^* \in H$ whenever $u \in H$; in fact

(6) $$\|u^*\| \leq \|u\|.$$

Equality holds in (6) if and only if $a_0 = 0$.

Finally, note that the sup-norm of u on T satisfies

(7) $$\|u\|_\infty \leq \sum_{-\infty}^{\infty} |a_k| \leq 2\|u\|,$$

by the Schwarz inequality.

10.6.5. Proposition (Bishop [2]). *Let Q be a convex neighborhood of 0 in R^n, with compact closure $\bar{Q}$. Let $h \colon \bar{Q} \to R^n$ be a C^2-map, with $h(0) = 0$ and $h'(0) = 0$. Then there is a constant $K < \infty$ such that*

(1) $$\|h \circ u - h \circ v\| \leq K(\|u\| + \|v\|)\|u - v\|$$

for all $u, v \in H$ whose range lies in Q. In particular,

(2) $$\|h \circ u\| \leq K\|u\|^2.$$

Proof. For each $x \in Q$, $h'(x)$ is a linear operator on R^n whose norm we denote by $|h'(x)|$. Since h is of class C^2 on the compact set $\bar{Q}$, there is a $c < \infty$ such that

(3) $$|h'(x) - h'(y)| \leq c|x - y| \qquad (x, y \in Q).$$

Thus $|h'(x)| \leq c|x|$, and therefore

(4) $$|h(x) - h(y)| \leq c(|x| + |y|)|x - y|.$$

Now fix $u, v \in H$, fix θ, put $x = u(\theta)$, $y = v(\theta)$ (writing θ in place of $e^{i\theta}$). By (4) and 10.6.4(7),

(5) $$|(h \circ u)(\theta) - (h \circ v)(\theta)| \leq 2c(\|u\| + \|v\|)|u(\theta) - v(\theta)|$$

so that

(6) $$\|h \circ u - h \circ v\|_2 \leq 2c(\|u\| + \|v\|)\|u - v\|.$$

Next,

$$|(h \circ u)'(\theta) - (h \circ v)'(\theta)| = |h'(x)u'(\theta) - h'(y)v'(\theta)|$$
$$\le |h'(x) - h'(y)| \, |u'(\theta)| + |h'(y)| \, |u'(\theta) - v'(\theta)|,$$

which, by (3) and 10.6.4(7), is at most

$$2c\|u - v\| \, |u'(\theta)| + 2c\|v\| \, |u'(\theta) - v'(\theta)|.$$

Minkowski's inequality shows therefore that

(7) $$\|(h \circ u)' - (h \circ v)'\|_2 \le 2c(\|u\| + \|v\|)\|u - v\|.$$

Now (1) follows from (6) and (7), with $K = 4c$, and (2) is the special case of (1) in which $v = 0$.

10.6.6. Generic Manifolds. Let M be a C^1-manifold in $\mathbb{C}^n$. Following Pinčuk [1], we say that M is *generic* if the $\mathbb{C}$-span of the tangent space $T_p(M)$ is all of $\mathbb{C}^n$, for every $p \in M$. In other words, it is required that

(1) $$T_p(M) + iT_p(M) = \mathbb{C}^n \qquad (p \in M).$$

It is clear that if (1) holds for *some* $p_0 \in M$, then it holds also for all $p \in M$ that are sufficiently close to p_0.

It is also clear that a totally real manifold is generic if and only if its dimension is n.

Here are some other simple properties:

(a) *Every generic vector space $V \subset \mathbb{C}^n$ contains a totally real generic subspace X.*

Proof. By assumption, V is a real vector space such that $V + iV = \mathbb{C}^n$. Let β be an R-basis of V. Since the $\mathbb{C}$-span of β is $\mathbb{C}^n$, β contains some n vectors that are linearly independent over $\mathbb{C}$. Their R-span X has the desired properties.

(b) *If X is as in (a), then there is an invertible $\mathbb{C}$-linear map Λ of $\mathbb{C}^n$ onto $\mathbb{C}^n$ such that $\Lambda X = R^n$, the R-span of the basis vectors $e_1, \ldots, e_n$.*

Proof. Let L be an R-linear map of X onto R^n, and define Λ by

$$\Lambda(x + iy) = Lx + iLy$$

for $x, y \in X$.

Note that it may not be possible to find a *unitary* Λ such that $\Lambda X = R^n$. For example, take $n = 2$, let X be the set of all

$$(\alpha + i\beta)e_1 + \beta e_2$$

with $\alpha, \beta \in R$. It is easily verified that X is totally real. Since X contains two vectors whose inner product is not real (namely e_1 and $ie_1 + e_2$), no unitary map carries X onto R^2.

(c) *Suppose V and X are as in* (a), *M is a manifold in $\mathbb{C}^n$, $p \in M$, and $V = T_p(M)$. Then M has a submanifold M_1, with $p \in M_1$, such that $T_p(M_1) = X$.*

Proof. Take $p = 0$, without loss of generality. Any R-linear projection P of $\mathbb{C}^n$ onto V is then a 1–1 map of some neighborhood N_1 of 0 in M onto a neighborhood N_2 of 0 in V. Let $M_1 = N_1 \cap P^{-1}(N_2 \cap X)$.

(d) *Suppose Ω is a connected open set in $\mathbb{C}^n$, M is a generic manifold in Ω, $f \in H(\Omega)$, and $f(z) = 0$ for every $z \in M$. Then $f \equiv 0$ in Ω.*

Proof. Fix $p \in M$. By (b) and (c) we may assume, without loss of generality, that $T_p(M) = R^n$. Thus $f(p + te_j)/t$ tends to 0 as t tends to 0 through real values. It follows that $(D_j f)(p) = 0$ for $1 \le j \le n$.

The same is true at every point of M. Thus each $D_j f$ vanishes identically on M. We can repeat the argument, with $D_j f$ in place of f, and so on, to reach the conclusion that $D^\alpha f \equiv 0$ on M, for every multi-index α. The power series expansion of f shows now that $f \equiv 0$ in Ω.

Here is an example, with $n = 2$, based on the fact that the space $\{\lambda e_1 + \bar{\lambda} e_2 : \lambda \in \mathbb{C}\}$ is totally real:

If Ω is a connected neighborhood of 0 in $\mathbb{C}^2$, $f \in H(\Omega)$, and $f(\lambda, \bar{\lambda}) = 0$ whenever $(\lambda, \bar{\lambda}) \in \Omega$, then $f \equiv 0$ in Ω.

10.6.7. We are now ready for the main result. Since it may not be possible to find a unitary transformation in the change of coordinates 10.6.6(b), we may as well state the theorem for arbitrary convex Ω, rather than just for B. In fact, convexity of Ω is not needed either, but then the proof (as given by Pinčuk [1]) requires some further background.

By analogy with $A(B)$, we let $A(\Omega)$ denote the class of all $f \in C(\overline{\Omega})$ that are holomorphic in Ω.

10.6.8. Theorem (Pinčuk [1]). *Suppose Ω is a convex open set in $\mathbb{C}^n$, M is a C^2-manifold in $\partial\Omega$, and there is a point $p \in M$ such that*

(1) $T_p(M) + iT_p(M) = \mathbb{C}^n.$

Then M is a (D)-set for $A(\Omega)$.

Proof. If (1) holds for one p, then it also holds for all points of M that are sufficiently close to p. By §10.6.6, we may therefore assume the following situation, without loss of generality:

$p = 0$, $T_0(M) = R^n$, and there is a ball $Q \subset R^n$, with center 0, and a C^2-map $h: \bar{Q} \to R^n$ with $h(0) = 0$, $h'(0) = 0$, such that

$$(2) \qquad\qquad M = \{x + ih(x): x \in Q\}.$$

Moreover, M is generic, and there is a vector $y \in R^n$, $y \neq 0$, such that the translates $M + ity$ of M lie in Ω whenever $0 < t \leq 1$.

Now choose a constant K, so large that the radius of Q is at least $1/K$, and so that the conclusion of Proposition 10.6.5 holds. Put $\delta = 1/(32\,K)$.

Next, choose $u \in H$ (see §10.6.4) with the following properties: $u(e^{i\theta}) = t(e^{i\theta})y$, where t is a real-valued function, $0 \leq t \leq 1$, t vanishes at every point of some arc $\gamma \subset T$, $t(1) > 0$, and $\|u\| < \delta$. Moreover, let u be an even function of $\theta(-\pi \leq \theta \leq \pi)$. Then u^* is an odd function of θ. In particular, $u^*(1) = 0$.

Our objective is now to solve the functional equation

$$(3) \qquad\qquad g = c - u^* - (h \circ g)^*$$

where $c \in R^n$, $|c| \leq 2\delta$, and $g \in H$.

To do this, let $X = \{g \in H: \|g\| \leq 4\delta\}$. Every $g \in X$ maps T into Q, so that we can define $\Psi: X \to H$ by

$$(4) \qquad\qquad \Psi g = c - u^* - (h \circ g)^*.$$

Then $\|(h \circ g)^*\| \leq \|h \circ g\| \leq K\|g\|^2 \leq \delta$, by Proposition 10.6.5. Since $|c| \leq 2\delta$ and $\|u^*\| \leq \|u\| < \delta$, we see that Ψ maps X into X.

If $g_1, g_2 \in X$, another application of Proposition 10.6.5 shows that

$$(5) \qquad \|\Psi g_1 - \Psi g_2\| \leq \|h \circ g_1 - h \circ g_2\|$$
$$\leq K(4\delta + 4\delta)\|g_1 - g_2\| = \tfrac{1}{4}\|g_1 - g_2\|.$$

Thus Ψ is a contraction of X into X, and since X is a complete metric space, Ψ has a unique fixed point in X.

This solves (3). More precisely, we have proved: *To every $c \in R^n$ with $|c| \leq 2\delta$ corresponds a unique $g_c \in H$ such that $\|g_c\| \leq 4\delta$ and*

$$(6) \qquad\qquad g_c = c - u^* - (h \circ g_c)^*.$$

We need one further property of $\{g_c\}$: *To every $x \in R^n$ with $|x| \leq \delta$ corresponds some $c \in R^n$ with $|c| \leq 2\delta$, such that*

$$(7) \qquad\qquad g_c(1) = x.$$

To prove (7), consider the map

(8) $$c \to g_c(1) = c - (h \circ g_c)*(1).$$

(Recall that $u*(1) = 0$.) A computation similar to (5) shows that

(9) $$\|g_{c_1} - g_{c_2}\| \le 2|c_1 - c_2|.$$

Also, if $|c| = 2\delta$, then

(10) $$|c - g_c(1)| \le 2\|h \circ g_c\| \le 2K\|g_c\|^2 \le \delta.$$

The continuous map (8) thus moves no point of the sphere $\{|c| = 2\delta\}$ by more than δ. Therefore the image of the ball $\{|c| \le 2\delta\}$ covers the ball $\{|x| \le \delta\}$. This establishes (7).

Now define $f_c : T \to \mathbb{C}^n$ by

(11) $$f_c = g_c + ih \circ g_c + iu$$

which, by (6) is the same as

(12) $$f_c = c + i[h \circ g_c + i(h \circ g_c)* + u + iu*].$$

By (12), each f_c has a continuous extension to $\bar{U}$ (which we still call f_c) whose restriction to U is a holomorphic map into $\mathbb{C}^n$. By (2), $g_c + i(h \circ g_c)$ maps T into $M \subset \partial\Omega$. Hence (11) and our choice of u show that $f_c(T) \subset \bar{\Omega}$ and that $f_c(1) \in \Omega$, so that $f_c(U) \subset \Omega$, by the maximum modulus principle and the convexity of Ω. Moreover, $f_c(\gamma) \subset M$, where γ is the arc in T on which $u = 0$.

Assume now that $F \in A(\Omega)$ and $F|_M = 0$. Then $F \circ f_c \in A(U)$ vanishes on γ, hence on U. In particular

(13) $$F(f_c(1)) = 0 \qquad (|c| \le 2\delta).$$

If $|x| \le \delta$ and (7) holds, then

(14) $$f_c(1) = x + ih(x) + iu(1).$$

Thus (13) implies that F vanishes on

(15) $$M_1 = \{x + ih(x) + iu(1): |x| < \delta\} \subset \Omega.$$

Since M_1 is a translate of a portion of M, M_1 is a generic manifold. By 10.6.6(d), $F \equiv 0$ in Ω. This proves that M is a (D)-set for $A(\Omega)$.

Here is an application, also due to Pinčuk [1]:

10.6.9. Theorem. *If $n > 1$ and M is a C^2-manifold of dimension $2n - 2$ in $S = \partial B_n$, then M is a (D)-set for $A(B)$.*

Proof. Since $2n - 2 > n - 1$, it follows from Theorem 10.5.6 that M is not complex-tangential. Hence there is a $p \in M$ at which the complex vector space

$$X_p = T_p(M) + iT_p(M)$$

is strictly larger than $T_p(M)$. The real dimension of X_p is thus $2n$. In other words, $X_p = \mathbb{C}^n$. Now refer to Theorem 10.6.8.

10.7. Peak Sets for Smooth Functions

10.7.1. When $A(B)$ is replaced by any of the algebras

$$(1) \qquad A^m(B) = A(B) \cap C^m(\bar{B}) \qquad (m = 1, 2, 3, \ldots, \infty)$$

then the peak sets, zero sets, and interpolation sets are no longer the same. This happens already when $n = 1$: Every peak set for $A^1(U)$ is finite, every finite subset of the unit circle is a peak set for $A^\infty(U)$ (even for a rational function), but there exist perfect sets that are zero sets and interpolation sets for $A^\infty(U)$, even in the strong sense of the possibility of interpolating all derivatives. These results are due to Carleson [1], Taylor–Williams [1], [2], and Alexander–Taylor–Williams [1].

When $n > 1$, complex-tangential conditions are again important, as in Theorem 10.5.4.

10.7.2. Definitions. For $1 \le m \le \infty$, a compact $K \subset S$ is a *peak set for* $A^m(B)$ if there exists $f \in A^m(B)$ such that $f = 1$ on K, $|f| < 1$ on $\bar{B} \backslash K$.

We say that K is *locally a peak set for* $A^m(B)$ if every point of K has a neighborhood V such that $K \cap \bar{V}$ is a peak set for $A^m(B)$.

If to every $g \in C^m(S)$ corresponds an $f \in A^m(B)$ such that $f = g$ on K, then K is said to be an *interpolation set for* $A^m(B)$.

A C^1-manifold $M \subset S$ is said to be *complex-tangential at a point* $\zeta \in M$ if

$$T_\zeta(M) \subset T_\zeta^{\mathbb{C}}(S).$$

This is a pointwise version of the condition that was discussed in §10.5.2.

As regards peak sets for $A^\infty(B)$, the following results are known.

10.7.3. Theorem. *Let* $n > 1$. *The following three properties of a compact* $K \subset S$ *are equivalent:*

(a) K *is locally a peak set for* $A^{\infty}(B)$.

(b)   *Every point of* K *has a neighborhood* V *such that* $K \cap V$ *lies in a totally real* C^{∞}*-manifold* $M \subset S$, *of dimension* $n - 1$, *that is complex-tangential at every point of* $K \cap V$.

(c)   *Same as* (b), *except that* M *is to be complex-tangential at every point of* M.

The implications (c) $\Rightarrow$ (a) $\Rightarrow$ (b) were proved by Hakim–Sibony [2]. That (b) $\Rightarrow$ (c) was added by Chaumat–Chollet [2], [3]; in the same paper, they prove that all compact subsets of peak sets for $A^{\infty}(B)$ are peak sets as well as interpolation sets for $A^{\infty}(B)$. In an earlier paper (Chaumat–Chollet [1]) they obtained a global version of (c) $\Rightarrow$ (a):

10.7.4. Theorem. *If* $n > 1$, K *is compact*, M *is a complex-tangential* C^{∞}*-manifold in* S, *and* $K \subset M$, *then* K *is a peak set for* $A^{\infty}(B)$.

Nagel [1] showed earlier that every complex-tangential closed C^3-manifold $M \subset S$ is the zero set of some $f \in A^{\infty}(B)$.

It is not known whether all sets that are locally peak sets for $A^{\infty}(B)$ are in fact peak sets for $A^{\infty}(B)$. The difficulty is that the class of all peak sets for $A^{\infty}(B)$ is not closed under the formation of finite unions; see §10.7.7.

We shall prove the implications (a) $\Rightarrow$ (b) $\Rightarrow$ (c) of Theorem 10.7.3, but only in the case $n = 2$, where the manifold M reduces to a curve. For the rest, we refer to the above-cited papers. That (b) implies (c) is quite easy in the case of curves:

10.7.5. Proposition. *Suppose that* $n = 2$, $1 < m \leq \infty$, $E \subset [-1, 1]$, *and* $\gamma : [-1, 1] \to S$ *is a nonsingular* C^m*-curve that satisfies*

(1) $\langle \gamma'(t), \gamma(t) \rangle = 0$ $(t \in E)$.

Then there is a $\delta > 0$ *and a nonsingular* C^{m-1}*-curve* $\Gamma : [-\delta, \delta] \to S$ *such that*

(2) $\Gamma(t) = \gamma(t)$ $(t \in E \cap [-\delta, \delta])$

and such that $\langle \Gamma'(t), \Gamma(t) \rangle = 0$ *for every* $t \in [-\delta, \delta]$.

(The term "nonsingular" means that $\gamma'(t) \neq 0$ for all t.)

Proof. Let γ_1, γ_2 be the components of γ. By a unitary change of variables we can arrange it so that none of $\gamma_1, \gamma_2, |\gamma_1|'$ have a zero on $[-\delta, \delta]$, for some

$\delta > 0$. This means that there are real C^m-functions x, u, v on $[-\delta, \delta]$, with $x > 0, x' > 0$, such that

(3)
$$\gamma_1 = \left(\frac{x}{1+x}\right)^{1/2} e^{iu}, \qquad \gamma_2 = \frac{e^{iv}}{(1+x)^{1/2}}$$

on $[-\delta, \delta]$. Define α by

(4)
$$x'\alpha = xu' + v'$$

and define $\Gamma = (\Gamma_1, \Gamma_2)$ by

(5)
$$\Gamma_1 = \gamma_1 e^{i\alpha}, \qquad \Gamma_2 = \gamma_2 e^{-ix\alpha}.$$

By (4), $\alpha \in C^{m-1}$, hence $\Gamma \in C^{m-1}$. Also, Γ is nonsingular, since the real part of $\Gamma'_1 \exp\{-i(u + \alpha)\}$ is positive.

A simple computation shows that $\langle \gamma', \gamma \rangle = 0$ if and only if $xu' + v' = 0$. Hence $\alpha = 0$ on E, which proves (2).

Finally, to prove that $\langle \Gamma', \Gamma \rangle = 0$ amounts to showing that

(6)
$$x(u + \alpha)' + (v - x\alpha)' = 0.$$

But (6) is an immediate consequence of (4).

10.7.6. Theorem. *Suppose* $n = 2$, $1 < m \le \infty$, $f \in A^m(B)$, f *peaks on* $K \subset S$, $\zeta_0 \in K$, *and* W *is a neighborhood of* ζ_0.

Then there is a neighborhood V *of* ζ_0, $\bar{V} \subset W$, *and there is a nonsingular* C^{m-1}-*curve* $\gamma : [-1, 1] \to S \cap \bar{V}$, *such that*

   (i) $K \cap V$ *lies in the range of* γ, *and*
   (ii) $\langle \gamma'(t), \gamma(t) \rangle = 0$ *whenever* $\gamma(t) \in K \cap V$.

Proof. Extend f to a C^m-function with domain $\mathbb{C}^2$. Assume that $\zeta_0 = e_2 = (0, 1)$, without loss of generality. The Hopf lemma (proved in §15.3.7), applied to the function $w \to f(0, w)$ in the unit disc, shows that $(D_2 f)(e_2) > 0$. By the implicit function theorem, there is a small polydisc $P \subset W$, centered at e_2, whose projection into the z-plane we call Ω, and there is a function $\alpha \in C^m(\Omega)$, such that $f(z, w) = 1$ in P if and only if $w = \alpha(z)$. In particular,

(1)
$$f(z, \alpha(z)) = 1 \qquad (z \in \Omega).$$

Let K_0 be the set of all $z \in \Omega$ such that $(z, \alpha(z)) \in K$. Thus K_0 is the projection of $K \cap P$ into the z-plane. *We claim that*

$$\frac{\partial \alpha}{\partial \bar{z}}, \frac{\partial^2 \alpha}{\partial z \, \partial \bar{z}} \quad and \quad \bar{z} + \overline{\alpha(z)} \frac{\partial \alpha}{\partial z}$$

are 0 *at every point of* K_0.

If $\partial/\partial\bar{z}$ is applied to (1), the result is

$$(2) \qquad (D_2 f)\frac{\partial\alpha}{\partial\bar{z}} + (\bar{D}_2 f)\frac{\partial\bar{\alpha}}{\partial\bar{z}} + \bar{D}_1 f = 0$$

in Ω; the derivatives $D_2 f$, $\bar{D}_2 f$, $\bar{D}_1 f$ are evaluated at $(z, \alpha(z))$. Since $f \in A^2(B)$, $\bar{D}_2 f$ and $\bar{D}_1 f$, as well as any of their first-order derivatives, are 0 on S.
Hence (2) gives $\partial\alpha/\partial\bar{z} = 0$ on K_0.
If $\partial/\partial z$ is applied to (2), one obtains that $\partial^2\alpha/\partial z\,\partial\bar{z} = 0$ on K_0.
Observe next that $u = \mathrm{Re}\,f$ is maximized (relative to S) on K, so that its directional derivatives along tangents to S are 0. Thus

$$(3) \qquad z\bar{D}_2 u - w\bar{D}_1 u = 0 \qquad \text{on } K.$$

Since $\bar{D}_1 f = \bar{D}_2 f = 0$ on $\bar{B}$, (3) is the same as

$$(4) \qquad z\overline{D_2 f} - w\overline{D_1 f} = 0 \quad \text{on } K.$$

Application of $\partial/\partial z$ to (1) shows that

$$(5) \qquad D_1 f + (D_2 f)\frac{\partial\alpha}{\partial z} = 0 \quad \text{on } K_0.$$

If we take conjugates in (4) and use (5), we obtain

$$(6) \qquad \bar{z} + \overline{\alpha(z)}\frac{\partial\alpha}{\partial z} = 0 \quad \text{on } K_0,$$

as claimed.
Next, define $h \in C^m(\Omega)$ by

$$(7) \qquad h(z) = |z|^2 + |\alpha(z)|^2 - 1.$$

Since $|f| < 1$ in B, we have $h \geq 0$ in Ω, and $h(z) = 0$ for all $z \in K_0$.
Since $\partial\alpha/\partial\bar{z} = \partial^2\alpha/\partial z\,\partial\bar{z} = 0$ on K_0, we have

$$(8) \qquad \frac{\partial^2 h}{\partial z\,\partial\bar{z}} = 1 + \left|\frac{\partial\alpha}{\partial z}\right|^2 \geq 1$$

on K_0, hence, in particular, at $z = 0$. One can therefore rotate coordinates in $\mathbb{C}$ so that

$$(9) \qquad h = ax^2 + by^2 + o(|z|^2),$$

with $b > 0$. Then $\partial h/\partial y = 2by + o(|z|)$, and another application of the implicit function theorem shows that there is a nonsingular C^{m-1}-curve γ_0 through 0 which is the zero set of $\partial h/\partial y$ in a (perhaps smaller) neighborhood Ω_0 of 0 in $\mathbb{C}$. (For the moment, we ignore the distinction between a curve and its range.) Since every point of K_0 is a local minimum of h, it follows that $K_0 \cap \Omega_0 \subset \gamma_0$.

Let $P_0 = \{(z, w) \in P : z \in \Omega_0\}$ and put

$$(10) \qquad \Gamma_0(t) = \gamma_0(t)e_1 + \alpha(\gamma_0(t))e_2$$

where t ranges over some parameter interval.

This Γ_0 runs through all points of $K \cap P_0$.

Since $\partial\alpha/\partial\bar{z} = 0$ on K_0,

$$(11) \qquad \Gamma_0'(t) = \gamma_0'(t)\left\{e_1 + \frac{\partial\alpha}{\partial z}(\gamma_0(t))e_2\right\}$$

if $\gamma_0(t) \in K_0$. By (10), (11), and (6),

$$(12) \qquad \langle\Gamma_0'(t), \Gamma_0(t)\rangle = 0 \quad \text{if} \quad \Gamma_0(t) \in K \cap P_0.$$

Finally, put $\gamma = \Gamma_0/|\Gamma_0|$. Then γ is a curve in S, $\gamma(t) = \Gamma_0(t)$ when $\Gamma_0(t) \in K$, and for these values of t we also have $\gamma'(t) = \Gamma_0'(t)$, since $|\Gamma_0|$ has a local minimum there. Thus (12) holds with γ in place of Γ_0. This proves the theorem.

10.7.7. Example (Hakim–Sibony [2]). Take $n = 2$, let K_1 and K_2 consist of all points $(\cos\theta, \sin\theta)$ and $(\cos\theta, i\sin\theta)$, respectively, where $-\pi \leq \theta \leq \pi$.

Then $\frac{1}{2}(z^2 + w^2 + 1)$ peaks on K_1, $\frac{1}{2}(z^2 - w^2 + 1)$ peaks on K_2, but $K_1 \cup K_2$ violates the conclusion of Theorem 10.7.6, with $m = 2$, at the points $(1, 0)$ and $(-1, 0)$.

Thus, although K_1 and K_2 are peak sets of polynomials, their union is not contained in the peak set of any $f \in A^2(B)$.

Chapter 11

Boundary Behavior of H^∞-Functions

11.1. A Fatou Theorem in One Variable

The objective of this preliminary section is Theorem 11.1.2, a one-variable Fatou-type theorem for *nonholomorphic* functions that will be needed in the proof of Theorem 11.2.4.

We begin with a simple lemma about functions on the real line.

11.1.1. Lemma. *Let* $f_1, \ldots, f_N$ *be nonnegative even functions on* $(-\infty, \infty)$ *that are nonincreasing on* $(0, \infty)$, *and define*

(1)
$$J(t_1, \ldots, t_N) = \int_{-\infty}^{\infty} f_1(x - t_1) \cdots f_N(x - t_N)dx,$$

for real numbers $t_1, \ldots, t_N$. *Then*

(2)
$$J(t_1, \ldots, t_N) \leq J(0, \ldots, 0).$$

Proof. It is enough to prove the lemma under the additional assumption that each f_j is continuously differentiable and has compact support.

The case $N = 1$ is trivial. Assume $N > 1$, and assume that the lemma is true with $N - 1$ in place of N. Then (2) holds whenever some two of the t_j are equal, since we can then replace the corresponding two functions f_j by their product, thus reducing the number of factors from N to $N - 1$. In the proof of (2) we may thus assume, without loss of generality, that $t_1 < \cdots < t_{N-1} < t_N$.

Differentiate (1) with respect to t_N, then replace x by $x + t_N$. Since f_N' is odd, we obtain

(3)
$$\frac{\partial J}{\partial t_N} = \int_0^\infty [g(-x) - g(x)]f_N'(x)dx$$

where $g(x) = f_1(x + t_N - t_1) \cdots f_{N-1}(x + t_N - t_{N-1})$. For $x > 0$, $g(x) \leq g(-x)$, since $t_N - t_j > 0$. Hence $\partial J/\partial t_N \leq 0$. This implies that

(4)
$$J(t_1, \ldots, t_{N-1}, t_N) \leq J(t_1, \ldots, t_{N-1}, t_{N-1})$$

234

whenever $t_1 < \cdots < t_{N-1} < t_N$. The right side of (4) is at most $J(0, \ldots, 0)$, by our induction hypothesis. The lemma follows.

11.1.2. Theorem. *Let $Q = (a, b) \times (0, c)$ be an open rectangle in the upper half of $\mathbb{C}$. Suppose that*

(a) $F: Q \to \mathbb{C}$ *is a bounded C^1-function, and*
(b) $\partial F/\partial \bar{z} \in L^p(Q)$ *for some $p > 1$.*
Then $\lim F(x + iy)$ exists for almost every $x \in (a, b)$, as $y \searrow 0$.

Note that (b) represents a considerable weakening of the classical hypothesis that $F \in H^\infty(Q)$, i.e., that $\partial F/\partial \bar{z} = 0$. It seems to be unknown whether the theorem fails when $p = 1$.

The original proof of the theorem (Nagel–Rudin [2]) involved an appeal to the theory of singular integrals. The more elementary proof that follows is patterned after pp. 60–61 of Carleson [3]. That such an elementary proof might exist was suggested by Ahern.

Proof. The hypothesis is preserved when p is replaced by any smaller value (>1). We may thus assume, without loss of generality, that the conjugate exponent q is an integer.

By shrinking (a, b) a little and making c somewhat smaller, we may also assume, without loss of generality, that F is defined and' C^1 on all of $\bar{Q}$, except, of course, on its lower edge $[a, b]$.

Since F is bounded, there is a sequence $\varepsilon_j \searrow 0$ such that the functions $x \to F(x + i\varepsilon_j)$ converge, in the weak*-topology of $L^\infty([a, b])$, to some $\varphi \in L^\infty([a, b])$. Extend F to $\bar{Q}$ by setting $F(x) = \varphi(x)$, $a \le x \le b$.

Let $Q_j = (a, b) \times (\varepsilon_j, c)$. If $z \in Q$ and j is large enough, then $z \in Q_j$. Since $F \in C^1(\bar{Q}_j)$, a standard application of Green's theorem (see, for example, p. 3 of Hörmander [2]) gives then

$$(1) \qquad F(z) = \frac{1}{2\pi i} \int_{\partial Q_j} \frac{F(\zeta)d\zeta}{\zeta - z} - \frac{1}{\pi} \int_{Q_j} \frac{h(w)}{w - z} \, du \, dv$$

where $w = u + iv$, $h = \partial F/\partial \bar{z}$.

The above-mentioned weak*-convergence, combined with the fact that $h \in L^p(Q) \subset L^1(Q)$, shows that we can let $j \to \infty$ in (1), to obtain $F = G - H$, where

$$(2) \qquad G(z) = \frac{1}{2\pi i} \int_{\partial Q} \frac{F(\zeta)d\zeta}{\zeta - z}$$

and

$$(3) \qquad H(z) = \frac{1}{\pi} \int_{Q} \frac{h(w)}{w - z} \, du \, dv.$$

Since G is the Cauchy integral of a bounded function, it is classical (see, for instance, Lemma 2.6 in Chap. V of Stein–Weiss [1] that $\lim G(x + iy)$ exists, as $y \searrow 0$, for almost every $x \in [a, b]$.

Define the oscillation of H at x to be

(4) $\operatorname{osc}(H, x) = \lim_{\delta \searrow 0} [\sup |H(x, y') - H(x, y'')| : 0 < y', y'' < \delta].$

The theorem will be proved as soon as we show that

(5) $\operatorname{osc}(H, x) = 0$ a.e.

The maximal function

(6) $\tilde{H}(x) = \sup\{|H(x + iy| : 0 < y < c\}$

can be used for this purpose. Let φ be any continuous function on $[a, b]$, $0 \le \varphi \le \tilde{H}$, that satisfies $\varphi(x) < \tilde{H}(x)$ wherever $\tilde{H}(x) > 0$. Since $\tilde{H}$ is lower semi-continuous, $\tilde{H}$ is the pointwise limit of an increasing sequence of such φ's. If x_0 is such that $\tilde{H}(x_0) > 0$, then $\varphi(x_0) < |H(x_0 + iy_0)|$ for some y_0, and therefore $\varphi(x) < |H(x + iy_0)|$ for all x sufficiently close to x_0. It follows that there is a Borel function $y : [a, b] \to (0, c)$ such that

(7) $0 \le \varphi(x) \le |H(x + iy(x))|$ $(a \le x \le b).$

By (3), Fubini's theorem, and Hölder's inequality, (7) implies

(8) $\displaystyle \int_a^b \varphi(x)dx \le \left\{ \int_Q |h|^p \right\}^{1/p} \cdot \left\{ \int_Q du\, dv \left| \int_a^b \frac{dx}{|w - x - iy(x)|} \right|^q \right\}^{1/q}.$

The qth power of the second factor in (8) is

(9) $\displaystyle \int_a^b \cdots \int_a^b dx_1 \cdots dx_q \int_a^b du \int_0^c \psi\, dv$

where

(10) $\psi = |u - x_1 + i(v - y(x_1))|^{-1} \cdots |u - x_q + i(v - y(x_q))|^{-1}.$

To symmetrize, let (9') be (9) with $(-c, c)$ in place of $(0, c)$. By Lemma 11.1.1, the v-integral over $(-c, c)$ is maximized when $y(x) \equiv 0$. There are absolute constants, $A_1, A_2 < \infty$ such that

(11) $\displaystyle \int_a^b \frac{dx}{|x - u + iv|} < A_1 + A_2 \log \frac{b - a}{|v|}.$

Hence (9′) is less than some $A_0 = A_0(a, b, q)$. Since (8) holds for every eligible φ, we conclude that

$$(12) \qquad \int_a^b \tilde{H}(x)dx \le A_0 \|h\|_p.$$

To finish, let Q_j be as in the beginning of this proof, let $h_j = 0$ in Q_j, $h_j = h$ in $Q \backslash Q_j$, and define H_j as in (3), with h_j in place of h. Then $H - H_j$ is continuous outside $\bar{Q}_j$, so that

$$(13) \qquad \mathrm{osc}(H, x) = \mathrm{osc}(H_j, x) \le 2\tilde{H}_j(x) \qquad (a \le x \le b).$$

Hence (12) implies

$$(14) \qquad \int_a^b \mathrm{osc}(H, x)dx \le 2A_0 \|h_j\|_p.$$

As $j \to \infty$, $\|h_j\|_p \to 0$, and (5) follows from (14). $\qquad \blacksquare$

11.2. Boundary Values on Curves in S

11.2.1. So far we have encountered two types of results concerning the boundary behavior of H^∞-functions: Korányi's generalization of Fatou's theorem, which asserts that every $f \in H^\infty(B)$ has K-limits at almost all points of S, and the Lindelöf–Čirka theorem that deals with limits at a single point of S. This leaves many questions. For instance, if γ is a smooth curve in S, does every $f \in H^\infty(B)$ have some sort of limit at almost every point of γ, relative to its arc-length measure? The answer turns out to be no or yes, depending on whether γ is or is not complex-tangential. Since complex-tangential curves are peak-sets, the first case is contained in the following simple fact:

11.2.2. Proposition. *If $K \subset S$ is a peak-set for $A(B)$, then there exists an $f \in H^\infty(B)$ which has no limit along any curve in B that ends at a point of K.*

Proof. Let $g \in A(B)$ peak on K. Then $\mathrm{Re}(1 - g) > 0$ on $\bar{B} \backslash K$, so that there is a function $h = \log(1 - g)$, holomorphic in B, with $|\mathrm{Im}\, h| < \pi/2$, such that

$$\mathrm{Re}\, h(z) = \log|1 - g(z)| \to -\infty$$

as $z \to K$. Put $f = \exp(ih)$.

Then $\exp(-\pi/2) < |f(z)| < \exp(\pi/2)$ for all $z \in B$. When z tends to K along any curve Γ, $f(z)$ spirals around the origin infinitely many times.

11.2.3. We now turn our attention to C^1-curves $\varphi: I \to S$ that are *nowhere* complex-tangential. Since $\text{Re}\langle\varphi', \varphi\rangle = 0$ for all C^1-curves in S (§6.4.2), this means that $\text{Im}\langle\varphi', \varphi\rangle \neq 0$ at every point of the parameter interval I. The continuity of φ' implies then that $\text{Im}\langle\varphi', \varphi\rangle$ is either positive on all of I, or negative on all of I. By reversing the parametrization, we can always achieve the first case. One can then find another parametrization which will simplify the proof of Theorem 11.2.4. To do this, let $I = [a, b]$, and define

$$(1) \qquad\qquad \beta(x) = -i \int_a^x \langle\varphi'(t), \varphi(t)\rangle dt$$

for $x \in I$. Then $\beta'(x) \geq \delta > 0$ for some δ and every $x \in I$, so that β has an inverse $\alpha \in C^1$, on $J = [0, \beta(b)]$. By the chain rule,

$$\langle(\varphi \circ \alpha)', \varphi \circ \alpha\rangle = \alpha'\langle\varphi' \circ \alpha, \varphi \circ \alpha\rangle = i(\beta' \circ \alpha)\alpha' = i$$

at every point of J. Setting $\psi = \varphi \circ \alpha$, ψ is a reparametrization of φ that satisfies

$$(2) \qquad\qquad\qquad \langle\psi', \psi\rangle = i.$$

We now come to the main result of this section. It was first proved by Nagel and Rudin [2], under the assumption that φ' satisfies a Lipschitz condition of some positive order. Nagel and Wainger [1] modified the proof so as to eliminate the need for this Lipschitz condition. They introduced the reparametrization (2), and the splitting into radial and tangential components that occurs in Step 3 of the proof of Theorem 11.2.4. The resulting proof applies then to any absolutely continuous φ, provided that $\text{Im}\langle\varphi', \varphi\rangle$ is positive a.e.; in fact, they weaken the hypotheses even further, putting rectifiability of φ in place of absolute continuity.

To avoid technicalities, we confine ourselves here to the case of continuous φ'. Both of the above-mentioned papers deal with arbitrary smoothly bounded domains, not just with B.

11.2.4. Theorem. *Suppose that* $\varphi: [a, b] \to S$ *is a* C^1-*curve that satisfies*

$$(1) \qquad\qquad \langle\varphi'(t), \varphi(t)\rangle \neq 0 \qquad (a \leq t \leq b).$$

Let $f \in H^\infty(B)$. *The restricted K-limit of f exists then at* $\varphi(t)$, *for almost every* $t \in [a, b]$.

"Restricted K-limits" are defined in §8.4.3. The word "restricted" cannot be omitted from the conclusion. Example 8.4.6 shows this, since the circle $\varphi(t) = (e^{it}, 0)$ satisfies (1).

Proof. The idea of the proof is to construct a C^1-map Φ that carries a rectangle $\bar{Q}$ (as in Theorem 11.1.2) into $\bar{B}$, in such a way that

    (i) $\Phi(x) = \varphi(x)$ for $a \le x \le b$,

    (ii) $\Phi(x + iy) \in B$ if $y > 0$,

    (iii) for every $x \in [a, b]$, the curve Γ_x defined by $\Gamma_x(y) = \Phi(x + iy)$ is a "special" approach curve (as defined in §8.4.3) to the point $\varphi(x) \in S$, and

    (iv) the composition $F = f \circ \Phi$ has $\partial F/\partial\bar{z}$ in $L^p(Q)$ for some $p > 1$.

Once we have this, it follows from Theorem 11.1.2 that the limit of f along Γ_x exists for almost every x. The Lindelöf–Čirka theorem 8.4.4 implies then that the restricted K-limits of f exist at the corresponding points $\varphi(x)$.

For convenience, we break the proof into three steps.

Step 1. The Map Φ. The remarks made in §11.2.3 show that it involves no loss of generality to assume that

$$\langle \varphi'(x), \varphi(x) \rangle = i \qquad (a \le x \le b). \tag{2}$$

(The reparametrization in question carries sets of measure 0 to sets of measure 0.)

Extend φ to a C^1-map of $(-\infty, \infty)$ into S, with $\|\varphi'\|_\infty < \infty$. Choose a positive function ψ on $(-1, 1)$, such that

$$\int_{-1}^{1} \psi(t)dt = 1, \qquad \int_{-1}^{1} |t\psi(t)|dt < \frac{1}{\|\varphi'\|_\infty}. \tag{3}$$

Define

$$u(x, y) = \int_{-1}^{1} \varphi(x + ty)\psi(t)dt, \tag{4}$$

$$\Phi(x + iy) = \varphi(x) - yu(x, y), \tag{5}$$

and put $Q = (a, b) \times (0, \tfrac{1}{3})$.

We claim that in Q *(writing* Φ *for* $\Phi(x + iy)$*),*

$$1 - |\Phi|^2 \ge y \tag{6}$$

and

$$\left|\frac{\partial\Phi}{\partial\bar{z}}\right| \le c, \qquad \left|\left\langle\frac{\partial\Phi}{\partial\bar{z}}, \Phi\right\rangle\right| \le c_1 y \tag{7}$$

for some constants $c < \infty, c_1 < \infty$.

Differentiation of (4) gives $|\partial u/\partial x| \leq \|\varphi'\|_\infty$, $|\partial u/\partial y| \leq 1$, so that, setting $\zeta = \varphi(x)$, we have

$$(8) \qquad\qquad |u(x, y) - \zeta| = \left| \int_0^y \frac{\partial u}{\partial y} \right| \leq y.$$

Thus $1 - y \leq \mathrm{Re}\langle \zeta, u \rangle$. By (5),

$$1 - |\Phi|^2 = 2y\, \mathrm{Re}\langle \zeta, u \rangle - y^2|u|^2 \geq 2y(1 - y) - y^2,$$

which gives (6) if $y \leq \frac{1}{3}$. Next, differentiation of (5) leads to

$$\frac{\partial \Phi}{\partial \bar{z}} = \frac{1}{2}[\varphi'(x) - iu(x, y)] - y\frac{\partial u}{\partial \bar{z}} = \frac{1}{2}[\varphi'(x) - i\varphi(x)] + O(y),$$

by another application of (8), since $\partial u/\partial \bar{z}$ is bounded. This proves (7), since (2) implies that

$$\langle \varphi' - i\varphi, \varphi \rangle = 0.$$

Step 2. The Curves Γ_x. Fix $x \in [a, b]$, put $\zeta = \varphi(x)$, and define $\Gamma_x(y) = \Phi(x + iy)$, $0 \leq y \leq \frac{1}{3}$. Write Γ in place of Γ_x, and let $\gamma = \langle \Gamma, \zeta \rangle \zeta$ be the projection of Γ into the complex line through 0 and ζ, as in §8.4.3. Then Γ is a ζ-curve that satisfies

$$|\Gamma - \gamma| \leq |\Gamma - \zeta| = y|u| \leq y,$$

by (4) and (5), so that (6) implies

$$\frac{|\Gamma - \gamma|}{1 - |\gamma|^2} \leq \frac{|\Gamma - \gamma|}{1 - |\Gamma|^2} \leq 1.$$

Hence (compare 8.4.3(4)) every Γ_x is a special ζ-curve, as $y \searrow 0$.

Step 3. Estimation of $(\partial/\partial \bar{z})(f \circ \Phi)$. Assume $f \in H^\infty(B)$, and define $F: Q \to \mathbb{C}$ by $F = f \circ \Phi$.

To each $z \in Q$, let $w = \Phi(z) = r\zeta$, with $\zeta \in S$, $r > 0$. By (6)

$$(9) \qquad\qquad 1 - r^2 = 1 - |w|^2 \geq y.$$

Consider the vectors

$$(10) \qquad\qquad \alpha = \frac{\partial \Phi}{\partial \bar{z}} = \left(\frac{\partial \Phi_1}{\partial \bar{z}}(z), \ldots, \frac{\partial \Phi_n}{\partial \bar{z}}(z) \right)$$

and

(11)
$$\beta = (\overline{(D_1 f)(w)}, \ldots, \overline{(D_n f)(w)}).$$

Since f is holomorphic, the chain rule gives

(12)
$$\frac{\partial F}{\partial \bar{z}} = \langle \alpha, \beta \rangle.$$

The vectors α, β have decompositions

(13)
$$\alpha = \langle \alpha, \zeta \rangle \zeta + \alpha_t, \qquad \beta = \langle \beta, \zeta \rangle \zeta + \beta_t$$

in which the "tangential" components α_t and β_t are orthogonal to ζ. Hence (12) becomes

(14)
$$\frac{\partial F}{\partial \bar{z}} = \langle \alpha, \zeta \rangle \langle \zeta, \beta \rangle + \langle \alpha_t, \beta_t \rangle.$$

Let us look at the three inner products in (14). First, (10) shows that

(15)
$$\langle \alpha, \zeta \rangle = \frac{1}{r} \left\langle \frac{\partial \Phi}{\partial \bar{z}}, \Phi \right\rangle = O(y),$$

by (7). Next, $|\langle \zeta, \beta \rangle| \le |\beta| \le \|f\|_\infty/(1 - |w|)$, by the Schwarz lemma. Hence (9) shows that

(16)
$$|\langle \zeta, \beta \rangle| = O\left(\frac{1}{y}\right).$$

The first summand on the right of (14) is thus bounded in Q.

To estimate $\langle \alpha_t, \beta_t \rangle$, note that $|\alpha_t| \le |\alpha| \le c$, where c is as in (7), and that $\langle w, \alpha_t \rangle = 0$. If $\lambda \in \mathbb{C}$ and $c|\lambda| < y^{1/2}$, it follows from (9) that

(17)
$$|w + \lambda \alpha_t|^2 = |w|^2 + |\lambda|^2 |\alpha_t|^2 < 1.$$

The function

(18)
$$h(\lambda) = f(w + \lambda \alpha_t)$$

is thus holomorphic in the disc $|\lambda| < c^{-1} y^{1/2}$. By the Schwarz lemma,

(19)
$$|h'(0)| \le \|f\|_\infty \cdot cy^{-1/2}.$$

By (10), (11), and the chain rule,

(20) $$h'(0) = \langle \alpha_t, \beta \rangle = \langle \alpha_t, \beta_t \rangle,$$

and comparison of (19) and (20) gives finally

(21) $$\frac{\partial F}{\partial \bar{z}} = O(y^{-1/2}).$$

Consequently, $\partial F/\partial \bar{z} \in L^p(Q)$ for every $p < 2$. This proves statement (iv) made at the start of this proof. As explained there, this implies the theorem.

The following consequence of Theorem 11.2.4 appears in Nagel [2], with a different proof.

11.2.5. Theorem. *If K is a (PI)-set in S and $\varphi: [a, b] \to S$ is a C^1-curve such that*

(1) $$\langle \varphi'(t), \varphi(t) \rangle \neq 0 \qquad (a \leq t \leq b),$$

then $\varphi^{-1}(K)$ has Lebesgue measure 0.

Corollary. *If (1) holds then $\varphi([a, b])$ is not a (PI)-set.*

Proof. Let $E = \varphi([a, b])$. Then $E \cap K$ is a peak-set. By Theorem 11.2.2 there is an $f \in H^\infty(B)$ whose radial limit exists at no point of $E \cap K$. By Theorem 11.2.4, $\varphi^{-1}(E \cap K) = \varphi^{-1}(K)$ has measure 0.

It is an open question whether the converse is true: If $K \subset S$ is compact and $\varphi^{-1}(K)$ has measure 0 for every φ that satisfies (1), is K a (PI)-set?

In view of Theorems 10.5.4 and 10.1.2, the following is a consequence of 11.2.5:

11.2.6. Theorem. *Suppose $\Phi: \Omega \to S$ is a nonsingular C^1-map of an open set $\Omega \subset R^m$ into S. Then $\Phi(\Omega)$ is totally null if and only if Φ is complex-tangential.*

11.2.7. While we are on the subject of curves in S that are nowhere complex-tangential, let us (following Nagel [2]) look at Cauchy integrals of certain measures concentrated on such curves.

To be specific, let $\varphi: [a, b] \to S$ be a C^2-curve such that $\langle \varphi', \varphi \rangle$ is nowhere 0, and let ψ be a complex C^1-function with support in (a, b). There is then a measure $\mu \in M(S)$, concentrated on the range of φ, such that

(1) $$\int_S f \, d\mu = \int_a^b f(\varphi(t))\psi(t)dt$$

for every $f \in C(S)$. Let $F = C[\mu]$, the Cauchy integral of μ. Then

$$(2) \qquad F(z) = \int_a^b \frac{\psi(t)dt}{\{1 - \langle z, \varphi(t)\rangle\}^n} \qquad (z \in B).$$

Since $\langle \varphi, \varphi' \rangle \neq 0$, we have $\langle z, \varphi'(t)\rangle \neq 0$ whenever z is close to $\varphi(t)$. If we express ψ as a sum $\psi_1 + \cdots + \psi_N$ so that the support of ψ_j lies in a small segment (a_j, b_j) (using a partition of unity), and then drop the subscripts j again, we achieve the following: There is a neighborhood V of $\varphi([a, b])$ in $\mathbb{C}^n$ and a $\delta > 0$ such that

$$(3) \qquad |\langle z, \varphi'(t)\rangle| \geq \delta \qquad (z \in V \cap B, a \leq t \leq b).$$

When $z \in V \cap B$, we can therefore rewrite (2) in the form

$$(4) \qquad F(z) = \int_a^b \frac{\langle z, \varphi'(t)\rangle}{\{1 - \langle z, \varphi(t)\rangle\}^n} \cdot \frac{\psi(t)}{\langle z, \varphi'(t)\rangle} dt$$

and integrate by parts. Since ψ and ψ' vanish at a and b,

$$(5) \qquad F(z) = \frac{1}{n-1} \int_a^b \frac{1}{\{1 - \langle z, \varphi(t)\rangle\}^{n-1}} \cdot \frac{d}{dt} \left\{ \frac{\psi(t)}{\langle z, \varphi'(t)\rangle} \right\} dt.$$

Now let $z = r\zeta$, $\zeta \in S$. For $\zeta \in S \setminus V$, $F(r\zeta)$ is bounded, since $\varphi([a, b]) \subset V$. For $\zeta \in V \cap S$ and r sufficiently close to 1, the derivative in (5) is bounded, by (3). Hence there are constants $c_1, c_2 < \infty$ such that

$$(6) \qquad \int_S |F(r\zeta)| d\sigma(\zeta) \leq c_1 + c_2 \int_S \frac{d\sigma(\zeta)}{|1 - r\langle \zeta, \eta\rangle|^{n-1}}.$$

The last integral stays bounded as $r \nearrow 1$ (Proposition 1.4.10).

We have thus proved the following:

The Cauchy integral (2) of the measure μ defined by (1) is in $H^1(B)$.

In particular, we see, when $n > 1$, that there exist measures on S that are singular with respect to σ but whose Cauchy integrals are in $H^1(B)$.

One can go much further if more differentiability is imposed on φ and ψ. For example, let $\varphi \in C^\infty$, $\psi \in C^\infty$, in addition to the preceding requirements. If α is any multi-index and F is given by (2), then $(D^\alpha F)(z)$ is given by an integral like (2), with $n + |\alpha|$ in place of n, and with some C^∞-numerator in place of $\psi(t)$. One can then integrate this by parts any number of times, each time multiplying and dividing by $\langle z, \varphi'(t)\rangle$. This decreases the exponent $n + |\alpha|$ by 1 each time. After a finite number of steps one finds that $D^\alpha F \in H^\infty(B)$. Thus:

If $\varphi: [a, b] \to S$ is a C^∞-curve such that $\langle \varphi', \varphi \rangle$ is nowhere 0, if ψ is a complex C^∞-function with support in (a, b), and if $\mu \in M(S)$ is the measure concentrated on $\varphi([a, b])$ that is defined by (1), then the Cauchy integral of μ lies in $A^\infty(B) = A(B) \cap C^\infty(\bar{B})$.

This topic was developed in more detail by Nagel [2] and Stout [5], [8].

11.2.8. Example. When φ is complex-tangential, the situation is entirely different. For example, take $n = 2$, $\varphi(t) = (\cos t, \sin t)$ on $[-\pi, \pi]$, $\psi(t) = 1/2\pi$. The Cauchy integral under consideration is now

$$(1) \qquad\qquad F(z, w) = \frac{1}{2\pi} \int_{-\pi}^{\pi} \frac{dt}{(1 - z \cos t - w \sin t)^2}.$$

By contour integration, one sees that

$$(2) \qquad\qquad F(z, w) = (1 - z^2 - w^2)^{-3/2}.$$

[In fact, it is enough to prove (2) when z and w are real, and this case reduces (by a translation) to the case $w = 0, 0 < z < 1$.]

There is a unitary transformation U such that, setting $G = F \circ U$, we have

$$(3) \qquad\qquad G(z, w) = (1 - 2zw)^{-3/2}.$$

If we expand $G^{1/2}$ by the binomial theorem and apply Parseval's theorem, we find that $G^{1/2}$ is not in H^2, hence F is not in H^1, in spite of the extreme smoothness of φ and ψ.

11.3. Weak*-Convergence

11.3.1. If $0 < r < 1, \zeta \in S$, and $f \in H^\infty(B)$, we use the familiar notation $f_r(\zeta) = f(r\zeta)$. In the present discussion, the symbol lim will always refer to $r \nearrow 1$.

Consider the following four properties that a measure $\mu \in M(S)$ may or may not have:

(a) For every $f \in H^\infty(B)$, lim $f_r(\zeta)$ exists pointwise a.e. $[|\mu|]$.
(b) For every $f \in H^\infty(B)$, lim f_r exists in the weak*-topology of $L^\infty(|\mu|)$, regarded as the dual of $L^1(|\mu|)$.
(c) For every $f \in H^\infty(B)$, lim $\int_S f_r \, d\mu$ exists.
(d) μ is a Henkin measure.

For example, (a) holds when $\mu = \sigma$ (Theorem 5.6.4), and when μ is arc-length measure on a C^1-curve in S that is nowhere complex-tangential (Theorem 11.2.4).

The following implications are known:

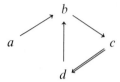

Of these, (a) → (b) follows from the dominated convergence theorem, (b) → (c) is trivial, and (d) → (b) is quite easy:

Suppose μ is a Henkin measure, $f \in H^\infty(B)$, $g \in L^1(|\mu|)$. Let $r_i \nearrow 1$, $t_i \nearrow 1$. Then $\{f_{r_i} - f_{t_i}\}$ is a Montel sequence. By Theorem 9.3.1, $g|\mu|$ is a Henkin measure. Thus

$$\int_S f_{r_i} g \, d|\mu| - \int_S f_{t_i} g \, d|\mu| \to 0$$

as $i \to \infty$. The arbitrariness of $\{r_i\}$ and $\{t_i\}$ implies therefore that $\lim \int f_r g \, d|\mu|$ exists. Thus (b) holds.

The implication (c) ⇒ (d) will be established in Theorem 11.3.4.

11.3.2. Lemma. *If $K \subset S$ is a peak-set for $A(B)$, then there is an $F \in H^\infty(B)$ and a sequence $r_p \nearrow 1$ such that $\|F\|_\infty = 1$ and*

(1) $$\lim_{p \to \infty} |F(r_p \zeta) - (-1)^p| = 0$$

uniformly on K.

Proof. Fix some $g \in A(B)$ that peaks on K. Choose $\varepsilon_p > 0$ so that $\Sigma \varepsilon_p < \infty$. Let Δ_p be the open triangle in $\mathbb{C}$ whose vertices are at $0, 1, i\varepsilon_p$. Put $r_0 = 0$ and let h_1 be a homeomorphism of $\bar{U}$ onto $\bar{\Delta}_1$ that is holomorphic in U (a Riemann map), with $h_1(1) = 1$, and define $f_1 = h_1 \circ g$.

Make the induction hypothesis, for some $p \geq 1$, that r_{p-1} and $f_1, \ldots, f_p \in A(B)$ are chosen, peaking on K. Then there is an r_p, $r_{p-1} < r_p < 1$, $1 - r_p < \varepsilon_p$, such that

(2) $$\sum_{j=1}^p |1 - f_j(r_p \zeta)| < \varepsilon_p \qquad (\zeta \in K),$$

and there is a Riemann map h_{p+1} of $\bar{U}$ onto $\bar{\Delta}_{p+1}$, with $h_{p+1}(1) = 1$, such that $|h_{p+1}| < \varepsilon_{p+1}$ on the compact set $g(r_p \bar{B})$. This follows from a normal family argument, since $h_{p+1}(0)$ can be moved arbitrarily close to 0 in applying the Riemann mapping theorem. Put $f_{p+1} = h_{p+1} \circ g$, to complete the induction.

This defines $\{r_p\}$ and $\{f_p\}$ so that f_p peaks on K, $\mathrm{Im}\, f_p > 0$ in B, and $|f_{p+1}| < \varepsilon_{p+1}$ on $r_p \bar{B}$. Put

$$(3) \qquad\qquad F = \exp\left\{ \pi i \sum_{p=1}^{\infty} f_p \right\}.$$

Then $F \in H(B)$ and $|F| < 1$ in B. For each p,

$$(4) \qquad \left| p - \sum_{j=1}^{\infty} f_j(r_p \zeta) \right| \le \sum_{j=1}^{p} |1 - f_j(r_p \zeta)| + \sum_{p+1}^{\infty} |f_j(r_p \zeta)|.$$

When $\zeta \in K$, the first sum on the right is $< \varepsilon_p$, by (2). In the second sum, $|f_j| < \varepsilon_j$ on $r_{j-1}\bar{B} \supset r_p\bar{B}$. Thus

$$(5) \qquad \left| p - \sum_{j=1}^{\infty} f_j(r_p \zeta) \right| < \sum_{p}^{\infty} \varepsilon_j \qquad (\zeta \in K).$$

Now (1) follows from (3) and (5).

11.3.3. Lemma. *If μ has property* (c) *(§11.3.1) and K is a peak-set, then* $\mu(K) = 0$.

Proof. Put $\mu(K) = \alpha$. Choose $\varepsilon > 0$. Associate F and $\{r_p\}$ to K, as in Lemma 11.3.2. If h is a sufficiently high power of some member of $A(B)$ that peaks on K, then

$$(1) \qquad\qquad \int_{S \setminus K} |h|\, d|\mu| < \varepsilon.$$

The same inequality holds then with h_r in place of h, if r is sufficiently close to 1. For such r, (1) implies

$$(2) \qquad\qquad \left| \int_S F_r h_r\, d\mu - \int_K F_r h_r\, d\mu \right| < \varepsilon.$$

Since $Fh \in H^\infty(B)$ and μ satisfies (c), the integrals over S converge to some limit L as $r \nearrow 1$. Lemma 11.3.2 shows that the integrals over K converge to α when $r = r_{2p}$ and $p \to \infty$, whereas they converge to $-\alpha$ when $r = r_{2p-1}$. Thus $|L - \alpha| \le \varepsilon$, $|L + \alpha| \le \varepsilon$, and therefore $|\alpha| \le \varepsilon$.

11.3.4. Theorem. *If $\mu \in M(S)$ and*

$$\lim_{r \to 1} \int_S f_r\, d\mu$$

exists for every $f \in H^\infty(B)$, then μ is a Henkin measure.

Proof. If K is any compact set in S that is totally null (i.e., $\rho(K) = 0$ for every representing measure $\rho \in M_0$) then K is a peak-set (Theorem 10.1.2), hence $\mu(K) = 0$, by Lemma 11.3.3. The totally singular part of μ in its GKS-decomposition relative to M_0 (see Theorem 9.4.4) is therefore 0. Consequently, $\mu \ll \rho$ for some $\rho \in M_0$, hence μ is a Henkin measure, by Theorem 9.3.1.

11.3.5. It is not known whether the missing arrow can be inserted into the diagram in §11.3.1, i.e., whether every $f \in H^\infty$ has radial limits a.e. with respect to every Henkin measure. Here is another way of asking the same question:

If $f \in H^\infty(B)$ and E_f is the exceptional set of points $\zeta \in S$ at which $\lim f(r\zeta)$ fails to exist, does it follow that E_f is totally null?

11.4. A Problem on Extreme Values

11.4.1. A Conjecture. Suppose $f \in A(B)$ and f is not constant. Let $E_1(f)$ be the set where $|f|$ attains its maximum. (Related sets E_2 and E_3 will be defined presently.) Thus $E_1(f)$ is the set of all $\zeta \in S$ at which $|f(\zeta)| = \|f\|_\infty$. It is probably true that

(*) $$\sigma(E_1) = 0 \quad \text{when} \quad n > 1.$$

This has some (perhaps only superficial) resemblance to the inner function problem which will be discussed in Chapter 19. The following definitions will make it easier to describe the progress that has been made so far toward proving (*).

11.4.2. Definitions. Again, $f \in A(B)$, f is not constant, and, from now on, $n > 1$. We associate three subsets of S to f:

$E_1 = E_1(f)$ is as in §11.4.1.

$E_2 = E_2(f)$ is the set of all $\zeta \in S$ that have a neighborhood $N_\zeta \subset S$ such that $f(\zeta)$ is a boundary point of $f(N_\zeta)$. In other words, $\zeta \in E_2$ if and only if the restriction of f to S fails to be an open mapping at ζ.

$E_3 = E_3(f)$ is the set of all $\zeta \in S$ that are not limit points of the associated variety

$$V_\zeta = \{z \in B : f(z) = f(\zeta)\}.$$

Note that it is not required, in the definition of E_3, that ζ be an isolated point of the set of all z in the *closed* ball $\bar{B}$ at which $f(z) = f(\zeta)$. For example, if f peaks on K, then $K \subset E_3(f)$, simply because V_ζ is empty for every $\zeta \in K$.

Since $f^{-1}(\partial(f(\bar{B}))) \subset E_2$, it is clear that $E_1 \subset E_2$ and that, in general, $E_1 \neq E_2$.

We shall see presently that $E_2 \subset E_3$.

For $k = 1, 2, 3, \ldots$, let $A^k(B) = A(B) \cap C^k(\bar{B})$.

Sibony [1] has obtained the following results (always assuming $n > 1$):

(i)   If $f \in A(B) \cap \mathrm{Lip}\ 1$, then $\sigma(E_1) = 0$.
(ii) If $f \in A^{2n-2}(B)$, then $\sigma(E_2) = 0$.
(iii) If $f \in A^2(B)$, then the n-dimensional Hausdorff measure of E_1 is finite.

The conclusion of (iii) is of course much stronger than $\sigma(E_1) = 0$, since $n < 2n - 1$. The n-torus that occurs in 10.1.5(e) is an example that is relevant to (iii). We shall not deal with (iii) here, but will weaken the hypotheses and strengthen the conclusions of (i) and (ii):

(iv) If $f \in A(B) \cap \mathrm{Lip}\ \alpha$ for some $\alpha > \frac{1}{2}$, then $\sigma(E_3) = 0$.

Theorem 11.4.7 will actually give a more precise conclusion. We first demonstrate some other properties of E_3.

11.4.3. Proposition. *Suppose* $n > 1, f \in A(B)$, *fix* t, $-1 < t < 1$, *and define*

$$\Omega = \{z \in B: x_1 > t\}, \qquad \omega = \{z \in S: x_1 > t\}.$$

If $p \in \Omega$ *and* $V_p = \{z \in \Omega: f(z) = f(p)\}$, *then the closure of* V_p *intersects* ω.

(As usual, $x_1 = \mathrm{Re}\ z_1$.) This proposition follows from a general maximum modulus theorem on varieties (Theorem 14.1.5) but a simple *ad hoc* proof can be given:

Proof. Write V for V_p and assume $f(p) = 0$, without loss of generality. Since $\bar{V}$ is compact, there is a point $a \in \bar{V}$ such that $\mathrm{Re}\ a_1 \geq \mathrm{Re}\ z_1$ for all $z \in \bar{V}$. If $a \in \omega$, we are done.

Assume that $a \in V$. Then there is an $r > 0$ such that the points $a + \varepsilon e_1 + \lambda e_2$ lie in B whenever $0 \leq \varepsilon \leq r, |\lambda| \leq r$. Define

$$g_\varepsilon(\lambda) = f(a + \varepsilon e_1 + \lambda e_2),$$

for $0 \leq \varepsilon \leq r, |\lambda| \leq r$. When $\varepsilon > 0$, our choice of a shows that g_ε has no zero in the disc $|\lambda| \leq \varepsilon$. Since $g_0(0) = 0$ and $g_\varepsilon \to g_0$ uniformly as $\varepsilon \searrow 0$, we conclude that $g_0(\lambda) = 0$ for all λ with $|\lambda| \leq r$. It follows that $V \supset B \cap L$, where L is the complex line $\{a + \lambda e_2: \lambda \in \mathbb{C}\}$. The closure of $B \cap L$ intersects ω in a circle.

Corollary 1. $E_3(f)$ *has empty interior.*

Indeed, the proposition asserts that ω contains points that are not in E_3, no matter how small $1 - t$ is.

Corollary 2. $f(\Omega) \subset f(\omega)$.

In particular, $f(B) \subset f(S)$ for every $f \in A(B)$.

Corollary 3. $E_2(f) \subset E_3(f)$ if f is not constant.

Proof. Suppose $e_1 \in E_2$. Then t can be chosen in the proposition so that $f(e_1)$ is a boundary point of $f(\omega)$. Since $f(\Omega)$ is open, it follows from Corollary 2 that $f(e_1)$ is not a point of $f(\Omega)$. Hence $e_1 \in E_3$.

The function $(1 - z_1)^3$ shows that E_2 can be a proper subset of E_3.

11.4.4. Proposition. $E_3(f)$ *is a set of type* F_σ.

Proof. Fix f, define V_ζ as in §11.4.2, and, for $k = 1, 2, 3, \ldots$, let X_k be the set of all $\zeta \in S$ whose distance from V_ζ is $< 1/k$. Since

$$S \backslash E_3 = \bigcap X_k,$$

it is enough to prove that each X_k is open.

Fix k, fix $\zeta \in X_k$. Then there is a $z \in V_\zeta$ and an $\varepsilon > 0$ such that

$$|z - \zeta| + 2\varepsilon < \frac{1}{k}.$$

Let W be the open ball with center z, radius ε. Then $f(W)$ is a neighborhood of $f(\zeta) = f(z)$ in $\mathbb{C}$. Hence there is a $\delta, 0 < \delta < \varepsilon$, such that $f(\eta) \in f(W)$ for all $\eta \in S$ with $|\eta - \zeta| < \delta$. Since V_η then intersects W, the distance from η to V_η is less than

$$\delta + |z - \zeta| + \varepsilon < \frac{1}{k},$$

so that $\eta \in X_k$. Thus X_k is open and the proof is complete.

11.4.5. Lemma. *If* $h \in H(U), 0 < |h(\lambda)| < 1$ *for all* $\lambda \in U$, *and* $h(0) = c$, *then*

$$|h'(0)| \leq 2|c| \log\left(\frac{1}{|c|}\right).$$

Proof. If $g = (\log h - \log c)/(\log h + \log c)$ then g maps U into U and $g'(0) = h'(0)/2c \log c$. Now apply the Schwarz lemma.

11.4.6. Carleson Sets. Given a compact K on a circle T, let J_i ($i = 1, 2, 3, \ldots$) be the components of $T \backslash K$ (the collection $\{J_i\}$ is finite or countable), and let δ_i be the length of the arc J_i. Then K is said to be a *Carleson set* if

   (a) K has Lebesgue measure 0, and
   (b) $\Sigma \delta_i \log(1/\delta_i) < \infty$.

It is easy to find sets K (even countable ones with only one limit point) that satisfy (a) but violate (b).

Suppose $f \in A(U) \cap \text{Lip } \varepsilon$ for some $\varepsilon > 0$, $f \not\equiv 0$, and K is the set of all boundary points of U where $f = 0$. The fact that

$$\int_{-\pi}^{\pi} \log|f(e^{i\theta})| d\theta > -\infty$$

leads then fairly directly to the conclusion that K is a Carleson set.

This was proved by Carleson [1]; he also obtained a converse, which will not concern us here: Every Carleson set is the set of zeros of some $f \in A^\infty(U)$.

We are finally ready for the result stated at the end of §11.4.2.

11.4.7. Theorem. *Suppose* $n > 1$, $\frac{1}{2} < \alpha \le 1$, *and*

   (i) $f \in A(B) \cap \text{Lip } \alpha$,
   (ii)   *the origin is not a critical point of* f.

Then there is at most one complex line L *through the origin such that the closure of* $E_3(f) \cap L$ *fails to be a Carleson set.*

The meaning of (ii) is that the gradient of f is not 0 at the origin: at least one of the number $(D_1 f)(0), \ldots, (D_n f)(0)$ is $\ne 0$.

Proof. Assume $\|f\|_\infty = 1$. Put $\alpha - \frac{1}{2} = \varepsilon$. Put

(1) $g(\lambda) = (D_2 f)(\lambda e_1)$ $(\lambda \in U)$.

As soon as we show that

   (I) g *extends to a function on* $\bar{U}$ *that belongs to* $A(U) \cap \text{Lip } \varepsilon$, *and*
   (II) $g(\lambda) = 0$ *if* $\lambda e_1 \in E_3(f)$,

the proof is essentially done.

To see this, let K be the closure of the set of all $\lambda \in T$ such that $\lambda e_1 \in E_3(f)$. Then $g(\lambda) = 0$ for all $\lambda \in K$, by (II) and the continuity of g on $\bar{U}$, and if K is not a Carleson set, then (I) implies that $g \equiv 0$. In particular, $(D_2 f)(0) = 0$. Of course, one proves in the same way that $(D_j f)(0) = 0$ for $j = 2, 3, \ldots, n$.

Now let L be any complex line through the origin, not necessarily through e_1. By means of a unitary transformation, the preceding paragraph implies the following: *If the closure of $L \cap E_3(f)$ is not a Carleson set, then all first-order directional derivatives of f at 0, in directions orthogonal to L, are 0.*

This can obviously not happen for two distinct L's, unless the gradient of f vanishes at the origin.

The proof of (I) uses the radial derivative $\mathscr{R}f$ defined in §6.4.4. The constants M_1, M_2 below depend only on α and the Lipschitz constant of f. Theorem 6.4.9 shows that

$$(2) \qquad |(\mathscr{R}f)(z)| \le M_1(1 - |z|)^{\alpha - 1} \qquad (z \in B)$$

and therefore Lemma 6.4.7 gives

$$(3) \qquad |(\mathscr{R}D_2 f)(\lambda e_1)| \le M_2(1 - |\lambda|)^{\alpha - 3/2} \qquad (\lambda \in U).$$

By (1), (3) is the same as

$$(4) \qquad |g'(\lambda)| \le M_2(1 - |\lambda|)^{\varepsilon - 1} \qquad (\lambda \in U).$$

The case $n = 1$ of Theorem 6.4.10 (this is the Hardy–Littlewood result mentioned in §6.4.3) shows that (4) implies statement (I).

To prove (II), suppose $\zeta = e^{i\theta}e_1 \in E_3(f)$, for some fixed θ. There is then a neighborhood N of ζ in $\mathbb{C}^n$ such that $f - f(\zeta)$ has no zero in $B \cap N$. If $0 < r < 1$, $1 - r$ is sufficiently small, and

$$(5) \qquad h(\lambda) = f(r\zeta + \lambda e_2) - f(\zeta) \qquad (|\lambda| < (1 - r^2)^{1/2})$$

it follows that $0 < |h(\lambda)| < 2$. Hence (1) and Lemma 11.4.5 imply

$$(6) \qquad |g(re^{i\theta})| = |(D_2 f)(r\zeta)| = |h'(0)| \le \frac{2|c| \log(2/|c|)}{(1 - r^2)^{1/2}}$$

where $c = h(0)$. Since $f \in \text{Lip}(\frac{1}{2} + \varepsilon)$,

$$(7) \qquad |c| = |h(0)| = |f(r\zeta) - f(\zeta)| \le M(1 - r)^{\varepsilon + 1/2}.$$

By (6) and (7), $g(re^{i\theta}) \to 0$ as $r \to 1$. This proves (II).

Corollary. *If $n > 1$, $\alpha > \frac{1}{2}$ and $f \in A(B) \cap \text{Lip } \alpha$, then $\sigma(E_3(f)) = 0$.*

If f is constant, there is nothing to prove. If not, then some point of B is not a critical point of f, hence 0 is not a critical point of $F = f \circ \psi$ for some $\psi \in \text{Aut}(B)$. The theorem applies to F and shows (by 1.4.7(1)) that $\sigma(E_3(F)) = 0$. Finally, $E_3(f) = \psi(E_3(F))$, and ψ preserves sets of measure 0.

11.4.8. Remark. The conclusion of Theorem 11.4.7 may fail if hypothesis (ii) is dropped:

When $n = 2$ and $f(z, w) = zw$, then $E_3(f)$ is the 2-torus $|z| = |w| = 1/\sqrt{2}$. (In this example, $E_1 = E_2 = E_3$.)

11.4.9. Remark. If $f \in A^1(B)$, f is not constant, and $n > 1$, let $E_4(f)$ be the set of all $\zeta \in S$ at which the equations

$$\bar{\zeta}_j(D_k f)(\zeta) = \bar{\zeta}_k(D_j f)(\zeta)$$

hold for $j, k = 1, \ldots, n$. (These are the conjugates of the tangential Cauchy–Riemann equations; see Chapter 18.)

The proof of Lemma 2 in Sibony [1] shows that if one of these equations fails at a point ζ, i.e., if $\zeta \notin E_4(f)$, then ζ is a limit point of V_ζ, hence $\zeta \notin E_3(f)$. Since $E_4(f)$ is closed, it follows that *the closure of $E_3(f)$ lies in $E_4(f)$.*

Chapter 12

Unitarily Invariant Function Spaces

This chapter deals with a subject that is basically a topic in harmonic analysis and which, at first glance, may seem to have little to do with our principal concern, namely with holomorphic functions. Nevertheless, one of its main results (Theorem 12.3.6) will be essential later in the classification of Moebius–invariant spaces, an obviously function-theoretic topic.

An interesting aspect of the second half of the chapter—a description of the $\mathscr{U}$-invariant subalgebras of $C(S)$—is that the structure of these algebras is more complicated in dimension 2 than in any other dimension.

Almost everything in this chapter (except for Section 12.1) comes from Nagel–Rudin [1] and Rudin [15].

12.1. Spherical Harmonics

This preparatory section contains proofs of two basic facts about spherical harmonics: their linear span is dense in $C(S)$, and harmonics of different degrees are orthogonal. Since the complex structure of C^n is irrelevant in this context, we shall temporarily use a euclidean space R^N as our setting.

12.1.1. Definitions. For $k = 0, 1, 2, \ldots$, $\mathscr{P}_k$ denotes the space of all homogeneous complex-valued polynomials on R^N of degree k, and $\mathscr{H}_k$ is the space of all $f \in \mathscr{P}_k$ that satisfy $\Delta f = 0$, where

$$\Delta = \frac{\partial^2}{\partial x_1^2} + \cdots + \frac{\partial^2}{\partial x_N^2}.$$

Naturally, the term "homogeneous" refers here to real scalars: if $f \in \mathscr{P}_k$ then $f(tx) = t^k f(x)$, for $x \in R^N$, $t \in R$.

Begin harmonic, each $f \in \mathscr{H}_k$ is uniquely determined by its restriction to the unit sphere S. These restrictions are the so-called *spherical harmonics* of degree k.

We shall freely identify $\mathscr{H}_k$ with its restriction to S.

12.1.2. Theorem. *If* $f \in \mathscr{H}_k$, $g \in \mathscr{H}_m$, *and* $k \neq m$, *then*

$$\int_S f \bar{g} \, d\sigma = 0.$$

Proof. The homogeneity of f and g shows that $\partial f / \partial r = kf$ and $\partial g / \partial r = mg$ on S, where $\partial / \partial r$ denotes the radial derivative. Since $\Delta f = 0 = \Delta g$, one of Green's identities gives

$$(m - k) \int_S f \bar{g} \, d\sigma = \int_S \left(f \frac{\partial \bar{g}}{\partial r} - \bar{g} \frac{\partial f}{\partial r} \right) d\sigma = \frac{1}{N} \int_B (f \Delta \bar{g} - \bar{g} \Delta f) dv = 0.$$

12.1.3. Theorem. *Each* $\mathscr{P}_k$ *is a direct sum*

(1) $$\mathscr{P}_k = \mathscr{H}_k \oplus |x|^2 \mathscr{H}_{k-2} \oplus |x|^4 \mathscr{H}_{k-4} \oplus \cdots$$

Here $|x|^2 = x_1^2 + \cdots + x_N^2$, and the sum stops when the subscript reaches 1 or 0.

Proof. Each $f \in \mathscr{P}_k$ has the form $f(x) = \Sigma f_\alpha x^\alpha$, where $f_\alpha \in \mathbb{C}$ and α ranges over the multi-indices with $|\alpha| = k$. The inner product

(2) $$\langle f, g \rangle_k = \sum_\alpha \alpha! f_\alpha \bar{g}_\alpha$$

turns $\mathscr{P}_k$ into a finite-dimensional Hilbert space.

Suppose $|\alpha| = |\beta| = k$. Then $D^\alpha x^\beta = \alpha!$ if $\alpha = \beta$, and is 0 if $\alpha \neq \beta$. Hence (2) is the same as

(3) $$\langle f, g \rangle_k = f(D)\bar{g} \qquad (f, g \in \mathscr{P}_k),$$

where $f(D) = \Sigma f_\alpha D^\alpha = \Sigma f_\alpha (\partial / \partial x_1)^{\alpha_1} \cdots (\partial / \partial x_N)^{\alpha_N}$.

If $h(x) = |x|^2 g(x)$ and $g \in \mathscr{P}_{k-2}$ then $h(D) = \Delta g(D)$, so that, for any $f \in \mathscr{P}_k$, (3) implies

(4) $$\langle |x|^2 g, f \rangle_k = \Delta g(D)\bar{f} = g(D)\Delta \bar{f} = \langle g, \Delta f \rangle_{k-2}$$

since $\Delta \cdot g(D) = g(D) \cdot \Delta$ and $\Delta \bar{f} = \overline{\Delta f}$. This shows that $f \perp |x|^2 \mathscr{P}_{k-2}$ (in the sense of the inner product (2)) if and only if $\Delta f \perp \mathscr{P}_{k-2}$, i.e., if and only if $\Delta f = 0$ (since $\Delta f \in \mathscr{P}_{k-2}$). In other words

(5) $$\mathscr{P}_k = \mathscr{H}_k \oplus |x|^2 \mathscr{P}_{k-2}.$$

Repeat the same argument with $\mathscr{P}_{k-2}$ in place of $\mathscr{P}_k$, and proceed. Finally, note that $\mathscr{P}_1 = \mathscr{H}_1$ and $\mathscr{P}_0 = \mathscr{H}_0$.

Corollary. *The linear span of* $\{\mathscr{H}_k: k = 0, 1, 2, \ldots\}$ *is dense in* $C(S)$.

Proof. On S, $|x| = 1$, so that every $f \in \mathscr{P}_k$ is a sum of spherical harmonics. The Stone–Weierstrass theorem implies that the linear span of $\{\mathscr{P}_k: k = 0, 1, 2, \ldots\}$ is dense in $C(S)$.

12.1.4. Although we shall not need it later, let us note one other consequence of 12.1.3(4):

If $g \in \mathscr{P}_{k-2}$ and $g \perp \Delta \mathscr{P}_k$, then $|x|^2 g \perp \mathscr{P}_k$. Since $|x|^2 g \in \mathscr{P}_k$, it follows that $|x|^2 g = 0$, hence $g = 0$. Conclusion: $\mathscr{P}_{k-2} = \Delta \mathscr{P}_k$.

Every polynomial is therefore the Laplacian of a polynomial.

Chapter IV of Stein and Weiss [1] contains more details about spherical harmonics, and their relation to Fourier analysis. The inner product 12.1.3(2) occurs there.

12.2. The Spaces $H(p, q)$

12.2.1. Definition. For nonnegative integers p and q (and fixed dimension n), $H(p, q)$ is the vector space of all *harmonic* homogeneous polynomials on $\mathbb{C}^n$ that have total degree p in the variables $z_1, \ldots, z_n$ and total degree q in the variable $\bar{z}_1, \ldots, \bar{z}_n$.

Thus every $f \in H(p, q)$ has *bidegree* (p, q).

Note that $H(p, 0)$ consists of holomorphic polynomials, and $H(0, q)$ consists of polynomials whose complex conjugates are holomorphic.

The case $n = 1$ is somewhat special. In that case, $\dim H(p, 0) = \dim H(0, q) = 1$, but $H(p, q) = \{0\}$ if both $p > 0$ and $q > 0$. These zero-dimensional spaces are henceforth excluded from consideration.

If we identify $\mathbb{C}^n$ with R^{2n}, it is clear that $H(p, q) \subset \mathscr{H}_k$ whenever $p + q = k$. Actually, more is true:

12.2.2. Proposition. $\mathscr{H}_k$ *is the sum of the pairwise orthogonal spaces* $H(p, q)$, *where* $p + q = k$.

Proof. Suppose $(p, q) \neq (r, s)$, $p + q = r + s = k$, $f \in H(p, q)$, and $g \in H(r, s)$. Then $p - q \neq r - s$, so that

$$\int_{-\pi}^{\pi} (f\bar{g})(e^{i\theta}\zeta)d\theta = (f\bar{g})(\zeta) \int_{-\pi}^{\pi} e^{i(p-q+s-r)\theta}\, d\theta = 0$$

for every $\zeta \in S$. This gives the orthogonality

$$\int_S f\bar{g}\, d\sigma = 0.$$

Next, fix $f \in \mathscr{H}_k$. Then $f = f_0 + f_1 + \cdots + f_k$, where f_i has bidegree $(i, k - i)$, so that Δf_i has bidegree $(i - 1, k - i - 1)$ (or is 0, when $i = 0$ or $i = k$). No cancellation can therefore account for the vanishing of the sum

$$\sum_{i=0}^{k} \Delta f_i = \Delta f = 0.$$

It follows that each f_i is harmonic. Thus $f_i \in H(i, k - i)$. This completes the proof.

12.2.3. Theroem. $L^2(\sigma)$ is the direct sum of the pairwise orthogonal spaces $H(p, q)$, $0 \le p < \infty$, $0 \le p < \infty$.

More explicitly, every $f \in L^2(\sigma)$ has a unique expansion $f = \Sigma f_{pq}$, with $f_{pq} \in H(p, q)$, that converges unconditionally to f in the L^2-norm topology.

Proof. Combine Proposition 12.2.2 with Theorem 12.1.2 and the Corollary to Theorem 12.1.3.

12.2.4. Definitions. A space Y of functions on S is said to be *unitarily invariant*, or simply $\mathscr{U}$-*invariant*, if $f \circ U \in Y$ whenever $f \in Y$ and $U \in \mathscr{U}$. We may also call such spaces $\mathscr{U}$-*spaces*.

For example, every $H(p, q)$ is a $\mathscr{U}$-space, since $\Delta(f \circ U) = (\Delta f) \circ U$.

If Y is a $\mathscr{U}$-space and T is an operator with domain Y that satisfies

$$T(f \circ U) = (Tf) \circ U$$

for all $f \in Y$ and $U \in \mathscr{U}$, we say that T *commutes with* $\mathscr{U}$.

For example, the Laplacian commutes with $\mathscr{U}$.

Here is another example: Let Y be a closed subspace of $L^2(\sigma)$ that is $\mathscr{U}$-invariant. Then $Y^\perp$ is also $\mathscr{U}$-invariant (since $f \to f \circ U$ is a unitary operator on $L^2(\sigma)$, for every $U \in \mathscr{U}$, by the $\mathscr{U}$-invariance of σ), and the projection $\pi: L^2 \to Y$ whose nullspace is $Y^\perp$ commutes with $\mathscr{U}$.

In particular, this is true of the orthogonal projection π_{pq} of $L^2(\sigma)$ onto $H(p, q)$.

To simplify notation, we shall sometimes use $[f, g]$ to denote the inner product in $L^2(\sigma)$:

$$[f, g] = \int_S f\bar{g} \, d\sigma.$$

12.2.5. Theorem. *Fix* (p, q). *To every* $z \in S$ *corresponds then a unique* $K_z \in H(p, q)$ *that satisfies*

(1) $(\pi_{pq} f)(z) = [f, K_z]$ $(f \in L^2(\sigma))$.

These kernel functions K_z have the following additional properties:

$$(2) \qquad\qquad K_z(w) = \overline{K_w(z)} \qquad (z, w \in S),$$

$$(3) \qquad\qquad \pi_{pq} f = \int_S f(\zeta) K_\zeta \, d\sigma(\zeta) \qquad (f \in L^2(\sigma))$$

$$(4) \qquad\qquad K_{Uz} = K_z \circ U^{-1} \qquad (U \in \mathcal{U}),$$

$$(5) \qquad\qquad K_z = K_z \circ V$$

for all $V \in \mathcal{U}$ that fix z, and

$$(6) \qquad\qquad K_z(z) = K_w(w) > 0 \qquad (z, w \in S).$$

Note: Since $K_\zeta \in C(S)$, the vector-valued integral (3) allows us to extend the domain of π_{pq} to $L^1(\sigma)$.

Proof. Let us write π for π_{pq}. Since

$$f \to (\pi f)(z)$$

is a bounded linear functional on L^2, there is a unique $K_z \in L^2$ that satisfies (1). Since $\pi f = 0$ for every $f \perp H(p, q)$, it follows that $K_z \in H(p, q)$. When $f \in H(p, q)$, (1) becomes

$$f(z) = [f, K_z].$$

In particular, $K_w(z) = [K_w, K_z]$. This proves (2), and (2) shows that (3) is just another way of writing (1). Since π commutes with $\mathcal{U}$,

$$[f, K_{Uz}] = (\pi f)(Uz) = \pi(f \circ U)(z) = [f \circ U, K_z] = [f, K_z \circ U^{-1}]$$

for every $f \in L^2(\sigma)$. (The last equality depends on the $\mathcal{U}$-invariance of σ.) This proves (4), hence also its special case (5). Finally,

$$K_{Uz}(Uz) = (K_z \circ U^{-1})(Uz) = K_z(z)$$

is another consequence of (4). It proves (6), because $K_z(z) = [K_z, K_z] > 0$.

The reason for mentioning (5) in the theorem just proved is that this property plays a role in the following proposition, which may be regarded as another characterization of K_z:

12.2.6. Proposition. *For each $z \in S$, $H(p, q)$ contains a unique f such that $f(z) = 1$ and $f = f \circ V$ for every $V \in \mathcal{U}$ that fixes z.*

Proof. The existence of f follows from 12.2.5(5). To prove the uniqueness, assume $z = e_1$, without loss of generality, and write points $w \in \mathbb{C}^n$ in the form $w = (w_1, w')$. The invariance $f = f \circ V$ shows then, for each w_1, that $f(w_1, w')$ is a radial polynomial in w', hence is a polynomial in $|w'|^2$. Having bidegree (p, q), f therefore has the form

$$(1) \qquad f(w) = \sum_{i=0}^{r} c_i |w'|^{2i} w_1^{p-i} (\overline{w}_1)^{q-i}$$

where $r = \min(p, q)$ and $c_0, \ldots, c_r$ are constants, $c_0 = 1$. Differentiation of (1) gives

$$(2) \qquad \sum_{k=1}^{n} \frac{\partial^2 f}{\partial w_k \, \partial \overline{w}_k} = \sum_{i=0}^{r-1} b_i |w'|^{2i} w_1^{p-i-1} (\overline{w}_1)^{q-i-1}$$

where

$$(3) \quad b_i = (p - i)(q - i)c_i + (i + 1)(n + i - 1)c_{i+1} \qquad (0 \le i < r).$$

Since f is harmonic, (2) vanishes, so that $b_i = 0$ for all i, and (3) successively determines $c_1, \ldots, c_r$.

12.2.7. Theorem. *Suppose $T: H(p, q) \to H(r, s)$ is linear and commutes with $\mathscr{U}$.*
When $(r, s) \ne (p, q)$, then $T = 0$.
When $(r, s) = (p, q)$, then $T = cI$, where c is a constant and I is the identity operator.

Proof. Let $K_z \in H(p, q)$ be as in Theorem 12.2.5 and let L_z be the corresponding function in $H(r, s)$. If $V \in \mathscr{U}$ fixes z, then

$$(1) \qquad (TK_z) \circ V = T(K_z \circ V) = TK_z$$

by 12.2.5(5), since T commutes with $\mathscr{U}$. By Proposition 12.2.6, (1) shows that there corresponds to each $z \in S$ a $c(z) \in \mathbb{C}$ such that $TK_z = c(z)L_z$. Hence

$$(2) \qquad (TK_z)(z) = c(z)L_z(z).$$

By 12.2.5(6), $L_z(z)$ is independent of z. If $w = Uz$, then, by 12.2.5(4),

$$(3) \qquad (TK_w)(w) = (TK_z \circ U^{-1})(Uz) = (TK_z)(z).$$

We conclude from this that $c(z) = c$, the same for all $z \in S$.

If $f \in H(p, q)$, 12.2.5(3) shows that

$$(4) \qquad\qquad f = \int_S f(\zeta) K_\zeta \, d\sigma(\zeta).$$

Apply T to (4), and use $TK_\zeta = cL_\zeta$:

$$(5) \qquad Tf = \int_S f(\zeta) TK_\zeta \, d\sigma(\zeta) = c \int_S f(\zeta) L_\zeta \, d\sigma(\zeta) = c\pi_{rs} f,$$

for every $f \in H(p, q)$.

If $(r, s) \neq (p, q)$, $\pi_{rs} f = 0$. If $(r, s) = (p, q)$, $\pi_{rs} f = f$. This proves the theorem.

12.2.8. Theorem. *Each $H(p, q)$ is $\mathscr{U}$-minimal.*

More explicitly, the assertion is that $H(p, q)$ has no proper $\mathscr{U}$-invariant subspace.

Proof. Since $H(p, q)$ is finite-dimensional, it is closed in $L^2(\sigma)$. Suppose Y is a $\mathscr{U}$-subspace of $H(p, q)$. Then so is $Y^\perp$, the orthogonal complement of Y relative to $H(p, q)$. The projection of $H(p, q)$ onto Y, with null-space $Y^\perp$, commutes with $\mathscr{U}$, and would violate Theorem 12.2.7 (the case $(p, q) = (r, s)$) unless $Y = H(p, q)$ or $Y = \{0\}$; for if cI is a projection, then $c = 1$ or $c = 0$.

12.3. $\mathscr{U}$-Invariant Spaces on S

12.3.1. Notation. In the rest of this chapter, Q will stand for the first quadrant of lattice points in the plane. Thus Q consists of all ordered pairs (p, q) in which p and q are nonnegative integers.

For $\Omega \subset Q$, the algebraic sum of all $H(p, q)$ with $(p, q) \in \Omega$ will be denoted by E_Ω. We adopt the convention that $E_\Omega = \{0\}$ when Ω is empty.

The letter X will stand for any of the Banach spaces $C(S)$ or $L^p(\sigma)$, $1 \leq p < \infty$. The X-closure of E_Ω will be denoted by X_Ω.

Trivially, every X_Ω is a closed $\mathscr{U}$-space in X. The proof of the converse is the main topic of this section.

We begin with the simplest case, $X = L^2 = L^2(\sigma)$.

12.3.2. Theorem. *If Y is a $\mathscr{U}$-invariant closed subspace of $L^2(\sigma)$, and if Ω is the set of all $(p, q) \in Q$ such that $\pi_{pq} Y \neq \{0\}$, then $Y = (L^2)_\Omega$.*

Proof. Pick $(p, q) \in \Omega$. Recall that π_{pq} is the orthogonal projection of L^2 onto $H(p, q)$. Since Y is $\mathscr{U}$-invariant and π_{pq} commutes with $\mathscr{U}$, $\pi_{pq} Y$ is a nontrivial $\mathscr{U}$-space in $H(p, q)$. The $\mathscr{U}$-minimality of $H(p, q)$ (Theorem 12.2.8) shows therefore that $\pi_{pq} Y = H(p, q)$.

Let $Y_0 = \{ f \in Y : \pi_{pq} f = 0 \}$, and let Y_1 be the orthogonal complement of Y_0 in Y. Then Y_0 is $\mathscr{U}$-invariant (being the null-space of a map that commutes with $\mathscr{U}$), hence so is Y_1. Moreover, $\pi_{pq} : Y_1 \to H(p, q)$ is an isomorphism, whose inverse we denote by Λ.

Choose $(r, s) \in Q$, $(r, s) \neq (p, q)$, and consider the linear map $T = \pi_{rs} \circ \Lambda$. It is clear that T commutes with $\mathscr{U}$ and that T carries $H(p, q)$ into $H(r, s)$. By Theorem 12.2.7, $T = 0$. Hence $\pi_{rs} Y_1 = \{0\}$ for every $(r, s) \neq (p, q)$. By Theorem 12.2.3, $Y_1 = H(p, q)$.

Thus $H(p, q) \subset Y$, for every $(p, q) \in \Omega$.

In other words, $(L^2)_\Omega \subset Y$.

Since $\pi_{rs} Y = \{0\}$ for every (r, s) not in Ω, another look at Theorem 12.2.3 completes the proof of the theorem.

The following lemmas will make it easy to pass from L^2 to X.

12.3.3. Lemma. *If $f \in X$ then $U \to f \circ U$ is a continuous map of $\mathscr{U}$ into X.*

Proof. If $\varepsilon > 0$, then $\| f - g \| < \varepsilon$ for some $g \in C(S)$. There is a neighborhood N of the identity in $\mathscr{U}$ such that $\| g - g \circ U \|_\infty < \varepsilon$ for every $U \in N$. Since

$$| f - f \circ U | \leq | f - g | + | g - g \circ U | + | (g - f) \circ U |,$$

we have $\| f - f \circ U \| < 3\varepsilon$ for every $U \in N$.

12.3.4. Lemma. *If Y is a closed $\mathscr{U}$-space in X, then $Y \cap C(S)$ is dense in Y.*

Proof. Pick $f \in Y$, choose N as in the proof of Lemma 12.3.3, let $\psi : \mathscr{U} \to [0, \infty)$ be continuous, with support in N and $\int \psi \, dU = 1$. Define

$$(1) \qquad\qquad g = \int_{\mathscr{U}} \psi(U) f \circ U \, dU.$$

The integrand is a continuous Y-valued function. Hence $g \in Y$. If $V \in \mathscr{U}$ and $Ve_1 = \zeta$, the invariance of the Haar measure dU shows that

$$(2) \qquad\qquad g(\zeta) = \int_{\mathscr{U}} \psi(UV^{-1}) f(Ue_1) dU.$$

Thus $g \in Y \cap C(S)$.

Finally, the relation

$$(3) \qquad\qquad f - g = \int_N \psi(U)(f - f \circ U) dU$$

gives $\| f - g \| < 3\varepsilon$, since $\| f - f \circ U \| < 3\varepsilon$ if $U \in N$.

12.3.5. Lemma. *If $Y \subset C(S)$, Y is a $\mathcal{U}$-space, and some $g \in C(S)$ is not in the uniform closure of Y, then g is not in the L^2-closure of Y.*

Proof. There is a $\mu \in M(S)$ such that $\int f\,d\mu = 0$ for all $f \in Y$, but $\int g\,d\mu = 1$. There is a neighborhood N of the identity in $\mathcal{U}$ such that Re $\int g \circ U\,d\mu > \frac{1}{2}$ for every $U \in N$. Associate ψ to N as in the proof of Lemma 12.3.4, and define $\Lambda \in C(S)^*$ by

$$(1) \qquad \Lambda h = \int_S d\mu(\zeta) \int_{\mathcal{U}} \psi(U)h(U\zeta)dU.$$

The Schwarz inequality shows that the square of the absolute value of the inner integral in (1) is at most

$$(2) \qquad \int_{\mathcal{U}} |\psi|^2\,dU \int_{\mathcal{U}} |h(U\zeta)|^2\,dU = \|\psi\|_2^2 \int_S |h|^2\,d\sigma,$$

by 1.4.7(3), so that

$$(3) \qquad |\Lambda h| \le \|\mu\|\,\|\psi\|_2\|h\|_2.$$

Thus Λ extends to a bounded linear functional on $L^2(\sigma)$.

If we interchange the two integrals in (1), we see that $\Lambda f = 0$ for every $f \in Y$, whereas Re $\Lambda g \ge \frac{1}{2}$. This completes the proof.

12.3.6. Theorem (Nagel–Rudin [1]). *If Y is a $\mathcal{U}$-invariant closed subspace of X, and if Ω is the set of all $(p, q) \in Q$ for which $\pi_{pq} Y \ne \{0\}$, then $Y = X_\Omega$.*

Proof. Recall that the domain of π_{pq} has been extended to $L^1(\sigma) \supset X$, by 12.2.5(3).

Define $\tilde{Y}$ to be the L^2-closure of $Y \cap C$, where $C = C(S)$.

Since Y is X-closed, $Y \cap C$ is uniformly closed, so that Lemma 12.3.5 gives

$$(1) \qquad \tilde{Y} \cap C = Y \cap C.$$

Observe next that $Y \cap C$ is L^2-dense in $\tilde{Y}$, by definition, and is X-dense in Y, by Lemma 12.3.4. Since each π_{pq} is X-continuous as well as L^2-continuous, it follows that $\pi_{pq} Y = \{0\}$ if and only if $\pi_{pq} \tilde{Y} = \{0\}$. Theorem 12.3.2 shows therefore that $\tilde{Y}$ is the L^2-closure of E_Ω. Since $E_\Omega \subset C$, another application of Lemma 12.3.5 gives

$$(2) \qquad \tilde{Y} \cap C = \text{uniform closure of } E_\Omega.$$

Since $Y \cap C$ is X-dense in Y (Lemma 12.3.4), (1) and (2) imply that Y is the X-closure of E_Ω. This is the assertion of the theorem.

The following consequence will be used later:

12.3.7. Theorem. *Let Y be a $\mathcal{U}$-invariant closed subspace of X. If there exists an $f \in Y$ and a $g \in H(p, q)$ such that*

$$(1) \qquad\qquad \int_S f\bar{g}\, d\sigma \neq 0$$

then $H(p, q) \subset Y$.

Note that (1) is particularly easy to verify (without computation) if $f = g\psi$ for some positive function ψ.

Proof. By 12.2.5(3) and Fubini's theorem,

$$\int_S \bar{g}\pi_{pq} f\, d\sigma = \int f\overline{\pi_{pq}g}\, d\sigma = \int f\bar{g}\, d\sigma \neq 0,$$

since $\pi_{pq}g = g$. Hence $\pi_{pq} f \neq 0$. Now apply Theorem 12.3.6.

Theorem 12.2.7 was used in the proof of Theorem 12.3.2. Here is another consequence:

12.3.8. Theorem. *Suppose $T: X_1 \to X_2$ is a continuous linear map that commutes with $\mathcal{U}$. Then there exist $c(p, q) \in \mathbb{C}$, for every $(p, q) \in Q$, such that*

$$(1) \qquad\qquad Tf = c(p, q)f \quad \text{for all } f \in H(p, q).$$

The numbers $c(p, q)$ determine T.

Formally, if $f = \Sigma f_{pq}$ with $f_{pq} \in H(p, q)$, then $Tf = \Sigma c(p, q)f_{pq}$. Thus T is a multiplier transformation.

Proof. $\pi_{rs} T$ maps $H(p, q)$ into $H(r, s)$ and commutes with $\mathcal{U}$, hence $\pi_{rs} Tf = 0$ if $f \in H(p, q)$ and $(r, s) \neq (p, q)$. Consequently, $TH(p, q)$ is a (finite-dimensional, hence closed) $\mathcal{U}$-invariant subspace of X_2 that is annihilated by π_{rs} whenever $(r, s) \neq (p, q)$. By Theorem 12.3.6, $TH(p, q) = H(p/q)$ or $\{0\}$. Now (1) follows from Theorem 12.2.7.

Here is an application:

12.3.9. Theorem. *There is no continuous linear projection of $L^1(\sigma)$ onto $H^1(S)$.*

In other words, H^1 is an uncomplemented subspace of L^1. The proof is almost the same as in the case $n = 1$; see Hoffman [1], p. 154, or Rudin [5].

Proof. Assume, to reach a contradiction, that there is a continuous linear projection P of L^1 onto H^1. Define $T: L^1 \to H^1$ by

$$(1) \qquad Tf = \int_{\mathscr{U}} [P(f \circ U^{-1})] \circ U \, dU.$$

It is easy to verify (see Rudin [5], or Theorem 5.18 in Rudin [2]) that T is then a continuous linear projection of L^1 onto H^1 that commutes with $\mathscr{U}$. Thus T is as in 12.3.8(1) (where $X_1 = X_2 = L^1$), with $c(p, 0) = 1$ for all p, and $c(p, q) = 0$ whenever $q > 0$. This implies that

$$(2) \qquad Tf = C[f]^*,$$

the boundary function of the Cauchy integral of f. But Theorems 6.3.3 and 6.3.5 show that the Cauchy integral of a positive $f \in L^1$ lies in H^1 if and only if $f \in L \log L$. Hence T cannot map all of L^1 into H^1.

12.3.10. We conclude this section with a rather curious characterization of $A(S)$, the restriction of the ball algebra $A(B)$ to S, in terms of the behavior on certain circles on S.

Fix t, $0 < t < 1$. Associate to each orthogonal pair of vectors ζ, η in S (i.e., $\langle \zeta, \eta \rangle = 0$) the circle $\Gamma_{\zeta, \eta}$, parametrized by

$$(1) \qquad \Gamma_{\zeta, \eta}(e^{i\theta}) = t\zeta + (1 - t^2)^{1/2} e^{i\theta} \eta.$$

This is the intersection of S and a complex line orthogonal to ζ, whose distance from the origin is t.

If $f \in A(B)$, then

$$(2) \qquad \lambda \to f(t\zeta + (1 - t^2)^{1/2} \lambda \eta)$$

is a function in the disc algebra whose boundary function is $f \circ \Gamma_{\zeta, \eta}$. This obviously necessary condition turns out to be sufficient as well:

12.3.11. Theorem. *Fix t, $0 < t < 1$, and define $\Gamma_{\zeta, \eta}$ as in §12.3.10.*

If $f \in C(S)$ and if $f \circ \Gamma_{\zeta, \eta}$ extends to a member of the disc algebra for every orthogonal pair of vectors ζ, η in S, then f has a continuous extension to $\bar{B}$ that is holomorphic in B.

Proof. Let Y be the class of all $f \in C(S)$ that satisfy the hypothesis. It is clear that Y is a closed subspace of $C(S)$. If $U \in \mathscr{U}$, then

$$(1) \qquad U\Gamma_{\zeta, \eta} = \Gamma_{U\zeta, U\eta}.$$

Thus Y is $\mathscr{U}$-invariant. By Theorem 12.3.6, it is enough to show that Y contains no $H(p, q)$ with $q > 0$.

Fix (p, q), $q > 0$, let $h(z) = z_1^p(\bar{z}_2)^q$, take $\zeta = e_1$, $\eta = e_2$. Then $h \in H(p, q)$ and

$$(2) \qquad (h \circ \Gamma_{\zeta, \eta})(e^{i\theta}) = t^p(1 - t^2)^{q/2}e^{-iq\theta},$$

which has no extension to the disc algebra. Thus $h \notin Y$, and the proof is complete.

With a stronger hypothesis (involving all complex lines that intersect B, not just those at distance t from the origin) this was first proved by Agranovskii–Valskii [1]. Stout [6] extended their result to arbitrary bounded regions in $\mathbb{C}^n$ with C^2-boundary.

The theorem fails when $t = 0$, i.e., when the complex lines through the origin are the only ones that occur in the hypothesis. To see this, take $f \in C(S)$ so that $f(e^{i\theta}\eta) = f(\eta)$ for all $\eta \in S$, $|\theta| \le \pi$, but which is otherwise arbitrary. For instance, f could be identically 0 on some circular open set in S without being 0 on all of S.

Theorem 12.3.11 is most interesting when t is near 1, since the complex lines that are involved do not enter the ball tB.

12.4. $\mathcal{U}$-Invariant Subalgebras of $C(S)$

12.4.1. If Ω is any subset of the quadrant Q, as in Section 12.3, the *uniform* closure of the linear span of the spaces $H(p, q)$ with $(p, q) \in \Omega$ will from now on be denoted by Y_Ω. Theorem 12.3.6 shows that every closed $\mathcal{U}$-invariant subspace of $C(S)$ is a Y_Ω; the converse is trivial.

The problem to which we now turn is to find those sets $\Omega \subset Q$ for which the corresponding $\mathcal{U}$-space Y_Ω is an algebra, relative to pointwise multiplication. We call such sets *algebra patterns*.

It will be helpful, throughout this section and the next, to think of Q as a geometric object, embedded in the plane, and as a semigroup relative to coordinatewise addition, rather than just as a collection of ordered pairs.

When $n \ge 3$, the algebra patterns (hence the $\mathcal{U}$-invariant algebras) can be completely characterized by a simple combinatorial criterion (Theorem 12.4.5). Some consequences of this are summarized in Theorem 12.4.7.

The case $n = 2$ is rather different and more complicated. It is taken up in the following section.

12.4.2. The case $n = 1$ is so simple that we dispose of it immediately:

If Ω is any additive semigroup of integers, then the set of all continuous functions on the unit circle whose Fourier coefficients vanish on the complement of Ω is a closed $\mathcal{U}$-algebra, and there are no others.

From now on we assume therefore that $n > 1$.

12.4.3. The Spaces $H(p, q) \cdot H(r, s)$. If $(p, q) \in Q$ and $(r, s) \in Q$, then $H(p, q) \cdot H(r, s)$ is defined to be the vector space of finite sums $\Sigma f_i g_i$ with $f_i \in H(p, q)$, $g_i \in H(r, s)$. It is clear that each $H(p, q) \cdot H(r, s)$ is a finite-dimensional 𝒰-space and that it is therefore a sum of finitely many $H(a, b)$'s, by Theorem 12.3.6. This theorem shows also that Y_Ω is an algebra if and only if $H(p, q) \cdot H(r, s) \subset Y_\Omega$ whenever $(p, q) \in \Omega$ and $(r, s) \in \Omega$.

We associate with each pair of points $(p, q) \in Q$ and $(r, s) \in Q$ the number

(1) $$\mu = \mu(p, q; r, s) = \min(p, s) + \min(r, q).$$

It is easy to check that μ is also given by

$$\mu = \min(p + q, r + s, p + r, q + s).$$

12.4.4. Theorem. *If* $n \geq 3, (p, q) \in Q, (r, s) \in Q$, and $\mu = \mu(p, q; r, s)$ *is as above, then*

(1) $$H(p, q) \cdot H(r, s) = \sum_{j=0}^{\mu} H(p + r - j, q + s - j).$$

When $n = 2$, *then* $H(p, q) \cdot H(r, s)$ *is a subspace of the sum in* (1).

As we shall see in the next section, equality can actually fail in (1) when $n = 2$.

When $n \geq 3$, the theorem has the following consequence: If Ω is an algebra pattern that contains (p, q) and (r, s), then, obviously, the left side of (1) lies in Y_Ω, hence so does every summand on the right of (1); this says that Ω contains the points

(2) $$(p + r - j, q + s - j) \qquad (0 \leq j \leq \mu).$$

These are $\mu + 1$ adjacent lattice points on a line of slope 1 (parallel to the diagonal of Q) whose highest point is $(p, q) + (r, s)$.

Proof. We begin by proving the inclusion

(3) $$H(p, q) \cdot H(r, s) \subset \sum_{j=1}^{\mu} H(p + r - j, q + s - j)$$

for any $n \geq 2$.

Let $f \in H(p, q), g \in H(r, s)$. Then fg is a polynomial that is homogeneous of degree $p + q + r + s$, with respect to real scalars. By Theorem 12.1.3, fg

coincides therefore on S with a finite sum of spherical harmonics of degree at most $p + q + r + s$. This yields an orthogonal decomposition (on S)

$$(4) \qquad\qquad fg = \sum h_{ab}$$

in which $h_{ab} \in H(a, b), a + b \le p + q + r + s$.

Let us write $[\varphi, \psi]$ for $\int_S \varphi\bar\psi \, d\sigma$.

Fix (a, b), put $h = h_{a,b}$. The orthogonality of the summands in (4) shows that $h \ne 0$ if and only if

$$(5) \qquad\qquad [fg, h] \ne 0.$$

The integrand in (5) is $fg\bar h$, a finite linear combination of monomials $z^\alpha \bar z^\beta$, where α and β are multi-indices with $|\alpha| = p + r + b, |\beta| = q + s + a$. Since $z^\alpha \perp z^\beta$ unless $\alpha = \beta$, (5) implies that $p + r + b = q + s + a$. Hence

$$(6) \qquad\qquad (a, b) = (p + r - j, q + s - j)$$

for some j. Since $a + b \le p + q + r + s, j \ge 0$.

Observe now that $[fg, h] = [f, \bar gh]$ and that $\bar gh$ is homogeneous of degree $r + s + a + b$. The reasoning used at the beginning of this proof shows therefore, because of (5), that $p + q \le r + s + a + b$. This, combined with (6), proves that $j \le r + s$.

The proof that $j \le p + q$ goes the same way.

The inequalities $j \le p + r, j \le q + s$ are obvious, by (6).

To sum up, we have proved the following:

If $h_{ab} \ne 0$ in (4), then (a, b) is given by (6), for some j such that $0 \le j \le \mu$. This proves (3).

In the second part of the proof, assume $n \ge 3$. Fix $j, 0 \le j \le \mu$. The definition of μ shows that $j = k + m$ for some integers k and m that satisfy

$$(7) \qquad\qquad 0 \le k \le \min(p, s), \qquad 0 \le m \le \min(r, q).$$

To simplify the notation, let us write u, v, w for the variables z_1, z_2, z_3. Define

$$(8) \qquad f = u^p \bar v^m \bar w^{q-m}, \qquad g = v^r \bar u^k \bar w^{s-k}, \qquad h = u^{p-k} v^{r-m} \bar w^{q+s-j}.$$

Because of (7), these are polynomials; they are obviously harmonic; hence $f \in H(p, q), g \in H(r, s), h \in H(p + r - j, q + s - j)$. Moreover, $fg\bar h > 0$ a.e. on S. Thus

$$(9) \qquad\qquad [fg, h] \ne 0.$$

Since $H(p, q) \cdot H(r, s)$ is $\mathcal{U}$-invariant, Theorem 12.3.7 implies that

(10) $$H(p + r - j, q + s - j) \subset H(p, q) \cdot H(r, s),$$

and now (1) follows from (10) and (3).

12.4.5. Theorem. *When* $n \geq 3$, *the following property* (*) *is necessary and sufficient for a set* $\Omega \subset Q$ *to be an algebra pattern:*
  (*) *If* $(p, q) \in \Omega$ *and* $(r, s) \in \Omega$, *then* Ω *contains all points* $(p + r - j, q + s - j)$ *with* $0 \leq j \leq \mu$, *where*

$$\mu = min(p, s) + min(r, q).$$

When $n = 2$, *then* (*) *implies that* Ω *is an algebra pattern.*

Proof. This is an immediate consequence of Theorem 12.4.4, since Y_Ω is an algebra if and only if $H(p, q) \cdot H(r, s) \subset Y_\Omega$ whenever $(p, q) \in \Omega$ and $(r, s) \in \Omega$.

12.4.6. The Semigroup $\Sigma(\Omega)$. For each integer k, let D_k be the set of all $(p, q) \in Q$ for which $p - q = k$. Thus D_0 is the diagonal of Q, and each D_k is a translate of D_0.

Let Ω be an algebra pattern, and let $\Sigma(\Omega)$ be the set of all k such that Ω intersects D_k. In other words, $k \in \Sigma(\Omega)$ if and only if $p - q = k$ for some $(p, q) \in \Omega$.

The relation $(z^p \overline{w}^q)(z^r \overline{w}^s) = z^{p+r} \overline{w}^{q+s}$ shows that every algebra pattern Ω is a subsemigroup of Q, for all $n \geq 2$. Consequently, $\Sigma(\Omega)$ *is an additive semigroup of integers.*

If Σ is any additive semigroup of integers, then the union of all D_k with $k \in \Sigma$ is easily seen to have property (*) of Theorem 12.4.5. In this way, every Σ gives rise to an algebra pattern. Our next theorem shows that under certain conditions there are no other possibilities.

12.4.7. Theorem. *Suppose* $n \geq 3$, *and let* Ω *be an algebra pattern.*
  (I) *If* Ω *contains some* (a, a) *with* $a > 0$, *then* Ω *is the union of all* D_k, $k \in \Sigma(\Omega)$.
  (II) *If* Ω *contains some* (p, q) *with* $p > q$ *and some* (r, s) *with* $r < s$, *then there is a positive integer* d *such that*

(1) $$\Omega = \bigcup_{m=-\infty}^{\infty} D_{md}.$$

  (III) *If* $p > q$, *then the smallest algebra pattern that contains* (p, q) *consists of the points*

(2) $(mp - j, mq - j)$ $(m = 2, 3, 4, \ldots, 0 \leq j \leq mq)$

plus, of course, (p, q) *itself.*

The points (x, y) described by (2) are those points on the lines $D_{m(p-q)}$ $(m = 2, 3, 4, \ldots)$ that satisfy $y \le (q/p)x$.

Proof. (I) If $a > 0$ and $(a, a) \in \Omega$, Theorem 12.4.5 implies that $(r, r) \in \Omega$ for $0 \le r \le 2a$. In particular, $(1, 1) \in \Omega$.

If $(p, q) \in \Omega$, it follows that $(p + 1, q + 1) \in \Omega$.

If also $p \ge 1$ and $q \ge 1$ (i.e., if (p, q) is not on an edge of Q), then $\mu(p, q; 1, 1) = 2$, and another application of 12.4.5 shows that $(p - 1, q - 1) \in \Omega$.

Repetition of these two steps shows that $D_{p-q} \subset \Omega$. This proves (I).

(II) In this case $\Sigma = \Sigma(\Omega)$ contains positive integers as well as negative ones, hence Σ is a subgroup of the integers (see §12.4.8), so that Σ consists of all integral multiples of d, where d is the smallest positive element of Σ.

Pick $(x, y) \in \Omega \cap D_d$, $(x', y') \in \Omega \cap D_{-d}$. Then $x > 0$, and thus $(x + x', y + y')$ is a point of $\Omega \cap D_0$ that is not $(0, 0)$. Now (II) follows from (I).

(III) The set consisting of (p, q) and all points listed in (2) evidently has property (*) of Theorem 12.4.5, and is therefore an algebra pattern.

For the converse, let Ω be the smallest subset of Q that contains (p, q) and has property (*). We have to show that Ω contains all points of the form (2).

Since $(p, q) \in \Omega$ and $\mu(p, q; p, q) = 2q$, (*) implies that Ω contains the points (2) with $m = 2$.

We proceed by induction, assuming that $m \ge 2$ and that Ω contains $(mp - j, mq - j)$ for $0 \le j \le mq$. Since $(p, q) \in \Omega$ and Ω is a semigroup, Ω contains the points

$$(3) \qquad ((m + 1)p - j, (m + 1)q - j) \qquad (0 \le j \le mq).$$

Take $j = (m - 1)q$ in our induction hypotheses. It follows that the point

$$(4) \qquad (r, s) = (mp - (m - 1)q, q)$$

lies in Ω. Since $\mu(p, q; r, s) = 2q$, (*) implies that Ω contains $(p + r - i, q + s - i)$ for $0 \le i \le 2q$. But these are exactly the points

$$(5) \qquad ((m + 1)p - j, (m + 1)q - 1) \qquad ((m - 1)q \le j \le (m + 1)q).$$

By (3) and (5), our induction hypothesis holds with $m + 1$ in place of m, and the proof is complete.

Remarks. As in the Theorem just proved, we continue to assume that $n \ge 3$, although some of the comments that follow are also true when $n = 2$, as will be seen later.

(a) If Ω is an algebra pattern that lies in D_0, then Ω is either empty, or $\{(0, 0)\}$, or D_0. This follows from (I). When $\Omega = D_0$, then Y_Ω is the algebra of all $f \in C(S)$ such that $f(e^{i\theta}\zeta) = f(\zeta)$ for all $\zeta \in S$, $-\pi \le \theta \le \pi$.

(b)   If Ω is as in (II), then Y_Ω is the algebra of all $f \in C(S)$ that are periodic in the sense that

$$f(e^{2\pi i/d}\zeta) = f(\zeta)$$

for all $\zeta \in S$.

(c)   If Ω is an algebra pattern that is symmetric, i.e., if $(p, q) \in \Omega$ implies $(q, p) \in \Omega$, then Ω is either empty, or $\{(0, 0)\}$, or D_0, or as in (II). The corresponding algebras Y_Ω are precisely those that are self-adjoint, in the sense that $f \in Y_\Omega$ implies $\bar{f} \in Y_\Omega$. We conclude: $C(S)$ contains only *countably many* $\mathscr{U}$-invariant closed self-adjoint sub-algebras.

(d)   In the same way, (I) shows that there are only *countably many* algebra patterns that intersect D_0 in a point other than $(0, 0)$, since there are only countably many semigroups Σ. (See §12.4.8.)

(e)   For $0 \le t < \infty$, let Ω_t be the set of all $(p, q) \in Q$ such that $q \le tp$. When $t \le 1$, then Ω_t has property (*). *The collection of all $\mathscr{U}$-invariant closed subalgebras of $C(S)$ has therefore the power of the continuum.*

   Ω_0 corresponds to the algebra $A(S)$.

   Ω_1 corresponds to the algebra of all $f \in C(S)$ that can be con-tinuously extended to $\bar{B}$ in such a way that the slice functions f_ζ lie in the disc algebra, for every $\zeta \in S$.

   When $t > 1$, Ω_t is not an algebra pattern.

(f)   Let us call a closed $\mathscr{U}$-invariant algebra $Y \subset C(S)$ *maximal* if no closed $\mathscr{U}$-invariant algebra lies strictly between Y and $C(S)$. The maximal $\mathscr{U}$-algebras can be easily determined from (I) and (II). The corresponding algebra patterns are:

(i) Ω_1, as defined in (e),
(ii)   the reflection of Ω_1 (the set of all (p, q) with $p \le q$),
(iii) the sets described in (II) for which d is a prime.

12.4.8. Semigroups of Integers. To make the conclusions of Theorem 12.4.7 more concrete, here is an explicit description of the semigroups that occur there.

First, suppose Σ has positive as well as negative elements. Let x and y be the smallest positive and largest negative one, respectively. Then $y = -x$, since otherwise $x + y$ is an element of Σ that lies strictly between x and y without being 0. From this it follows easily that Σ is the group generated by x.

Secondly, suppose $s \ge 0$ for all $s \in \Sigma$, and that Σ contains a positive number. Let d be the greatest common divisor of the positive elements of Σ, and let A_d be the arithmetic progression $\{0, d, 2d, \ldots\}$. Obviously, $\Sigma \subset A_d$. We claim that, conversely, *at most finitely many members of A_d are not in Σ.*

It is enough to prove this for $d = 1$. Then Σ contains positive integers $m_1 < m_2 < \cdots < m_k$ whose g.c.d. is 1. Hence there are integers a_j such that

$$1 = a_1 m_1 + \cdots + a_k m_k.$$

Put $c = |a_1| m_1 + \cdots + |a_k| m_k$. Since $m_1 |a_j| + t a_j \geq 0$ when $0 \leq t < m_1$, it follows that $m_1 c + t$ lies in Σ for these t. Thus Σ contains m_1 consecutive integers. Adding multiples of m_1 to these, we find that Σ contains all integers $\geq m_1 c$.

12.5. The Case $n = 2$

We begin this section by showing how part (I) of Theorem 12.4.7 can fail when $n = 2$ (Theorem 12.5.3). We then develop enough information to prove the analogues of 12.4.7(I) and (II). (Theorem 12.5.6); the statement of (II) is actually the same as before, but the proof is a little different. The singly generated case (part (III) of Theorem 12.4.7) is more difficult and will be done last.

Throughout this section, $n = 2$, and points of $\mathbb{C}^2$ will be denoted by (z, w). Here is a convenient description of $H(p, q)$:

12.5.1. Proposition. When $n = 2$, then every $H(p, q)$ is the linear span of the functions g_α defined by

$$(1) \qquad\qquad g_\alpha(z, w) = (z + \alpha w)^p (\bar{w} - \alpha \bar{z})^q \qquad (\alpha \in \mathbb{C}).$$

Proof. Clearly, $g_0 \in H(p, q)$. Since $H(p, q)$ is $\mathscr{U}$-minimal, $H(p, q)$ is spanned by the functions

$$(2) \qquad\qquad g_0(az + bw, cz + dw) = (az + bw)^p (\overline{cz} + \overline{dw})^q,$$

where $\begin{pmatrix} a & b \\ c & d \end{pmatrix}$ runs over the unitary matrices. We can restrict ourselves to the case $a \neq 0$ (hence $d \neq 0$), since $\dim H(p, q) < \infty$, so that every dense subspace of $H(p, q)$ is in fact $H(p, q)$. Since $a\bar{c} + b\bar{d} = 0$, one checks easily that every function of the form (2) is a scalar multiple of $g_{b/a}$.

12.5.2. Let Ω be an algebra pattern such that $p \geq q$ for all $(p, q) \in \Omega$. If $f \in Y_\Omega$ and $\zeta \in S$, the slice function f_ζ (defined on the unit circle) has therefore a continuous extension to the closed unit disc that is holomorphic in its interior; for each $\zeta \in S$, the map

$$(1) \qquad\qquad\qquad f \to f_\zeta(0)$$

is thus a complex homomorphism of Y_Ω. Since

$$(2) \qquad f_\zeta(0) = \frac{1}{2\pi} \int_{-\pi}^{\pi} f(e^{it}\zeta)dt,$$

$f_\zeta(0)$ is not changed if ζ is replaced by $e^{i\theta}\zeta$.

The orthogonal complement of the vector ζ in $\mathbb{C}^2$ has complex dimension 1. Let $\zeta^\perp$ be a unit vector, orthogonal to ζ. Then $\zeta^\perp$ is determined by ζ, up to multiplication by scalars of absolute value 1, so that the number $f_{\zeta^\perp}(0)$ is uniquely determined by ζ, for every $f \in Y_\Omega$.

Define Y to be the set of all $f \in Y_\Omega$ that satisfy

$$(3) \qquad f_\zeta(0) = f_{\zeta^\perp}(0)$$

for every $\zeta \in S$. It is clear that Y is a closed $\mathcal{U}$-invariant subalgebra of Y_Ω. Thus $Y = Y_{\Omega_0}$ for some $\Omega_0 \subset \Omega$. Let us determine Ω_0.

If $f \in H(p, q)$ then $f_\zeta(\lambda) = f(\zeta)\lambda^p \bar{\lambda}^q$.

If $f \in H(p, q)$ and $p > q$, it follows that $f_\zeta(0) = 0$, so that (3) holds.

If $f \in H(r, r)$ then each f_ζ is a constant, and we claim that

$$(4) \qquad f_\zeta = (-1)^r f_{\zeta^\perp}.$$

Because of the $\mathcal{U}$-invariance of $H(r, r)$, it is enough to check (4) when $\zeta = (1, 0)$, $\zeta^\perp = (0, 1)$, and this follows immediately from Proposition 12.5.1, with $(p, q) = (r, r)$.

By (4), every $f \in H(r, r)$ satisfies (3) if and only if r is even.

This determines Ω_0. For easier reference, let us state the result more formally:

12.5.3. Theorem. *Let* $n = 2$. *If* Ω *is an algebra pattern such that* $p \geq q$ *for all* $(p, q) \in \Omega$, *and if* Ω_0 *is the set of all points of* Ω *that are not of the form*

$$(1, 1), (3, 3), (5, 5), \ldots,$$

then Ω_0 *is also an algebra pattern.*

For example, the set of all points $(2k, 2k)$, $k = 0, 1, 2, \ldots$, is an algebra pattern when $n = 2$, but not for any $n \geq 3$.

The next two lemmas give part of Theorem 12.4.4 for $n = 2$; they are actually sharp, as shown by Theorems 12.5.9 and 12.5.10.

12.5.4. Lemma. *If* Ω *is an algebra pattern that contains two points* (p, q), (r, s), *with* $ps \neq rq$, *then* $(p + r - 1, q + s - 1) \in \Omega$.

Geometrically, $ps \neq rq$ means that (p, q), (r, s), and $(0, 0)$ are not collinear.

Proof. Define

$$f(z, w) = pz^{p-1}w\overline{w}^q - qz^p\overline{z}\,\overline{w}^{q-1}.$$

Since $\partial^2 f/\partial z\,\partial\overline{z} + \partial^2 f/\partial w\,\partial\overline{w} = 0$, $f \in H(p, q)$.

If $g(z, w) = z^r\overline{w}^s$ and $h(z, w) = z^{p+r-1}\overline{w}^{q+s-1}$, a computation using Proposition 1.4.9 gives

$$\int_S f g\overline{h}\, d\sigma = \frac{(p+r-1)!(q+s-1)!}{(p+q+r+s)!}(ps - rq) \neq 0.$$

Thus $H(p+r-1, q+s-1) \subset H(p, q) \cdot H(r, s)$, by Theorem 12.3.7.

12.5.5. Lemma. *If $p \geq q$ then*

(1) $$H(2p - 2k, 2q - 2k) \subset H(p, q) \cdot H(p, q)$$

for $k = 0, 1, \ldots, q$.

Proof. Let f be the coefficient of α^k in

(2) $$g_\alpha(z, w) = (z + \alpha w)^p(\overline{w} - \alpha\overline{z})^q,$$

the function that occurs in Proposition 12.5.1. Restrict α to the unit circle, multiply g_α by α^{-k}, and integrate the product over T. This shows that $f \in H(p, q)$. By the binomial theorem,

$$f(z, w) = z^{p-k}\overline{w}^{q-k} \sum_{u+v=k} (-1)^v \binom{p}{u}\binom{q}{v}|z^v w^u|^2.$$

The sum is real. If $h(z, w) = z^{2p-2k}\overline{w}^{2q-2k}$, it follows that $\int_S f^2\overline{h}\, d\sigma > 0$. Theorem 12.3.7 implies therefore that (1) holds.

The following theorem should be compared to 12.4.7.

12.5.6. Theorem. *Suppose $n = 2$ and Ω is an algebra pattern.*

(I)   *If Ω contains some (a, a) with $a > 0$, and $p \geq q$ for every $(p, q) \in \Omega$, then Ω is either the union of all D_k, $k \in \Sigma(\Omega)$, or Ω is the set obtained by removing $(1, 1), (3, 3), (5, 5), \ldots$ from this union.*

(II) *If Ω contains some (p, q) with $p > q$ and some (r, s) with $r < s$, then there is a positive integer d such that*

$$\Omega = \bigcup_{m=-\infty}^{\infty} D_{md}.$$

Proof. (I) Apply Lemma 12.5.5 with $(p, q) = (a, a)$. Since Ω is a semigroup, it follows that Ω contains the "even points" of D_0, i.e., the points $(2i, 2i)$, $i = 0$, $1, 2, \ldots$. In particular, $(2, 2) \in \Omega$.

This has two consequences:

(i) If $(p, q) \in \Omega$, $p \geq 2$, $q \geq 2$, then $(p - 2, q - 2) \in \Omega$.
(ii) If $(p, q) \in \Omega$, $p \neq q$, then $(p + 1, q + 1) \in \Omega$.

The first of these follows from

$$(z^p\overline{w}^q) \cdot (w^2\overline{z}^2) = z^{p-2}\overline{w}^{q-2}|zw|^4$$

and Theorem 12.3.7. To obtain the second, refer to Lemma 12.5.4, with $r = s = 2$.

If $(b, b) \in \Omega$ for some odd b, repeated application of (i) gives $(1, 1) \in \Omega$, hence $D_0 \subset \Omega$.

As far as D_0 is concerned, there are thus only two possibilities: Either $D_0 \subset \Omega$, or $\Omega \cap D_0$ consists of the even points of D_0.

If now $(p, q) \in \Omega$ and $p \neq q$, (ii) shows that $(p + 1, q + 1) \in \Omega$, and (provided that $p > 0$ and $q > 0$) another application of (i) puts $(p - 1, q - 1)$ into Ω. Thus $D_{p-q} \subset \Omega$, and (I) is proved.

(II) As in the proof of Theorem 12.4.7(II), $\Sigma(\Omega)$ is now a group, generated by some $d > 0$. Let $(p, q) \in D_d$ and $(r, s) \in D_{-d}$ belong to Ω. By the semigroup property of Ω and Lemma 12.5.4, the two adjacent points

$$(p + r, q + s), \quad (p + r - 1, q + s - 1)$$

of D_0 lie in Ω. The rest of the proof is exactly like that of (I), with $D_0 \subset \Omega$.

Remark. The remarks that follow Theorem 12.4.7 apply now equally well to the case $n = 2$, except that in (a) and (c) there is the additional possibility that $\Omega \cap D_0$ can consist of just the even points of D_0. (See Theorem 12.5.3.)

12.5.7. Lemma. *If $n = 2$ and $ps = rq$, then the map*

(1) $$u \to \Delta u|_S$$

takes $H(p, q) \cdot H(r, s)$ onto $H(p - 1, q - 1) \cdot H(r - 1, s - 1)$.

The symbol $\Delta u|_S$ is the Laplacian of u, restricted to S. We adopt the convention that $H(x, y) = \{0\}$ when $x < 0$ or $y < 0$.

Proof. By Proposition 12.5.1, $H(p, q) \cdot H(r, s)$ is the linear span of the functions

$$u = (\hat{z} + \alpha w)^p(\overline{w} - \alpha\overline{z})^q \cdot (z + \beta w)^r(\overline{w} - \beta\overline{z})^s = f^p g^q h^r k^s,$$

using obvious abbreviations. Here α and β are arbitrary complex numbers. Straightforward differentiation of u gives

$$\Delta u = 4(\beta - \alpha)f^{p-1}g^{q-1}h^{r-1}k^{s-1}(rqfk - psgh).$$

Note that $fk - gh = (\alpha - \beta)(z\bar{z} + w\bar{w}) = \alpha - \beta$ on S. If $ps = rq$, it follows that

$$\Delta u = -4ps(\beta - \alpha)^2(f^{p-1}g^{q-1}) \cdot (h^{r-1}k^{s-1})$$

on S.

12.5.8. Lemma. *If $u \in H(p, q) \cdot H(r, s)$ and $h \in H(a, b)$ then*

(1)
$$\int_S (\Delta u)\bar{h}\, d\sigma = \gamma \int_S u\bar{h}\, d\sigma$$

where $\gamma = \gamma(p, q, r, s, a, b) \neq 0$ unless $a + b = p + q + r + s$.

Note. As the proof will show, the lemma depends only on the homogeneity of u and h, not on any considerations that involve bidegrees. We shall prove it for $n = 2$ although the computation works equally well for any n, yielding a different γ.

Proof. Put $c = p + q + r + s$. Then u and $\bar{h}$ are homogeneous of degree c and $a + b$, respectively. On S, their outward normal derivatives are therefore $\partial u/\partial n = cu$, $\partial\bar{h}/\partial n = (a + b)\bar{h}$. This gives the first equality in

$$(c - a - b)\int_S u\bar{h}\, d\sigma = \int_S \left(\bar{h}\frac{\partial u}{\partial n} - u\frac{\partial\bar{h}}{\partial n}\right)d\sigma = \frac{1}{4}\int_B \bar{h}\Delta u\, dv$$

$$= \frac{1}{a + b + c + 2}\int_S \bar{h}\Delta u\, d\sigma,$$

where v is Lebesgue measure on $\mathbb{C}^2$, as in §1.4.1. The second equality is Green's theorem, and the third is obtained by using polar coordinates and taking the homogeneity of $\bar{h}$ and Δu into account.

12.5.9. Theorem. *If $n = 2$, then a necessary and sufficient condition for the inclusion*

(1)
$$H(p + r - 1, q + s - 1) \subset H(p, q) \cdot H(r, s)$$

is $ps \neq rq$.

Proof. The sufficiency is Lemma 12.5.4. To prove the necessity, we will show that the assumption $ps = rq$ implies

(2) $$H(p, q) \cdot H(r, s) \perp H(p + r - 1, q + s - 1).$$

By Lemma 12.5.7,

(3) $$\Delta[H(p, q) \cdot H(r, s)]|_s \subset H(p - 1, q - 1) \cdot H(r - 1, s - 1).$$

Since $(p - 1) + (r - 1) < p + r - 1$, it follows that the left side of (3) is orthogonal to $H(p + r - 1, q + s - 1)$. Hence (2) follows from Lemma 12.5.8.

12.5.10. Theorem. If $n = 2$ and $p \geq q$, then

(1) $$H(p, q) \cdot H(p, q) = \sum_{k=0}^{q} H(2p - 2k, 2q - 2k).$$

Proof. By Lemma 12.5.5, it is enough to show that

(2) $$H(p, q) \cdot H(p, q) \perp H(2p - j, 2q - j)$$

whenever j is odd. Theorem 12.5.9 proves (2) when $j = 1$. Assume that $j \geq 3$ and that (2) is proved with $j - 2$ in place of j (and for all p, q). Take

(3) $$u \in H(p, q) \cdot H(p, q), \qquad h \in H(2p - j, 2q - j).$$

By 12.5.7, $\Delta u|_s \in H(p - 1, q - 1) \cdot H(p - 1, q - 1)$.
Since $h \in H(2(p - 1) - (j - 2), 2(q - 1) - (j - 2))$, our induction hypothesis implies that $[\Delta u, h] = 0$. Hence $[u, h] = 0$, by Lemma 12.5.8. This proves (2).

In the following lemma, (A) and (B) are already known; they are included for convenient reference.

12.5.11. Lemma. *Suppose $p > q$, $r > s$, and Ω is an algebra pattern that contains (p, q) and (r, s). Then Ω also contains*

(A) $(p + r, q + s)$,
(B) $(p + r - 1, q + s - 1)$ *if $ps \neq rq$*,
(C) $(r - q, s - p)$ *if $s \geq p$*,
(D) $(r - q + 1, s - p + 1)$ *if $s \geq p - 1$*,
(E) $(p + r - k, q - k)$ *if $s = 0$ and $1 \leq k \leq \min(q, r)$*.

Proof. (A) says simply that Ω is a semigroup, (B) is 12.5.4, and (C) follows from

$$(z^p\overline{w}^q)(w^r\overline{z}^s) = w^{r-q}\overline{z}^{s-p}|z^pw^q|^2,$$

and Theorem 12.3.7. For (D) and (E) use the function $f \in H(p, q)$ given by

$$f(z, w) = pz^{p-1}w\overline{w}^q - qz^p\overline{z}\overline{w}^{q-1},$$

as in 12.5.4.

If $g = w^r\overline{z}^s$ and $h = w^{r-q+1}\overline{z}^{s-p+1}$, a computation, based on Proposition 1.4.9, gives

$$\int_S fg\overline{h}\, d\sigma = \frac{r!s![p(r+1) - q(s+1)]}{(r+s+2)!} \neq 0,$$

hence (D), by Theorem 12.3.7.

For (E), take $g = z^{r+1-k}w^{k-1}$, $h = z^{p+r-k}\overline{w}^{q-k}$, in which case

$$-\int_S fg\overline{h}\, d\sigma = \frac{(p+r-k)!q!(r+1-k)}{(p+q+r+s+1-k)!} \neq 0.$$

12.5.12. Theorem. *Suppose* $n = 2$, $(p, q) \in Q$, $p > q$. *Put*

$$E_1 = \{(kp, kq): 1 \le k < \infty\}$$

$$E_2 = \{(2p - 2q + 2i, 2i): 0 \le i \le q - 1\}$$

and, for $m \ge 3$,

$$E_m = \{(mp - mq + j, j): 0 \le j \le mq - 2\}.$$

Let Ω *be the smallest algebra pattern that contains* (p, q). *Then*

$$\Omega = E_1 \cup E_2 \cup E_3 \cup \cdots.$$

Note. To get a clearer picture of this set, let $\tilde{\Omega}$ be the algebra pattern described in part (III) of Theorem 12.4.7. The union $E_1 \cup E_2 \cup E_3 \cup \cdots$ is then the subset of $\tilde{\Omega}$ that is left after deletion of the points $(2p - j, 2q - j)$ with odd j, and of the points $(mp - 1, mq - 1)$, $m = 3, 4, 5, \ldots$.

Proof. The above-mentioned set $\tilde{\Omega}$ has property (*) of Theorem 12.4.5, hence is an algebra pattern even when $n = 2$. Thus $\Omega \subset \tilde{\Omega}$. It follows from Theorems 12.5.9 and 12.5.10 that the points of $\tilde{\Omega}$ that are not in $\bigcup E_m$ are actually missing from Ω. What remains to be proved is that $E_m \subset \Omega$, for $m = 1, 2, 3, \ldots$.

Since Ω is a semigroup, $E_1 \subset \Omega$.

By Theorem 12.5.10, $E_2 \subset \Omega$.

To prove $E_3 \subset \Omega$, we apply the various parts of Lemma 12.5.11 to (p, q) and

$$(r, s) = (2p - 2q + 2i, 2i), \qquad i = 0, 1, \ldots, q - 1.$$

(A) and (B) produce those points of E_3 that have $q - 1 \leq j \leq 3q - 2$. If $3q \leq 2p$, and $(r, s) = (2p - 2q, 0)$, then

$$\min(q, r) = q,$$

so that (E) produces all points of E_3 that have $0 \leq j \leq q - 1$. In that case, $E_3 \subset \Omega$.

Assume now that $2p < 3q$. Then (E) produces only those points of E_3 that have $3q - 2p \leq j \leq q - 1$, and we still have to account for $0 \leq j < 3q - 2p$.

Fix j in this range. Then $p + j < 3q - p < 2q$. If $p + j$ is even, we can use (C) with $s = 2i = p + j$. If $p + j$ is odd, use (D) with $s = 2i = p + j - 1$.

This completes the proof that $E_3 \subset \Omega$.

The rest is easy: E_4 and E_5 can be obtained from the preceding ones, using only (A) and (B); for $E_6, E_7, \ldots$, (A) alone is enough.

Chapter 13

Moebius-Invariant Function Spaces

A space Y of functions with domain S, or B, or $\bar{B}$, is said to be *Moebius-invariant*, or simply $\mathcal{M}$-*invariant*, if $f \circ \psi \in Y$ for all $f \in Y$ and all $\psi \in \mathrm{Aut}(B)$.

The closed $\mathcal{M}$-invariant subspaces of various function spaces are determined in this chapter. Most of the theorems say that the obvious possibilities are in fact the only ones.

13.1. $\mathcal{M}$-Invariant Spaces on S

13.1.1. As in Section 12.3, X will now stand for either $C(S)$ or $L^p(\sigma)$ $(1 \le p < \infty)$. If Y is a closed $\mathcal{M}$-invariant subspace of X, then Y is obviously also $\mathcal{U}$-invariant, and Theorem 12.3.6 shows that $Y = X_\Omega$, where Ω is the set of all lattice points (p, q) such that $\pi_{pq} Y \ne \{0\}$. The problem is thus to pick out those sets Ω for which X_Ω is $\mathcal{M}$-invariant. The following lemma does this.

13.1.2. Lemma. *Suppose Y is a closed $\mathcal{M}$-invariant subspace of X, $p > 0$, and $H(p, q) \subset Y$ for some $q \ge 0$.*

Then $H(p - 1, q) \subset Y$, and $H(p + 1, q) \subset Y$.

Proof. Put $g(z) = z_1^p \bar{z}_n^q$. Then $g \in H(p, q) \subset Y$. (Note that this makes sense even when $n = 1$, since then $q = 0$; see §12.2.1.)

For $-1 < t < 1$, put $s = (1 - t^2)^{1/2}$, and let

$$(1) \qquad \psi_t(z) = \left(\frac{z_1 + t}{1 + tz_1}, \frac{sz_2}{1 + tz_1}, \dots, \frac{sz_n}{1 + tz_1} \right).$$

Then $\psi_t \in \mathrm{Aut}(B)$, hence $g \circ \psi_t \in Y$. This says that $h_t \in Y$, where

$$(2) \qquad h_t(z) = \left(\frac{z_1 + t}{1 + tz_1} \right)^p \left(\frac{\bar{z}_n}{1 + t\bar{z}_1} \right)^q.$$

As $t \to 0$, the quotients $(h_t - h_0)/t$ converge, uniformly on S, to

$$(3) \qquad f(z) = (pz_1^{p-1} - pz_1^{p+1} - qz_1^p \bar{z}_1) \bar{z}_n^q.$$

278

Thus $f \in Y$. Setting

(4) $$f_1 = z_1^{p-1} \bar{z}_n^q, \qquad f_2 = z_1^{p+1} \bar{z}_n^q, \qquad f_3 = z_1^p \bar{z}_1 \bar{z}_n^q,$$

we have

(5) $$f = pf_1 - pf_2 - qf_3.$$

The function $u = f_3 - c(z_1\bar{z}_1 + \cdots + z_n\bar{z}_n)f_1$ is harmonic, hence lies in $H(p, q+1)$, when $c = p/(p + q + n - 1)$. On S, $u = f_3 - cf_1$. Hence

(6) $$f = (p - cq)f_1 - pf_2 - qu$$

is the orthogonal decomposition of f (on S) into components that lie in $H(p-1, q)$, $H(p+1, q)$, $H(p, q+1)$, respectively. Since $p > 0$, $p - cq \neq 0$. Thus

(7) $$\pi_{p-1, q} Y \neq \{0\}, \qquad \pi_{p+1, q} Y \neq \{0\},$$

and the lemma follows from Theorem 12.3.6.

13.1.3. The roles of p and q can of course be switched in the preceding lemma. If X_Ω is $\mathscr{M}$-invariant, repeated application of the lemma shows therefore that only the following six cases can occur:

    (i) Ω is empty.
    (ii) $\Omega = \{(0, 0)\}$.
    (iii) $\Omega = \{(p, 0): p = 0, 1, 2, \ldots\}$.
    (iv) $\Omega = \{(0, q): q = 0, 1, 2, \ldots\}$.
    (v) $\Omega = \{(p, q): pq = 0\}$.
    (vi) $\Omega = Q$.

The sets described by (iii) and (iv) are the "edges" of the quadrant Q, and (v) is their union. The six subspaces of $C(S)$ that correspond to these sets Ω are displayed in the following diagram, which also indicates the inclusions that exist among them:

Here (a) $\mathbb{C}$ denotes the constant functions,
  (b) conj $A(S)$ is the space of all $f \in C(S)$ whose complex conjugates lie in $A(S)$, and
  (c) plh(S) consists of all $f \in C(S)$ whose Poisson integrals are pluriharmonic in B. Thus $f \in$ plh(S) if and only if f has a continuous extension

to $\bar{B}$ that is pluriharmonic in B. The spade plh(S) can also be described as the uniform closure of $A(S)$ + conj $A(S)$.

[When $n = 1$, it should be noted that plh$(S) = C(S)$.]

For ease of reference, here is a more formal summary of this discussion:

13.1.4. Theorem. *The above diagram lists every $\mathscr{M}$-invariant closed subspace of $C(S)$.*

Naturally, the list of sets Ω given in §13.1.3 also describes the $\mathscr{M}$-invariant closed subspaces of $L^p(\sigma)$, $1 \le p < \infty$. They are

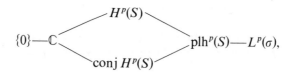

where (plh)$^p(S)$ is the space of all $f \in L^p(\sigma)$ whose Poisson integrals are pluriharmonic in B.

13.2. $\mathscr{M}$-Invariant Subalgebras of $C_0(B)$

13.2.1. If $4n^2$ Re λ + (Im λ)^2 < 0 and X_λ is the corresponding eigenspace of the invariant Laplacian $\widetilde{\Delta}$ (§4.2.1) then

$$(1) \qquad\qquad\qquad X_\lambda \cap C_0(B)$$

is obviously $\mathscr{M}$-invariant, is a closed subspace of $C_0(B)$ (Corollary 1 to Theorem 4.2.4), and is neither $C_0(B)$ nor $\{0\}$ (Theorem 4.2.7).

Spaces of the form (1) and their closed sums thus furnish a large variety of closed $\mathscr{M}$-invariant subspaces of $C_0(B)$; these have not been classified. But the situation becomes extremely simple if we specialize to algebras:

13.2.2. Theorem. *$C_0(B)$ has no nontrivial closed $\mathscr{M}$-invariant subalgebra.*

Proof. Let Y be a closed $\mathscr{M}$-invariant subalgebra of $C_0(B)$, $Y \ne \{0\}$. We must prove that $Y = C_0(B)$.

Let μ be a complex Borel measure on B that annihilates Y.

Let $\widetilde{Y}$ be the class of all functions on B that are pointwise limits of uniformly bounded sequences of members of Y. Then $\widetilde{Y}$ is an $\mathscr{M}$-invariant algebra of bounded Borel functions, and the dominated convergence theorem shows that μ annihilates $\widetilde{Y}$.

Suppose we have proved the following statement:

(∗)   *There is a $\delta > 0$ such that $\widetilde{Y}$ contains the characteristic function of every ball rB, for $0 < r < \delta$.*

Since $\tilde{Y}$ is *M*-invariant, (∗) implies that $\tilde{Y}$ contains the characteristic function of every ellipsoid $E(a, \varepsilon) = \varphi_a(\varepsilon B)$ (§2.2.7) where $a \in B$, $0 < \varepsilon < \delta$. If W is a union of finitely many such ellipsoids, the fact that $\tilde{Y}$ is an algebra implies that $\tilde{Y}$ contains the characteristic function of W. Thus $\mu(W) = 0$. Every open set $\Omega \subset B$ is the union of an increasing sequence of W's. Consequently, $\mu(\Omega) = 0$, $\mu = 0$, and $Y = C_0(B)$, by the Hahn–Banach theorem.

We turn to the proof of (∗).

Since Y is *M*-invariant and $Y \neq \{0\}$, some $g \in Y$ has $g(0) \neq 0$. Its radialization (see §4.2.1)

$$ (1) \qquad\qquad g^{\#} = \int_{\mathcal{U}} g \circ U \, dU $$

lies in Y, and $g^{\#}(0) = g(0) \neq 0$.

For $m > n$, define

$$ (2) \qquad\qquad K_m(z) = c_m(1 - |z|^2)^m; $$

the constants c_m are so adjusted that

$$ (3) \qquad\qquad \int_B K_m \, d\tau = 1 $$

where τ is the *M*-invariant measure defined in Theorem 2.2.6. Put

$$ (4) \qquad\qquad h_m(z) = \int_B (g^{\#} \circ \varphi_w)(z) K_m(w) d\tau(w), $$

where φ_w is the involution defined in §2.2.1.

Since $g^{\#} \circ \varphi_w \in Y$ for all $w \in B$, $h_m \in Y$. When $z = 0$ and $m \to \infty$, the integral (4) converges to $g^{\#}(0) \neq 0$. Thus for some large m (fixed from now on), $h_m(0) \neq 0$.

Part (iv) of Theorem 2.2.2 implies that

$$ (5) \qquad\qquad |\varphi_w(z)| = |\varphi_z(w)|. $$

Since $g^{\#}$ is radial, (4) can therefore be rewritten in the form

$$ (6) \qquad\qquad h_m(z) = \int_B g^{\#}(\varphi_z(w)) K_m(w) d\tau(w) $$

which, because of the invariance of τ, is the same as

$$(7) \qquad h_m(z) = \int_B g^*(w) K_m(\varphi_z(w)) d\tau(w),$$

or

$$(8) \qquad h_m(z) = c_m(1 - |z|^2)^m \int_B g^*(w) \frac{(1 - |w|^2)^m}{|1 - \langle z, w \rangle|^{2m}} d\tau(w),$$

by (2) and another application of 2.2.2(iv). It is clear from (8) that h_m is real-analytic.

Now choose a point $a \in B$ so that $|h_m(a)| = \|h_m\|_\infty$ but $|h_m(z)| < \|h_m\|_\infty$ if $|a| < |z| \le 1$. Put $h = \mathrm{ch}_m \circ \varphi_a$, where the constant c is so chosen that $h(0) = \mathrm{ch}_m(a) = 1$, and put $f = h^*$.

This f is a radial real-analytic function, $f \in Y$, $f(0) = 1$, and $|f(z)| < 1$ if $0 < |z| \le 1$.

Consequently, there is a $\delta > 0$ such that

(a) $|f(z)|$ is a *strictly decreasing* function of $|z|$ in the ball δB, and
(b) $|f(w)| < |f(z)|$ whenever $|z| \le \delta < |w| \le 1$.

In other words, f maps $\delta \bar{B}$ onto an arc L with one endpoint at 1; f maps the rest of B into a closed disc D with center at 0, whose boundary contains the other endpoint of L; and no circle with center at 0 intersects L in more than one point.

Let $L_{u,v}$ be the subarc of L that has u, v as endpoints.

Choose r, s, so that $0 < s < r < \delta$. There exist $\alpha \in L$, $\beta \in L$, so that

$$(9) \qquad f(sB) = L_{1,\alpha}, \qquad f(rB) = L_{1,\beta}.$$

Define a continuous function q on $D \cup L$ by setting $q(\lambda) = 1$ on $L_{1,\alpha}$,

$$(10) \qquad q(\lambda) = \frac{|\lambda| - |\beta|}{|\alpha| - |\beta|} \quad \text{on} \quad L_{\alpha,\beta},$$

and $q(\lambda) = 0$ on the rest of $D \cup L$.

The compact set $D \cup L$ does not separate the plane, and q is holomorphic in its interior. Mergelyan's theorem (see, for instance, Chap. 20 in Rudin [3]) shows therefore that there are polynomials Q_i, with $Q_i(0) = 0$, such that $Q_i \to q$ uniformly on $D \cup L$. Since Y is an algebra, $Q_i \circ f \in Y$. Since Y is uniformly closed, it follows that $q \circ f \in Y$.

Note also that $\|q \circ f\|_\infty = 1$.

If we now fix r and let $s \nearrow r$, we obtain a uniformly bounded sequence of functions $q_s \circ f \in Y$ that converges to 1 in rB and to 0 outside rB. This proves (*).

13.3. $\mathscr{M}$-Invariant Subspaces of $C(\bar{B})$

In this section, the closed $\mathscr{M}$-invariant *subalgebras* of $C(\bar{B})$ are completely described. For *subspaces* the classification is not yet complete.

13.3.1. Definition. $P(B)$ denotes the space of all (invariant) Poisson integrals of members of $C(S)$.

Thus $P(B)$ is a closed subspace of $C(\bar{B})$ which is $\mathscr{M}$-invariant, by Theorem 3.3.8.

The space of all $f \in C(\bar{B})$ that are pluriharmonic in B will be denoted by $\mathrm{plh}(B)$.

13.3.2. Theorem. *The following diagram lists every $\mathscr{M}$-invariant closed subspace of $C(\bar{B})$ whose intersection with $C_0(B)$ is $\{0\}$:*

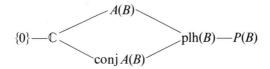

Proof. Let Y be an $\mathscr{M}$-invariant closed subspace of $C(\bar{B})$, with $Y \cap C_0(B) = \{0\}$. Let Λ_0 and Λ_1 be the linear functionals defined on Y by

$$(1) \qquad \Lambda_0 f = f(0), \qquad \Lambda_1 f = \int_S f \, d\sigma.$$

If there were an $f \in Y$ with $\Lambda_1 f = 0$ but $\Lambda_0 f \neq 0$, the radialization $f^{\#}$ of f would be a nontrivial member of $Y \cap C_0(B)$, in contradiction to our hypothesis.

The null-space of Λ_1 is thus contained in that of Λ_0. Hence there is a constant c such that $\Lambda_0 f = c\Lambda_1 f$ for every $f \in Y$, or

$$(2) \qquad f(0) = c \int_S f \, d\sigma \qquad (f \in Y).$$

Choose $z \in B$, choose $\psi \in \mathrm{Aut}(B)$ so that $\psi(0) = z$, let $P[f]$ denote the Poisson integral of the restriction of f to S, and apply (2) to $f \circ \psi$ in place of f, to obtain

$$(3) \qquad f(z) = (f \circ \psi)(0) = c \int_S (f \circ \psi) \, d\sigma$$

$$= cP[f \circ \psi](0) = cP[f](\psi(0)) = cP[f](z),$$

by Theorem 3.3.8. If z converges to some $\zeta \in S$, it follows that $f(\zeta) = cf(\zeta)$. Thus $c = 1$, and (3) becomes

(4) $f(z) = P[f](z)$ $(z \in B, f \in Y)$.

Every $f \in Y$ is thus the Poisson integral of its restriction to S. The supremum of $|f|$ is therefore the same on B as on S, so that the restriction of Y to S is a closed $\mathcal{M}$-invariant subspace of $C(S)$, i.e., it is one of the six spaces described by Theorem 13.1.4. Hence Y consists of the Poisson integrals of all functions in one of these spaces.

13.3.3. Theorem. *If X is an $\mathcal{M}$-invariant closed subspace of $C(\bar{B})$, and $X \supset C_0(B)$, then*

$$X = C_0(B) + Y,$$

where Y is one of the six spaces listed in Theorem 13.3.2.

Note that $C_0(B) + P(B) = C(\bar{B})$.

Proof. Since $X \supset C_0(B)$, the restriction of X to S is a *closed* $\mathcal{M}$-invariant subspace of $C(S)$, and $P[f] \in X$ for every $f \in X$. The conclusion of the theorem follows now from Theorem 13.1.4.

13.3.4. Remark. If Y is a closed $\mathcal{M}$-invariant subspace of $C(\bar{B})$ such that $Y_0 = Y \cap C_0(B)$ is neither $\{0\}$ nor $C_0(B)$, it is not known whether Y must be the direct sum of Y_0 and one of the spaces of Poisson integrals listed in Theorem 13.3.2. (Each of these direct sums is of course closed and $\mathcal{M}$-invariant.)

13.3.5. Theorem. *The closed $\mathcal{M}$-invariant subalgebras of $C(\bar{B})$ are*

   (a) $\{0\}$, $\mathbb{C}$, $A(B)$, conj $A(B)$,
   (b) $C_0(B) + Y$, *where Y is as in* (a),
   (c) $C(\bar{B})$.

Proof. If Y is a closed $\mathcal{M}$-invariant subalgebra of $C(\bar{B})$, Theorem 13.2.2 shows that $Y \cap C_0(B)$ is either $\{0\}$ or $C_0(B)$. Thus Y is one of the 12 spaces listed in Theorems 13.3.2 and 13.3.3. Of these, only

$$\text{plh}(B), P(B), C_0(B) + \text{plh}(B)$$

fail to be algebras.

As regards $\text{plh}(B)$, the following more precise result holds:

13.3.6. Theorem. *If both u and u^2 are pluriharmonic in a region $\Omega \subset \mathbb{C}^n$, then either $u \in H(\Omega)$ or $\bar{u} \in H(\Omega)$.*

Proof. By assumption, $\partial^2 u / \partial \bar{z}_j \, \partial z_k = 0$ for all j, k. Hence

$$(1) \qquad \qquad \frac{\partial u}{\partial z_k} \in H(\Omega) \qquad (1 \le k \le n).$$

The same is true for u^2, so that

$$(2) \qquad \qquad u \frac{\partial u}{\partial z_k} \in H(\Omega) \qquad (1 \le k \le n).$$

Either $\partial u / \partial z_k = 0$ for all k, in which case $\bar{u} \in H(\Omega)$, or there is an index k and a dense open subset Ω_0 of Ω such that $\partial u / \partial z_k$ has no zero in Ω_0. Division of (2) by (1) shows then that $u \in H(\Omega_0)$. Since u is continuous in Ω, the corollary to Theorem 4.4.7 implies that $u \in H(\Omega)$.

13.4. Some Applications

The preceding theorems enable us to draw conclusions about functions f, from certain hypotheses imposed on the family $\{f \circ \psi\}$, where ψ ranges over $\text{Aut}(B)$.

13.4.1. Proposition. *If $f \in C(S)$ and*

$$\int_{-\pi}^{\pi} f(\psi(e^{i\theta}, 0, \dots, 0)) e^{i\theta} \, d\theta = 0$$

for every $\psi \in \text{Aut}(B)$, then $f \in A(S)$.

Proof. The set Y of all f that satisfy the hypothesis forms an $\mathcal{M}$-invariant closed subspace of $C(S)$ that does not contain the function $\bar{z}_1$. Since $Y \supset A(S)$, Theorem 13.1.4 shows that $Y = A(S)$.

13.4.2. Proposition. *Suppose $f \in C(S)$, and f^2 cannot be uniformly approximated on S by finite linear combinations of functions $f \circ \psi$, where $\psi \in \text{Aut}(B)$.*
 Then $f \in \text{plh}(S)$, but f is not in $A(S)$, nor in conj $A(S)$.

Proof. Let Y be the $\mathcal{M}$-invariant closed subspace of $C(S)$ generated by f. The hypothesis says that Y is not an algebra. By Theorem 13.1.4, $Y = \text{plh}(S)$.

Note: When $n = 1$, then $\text{plh}(S) = C(S)$, an algebra. The hypothesis of Proposition 13.4.2 can therefore not be satisfied when $n = 1$. In other words: If $f \in C(T)$, then f^2 can be uniformly approximated on T by finite linear combinations of $f \circ \psi$, $\psi \in \text{Aut}(U)$.

13.4.3. Proposition. *Suppose* $f \in L^1(\sigma)$, *and the Cauchy integrals of* $f \circ \psi$ *satisfy*

(1) $$C[f \circ \psi] = C[f] \circ \psi$$

for every $\psi \in \text{Aut}(B)$. *Then* $f \in H^1(S)$.

Proof. Let Y be the (obviously closed) subspace of $L^1(\sigma)$ consisting of all f that satisfy the hypothesis. If $f \in Y$, and $\psi, \varphi \in \text{Aut}(B)$, two applications of (1) shows that

$$C[f \circ \psi \circ \varphi] = C[f] \circ \psi \circ \varphi = C[f \circ \psi] \circ \varphi.$$

Thus $f \circ \psi \in Y$, so that Y is $\mathcal{M}$-invariant. Clearly, $H^1(S) \subset Y$. On the other hand, Example 3.3.9 shows that ζ_1 is not in Y. Hence $Y = H^1(S)$, by the L^1-analogue of Theorem 13.1.4.

13.4.4. Proposition. *If* $f \in C(\bar{B})$ *has the mean-value property*

(1) $$f(\psi(0)) = \int_B f \circ \psi \, dv$$

for every $\psi \in \text{Aut}(B)$, *then* $\tilde{\Delta} f = 0$ *in* B.

Proof. The set Y of all f that satisfy the hypothesis forms a closed $\mathcal{M}$-invariant subspace of $C(\bar{B})$. Assume, to reach a contradiction, that there is a $g \in Y \cap C_0(B)$, $g \not\equiv 0$. Choose ψ so that $g(\psi(0)) = \|g\|_\infty$. Since $v(B) = 1$, g cannot satisfy (1).

Thus $\bar{Y} \cap C_0(B) = \{0\}$, and Theorem 13.3.2 implies that $Y = P(B)$. (Compare with Theorem 3.3.7.)

13.4.5. Proposition *Let* $K \subset B$ *be a compact set such that some* $h \in C(K)$ *cannot be uniformly approximated on* K *by holomorphic polynomials.*

If $f \in C(\bar{B})$ *and* $f \circ \psi$ *can be uniformly approximated on* K *by holomorphic polynomials, for every* $\psi \in \text{Aut}(B)$, *then* f *is holomorphic in* B.

Proof. The set of all such f forms an $\mathcal{M}$-invariant closed subalgebra of $C(\bar{B})$ that contains $A(B)$ but does not contain $C_0(B)$. By Theorem 13.3.5, this algebra is $A(B)$.

Note that the set K may be quite small. For example, if $r > 0$ and K is the circle

$$\{(re^{i\theta}, 0, \dots, 0): -\pi \le \theta \le \pi\},$$

the hypothesis of the theorem is satisfied.

This proposition is related to some aspects of the Pompeiu problem (Zalcman [1]).

13.4.6. A Problem. There seems to be no good reason why continuity on the boundary should play any role in the preceding result. The question thus arises whether Proposition 13.4.5 remains true if $C(\bar{B})$ is replaced by $C(B)$, the set of all (possibly unbounded) continuous complex functions with domain B.

Equipped with the topology of uniform convergence on compact subsets of B, $C(B)$ is a well-known Fréchet algebra. It would be interesting to know the $\mathcal{M}$-invariant closed subalgebras of $C(B)$. Probably there are only two nontrivial ones, namely $H(B)$ and conj $H(B)$.

Here is a small partial result that points in this direction:

13.4.7. Proposition. *If X is an $\mathcal{M}$-invariant closed subspace of $C(B)$, and if X contains a nonconstant $f \in H(B)$, then $X \supset H(B)$.*

Proof. Let $f = \Sigma F_p$ be the homogeneous expansion of f. There is a $p > 0$ such that $F_p \neq 0$. The $\mathcal{U}$-invariance of X, combined with the formula

$$F_p(z) = \frac{1}{2\pi} \int_{-\pi}^{\pi} f(e^{i\theta}z)e^{-ip\theta} \, d\theta,$$

shows that $F_p \in X$, since X is closed. Thus $H(p, 0) \subset X$, by the $\mathcal{U}$-minimality of the spaces $H(p, q)$. In particular, $z_1^p \in X$, and the proof of Lemma 13.1.2 leads to the conclusion that $pz_1^{p-1} - pz_1^{p+1} \in X$. Thus X contains $H(p - 1, 0)$ and $H(p + 1, 0)$. Repetition of this argument shows that X contains $H(p, 0)$ for every $p \geq 0$. Hence $X \supset H(B)$.

13.4.8. Note. With the exception of §13.3.6 and §13.4.7, the results of this chapter are contained in Nagel–Rudin [1]. Some earlier work on these topics was done by Agranovskii [1], [2], and by Agranovskii–Valskii [1].

Chapter 14

Analytic Varieties

This chapter contains a brief introduction to analytic varieties. It is quite elementary, but will be sufficient for the material that follows.

14.1. The Weierstrass Preparation Theorem

14.1.1. The Order of a Zero. Suppose Ω is a region in $\mathbb{C}^n$, $f \in H(\Omega)$, $a \in \Omega$, and $f(a) = 0$. If f is not identically 0 in Ω, then there are vectors $b \in \mathbb{C}^n$ such that the one-variable function

$$(1) \qquad \qquad \lambda \to f(a + \lambda b)$$

does not vanish identically in any neighborhood of $\lambda = 0$. Thus (1) has a zero of some positive integral order k at $\lambda = 0$. Of course, k may depend on the choice of b. (Example: $f = z^2 - w^3$.) The smallest k that can be obtained in this way, by varying b, is said to be the *order* of the zero that f has at a.

If f has a zero of order m at 0 (for simplicity, we replace a by 0), it follows that coordinates can be so chosen in $\mathbb{C}^n$, by an invertible linear change of variables, that the function $f(0', z_n)$ has a zero of order m at $z_n = 0$, using our customary notation $z = (z', z_n)$, with $z' \in \mathbb{C}^{n-1}$, $z_n \in \mathbb{C}$.

Polydiscs Δ in $\mathbb{C}^n$ will be written in the form $\Delta = \Delta' \times \Delta_n$, where Δ' is a polydisc in $\mathbb{C}^{n-1}$ and Δ_n is a disc in $\mathbb{C}$. Throughout this section it is assumed that $n > 1$.

The Weierstrass theorem (14.1.3) is, as we shall see, a simple consequence of the following lemma, which will also have some other applications.

14.1.2. The Two-Function Lemma. *Suppose Ω is a neighborhood of 0 in $\mathbb{C}^n$, $f \in H(\Omega)$, $g \in H(\Omega)$, and $f(0', z_n)$ has a zero of multiplicity m at $z_n = 0$.*

    (i)   *There is then a polydisc $\Delta = \Delta' \times \Delta_n \subset \Omega$, with center at 0, such that $f(z', \cdot)$ has, for each $z' \in \Delta'$, exactly m zeros in Δ_n, counted according to their multiplicities.*

(ii) *If these zeros are denoted by* $\alpha_1(z'), \ldots, \alpha_m(z')$, *then the coefficients* $c_0, \ldots, c_{m-1}$, *determined by*

$$\prod_{j=1}^{m} [\lambda - g(z', \alpha_j(z'))] = \lambda^m + \sum_{j=0}^{m-1} c_j(z')\lambda^j,$$

are holomorphic functions in Δ'.

Note that the coefficients $c_j(z')$ are precisely the elementary symmetric functions of the unordered m-tuple

$$\{g(z', \alpha_j(z')): 1 \le j \le m\}.$$

The proof will, however, not rely on any knowledge of symmetric functions.

Proof. Since $f(0', \cdot)$ has a zero of order m at 0 and since the zeros of holomorphic functions of one complex variable are isolated, there is a number $r > 0$ such that $f(0', \cdot)$ has no other zeros in the closure of the disc $\Delta_n = \{\lambda: |\lambda| < r\}$. Hence there exist $\delta > 0$ and a polydisc Δ' in $\mathbb{C}^{n-1}$, centered at $0'$, such that $|f(z', \lambda)| > \delta$ whenever $z' \in \Delta'$ and $|\lambda| = r$, and such that the closure of $\Delta = \Delta' \times \Delta_n$ lies in Ω.

We now associate to every $h \in H(\Omega)$ and $z' \in \Delta'$ the integral

(1)
$$J_h(z') = \frac{1}{2\pi i} \int_{|\lambda|=r} \left(\frac{hD_n f}{f}\right)(z', \lambda)d\lambda,$$

where $D_n f = \partial f / \partial z_n$. The denominator is bounded from 0 on the path of integration. Thus J_h is continuous in Δ', and Morera's theorem, applied in any of the variables $z_1, \ldots, z_{n-1}$, shows that J_h is in fact *holomorphic* in Δ'.

When $h \equiv 1$, then $J_h(z')$ is the number of zeros of $f(z', \cdot)$ in Δ_n. In particular, $J_h(0') = m$. Being a continuous integer-valued function in the connected set Δ', J_h is constant. This proves (i).

Returning to an arbitrary $h \in H(\Omega)$, another application of the residue theorem shows that

(2)
$$J_h(z') = \sum_{j=1}^{m} h(z', \alpha_j(z'))$$

where $\alpha_1(z'), \ldots, \alpha_m(z')$ are the zeros of $f(z', \cdot)$ in Δ_n. *The sum* (2) *is thus holomorphic in* Δ'.

If $\zeta \in \mathbb{C}$ and $|\zeta|$ is sufficiently small, then $|\zeta g| < 1$ on some neighborhood of $\bar{\Delta}$, and the preceding reasoning can be applied to $h = \log(1 - \zeta g)$. In particular, the sum (2) is holomorphic for this h, and if we exponentiate, we see that $G_\zeta \in H(\Delta')$, where

$$(3) \qquad\qquad G_\zeta(z') = \prod_{j=1}^{m} [1 - \zeta g(z', a_j(z'))].$$

For each $z' \in \Delta'$, $G_\zeta(z')$ is a polynomial in ζ. The coefficient of ζ^k is

$$(4) \qquad\qquad \frac{1}{2\pi i} \int_\Gamma G_\zeta(z') \zeta^{-k-1} \, d\zeta,$$

where Γ is a small circle around the origin. Since $G_\zeta \in H(\Delta')$, (4) defines a holomorphic function in Δ'. The coefficient of ζ^k is the coefficient of λ^{m-k} in (ii). The lemma is therefore proved.

14.1.3. The Weierstrass Theorem. *Suppose Ω is a neighborhood of 0 in $\mathbb{C}^n$, $f \in H(\Omega)$, $f(0', z_n)$ has a zero of multiplicity m at $z_n = 0$, and $\Delta = \Delta' \times \Delta_n$ is as in 14.1.2(i). Then*

$$(1) \qquad\qquad f(z) = W(z)h(z) \qquad (z \in \Delta)$$

where $h \in H(\Delta)$, h has no zero in Δ, and

$$(2) \qquad\qquad W(z) = z_n^m + b_1(z')z_n^{m-1} + \cdots + b_m(z'),$$

with $b_j \in H(\Delta')$, $b_j(0') = 0$.

The function W is called a *Weierstrass polynomial.* The factorization (1) is clearly unique, since $W(z)$ is, for each $z' \in \Delta'$, a monic polynomial in z_n, of degree m, whose m zeros are determined by f.

Proof. If we apply the two-function lemma with $g(z) = z_n$, and if we define

$$(3) \qquad\qquad W(z) = W(z', z_n) = \prod_{j=1}^{m} [z_n - \alpha_j(z')],$$

we see that W is a monic polynomial in z_n whose coefficients b_j are holomorphic in Δ'. Also, $\alpha_j(0') = 0$, for $1 \le j \le m$. Thus $W(0', z_n) = z_n^m$, and $b_j(0') = 0$. To complete the proof, define

$$(4) \qquad\qquad h(z) = \frac{1}{2\pi i} \int_{|\lambda|=r} \left(\frac{f}{W}\right)(z', \lambda) \frac{d\lambda}{\lambda - z_n} \qquad (z \in \Delta),$$

where r is as in the proof of 14.1.2. Then $h \in H(\Delta)$ (note that W has no zero on the path of integration), and since $W(z', \cdot)$ is, for fixed z', a polynomial with the same zeros as $f(z', \cdot)$, Riemann's one-variable theorem about removable singularities shows that $f = Wh$ in Δ.

14.1.4. Remarks. Part (i) of 14.1.2 shows that to every $z' \in \Delta'$ correspond exactly m values of $z_n \in \Delta_n$ (possibly with repetitions) such that $f(z', z_n) = 0$. Letting $Z(f)$ be the zero-variety of f, this says, roughly speaking, that $Z(f) \cap \Delta$ consists of m "sheets" over Δ' which come together at the origin (and possibly at other points).

In particular, no holomorphic function of more than one complex variable has any isolated zeros.

The case $m = 1$ is particularly simple: Theorem 14.1.3 shows then that $Z(f) \cap \Delta$ is the graph of a holomorphic function of $n - 1$ variables (with domain Δ').

Here is another application of the two-function lemma, to a maximum modulus theorem.

14.1.5. Theorem. *Let f, g, Δ be as in Lemma* 14.1.2. *Assume g is not constant on $Z(f) \cap \Delta$. Then $|g(0)| < |g(z)|$ for some $z \in Z(f) \cap \Delta$.*

Proof. Assume $g(0) = 1$, without loss of generality, and assume (to reach a contradiction) that $|g(z', \alpha_j(z'))| \leq 1$ for all $z' \in \Delta', 1 \leq j \leq m$. Put

$$(1) \qquad\qquad G(z') = \sum_{j=1}^{m} g(z', \alpha_j(z')) \qquad (z' \in \Delta').$$

Then $G \in H(\Delta')$, $G(0') = m$, $|G(z')| \leq m$ for all $z' \in \Delta'$. Since the maximum modulus theorem holds in $H(\Delta')$, it follows that $G(z') = m$ for all $z' \in \Delta'$. Since no summand in (1) exceeds 1 in absolute value, all must equal 1. Thus $g(z) = 1$ for all $z \in Z(f) \cap \Delta$, a contradiction.

14.2. Projections of Varieties

We shall now introduce the general notion of an analytic variety (so far, we have only encountered zero-varieties of single holomorphic functions) and prove a projection theorem which will be used in Section 14.3 to establish the finiteness of all compact subvarieties of $\mathbb{C}^n$.

14.2.1. Definition. Let Ω be an open set in $\mathbb{C}^n$. A set $V \subset \Omega$ is said to be an *analytic subvariety* of Ω if

(a) V is (relatively) closed in Ω, and
(b) every point $p \in \Omega$ has a neighborhood $N(p)$ such that

$$(1) \qquad\qquad V \cap N(p) = Z(f_1) \cap \cdots \cap Z(f_r)$$

for some $f_1, \ldots, f_r \in H(N(p))$.

Here, as before, $Z(f_i)$ is the zero-variety of f_i.

We may describe (1) by saying that "V is defined in $N(p)$ by $f_1, \ldots, f_r$."

The adjective "analytic" will occasionally be dropped, in which case we will just talk about *subvarieties* of Ω.

Note that (a) is really redundant, being a consequence of (b). We include (a) for emphasis. Moreover, if (a) is known to hold, then (b) needs only to be verified for $p \in V$.

14.2.2. Remarks. (i) The empty set is a subvariety of Ω. (Take $r = 1, f_1 = 1$.)

(ii) Ω is a subvariety of Ω. (Take $r = 1, f_1 = 0$.)

(iii)   If Ω is a region in $\mathbb{C}$ (the case $n = 1$) then the subvarieties of Ω (nonempty, $\neq \Omega$) are precisely the discrete subsets of Ω.

(iv)   If V_1 and V_2 are subvarieties of Ω, so are $V_1 \cap V_2$ and $V_1 \cup V_2$.

Proof. If $\{f_i\}$ defines V_1 and $\{g_k\}$ defines V_2 in $N(p)$, then $\{f_i\} \cup \{g_k\}$ defines $V_1 \cap V_2$ and $\{f_i g_h\}$ defines $V_1 \cup V_2$.

(v)   If a subvariety V of Ω is compact, then V is also a subvariety of $\mathbb{C}^n$.

14.2.3. Notation. When $n > 1$, we shall write $z = (z', z_n)$, $\Delta = \Delta' \times \Delta_n$, as in Section 14.1, and shall let π denote the projection of $\mathbb{C}^n$ onto $\mathbb{C}^{n-1}$ defined by $\pi(z', z_n) = z'$.

In particular, $\pi(\Delta) = \Delta'$.

14.2.4. The Projection Theorem. *Let V be an analytic subvariety of a region $\Omega \subset \mathbb{C}^n$, $n > 1$, let $p = (p', p_n)$ be a point of V, and let*

$$(1) \qquad\qquad L = \{(p', \lambda) \colon \lambda \in \mathbb{C}\}.$$

If p is an isolated point of $L \cap V$, then p is the center of a polydisc $\Delta \subset \Omega$ such that $\pi(V \cap \Delta)$ is an analytic subvariety of $\pi(\Delta)$.

Proof. Without loss of generality, assume that p is the origin of $\mathbb{C}^n$ and that Ω is a polydisc in which V is defined by holomorphic functions $f_1, \ldots, f_r$. Our assumption about $L \cap V$ shows that for (at least) one f_i, say for f_r, the origin in $\mathbb{C}$ is an isolated zero of $f_r(0', \cdot)$. To emphasize the special role played by this property of f_r, let us write F for f_r.

There is a polydisc $\Delta = \Delta' \times \Delta_n \subset \Omega$, with center at 0, such that the conclusion of the two-function lemma (14.1.2) holds for Δ, F, and any $g \in H(\Omega)$. In particular, the product P defined by

$$(1) \qquad\qquad P(z') = \prod_{j=1}^{m} g(z', \alpha_j(z')) \qquad (z' \in \Delta')$$

is holomorphic in Δ'; recall that $\alpha_j(z')$ $(1 \le j \le m)$ are the zeros of $F(z', \cdot)$.

Now fix some $z' \in \Delta'$. It is clear that $P(z') = 0$ if and only if some $\alpha_j(z')$ is also a zero of $g(z', \cdot)$, i.e., if and only if F and g have a common zero in Δ that lies "above" z'. Hence

$$(2) \qquad\qquad \pi(\Delta \cap Z(F) \cap Z(g)) = Z(P).$$

Since $P \in H(\Delta')$, *we conclude that* $\pi(\Delta \cap Z(F) \cap Z(g))$ *is an analytic subvariety of* $\Delta' = \pi(\Delta)$.

We now return to the functions $f_1, \ldots, f_r$ (with $f_r = F$) that define V in Ω. (Note that we can assume $r > 1$, without loss of generality, by repetition, if necessary.)

Let (c_{ij}) be a rectangular matrix of complex numbers, with $(r - 1)m$ rows and $r - 1$ columns, in which every square matrix of size $(r - 1) \times (r - 1)$ has non-zero determinant. In other words, every set of $r - 1$ rows is linearly independent. Define

$$(3) \qquad\qquad g_i = \sum_{j=1}^{r-1} c_{ij} f_j \qquad (1 \le i \le rm - m).$$

Applying (2) to g_i in place of g, we see that each of the sets

$$(4) \qquad\qquad E_i = \pi(\Delta \cap Z(F) \cap Z(g_i))$$

is a subvariety of Δ'. We claim that

$$(5) \qquad\qquad \pi(\Delta \cap V) = \bigcap_i E_i.$$

To prove one half of (5), let $z \in \Delta \cap V$. Then $z \in Z(g_i)$ for all i, and $z \in Z(F)$. Hence $\pi(z) \in E_i$ for all i. The left side of (5) is thus a subset of the right.

For the opposite inclusion, take $z' \in \bigcap E_i$. To each of the $(r - 1)m$ values of i corresponds then an $\alpha_k(z')$ such that

$$(6) \qquad\qquad g_i(z', \alpha_k(z')) = 0.$$

This follows from (4). Since k runs over only m values, there is some k (fixed from now on) and some set I of $r - 1$ distinct i's, for which (6) holds. The corresponding system of equations

$$(7) \qquad \sum_{j=1}^{r-1} c_{ij} f_j(z', \alpha_k(z')) = g_i(z', \alpha_k(z')) = 0 \qquad (i \in I)$$

has a unique solution, by our choice of (c_{ij}). Thus $f_j(z', \alpha_k(z')) = 0$ for all j, and therefore $z' \in \pi(\Delta \cap V)$.

This proves (5). Since each E_i is a subvariety of Δ', the same is true of their intersection, and the theorem is proved.

14.2.5. Remarks. (i) That the hypothesis on $L \cap V$ cannot be removed from the projection theorem is shown by the variety

$$V = \{z \in \mathbb{C}^3 : z_1 = z_2 z_3\},$$

at $p = (0, 0, 0)$. For if Δ is any polydisc with center at p, then $\pi(V \cap \Delta)$ is not even a closed subset of $\pi(\Delta)$: the points $(z_1, 0)$ with $z_1 \neq 0$ are missing.

(ii)   The variety $V = \{z \in \mathbb{C}^2 : z_1 z_2 = 1\}$ satisfies the hypotheses of the projection theorem at every point, but nevertheless the conclusion holds only locally, not globally: $\pi(V) = \mathbb{C} \setminus \{0\}$ is not a subvariety of $\mathbb{C}$.

14.3. Compact Varieties in $\mathbb{C}^n$

14.3.1. Theorem. *Every compact analytic subvariety of $\mathbb{C}^n$ is a finite set of points.*

Proof. When $n = 1$, the theorem is true because zero-sets of nonconstant holomorphic functions of one variable are discrete. Assume that $n \geq 2$ and that the theorem is true in $\mathbb{C}^{n-1}$. Let V be a compact subvariety of $\mathbb{C}^n$.

Pick $z' \in \pi(V)$, where $\pi: \mathbb{C}^n \to \mathbb{C}^{n-1}$ is as in §14.2.3, and define

$$L = \{(z', \lambda): \lambda \in \mathbb{C}\}.$$

After an obvious identification of L with $\mathbb{C}$, we see that $L \cap V$ is a compact subvariety of $\mathbb{C}$, hence is finite. Let $p^{(i)}$ $(1 \leq i \leq m)$ be the points of $L \cap V$. By the projection theorem, each $p^{(i)}$ is the center of a polydisc Δ_i in $\mathbb{C}^n$ such that $\pi(V \cap \Delta_i)$ is a subvariety of $\pi(\Delta_i)$. The part of V that is not covered by $\Delta_1 \cup \cdots \cup \Delta_m$ is compact and has positive distance from L. Hence z' is the center of a polydisc

$$\Delta' \subset \pi(\Delta_1) \cap \cdots \cap \pi(\Delta_m),$$

so small that all points of V that project into Δ' lie in $\Delta_1 \cup \cdots \cup \Delta_m$. In other words,

$$\Delta' \cap \pi(V) = \Delta' \cap \bigcup_{i=1}^{m} \pi(V \cap \Delta_i).$$

Thus $\Delta' \cap \pi(V)$ is a subvariety of Δ'. Since Δ' is a neighborhood of the arbitrarily chosen point $z' \in \pi(V)$, and since $\pi(V)$ is compact (hence closed), it follows that $\pi(V)$ is a subvariety of $\mathbb{C}^{n-1}$.

Hence $\pi(V)$ is a finite set, by our induction hypothesis. Since each point of $\pi(V)$ is the π-image of only finitely many points of V, we conclude that V is the union of finitely many finite sets.

14.3.2. The preceding theorem plays an important role in the study of proper holomorphic maps (Chapter 15). Here is another application of it, to functions in

$$A(\Omega) = C(\bar{\Omega}) \cap H(\Omega):$$

14.3.3. Theorem. *Assume $n > 1$. Let Ω be a bounded region in $\mathbb{C}^n$. If $f \in A(\Omega)$, $g \in A(\Omega)$, and $|f(\zeta)| \le |g(\zeta)|$ for every boundary point ζ of Ω, then also $|f(z)| \le |g(z)|$ for every $z \in \Omega$.*

To see that this is false when $n = 1$, let Ω be the unit disc in $\mathbb{C}$ and consider $(\alpha - z)/(1 - \bar{\alpha}z)$ for two distinct values of $\alpha \in U$.

Proof. The theorem is trivial when $g \equiv 0$. So assume $g \not\equiv 0$. Since $Z(g)$ has then no interior, it suffices to prove that $|f(p)| \le |g(p)|$ for every $p \in \Omega$ where $g(p) \neq 0$. Pick such a point p, put $\alpha = f(p)/g(p)$, and let V be the zero-variety of $f - \alpha g$ (in Ω).

Since $\bar{V}$ (the closure of V in $\mathbb{C}^n$) is compact and g is continuous, there is a $\beta \in \mathbb{C}$ such that $g(z_0) = \beta$ for some $z_0 \in \bar{V}$, but $|g(z)| \le |\beta|$ for every $z \in \bar{V}$. Clearly, $|\beta| \ge |g(p)| > 0$. Let

$$E = \{z \in \bar{V}: g(z) = \beta\}.$$

Then E is compact, and nonempty because $z_0 \in E$.

If E contains no boundary point of Ω, then E is a compact subvariety of Ω, hence E is a finite set, by Theorem 14.3.1. But $|g|$ cannot have an isolated maximum on $V = Z(f - \alpha g)$, by Theorem 14.1.5.

Consequently, there exists $\zeta \in E \cap \partial\Omega$.

Since $\zeta \in E$, $|\alpha\beta| = |\alpha g(\zeta)| = |f(\zeta)|$.

Since $\zeta \in \partial\Omega$, $|f(\zeta)| \le |g(\zeta)| = |\beta|$.

Since $\beta \neq 0$, it follows that $|\alpha| \le 1$, hence $|f(p)| \le |g(p)|$.

14.4. Hausdorff Measures

Hausdorff measures furnish just the right framework for deriving certain topological properties of analytic varieties from size estimates. The results of the present section will lead to an easy proof of Proposition 15.1.5, which establishes an important feature of proper holomorphic maps.

Since Hausdorff measures have nothing to do with the complex structure of $\mathbb{C}^n$, it seems best to study them in euclidean spaces R^N, rather than in $\mathbb{C}^n$. A large part of the subject can in fact be developed in arbitrary metric spaces.

14.4.1. Definitions. Let $A \subset R^N, \varepsilon > 0$. An ε-*cover* of A is an at most countable collection of sets $A_i \subset R^N$ such that diam $A_i < \varepsilon$ for all i, and $A = \bigcup A_i$. For any $t > 0$, define

(1) $$h_{\varepsilon, t}(A) = \inf \sum_i (\operatorname{diam} A_i)^t,$$

the infimum being taken over all ε-covers $\{A_i\}$ of A, and define

(2)
$$h_t(A) = \lim_{\varepsilon \searrow 0} h_{\varepsilon, t}(A).$$

The monotonicity of $h_{\varepsilon, t}(A)$ as a function of ε shows that lim could be replaced by sup in (2).

The number $h_t(A)$ is the *t-dimensional Hausdorff measure of A*.

What usually matters most is not the precise value of $h_t(A)$, but whether $h_t(A)$ is or is not 0, positive, finite, or ∞.

Customarily, h_0 is counting measure.

We note, in passing, that the *Hausdorff dimension* of A is the supremum of the set of all t for which $h_t(A) > 0$.

14.4.2. Proposition. *Let Q be the unit cube of R^N. Then*

(i) $h_t(Q) = \infty$ *when* $t < N$,
(ii) $0 < h_N(Q) < \infty$, *and*
(iii) $h_t(Q) = 0$ *when* $t > N$.

Proof. Given $\varepsilon > 0$, let k be an integer, $k > \varepsilon^{-1}N^{1/2}$. Cover Q by k^N cubes of edge $1/k$; their diameter is $< \varepsilon$. Hence

$$h_{\varepsilon, t}(Q) \le k^N (N^{1/2} k^{-1})^t = N^{t/2} k^{N-t}.$$

Thus $h_N(Q) \le N^{N/2}$, and (letting $k \to \infty$), $h_t(Q) = 0$ if $t > N$.

Next, suppose $t \le N$. Let $\{A_i\}$ be an ε-cover of Q, let δ_i be the diameter of A_i, cover each A_i by a ball B_i of radius δ_i. Then $m(B_i) = c\delta_i^N$, where m denotes Lebesgue measure in R^N and c is a positive real number that depends only on N. Since $\delta_i \le \varepsilon$,

$$m(B_i) \le c\varepsilon^{N-t}\delta_i^t,$$

so that

$$\sum_i \delta_i^t \ge c^{-1}\varepsilon^{t-N} \sum_i m(B_i) \ge c^{-1}\varepsilon^{t-N},$$

since $\{B_i\}$ is a cover of Q and $m(Q) = 1$.

Hence $h_N(Q) \ge c^{-1}$, and $h_t(Q) = \infty$ if $t < N$.

14.4.3. Proposition. *If $t < N$, and the N-dimensional cube Q is a union of countably many compact sets K_i, then $h_t(K_i) = \infty$ for at least one i.*

Proof. By Baire's theorem, some K_i contains an N-cube, hence $h_t(K_i) = \infty$, by Proposition 14.4.2.

14.4.4. Proposition. *If $h_t(A) < \infty$ and $F \in \text{Lip } 1$, then $h_t(F(A)) < \infty$.*

This follows immediately from the definitions.

14.4.5. Theorem. *Suppose Ω is a connected open set in R^N, $0 \le t < N - 1$, E is a (relatively) closed subset of Ω, and E is a union of countably many compact sets K_i with $h_t(K_i) < \infty$.*
Then $\Omega \backslash E$ is connected.

Proof. It suffices to prove that $V \backslash E$ is connected, for every nonempty convex open set $V \subset \Omega$. (Note that E has no interior, by 14.4.3, so that $V \backslash E$ is not empty.)

Fix such a set V and let x, y be distinct points of $V \backslash E$. Since $V \backslash E$ is open, there is an $(N - 1)$-cube $Q \subset V \backslash E$, centered at y, perpendicular to the interval $[x, y]$. If x and y were in different components of $V \backslash E$, then, for every $z \in Q$, the interval $[x, z]$ would meet E, hence some K_i. If π denotes the conical projection into Q with x as origin, it follows that Q is covered by the sets $\pi(K_i)$. Since $\pi \in \text{Lip } 1$, $h_t(\pi(K_i)) < \infty$. This contradicts Proposition 14.4.3 (with $N - 1$ in place of N). Hence $V \backslash E$ is connected.

14.4.6. Proposition. *Suppose $A \subset R^N$ and I is the unit interval in R, so that $A \times I \subset R^{N+1}$.*
If $h_t(A) < \infty$ then $h_{t+1}(A \times I) < \infty$.

Proof. Pick ε, $0 < \varepsilon < 1$, pick $\eta > 0$. There is an ε-cover $\{A_i\}$ of A with

$$\sum_i \delta_i^t \le \eta + h_t(A)$$

where $0 < \delta_i = \text{diam } A_i < \varepsilon$.

Let k_i be the smallest integer such that $k_i \delta_i > 1$. Associate to each A_i the sets $A_i \times I_{ij}$, $1 \le j \le k_i$, where the I_{ij} are intervals of length δ_i whose union covers I. Then

$$\text{diam}(A_i \times I_{ij}) < 2\delta_i$$

so that $\{A_i \times I_{ij}\}_{i,j}$ is a (2ε)-cover of $A \times I$. Also,

$$\sum_{i,j} (2\delta_i)^{t+1} = 2^{t+1} \sum_i k_i \delta_i \cdot \delta_i^t \le 2^{t+2} \sum_i \delta_i^t \le 2^{t+2}[\eta + h_t(A)].$$

Hence $h_{t+1}(A \times I) \le 2^{t+2} h_t(A)$.

14.4.7. In the proof of Theorem 14.4.9 we shall need some information about factorings of Weierstrass polynomials.

Let Δ' be a polydisc in $\mathbb{C}^{n-1}$, as in §14.1.1. Let $\mathscr{R} = H(\Delta')$ be the ring of all holomorphic functions in Δ'. Since Δ' is connected, $\mathscr{R}$ has no zero-divisors (i.e., $\mathscr{R}$ is an integral domain), and therefore $\mathscr{R}$ has a field of quotients which we call $\mathscr{F}$.

As usual, $\mathscr{R}[z_n]$ and $\mathscr{F}[z_n]$ denote the rings of polynomials in z_n whose coefficients lie in $\mathscr{R}$ and $\mathscr{F}$, respectively.

For example, each Weierstrass polynomial (Theorem 14.1.3) is a monic member of $\mathscr{R}[z_n]$.

14.4.8. Lemma. *Suppose Q_1 and Q_2 are monic polynomials in $\mathscr{F}[z_n]$ whose product P lies in $\mathscr{R}[z_n]$. Then $Q_i \in \mathscr{R}[z_n]$, for $i = 1, 2$.*

Proof. Let m be the degree of P, and let the zeros of $P(z', \cdot)$ be $\alpha_1(z'), \ldots, \alpha_m(z')$, for $z' \in \Delta'$. Then

$$(1) \qquad\qquad P(z', z_n) = \prod_{j=1}^{m} (z_n - \alpha_j(z')).$$

It involves no loss of generality to assume that the coefficients of P are bounded in Δ'. (If this is not the case, replace Δ' by a slightly smaller polydisc.) Hence there is an $M < \infty$ such that $|\alpha_j(z')| < M$ for all $z' \in \Delta'$, $1 \le j \le m$.

The factors Q have the form

$$(2) \qquad\qquad Q_1(z) = \sum_{i=1}^{r} b_i(z') z_n^i, \qquad Q_2(z) = \sum_{i=1}^{s} c_i(z') z_n^i$$

with $b_i, c_i \in \mathscr{F}$, $b_r = c_s = 1$, $r + s = m$. Let $g \in \mathscr{R} = H(\Delta')$ be a common denominator of the coefficients b_i, c_i. In $\Delta' \backslash Z(g)$, all b_i and c_i are holomorphic. For any $z' \in \Delta' \backslash Z(g)$ it is clear that $Q_1(z', z_n)$ and $Q_2(z', z_n)$ are subproducts of (1). Since $|\alpha_j(z')| < M$, the coefficients b_i, c_i are bounded holomorphic function in $\Delta' \backslash Z(g)$. But $g \not\equiv 0$, so that $Z(g)$ is H^∞-removable (Theorem 4.4.7). Hence $b_i \in \mathscr{R}$, $c_i \in \mathscr{R}$.

14.4.9. Theorem. *Suppose Ω is a region in $\mathbb{C}^n$, $f \in H(\Omega)$, and $f \not\equiv 0$. Then $Z(f)$ is a union of countably many compact sets K_j with $h_{2n-2}(K_j) < \infty$.*

Proof. The theorem is obviously true when $n = 1$. Assume $n > 1$, and make the induction hypothesis that it is proved in $\mathbb{C}^{n-1}$. It is then clearly sufficient to prove the theorem locally, i.e., for the zero-variety of a Weierstrass polynomial $W(z) = W(z', z_n)$.

Let $\mathscr{R}$ and $\mathscr{F}$ be as in §14.4.7. Factor

$$W = P_1 \cdots P_k$$

where each $P_i \in \mathcal{R}[z_n]$ is monic, and no P_i can be further factored in $\mathcal{R}[z_n]$ (so that each factor has positive degree in z_n). Lemma 14.4.8 shows that each P_i is then irreducible in $\mathcal{F}[z_n]$.

Fix some i, write P in place of P_i, let $P' = \partial P / \partial z_n$. Since P is irreducible, P and P' are relatively prime in $\mathcal{F}[z_n]$. The division algorithm (Euclid's algorithm) for polynomials in one variable with coefficients in a field furnishes now elements a, $b \in \mathcal{F}[z_n]$ such that $aP + bP' = 1$.

Multiply this by $D \in H(\Delta')$, a common denominator of the coefficients of a and b. Then $D \not\equiv 0$, and $A = Da$, $B = Db$ are in $H(\Delta')$. Also

$$A(z', z_n)P(z', z_n) + B(z', z_n)P'(z', z_n) = D(z').$$

Let $V = Z(P)$. Call a point $z \in V$ *singular* if $P'(z) = 0$; the other points of V are *regular*. Correspondingly, $V = V_s \cup V_r$. The points of V_r are those where P has a zero of order 1.

If (z', z_n) is singular, it follows that $D(z') = 0$, i.e., that $z' \in Z(D)$. Our induction hypothesis says that $Z(D)$ is a countable union of compact sets K with $h_{2n-4}(K) < \infty$. Two applications of Proposition 14.4.6 show now that V_s lies in a countable union of compact sets of finite $(2n - 2)$-dimensional Hausdorff measure.

Finally, V_r is a countable union of compact sets K, each of which is the graph of a holomorphic function g on a set $E \subset \Delta'$ (see §14.1.4), such that the gradient of g is bounded on E. Thus K is a Lip 1 image of E. By Proposition 14.4.4, $h_{2n-2}(K) < \infty$.

This completes the proof.

Chapter 15

Proper Holomorphic Maps

15.1. The Structure of Proper Maps

15.1.1. Definition. Let X and Y be topological spaces. A continuous map $f: X \to Y$ is said to be *proper* if $f^{-1}(K)$ is compact in X for every compact $K \subset Y$.

We shall study proper holomorphic maps $F: \Omega \to \Omega'$, where Ω and Ω' are regions in $\mathbb{C}^n$ and $\mathbb{C}^k$, respectively. (The case $k = n$ will be the one of greatest interest.) In this context, the compactness of $F^{-1}(K)$ for every compact $K \subset \Omega'$ is equivalent to the following requirement: If $\{p_i\}$ is a sequence in Ω that has no limit point in Ω, then $\{F(p_i)\}$ has no limit point in Ω'.

15.1.2. Remarks. (i) The role played by the target space Ω' in the definition of "proper" should be stressed: The identity map, applied to the open unit disc U in $\mathbb{C}$, is a proper map of U into U, but is not proper as a map of U into $\mathbb{C}$.

(ii) The finite Blaschke products

$$B(z) = c \prod_{i=1}^{m} \frac{\alpha_i - z}{1 - \bar{\alpha}_i z} \qquad (|\alpha_i| < 1, |c| = 1)$$

are the only proper holomorphic maps of U into U. This fact, whose easy proof we omit, indicates that proper holomorphic maps are quite special.

(iii) The map $(z, w) \to (z, zw)$ of U^2 into U^2 is a simple example of one that is not proper.

15.1.3. Some Elementary Facts. Let Ω and Ω' be regions in $\mathbb{C}^n$ and $\mathbb{C}^k$, respectively, and suppose that $F: \Omega \to \Omega'$ is holomorphic and proper.

If $w = (w_1, \ldots, w_k) \in \Omega'$, then $F^{-1}(w)$ is a subvariety of Ω, being the intersection of the zero-sets of $f_i - w_i$, where f_i is the ith component of F, and $F^{-1}(w)$ is compact, since F is proper. By Theorem 14.3.1, $F^{-1}(w)$ *is a finite set*.

The number of points in $F^{-1}(w)$ will be denoted by $\#(w)$. This count ignores multiplicities: for example, if $F(z) = z^2$ $(n = k = 1)$ then $\#(w) = 2$ if $w \neq 0$, $\#(0) = 1$.

If $n = k$, it follows that the Jacobian J of F *cannot vanish at all points of* Ω. Otherwise, the rank of the linear operator $F'(z)$ would be at most $2n - 1$ at every $z \in \Omega$ (with respect to real coordinates) and the rank theorem (see, for example, Rudin [16], p. 228) would show that $F^{-1}(w)$ contains a 1-dimensional manifold, hence an infinite set, for some $w \in \Omega'$.

The same reasoning shows that one must have $n \le k$, i.e., that proper holomorphic maps cannot decrease dimension.

15.1.4. Critical Values and Regular Values. Let Ω and Ω' be regions in $\mathbb{C}^n$ and suppose that $F: \Omega \to \Omega'$ is holomorphic and proper.

Let $M = Z(J)$, J being the Jacobian of F. Its image $F(M)$ is called the *critical set* of F. Each $w \in F(M)$ is a *critical value* of F. Every other point of $F(\Omega)$ is a *regular value* of F.

Since F is proper, it is easy to see that F is a closed map: if E is closed in Ω then $F(E)$ is closed in Ω'. In particular, $F(M)$ and $F(\Omega)$ are closed in Ω', and the regular values of F form an open set.

Theorem 14.4.9 shows that M is a countable union of compact sets K_j with $h_{2n-2}(K_j) < \infty$. On each K_j, F satisfies a Lipschitz condition. By 14.4.4, $F(M)$ is a countable union of compact sets whose $(2n - 2)$-dimensional Hausdorff measure is finite. Proposition 14.4.3 implies therefore that $F(M)$ has no interior, and Theorem 14.4.5 shows that $\Omega' \backslash F(M)$ is connected.

Any point of Ω' that is a boundary point of $F(\Omega)$ must lie in $F(M)$, since $F(\Omega)$ is closed in Ω' and the regular values form an open set. But we just saw that $F(M)$ is too small to separate Ω'. Thus $F(\Omega) = \Omega'$.

Let us summarize:

15.1.5. Proposition. *If* Ω *and* Ω' *are regions in* $\mathbb{C}^n$ *and* F *is a proper holomorphic map of* Ω *into* Ω', *then*

(a) $F(\Omega) = \Omega'$ *and*
(b)   *the regular values of* F *form a connected open set that is dense in* Ω'.

In the next theorem it is not assumed that the holomorphic map F is proper, but only that the inverse image of every *point* be compact (hence finite). The conclusion is that F is then locally proper:

15.1.6. Theorem. *Suppose* Ω *is a region in* $\mathbb{C}^n$, $F: \Omega \to \mathbb{C}^n$ *is holomorphic, and* $F^{-1}(w)$ *is compact for every* $w \in \mathbb{C}^n$. *Pick* $p \in \Omega$.

Every neighborhood of p *contains then a connected neighborhood* D *of* p *such that the restriction of* F *to* D *is a proper map of* D *onto the region* $F(D)$.

Consequently, F *is an open map.*

Proof. Put $F(p) = w$. Since $F^{-1}(w)$ is a compact variety in Ω, $F^{-1}(w)$ is a finite set, hence p lies in an open ball Q such that p is the only point of the closed ball $\bar{Q} \subset \Omega$ that F maps to w. Put $E = F(\partial Q)$. Then E is compact, $w \notin E$, hence w lies in an open ball N that does not intersect E.

Put $\Omega_0 = Q \cap F^{-1}(N)$, and let K be a compact subset of N. Since no boundary point of Q maps into N,

$$\Omega_0 \cap F^{-1}(K) = \bar{Q} \cap F^{-1}(K),$$

and the latter set is compact. The restriction of F to Ω_0 is therefore a proper map of the open set Ω_0 into N.

If D is the component of Ω_0 that contains p, it follows that the restriction of F to D is a proper map of D into N. By Proposition 5.1.5, $F(D) = N$, and the theorem is proved.

Note: Nonholomorphic proper maps need not be open. The map $(x, y) \to (|x|, y)$ of R^2 into R^2 is an example of this.

The following theorem about removable singularities will be used in the proofs of Theorems 15.1.8 and 15.1.9.

15.1.7. Radó's Theorem. *Suppose Ω is a region in $\mathbb{C}^n$, $f: \Omega \to \mathbb{C}$ is continuous, and f is holomorphic in the open subset of Ω in which $f(z) \neq 0$. Then $f \in H(\Omega)$.*

Proof. Definition 1.1.3 shows that it is enough to prove this in the case $n = 1$, hence when $\Omega \supset \bar{U}$.

So assume $f \in C(\bar{U})$, $E = \{f = 0\}$, $f \in H(U \backslash E)$, and $|f| < 1$.

Let $g = P[f]$, the Poisson integral of the restriction of f to $T = \partial U$, let α be a positive constant, and define

$$(1) \qquad \qquad \varphi = \text{Re}(f - g) + \alpha \log|f|.$$

Then φ is harmonic in $U \backslash E$.

When $z \to z_0 \in E$ ($z \in U \backslash E$), then $\varphi(z) \to -\infty$.

When $z \to e^{i\theta} \in T$, then $\varphi(z) \to \alpha \log|f(e^{i\theta})| < 0$.

The maximum principle for harmonic functions shows therefore that $\varphi(z) < 0$ in $U \backslash E$. Letting $\alpha \searrow 0$, we conclude that

$$(2) \qquad \qquad \text{Re}(f - g) \leq 0 \quad \text{in} \quad U \backslash E.$$

The same argument, with $\alpha < 0$, leads to the conclusion that $\text{Re}(f - g) \geq 0$ in $U - E$. The same holds for the imaginary part. Hence $f(z) = g(z)$ for every $z \in T \cup (U \backslash E)$.

If $z \in \partial E$, then $f(z) = 0$, hence $g(z) = 0$. Since g is harmonic, it follows that $g \equiv 0$ on E. Thus $f \equiv g$ on $\bar{U}$. In particular, $f \in C^1(U)$, and $\partial f / \partial \bar{z} = 0$ in U/E as well as at every interior point of E (if there are any). By continuity, $\partial f / \partial \bar{z} \equiv 0$ in U, hence $f \in H(U)$, which was to be proved.

The following is a converse of the inverse function theorem:

15.1.8. Theorem. *If Ω is a region in $\mathbb{C}^n$ and $F: \Omega \to \mathbb{C}^n$ is holomorphic and one-to-one, then the Jacobian J of F has no zero in Ω.*

Consequently (see Theorem 1.3.7) F is a *biholomorphic* map of Ω onto $F(\Omega)$.

Proof. F satisfies the hypotheses of Theorem 15.1.6. Hence F is an open map of Ω onto $\Omega' = F(\Omega)$, so that F is in fact a homeomorphism of Ω onto Ω'. Define

$$g(w) = J(F^{-1}(w)) \qquad (w \in \Omega').$$

By Theorem 1.3.7, $g \in H(\Omega'\backslash F(M))$, where $M = Z(J)$. Since g is continuous in Ω', and $g(w) = 0$ exactly when $w \in F(M)$, Radó's theorem shows that $g \in H(\Omega')$. Thus $F(M) = Z(g)$, a zero-variety, hence H^∞-removable (Theorem 4.4.7). It follows that $F^{-1} \in H(\Omega')$. The chain rule, applied to $F^{-1}(F(z)) = z$, shows now that $J(z) \neq 0$, so that M is in fact empty.

We now come to the main result of this section. Note that the hypotheses are exactly those of Proposition 15.1.5.

15.1.9. Theorem. *Suppose that Ω and Ω' are regions in $\mathbb{C}^n$, and that $F: \Omega \to \Omega'$ is holomorphic and proper.*
  Let $\#(w)$ denote the number of points in the set $F^{-1}(w)$, for $w \in \Omega'$.

  (a) *There is then an integer m (the so-called multiplicity of F) such that*

$$\#(w) = m \text{ for every regular value of } F,$$

$$\#(w) < m \text{ for every critical value of } F.$$

  (b) *The critical set of F is a zero-variety in Ω'.*
  (c) *More generally, $F(V)$ is an analytic subvariety of Ω' whenever V is an analytic subvariety of Ω.*

Proof. Pick $w_0 \in \Omega'$, let $z_1, \ldots, z_k$ be the points in $F^{-1}(w_0)$; thus $k = \#(w_0)$. There are open balls Q_i with center z_i whose closures $\overline{Q}_i$ are disjoint and lie in Ω. Put

(1) $$E = \Omega\backslash(Q_1 \cup \cdots \cup Q_k).$$

Then E is closed in Ω. Since F is proper, F is a closed map, hence $F(E)$ is closed in Ω', so that w_0 is the center of an open ball $N \subset \Omega'\backslash F(E)$. Define

(2) $$D_i = Q_i \cap F^{-1}(N) \qquad (i = 1, \ldots, k).$$

As in the proof of Theorem 15.1.6, $F: D_i \to N$ is proper, for each i. Note that each D_i is connected, since (as is easily seen) the restriction of F to any component Δ of D_i is proper, so that $F(\Delta) = N$; but w_0 has only one inverse image in D_i. Moreover, F maps no point outside $D_1 \cup \cdots \cup D_k$ into N, since N does not intersect $F(E)$.

So far, we have proved the following:

(*) *If* $w_0 \in \Omega'$, $\#(w_0) = k$, $F^{-1}(w_0) = \{z_1, \ldots, z_k\}$, *then* w_0 *has a neighborhood* N *and the* z_i's *have disjoint connected neighborhoods* D_i *such that* $F(D_i) = N$ *for* $1 \le i \le k$, *and* $F^{-1}(N) = D_1 \cup \cdots \cup D_k$.

Moreover, the D_i's *can be taken so as to lie within prescribed neighborhoods of the points* z_i.

Now let w_0 be a regular value of F. By the inverse function theorem, the D_i's can be so chosen that F is one-to-one in each D_i. Thus (*) shows that $\#(w) = \#(w_0)$ for every $w \in N$. Since the set of all regular values is connected (by 15.1.5), there is an m that satisfies the first half of conclusion (a).

Returning to (*), for arbitrary $w_0 \in \Omega'$, we see that N contains a regular value w, again by 15.1.5. Hence (*) implies that $\#(w_0) \le m$ for every $w_0 \in \Omega'$. If $\#(w_0) = m$, it follows that F is one-to-one in each of the regions D_i. By Theorem 15.1.8, J has then no zero in D_i, so that w_0 is a regular value of F.

This completes the proof of (a).

If w_0 is regular, (*) and the inverse function theorem show that there are holomorphic maps

$$(3) \qquad\qquad p_i: N \to D_i \qquad (1 \le i \le m)$$

that invert F. The product

$$(4) \qquad\qquad \psi(w) = \prod_{i=1}^{m} J(p_i(w))$$

is thus holomorphic in $\Omega' \setminus F(M)$, where $M = Z(J)$, and obviously has no zero in this region. Put $\psi(w) = 0$ for $w \in F(M)$. If we can show that ψ is then continuous in Ω', Radó's theorem will imply that $\psi \in H(\Omega')$; since $F(M)$ is the zero-variety of ψ, this will prove (b).

Choose $w_0 \in F(M)$, $z_1 \in M$ so that $F(z_1) = w_0$, $\varepsilon > 0$. Apply (*) with the neighborhood D_1 of z_1 chosen so small that $|J| < \varepsilon$ in D_1. At least one factor in (4) has then absolute value $< \varepsilon$ in N; the others are bounded in N. This proves that ψ is continuous at w_0, and thus gives (b).

We turn to the proof of (c). Let $g \in H(\Omega)$ and let the maps $p_1, \ldots, p_m$ be as in (3). The product

$$(5) \qquad\qquad h(w) = \prod_{i=1}^{m} g(p_i(w))$$

is then holomorphic in $\Omega' \setminus F(M)$. If K is compact in Ω' and $w \in K \setminus F(M)$, then $p_i(w)$ lies in the compact set $F^{-1}(K)$ for each i. Thus h is bounded on $K \setminus F(M)$. Since $F(M)$ is a zero-variety, it is H^∞-removable (Theorem 4.7.7). Consequently, h extends so as to be holomorphic in Ω', and (5) shows that

$$(6) \qquad\qquad F(Z(g)) = Z(h).$$

Thus $F(Z(g))$ *is a subvariety of* Ω'.

Assume next that $V = Z(f_1) \cap \cdots \cap Z(f_{r-1})$, where $f_1, \ldots, f_{r-1} \in H(\Omega)$, and define

(7) $$g_i = \sum_{j=1}^{r-1} c_{ij} f_j \qquad (1 \le i \le rm - m),$$

where (c_{ij}) is a matrix exactly as in the proof of Theorem 14.2.4. The argument used there can be repeated almost word for word to show that

(8) $$F(V) = \bigcap_i F(Z(g_i)).$$

This proves that $F(V)$ is a subvariety of Ω', in the special case in which V is globally defined in Ω as an intersection of zero-varieties.

In the general case, pick $w_0 \in F(V)$, and choose $D_1, \ldots, D_k, N$ as in $(*)$, making sure that each D_i is so small that the preceding special case can be applied to show that $F(V \cap D_i)$ is a subvariety of N, for each i. Since

(9) $$N \cap F(V) = \bigcup_{i=1}^{k} F(V \cap D_i),$$

the proof of (c) is complete.

15.1.10. In the context of the preceding theorem, $F'(z)$ is invertible for all $z \in \Omega$ such that $F(z)$ is a regular value. The restriction of F to $\Omega \backslash F^{-1}(F(M))$ is therefore an m-to-1 *covering map* (a local homeomorphism) whose range is the set of regular values of F; branching occurs at the points of M.

The triple (Ω, F, Ω') is a special case of what has been called an *analytic cover*. This topic is discussed in Gunning–Rossi [1], in more detail and in a more general setting.

15.2. Balls vs. Polydiscs

Theorem 2.2.4 showed that U^n and B_n are not biholomorphically equivalent when $n > 1$. Actually, as we shall now see, there do not even exist any proper holomorphic maps from U^n to B_n, nor from B_n to U^n, when $n > 1$. This is a very special case of Theorem 15.2.4 whose statement and proof show a basic reason for this nonexistence: a large part of the boundary of U^n contains analytic discs, whereas every boundary point of B is a peak point for $A(B)$.

15.2.1. Local Peak Points. Let Ω be a region in $\mathbb{C}^n$. A point $\zeta \in \partial\Omega$ is said to be a *local peak point for $A(\Omega)$* if there is an $r > 0$ and a function h, continuous on the closure of

$$\Omega_0 = \Omega \cap (\zeta + rB),$$

holomorphic in Ω_0, such that $h(\zeta) = 1$ but $|h(z)| < 1$ for every $z \in \bar{\Omega}_0 \backslash \{\zeta\}$.

15.2.2. Lemma. *Assume that*

    (i)   *D is a region in $\mathbb{C}^k$, $z_0 \in D$;*

    (ii) *Ω is a bounded region in $\mathbb{C}^n$, ζ is a local peak point for $A(\Omega)$;*

    (iii) *$\{F_i\}$ is a sequence of holomorphic maps, taking D into Ω, and $F_i(z_0) \to \zeta$ as $i \to \infty$.*

Then $F_i(z) \to \zeta$, uniformly on every compact subset of D.

Proof. Choose h and r as in §15.2.1. Since Ω is bounded, $\{F_i\}$ is equicontinuous on every compact subset of D. The closure of $\{F_i\}$ is therefore compact, in the topology of uniform convergence on compact subsets of D. Let F be any limit point of $\{F_i\}$. Then $F(z_0) = \zeta$, and z_0 has a connected neighborhood $N \subset D$ such that $|F(z) - \zeta| < r$ for all $z \in N$. For large i, $h \circ F_i \in H(N)$.

Put $g = h \circ F$. Then $g \in H(N)$, $g(z_0) = 1$, and $|g(z)| \le 1$ for all $z \in N$. By the maximum modulus theorem, $g(z) = 1$, hence $F(z) = \zeta$, for all $z \in N$. Since D is connected, $F(z) = \zeta$ for all $z \in D$.

The constant ζ is thus the only limit point of $\{F_i\}$. This proves the lemma.

15.2.3. Lemma. *Suppose $n > 1$, $f \in H^\infty(U^n)$. Then there is a set $E \subset [0, 2\pi]$, $m_1(E) = 2\pi$, such that*

$$(1) \qquad\qquad f_\theta(z') = \lim_{r \nearrow 1} f(z', re^{i\theta})$$

exists for all $(z', \theta) \in U^{n-1} \times E$.

Moreover, $f_\theta \in H^\infty(U^{n-1})$ for every $\theta \in E$.

Let E_0 be the set of all $\theta \in E$ for which f_θ is constant. If $m_1(E_0) > 0$, then f is a function of z_n alone.

(Recall that m_1 denotes one-dimensional Lebesgue measure.)

Proof. For $z' \in U^{n-1}$, let $E(z')$ be the set of all $\theta \in [0, 2\pi]$ for which the limit (1) exists. By Fatou's theorem, $m_1(E(z')) = 2\pi$. Let X be a countable dense subset of U^{n-1}. Put $E = \bigcap E(z')$, where z' ranges over X.

Since X is countable, $m_1(E) = 2\pi$. The functions $f(\cdot, \lambda)(\lambda \in U)$ form a normal family in $H^\infty(U^{n-1})$. Since X is dense in U^{n-1}, the existence of (1) for all $z' \in X$ implies its existence for all $z' \in U^{n-1}$. This gives the first two conclusions of the lemma.

If $z', w' \in U^{n-1}, \theta \in E_0$, and

$$(2) \qquad\qquad \varphi(\lambda) = f(z', \lambda) - f(w', \lambda) \qquad (\lambda \in U)$$

then $\varphi \in H^\infty(U)$ and $\varphi(re^{i\theta}) \to 0$ as $r \nearrow 1$. If $m_1(E_0) > 0$, it follows that $\varphi(\lambda) \equiv 0$, hence $f(z', \lambda) = f(w', \lambda)$ for all $\lambda \in U$.

15.2.4. Theorem. *Let Ω be a bounded region in $\mathbb{C}^n$ such that every boundary point of Ω is a local peak point for $A(\Omega)$.*

(i) If $k > 1$, then there is no proper holomorphic map of U^k into Ω.

(ii) If $n > 1$, then there is no proper holomorphic map of Ω into U^n.

Proof. (i) Assume, to reach a contradiction, that $F: U^k \to \Omega$ is holomorphic and proper. Write points $z \in U^k$ in the form (z', z_k). The components of $F(z', \cdot)$ belong then to $H^\infty(U)$, for every $z' \in U^{k-1}$.

Let $E \subset [0, 2\pi]$ be the set of all θ such that

(1)
$$\lim_{r \nearrow 1} F(0', re^{i\theta}) = \zeta(\theta)$$

exists. Since F is proper, $\zeta(\theta) \in \partial\Omega$, and hence Lemma 15.2.2 (applied to maps from U^{k-1} into Ω) show that

(2)
$$\lim_{r \nearrow 1} F(z', re^{i\theta}) = \zeta(\theta)$$

for all $z' \in U^{k-1}$, $\theta \in E$. By Lemma 15.2.3, F depends on z_k alone.

By the same reasoning, F depends only on z_1. Hence F is constant, which cannot happen if F is proper.

(ii) Again, assume that $F: \Omega \to U^n$ is proper and holomorphic. Let m be the multiplicity of F, as in Theorem 15.1.9. If $m = 1$, follow F by the 2-to-1 map

$$(z_1, z_2, \ldots, z_n) \to (z_1^2, z_2, \ldots, z_n)$$

of U^n onto U^n. We may thus assume that $m > 1$.

Let $V \subset U^n$ be the critical set of F, as in Theorem 15.1.9. Every regular value of F, i.e., every point of $U^n \setminus V$, has then a neighborhood N in which m holomorphic maps $p_1, \ldots, p_m$ into Ω are defined, such that $F(p_i(z)) = z$ and $p_i(z) \neq p_j(z)$ if $i \neq j$ and $z \in N$.

Choose a polynomial $Q: \mathbb{C}^n \to \mathbb{C}$ that separates the points $p_1(a), \ldots, p_m(a)$ for some particular $a \in U^n \setminus V$ but fails to separate $p_1(b), \ldots, p_m(b)$ for some $b \in U^n \setminus V$. The product

(3)
$$h(z) = \prod \{Q(p_i(z)) - Q(p_j(z))\}^2,$$

extended over all (i, j) such that $1 \leq i < j \leq m$, is then a *nonconstant* bounded holomorphic function in $U^n \setminus V$, since $h(a) \neq 0 = h(b)$. Being a zero-variety (Theorem 15.1.9), V is H^∞-removable (Theorem 4.4.7), so that h extends to a member of $H^\infty(U^n)$.

Since h is not constant, we may assume (by permuting coordinates if necessary) that $\partial h / \partial z_1 \not\equiv 0$.

By Lemma 15.2.3 there is then a set $E \subset [0, 2\pi]$, $m_1(E) = 2\pi$, such that

(4)
$$h_\theta(z') = \lim_{r \nearrow 1} h(z', re^{i\theta}) \qquad (z' \in U^{n-1})$$

exists for all $\theta \in E$, and such that h_θ is not constant in U^{n-1} if $\theta \in E$.

Fix $\theta \in E$. Since h_θ is not constant, there is a polydisc P, with compact closure $\bar{P} \subset U^{n-1}$, and a $\delta > 0$, such that $|h_\theta(z')| > \delta$ on $\bar{P}$. By Lindelöf's theorem, the existence of the radial limit (4) implies that the nontangential limit exists also. The convergence is uniform on $\bar{P}$, by equicontinuity. Consequently, there is a triangle $\Delta \subset U$, with one vertex at $e^{i\theta}$, that contains part of the radius ending at $e^{i\theta}$, such that

$$(5) \qquad h(z', z_n) \neq 0 \quad \text{if} \quad (z', z_n) \in \bar{P} \times \Delta.$$

The (locally defined) p_j's are thus distinct in the simply connected region $P \times \Delta$. Hence they determine m holomorphic maps $p_j: P \times \Delta \to \Omega$.

As $r_k \nearrow 1$ through a suitable sequence,

$$(6) \qquad \lim_{k \to \infty} p_j(z', r_k e^{i\theta}) = g_j(z')$$

exists for each j and each $z' \in P$. Since $(z', r_k e^{i\theta})$ tends to ∂U^n, $p_j(z', r_k e^{i\theta})$ tends to $\partial \Omega$. By Lemma 15.2.2, each g_j is constant in P.

It now follows from (3), (4), and (6) that h_θ is constant in P. Since $h_\theta \in H^\infty(U^{n-1})$, h_θ is constant in U^{n-1}. This is a contradiction, since $\theta \in E$.

15.2.5. A variety of further results in this direction may be found in Chapter 5 of Narasimhan [1].

To conclude this section, here is a theorem of Alexander (he only considered the case $k = n$) that gives a *quantitative* reason for the nonexistence of proper holomorphic maps from U^n to B_k:

15.2.6. Theorem (Alexander [4]). *Suppose $n > 1$, $k \geq 1$, $F: U^n \to B_k$ is holomorphic, and $0 \in F(U^n)$.*

Then there is a sequence in U^n that converges to a boundary point of U^n, but which is mapped into the ball $n^{-1/2} B_k$ by F.

Proof. Assume $F(0) = 0$, without loss of generality. (We can precede F by an automorphisms of U^n, i.e., by a Moebius transformation in each of the n coordinates.) Writing $F = (f_1, \ldots, f_k)$, each component f_j is then the sum of a power series

$$(1) \qquad f_j(z) = \sum{}' c(j, \alpha) z^\alpha \qquad (z \in U^n).$$

The symbol $\sum'$ indicates that the multi-index 0 is missing.

For $p = 1, \ldots, n$, let M_p be the set of all multi-indices α such that $\alpha_p > 0$ but $\alpha_i = 0$ for all $i \neq p$.

For $0 < r < 1$, define

$$(2) \qquad .\psi(r) = \min\{|F(r\zeta)| : \zeta \in \partial U^n\}.$$

Note that (2) involves the entire boundary of U^n in $\mathbb{C}^n$, not just the distinguished boundary T^n. For every θ,

(3) $$\psi^2(r) \leq |F(re^{i\theta}, 0, \ldots, 0)|^2 = \sum_{j=1}^{k} |f_j(re^{i\theta}, 0, \ldots, 0)|^2.$$

If we integrate (3) over the unit circle, and use (1) and the Parseval theorem, we obtain

(4) $$\psi^2(r) \leq \sum_{j=1}^{k} \sum_{\alpha \in M_1} |c(j, \alpha)|^2 r^{2|\alpha|}.$$

The same estimate holds with $M_2, \ldots, M_n$ in place of M_1. Adding these, we see that

$$n\psi^2(r) \leq \sum_{j, \alpha} |c(j, \alpha)|^2 r^{2|\alpha|} = \int_{T^n} |F_r|^2 \, d\lambda_n < 1.$$

Thus $\psi(r) < n^{-1/2}$, which proves the theorem.

Note: If $1 \leq n \leq k$ and

$$F(z_1, \ldots, z_n) = n^{-1/2}(z_1, \ldots, z_n, 0, \ldots, 0),$$

then F maps U^n into B_k. The factor $n^{-1/2}$ can therefore not be replaced by any smaller one in Theorem 15.2.6.

The theorem does not assert that there is no holomorphic map of U^n onto B_n. In fact, according to Fornaess and Stout [1], [2], there do exist locally biholomorphic maps of U^n onto every n-dimensional complex manifold M such that, furthermore, no point of M has more than $(2n + 1)4^n + 2$ preimages in U^n.

15.3. Local Theorems

The two principal results of this section (Theorems 15.3.4 and 15.3.8) state that certain kinds of holomorphic maps defined near boundary points of B must extend to automorphisms of B, when $n > 1$. Both of them are obviously false when $n = 1$. The first one will be used in Section 15.4, in the proof that the automorphisms of B are the only proper holomorphic maps from B to B when $n > 1$.

The essence of Theorem 15.3.4 is contained in the following simple consequence of the Schwarz lemma. To state it, we introduce the notation

$$D_z = \{\lambda z : \lambda \in \mathbb{C}, |\lambda z| < 1\},$$

for any $z \in \mathbb{C}^n, z \neq 0$. Thus D_z is the disc of radius 1 in the complex line through 0 and z that lies in B.

15.3.1. Theorem. *Suppose*

(a) Ω_1 *and* Ω_2 *are regions in* B, $0 \in \Omega_1 \cap \Omega_2$,
(b) F *is a biholomorphic map of* Ω_1 *onto* Ω_2, *with* $F(0) = 0$, *and*
(c) *some point* $p \in \Omega_1$, $p \neq 0$, *has a neighborhood* $N_p \subset \Omega_1$ *with the following property: For every* $z \in N_p$, $D_z \subset \Omega_1$ *and* $D_{F(z)} \subset \Omega_2$.

Then F extends to a unitary operator on $\mathbb{C}^n$.

Proof. Let $z \in N_p$, put $w = F(z)$. Then F maps D_z into B, and F^{-1} maps D_w into B. Theorem 8.1.2, applied to the restriction of F to D_z, shows therefore that $|F(z)| \leq |z|$, i.e., that $|w| \leq |z|$. The same reasoning, applied to F^{-1} in place of F, shows that $|z| \leq |w|$.

Thus $|F(z)|^2 = |z|^2$ for all $z \in N_p$. These functions are real-analytic. Hence $|F(z)|^2 = |z|^2$ for every $z \in \Omega_1$, and in particular for every $z \in rB$, where $r > 0$ is sufficiently small so that $rB \subset \Omega_1$. The restriction of F to rB is thus an automorphism of rB that fixes 0. Hence F is unitary.

15.3.2. Lemma. *If $F: B \to B$ is holomorphic, $F(0) = 0$, and $|(JF)(0)| = 1$, then F is unitary.*

Here, as usual, JF is the Jacobian of F.

Proof. Let $A = F'(0)$, so that $(JF)(0) = \det A$. Since $|\det A| = 1$, A preserves volume. By Theorem 8.1.2, A maps B into B. Hence A maps B onto B, i.e., A is unitary. Now apply Theorem 8.1.3.

15.3.3. Lemma. *Assume $n > 1$. If $0 < t < r < 1$, $\zeta \in S$, $a = r\zeta$, Ω is a region such that*

$$\{z \in B: t < \mathrm{Re}\langle z, \zeta \rangle\} \subset \Omega \subset B,$$

and $\delta = (1 - r)(1 + t)/(1 - rt)$, then $\varphi_a(\Omega)$ contains

(i) *all $w \in B$ with $|\langle w, \zeta \rangle| < 1 - \delta$, and*
(ii) *all D_η, where $\eta \in S$ and $|\langle \eta, \zeta \rangle| < 1 - \delta$.*

Here φ_a is the involution defined in §2.2.1, and D_η is a disc as in Theorem 15.3.1. In the application of the lemma, t will be fixed, while $r \nearrow 1$, so that $\delta \searrow 0$.

Proof. Let $\zeta = e_1$, without loss of generality. If $\mathrm{Re}\, z_1 = t$ and $w = \varphi_a(z)$, it follows from §2.2.1 that

$$\mathrm{Re}\, w_1 = \mathrm{Re}\, \frac{r - z_1}{1 - rz_1} \geq \frac{r - t}{1 - rt} = 1 - \delta.$$

Since $\varphi_a(e_1) = -e_1$, it follows that $\varphi_a(\Omega)$ contains every $w \in B$ with Re $w_1 < 1 - \delta$. This gives (i), and it is clear that (i) implies (ii).

15.3.4. Theorem. *Let $n > 1$. Suppose, for $i = 1, 2$, that Ω_i is a region in B whose boundary $\partial\Omega_i$ contains an open subset Γ_i of S, and that F is a biholomorphic map of Ω_1 onto Ω_2.*

If there is a sequence $\{a_k\}$ in Ω_1, converging to a point $\alpha \in \Gamma_1$ which is not a limit point of $B \cap \partial\Omega_1$, such that the points $b_k = F(a_k)$ converge to a point $\beta \in \Gamma_2$ which is not a limit point of $B \cap \partial\Omega_2$, then F extends to an automorphism of B.

Proof. The hypotheses concerning α and β show that there exists $t, 0 < t < 1$, with the following property: If k is sufficiently large, if $a_k = |a_k|\tilde{a}_k$, $b_k = |b_k|\tilde{b}_k$, so that $\tilde{a}_k \in S$, $\tilde{b}_k \in S$, then Ω_1 contains all $z \in B$ with $t < \text{Re}\langle z, \tilde{a}_k\rangle$, and Ω_2 contains all $w \in B$ with $t < \text{Re}\langle w, \tilde{b}_k\rangle$. Also, $t < |a_k|, t < |b_k|$.

Define

$$(1) \qquad\qquad G_k = \varphi_{b_k} \circ F \circ \varphi_{a_k}.$$

Then G_k is a biholomorphic map of Ω_1^k onto Ω_2^k, where

$$(2) \qquad\qquad \Omega_1^k = \varphi_{a_k}(\Omega_1), \qquad \Omega_2^k = \varphi_{b_k}(\Omega_2).$$

Also, $G_k(0) = 0$.

By Lemma 15.3.3, Ω_1^k and Ω_2^k contain a ball $(1 - \delta_k)B$, where $\delta_k \to 0$ as $k \to \infty$. Applying properly scaled versions of Theorem 8.1.2 to G_k and G_k^{-1}, it follows that

$$(3) \qquad |(JG_k)(0)| \leq (1 - \delta_k)^{-n}, \qquad |(JG_k^{-1})(0)| \leq (1 - \delta_k)^{-n},$$

or $1 - \delta_k \leq |(JG_k)(0)|^{1/n} \leq (1 - \delta_k)^{-1}$.

A subsequence of $\{G_k\}$ converges therefore, uniformly on compact subsets of B, to a holomorphic $G: B \to B$, with $G(0) = 0, |(JG)(0)| = 1$. By Lemma 15.3.2, $G = U$, a unitary operator.

Fix $c, 0 < c < \frac{1}{100}$. For each k, let Y_k be the set of all $z, 0 < |z| < 1 - c$ such that

$$(4) \qquad\qquad D_z \subset \Omega_1^k \quad \text{and} \quad D_{Uz} \subset \Omega_2^k.$$

By Lemma 15.3.3, Y_k is a large set when k is large, i.e., when δ_k is small. Hence there exists k, fixed from now on, such that

$$(5) \qquad\qquad |G_k(z) - Uz| < c \quad \text{if} \quad |z| \leq 1 - c,$$

and such that Y_k contains an open ball of radius $2c$. Let p be its center.

If $|z - p| < c$ and $w = G_k(z)$, then $D_z \subset \Omega_1^k$, and since

(6) $|U^{-1}w - p| = |w - Up| = |G_k(z) - Uz + Uz - Up| < 2c$,

we have $U^{-1}w \in Y_k$, hence $D_w \subset \Omega_2^k$.

Theorem 15.3.1 can therefore be applied to G_k. It shows that G_k is unitary. Since $F = \varphi_{b_k} \circ G_k \circ \varphi_{a_k}$, we conclude that $F \in \text{Aut}(B)$.

15.3.5. Example. Let Ω consist of all $z \in \mathbb{C}^n$ that satisfy $|z| < 1$ and $|e_1 - z| < 1$. Then $z \to e_1 - z$ is a biholomorphic map of Ω onto Ω which does not extend to an automorphism of B, although $\partial\Omega$ contains an open subset of B.

This illustrates the importance of the hypotheses about the location of α and β in Theorem 15.3.4.

Our next objective is Theorem 15.3.8, where F is not assumed to be one-to-one, but is instead required to be continuously differentiable up to the boundary. Whenever this holds, F can be extended to a C^1-map whose domain is a larger open set.

15.3.6. Extensions of C^1-Functions. In R^N, with points $x = (x_1, x_2, \ldots, x_N)$ $= (x_1, x')$, let Q be the cube defined by $|x_i| < 1$ for $i = 1, \ldots, N$, let Q_+ be the part of Q in which $x_1 > 0$, and consider a function $f \in C^1(\bar{Q}_+)$, i.e., assume that every $\partial f/\partial x_i$ exists in Q_+ and has a continuous extension to $\bar{Q}_+$. In this situation, there are functions $g \in C^1(\bar{Q})$ that coincide with f on $\bar{Q}_+$. An explicit example of such an extension is obtained by defining $g(x_1, x') = f(x_1, x')$ in $\bar{Q}_+$ and

$$g(x_1, x') = 2f(0, x') - f(-x_1, x')$$

when $-1 \le x_1 < 0$. It is easy to check that this g lies in $C^1(\bar{Q})$.

A stereographic projection, or any other suitable diffeomorphism, transfers this result to regions whose boundary contains portions of a sphere rather than a hyperplane. It is in this setting that the existence of C^1-extensions will be used (sometimes tacitly), in order to be able to talk about derivatives *on* the boundary.

The following simple fact is often referred to as the Hopf lemma.

15.3.7. Lemma. *If $f \in A^1(U)$, $f(1) = 1$, and $\text{Re } f(z) < 1$ for all $z \in U$, then* $f'(1) > 0$.

Recall that $A^1(U) = C^1(\bar{U}) \cap A(U)$.

Proof. Any C^1-extension of $f = u + iv$ to a neighborhood of $\bar{U}$ satisfies the Cauchy–Riemann equations on $\bar{U}$. Since u attains its maximum (relative to

$\overline{U}$) at $z = 1$, we have $\partial v/\partial x = -\partial u/\partial y = 0$ at that point. Thus

$$f'(1) = \frac{\partial u}{\partial x}(1) = \lim_{r \nearrow 1} \frac{1 - u(r)}{1 - r}.$$

Since the Poisson kernel for U satisfies

$$P_r(\theta) \geq \frac{1 - r}{1 + r},$$

the Poisson integral representation of $1 - u$ shows that

$$\frac{1 - u(r)}{1 - r} \geq \frac{1 - u(0)}{1 + r}.$$

Thus $2f'(1) \geq 1 - u(0) > 0$.

Part (a) of the following theorem is due to Fornaess [1; p. 549] and Pinčuk [2]. Part (c) was first proved by Alexander [2], with C^∞ in place of C^1, then by Pinčuk [3], [4] in its present form. Pinčuk actually established the result for larger classes of domains.

15.3.8. Theorem. *Let n and N be positive integers, $n > 1$. Let β be an open ball in $\mathbb{C}^n$ that intersects ∂B_n. Put $\Omega = \beta \cap B_n$.*

Assume that $F: \overline{\Omega} \to \mathbb{C}^N$ is nonconstant, of class C^1, holomorphic in Ω, and that F sends $\overline{\Omega} \cap \partial B_n$ into ∂B_N. Then

(a)   *the linear operator $F'(\zeta)$ has rank n for every $\zeta \in \beta \cap \partial B_n$,*
(b)   *$n \leq N$, and*
(c)   *in the case $n = N$, F extends to an automorphism of B_n.*

Proof. Since $F'(\zeta)$ maps $\mathbb{C}^n$ to $\mathbb{C}^N$, it is clear that (a) implies (b).

But (a) also implies (c). To see this, assume $n = N$, fix $\zeta \in \beta \cap \partial B_n$. By (a), the Jacobian of any C^1-extension $\tilde{F}$ of F (as in §15.3.6) is different from 0 at ζ. The standard real-variable version of the inverse function theorem implies therefore that $\tilde{F}$ is one-to-one in some neighborhood Ω_0 of ζ. Choose $t < 1$ so that $\beta \cap \Omega_0$ contains the set Ω_1 of all $z \in B_n$ for which $t < \text{Re}\langle z, \zeta \rangle$. Since F maps $\overline{\Omega}_1 \cap \partial B_n$ into ∂B_n, the maximum modulus theorem shows that $F(\Omega_1) \subset B$. Theorem 15.3.4 can now be applied to the restriction of F to Ω_1. This proves (c).

In the proof of (a) it involves no loss of generality to assume that $e_1 \in \beta$ and that $F(e_1) = e_1$. Let $f_1, \ldots, f_N$ be the components of F, and extend them to C^1-functions in β, as in §15.3.6. Since $u_1 = \text{Re } f_1$ has a local maximum (relative to ∂B_n) at e_1, we have $\partial u_1/\partial x_j = 0$ and $\partial u_1/\partial y_j = 0$ at e_1, for $j = 2, \ldots, n$, and since the Cauchy–Riemann equations hold at e_1, we conclude that $\partial f_1/\partial z_j = 0$ at e_1 if $j \geq 2$. Therefore the Taylor expansion of

F at e_1 has the form

(1) $$f_1(z) - 1 = \alpha_{11}(z_1 - 1) + \varepsilon_1(z)$$

and, for $2 \leq i \leq N$,

(2) $$f_i(z) = \alpha_{i1}(z_1 - 1) + \sum_{j=2}^{n} \alpha_{ij}z_j + \varepsilon_i(z),$$

where $\alpha_{ij} = (D_j f_i)(e_1)$ and $\varepsilon_i(z) = o(|z - e_1|)$ for $1 \leq i \leq N$.

Note that $\alpha_{11} > 0$, by the Hopf lemma.

Let A be the matrix (α_{ij}), with N rows and n columns. We have to prove that A has rank n. Except for α_{11}, the top row of A has only zero entries. Assume, to reach a contradiction, that the rank of A is $< n$. The rank of the matrix obtained from A by deleting the top row and the leftmost column is then $< n - 1$. Hence there is a unit vector $u \in \mathbb{C}^n$, $u \perp e_1$, such that $Au = 0$.

If $0 < t < 1$, put $s = (1 - t^2)^{1/2}$. When t is sufficiently close to 1, the points $te_1 + \lambda u$ lie in $\bar{\Omega}$ for all $\lambda \in \mathbb{C}$ with $|\lambda| \leq s$. For such t, put

(3) $$z(\theta) = te_1 + se^{i\theta}u \qquad (-\pi \leq \theta \leq \pi)$$

and replace z by $z(\theta)$ in (1) and (2). The sum in (2) is then 0, since $Au = 0$. Note that

(4) $$|z(\theta) - e_1|^2 = (1 - t)^2 + s^2 = 2(1 - t),$$

so that $\varepsilon_i(z(\theta)) = o((1 - t)^{1/2})$.

Thus $|f_1(z(\theta)) - 1|$ and $|f_i(z(\theta))|$ are $o((1 - t)^{1/2})$, by (1) and (2), for $2 \leq i \leq N$. Since $|F(z(\theta))|^2 = 1$,

(5) $$2\,\mathrm{Re}[1 - f_1(z(\theta))] = |1 - f_1|^2 + \sum_{2}^{N}|f_i|^2 = o(1 - t).$$

But $\mathrm{Re}\, f_1(te_1 + \lambda u)$ is a harmonic function of λ, in $|\lambda| < s$. The boundary estimate (5) holds therefore also when $\lambda = 0$. Thus

(6) $$\lim_{t \nearrow 1} \frac{\mathrm{Re}\{1 - f_1(te_1)\}}{1 - t} = 0,$$

contradicting the fact that $\alpha_{11} > 0$.

15.4. Proper Maps from B to B

As was pointed out in §15.1.2, there exist proper holomorphic maps of U onto U that have any prescribed multiplicity $m \geq 1$. The simplest of these is the map that takes z to z^m. When U is replaced by $B_n(n > 1)$ the situation changes completely. In that case, no branching can occur, the multiplicity must be 1, and the automorphisms are in fact the only possibilities. This is the content of Theorem 15.4.2.

The following lemma is due to Henkin [6]. It involves the approach regions D_α (§5.4.1) that play a role in the definition of K-limits (§5.4.6) and in Korányi's theorem 5.4.9.

15.4.1. Lemma. *Suppose that* $F: B \to B$ *is a proper holomorphic map, with* $F(0) = 0$. *Then* F *has a continuous extension to* $\bar{B}$, *and there is a constant* $A < \infty$ *such that*

$$(1) \qquad F(D_\alpha(\zeta)) \subset D_{A\alpha}(F(\zeta))$$

for every $\zeta \in S$ *and every* $\alpha > 1$.

Proof. For $w \in B$, let $\{p_i(w)\}$ be the points of B such that $F(p_i(w)) = w$. Define

$$(2) \qquad \mu(w) = \max_i |p_i(w)|^2 \qquad (w \in B).$$

Then $\mu \in C(B)$, $\mu < 1$.

Let V be the critical set of F.

Choose $a \in B$, $b \in \mathbb{C}^n$, let D be the disc of all $\lambda \in \mathbb{C}$ with $a + \lambda b \in B$, and put

$$(3) \qquad u(\lambda) = \mu(a + \lambda b) \qquad (\lambda \in D).$$

If $a \in B \backslash V$, then $E = \{\lambda \in D : a + \lambda b \in V\}$ is *discrete*, since V is an analytic variety in B. Each p_i is locally holomorphic in $B \backslash V$. Hence u is subharmonic in $D \backslash E$. Being continuous in D, it follows that u is subharmonic in D. If $a \in V$, approximate a by points of $B \backslash V$. Then u is a limit of subharmonic functions, hence u is subharmonic.

This proves that μ is plurisubharmonic in B.

Since F is proper, there exists $c < 1$ such that $|z| \leq c$ whenever $|F(z)| \leq \frac{1}{2}$. Thus $\mu(w) \leq c^2$ if $|w| \leq \frac{1}{2}$, and $\mu(w) \leq h(w)$ if $\frac{1}{2} < |w| < 1$, where h is the (radial) harmonic function that is 1 when $|w| = 1$ and is c^2 when $|w| = \frac{1}{2}$. It follows that there is a $c_1 > 0$ such that

$$(4) \qquad |p_i(w)|^2 \leq \mu(w) \leq 1 - c_1(1 - |w|^2) \qquad (w \in B),$$

or, equivalently, $c_1(1 - |F(z)|^2) \leq 1 - |z|^2$ if $z \in B$.

The upshot of all this is the existence of a constant $A < \infty$ such that

$$(5) \qquad 1 - |F(z)|^2 \leq A(1 - |z|^2) \qquad (z \in B).$$

By (5) and Theorem 8.1.4, F satisfies

$$(6) \qquad |1 - \langle F(z), F(a) \rangle| \leq A|1 - \langle z, a \rangle|$$

for all $z \in B$, $a \in B$. When z and a tend to the same boundary point $\zeta \in S$, the right side of (6) tends to 0, hence so does the left. This proves that F extends continuously to $\bar{B}$.

Since $F(0) = 0$, $|F(z)|^2 \le |z|^2$, hence

$$(7) \qquad\qquad \{1 - |F(z)|^2\}^{-1} \le \{1 - |z|^2\}^{-1},$$

which, when multiplied by (6), gives

$$(8) \qquad\qquad \frac{|1 - \langle F(z), F(\zeta) \rangle|}{1 - |F(z)|^2} \le A \frac{|1 - \langle z, \zeta \rangle|}{1 - |z|^2}$$

for $z \in B$, $\zeta \in S$. But (8) is just another way of writing (1). This proves the lemma.

The assumption $F(0) = 0$ is of course inessential, but it simplifies the proof a little.

15.4.2. Theorem (Alexander [3]). *If $n > 1$, and F is a proper holomorphic map of B into B, then $F \in \mathrm{Aut}(B)$.*

Proof. Assume $F(0) = 0$, without loss of generality. Denote the multiplicity of F by m, and let $p_1(w), \ldots, p_m(w)$ have the same meaning as in the proof of Lemma 15.4.1. By that lemma, F is a continuous map of $\bar{B}$ onto $\bar{B}$. We can therefore define $\#(w)$ to be the cardinality of the set $F^{-1}(w)$, as in Theorem 15.1.9, but now for all $w \in \bar{B}$, not just in B. A priori, $\#(w) = \infty$ is possible, if $w \in S$. However, we will show that $\#(w) = m$ for some $w \in S$; this will put us in a position where Theorem 15.3.4 can be applied.

Step 1. $\#(\eta) \ge m$ *for almost all* $\eta \in S$. Let Λ be a linear functional on $\mathbb{C}^n$ that separates the points $p_i(\tilde{w})$ for some $\tilde{w} \in B$ with $\#(\tilde{w}) = m$, i.e.,

$$(1) \qquad\qquad \Lambda p_i(\tilde{w}) \ne \Lambda p_j(\tilde{w}) \quad \text{if} \quad i \ne j.$$

The same argument that was used in the proof of Theorem 15.2.4 shows that there is an $h \in H^\infty(B)$ such that

$$(2) \qquad\qquad h(w) = \prod_{i<j} \{\Lambda p_i(w) - \Lambda p_j(w)\}^2$$

for every regular value w of F; the way in which $\{p_1(w), \ldots, p_m(w)\}$ is ordered is clearly irrelevant for (2).

By (1), $h \not\equiv 0$. Let E_1 be the set of all $\eta \in S$ at which the K-limit of h exists and is $\ne 0$. Then $\sigma(E_1) = 1$. Fix $\eta \in E_1$. We will see that $\#(\eta) \ge m$.

There is an approach region $D_\alpha(\eta)$ and a $\delta > 0$ such that

(3)
$$|h(w)| > \delta \qquad (w \in D_\alpha(\eta)).$$

Every point of $D_\alpha(\eta)$ is thus a regular value of F, and there exists $\varepsilon > 0$ and functions $p_i \in H(D_\alpha(\eta))$ such that

(4)
$$|p_i(w) - p_j(w)| > \varepsilon \qquad (w \in D_\alpha(\eta), i \neq j).$$

Let $\{w_k\}$ be a sequence in $D_\alpha(\eta)$ that converges to η. It has a subsequence, again denoted by $\{w_k\}$, such that $\lim p_i(w_k)$ exists, as $k \to \infty$, for $i = 1, \ldots, m$. By (4), this gives m distinct points

(5)
$$\zeta_i = \lim_{k \to \infty} p_i(w_k) \qquad (1 \leq i \leq m)$$

on S, such that $F(\zeta_i) = \eta$. Thus $\#(\eta) \geq m$, and Step 1 is completed.

Step 2. $\#(\eta) \leq m$ *for almost all* $\eta \in S$. There is a countable collection Φ of linear functionals Λ on $\mathbb{C}^n$ such that every finite subset of $\mathbb{C}^n$ is separated by some $\Lambda \in \Phi$. Define

(6)
$$Q_\Lambda(t, w) = \prod_{i=1}^{m} (t - \Lambda p_i(w))$$

for $\Lambda \in \Phi$, $t \in \mathbb{C}$, and w a regular value of F. The coefficients $g_{k, \Lambda}(w)$ in the expansion

(7)
$$Q_\Lambda(t, w) = t^m + \sum_{k=0}^{m-1} g_{k, \Lambda}(w) t^k$$

are polynomials in the $\Lambda p_i(w)$, hence are bounded, and therefore extend to members of $H^\infty(B)$, since the critical values of F form an H^∞-removable set. Thus (7) holds for all $w \in B$, and so does (6) (with possible repetition of factors).

Let E_2 be the set of all $\eta \in S$ at which the K-limit of every $g_{k, \Lambda}$ exists. Since Φ is countable, $\sigma(E_2) = 1$.

Fix $\eta \in E_2$. We will see that $\#(\eta) \leq m$.

Choose $\zeta \in S$ so that $F(\zeta) = \eta$. As $r \nearrow 1$, Lemma 15.4.1 shows that $F(r\zeta)$ tends to η within some region $D_\alpha(\eta)$, so that

(8)
$$\lim_{r \nearrow 1} g_{k, \Lambda}(F(r\zeta)) = g_{k, \Lambda}(\eta)$$

exists for all k and all Λ.

Since $r\zeta = p_i(F(r\zeta))$ for some i, (6) shows that

(9)
$$Q_\Lambda(\Lambda r\zeta, F(r\zeta)) = 0,$$

and (7) transforms this to

$$(10) \qquad (r\Lambda\zeta)^m + \sum_{k=0}^{m-1} g_{k,\Lambda}(F(r\zeta))(r\Lambda\zeta)^k = 0.$$

Letting $r \nearrow 1$, (8) and (10) show that $\Lambda\zeta$ *is a root of the monic polynomial* $Q_\Lambda(\cdot, \eta)$ *of degree m.*

If η had $m + 1$ inverse images, some $\Lambda \in \Phi$ would separate them, and the corresponding polynomial $Q_\Lambda(\cdot, \eta)$ would have $m + 1$ distinct roots, which is impossible.

This completes Step 2.

Step 3. Completion of proof. By Steps 1 and 2 there is an $\eta \in S$ with $\#(\eta) = m$. Let $\zeta_1, \ldots, \zeta_m$ be the points of $F^{-1}(\eta)$. Choose $r > 0$ so that $|\zeta_i - \zeta_k| > 3r$ if $i \neq j$. Let β_i be the set of all $z \in B$ with $|z - \zeta_i| < r$. The β_i's are disjoint open sets. None of the compact sets

$$(11) \qquad F(\overline{B \cap \partial\beta_i}) \qquad (1 \le i \le m)$$

contains η. Therefore there exists $\delta > 0$ such that the set

$$(12) \qquad V = \{w \in B : |w - \eta| < \delta\}$$

intersects none of the sets (11).

We claim that $F(\beta_i) \supset V$, for $i = 1, \ldots, m$.

Since F is an open mapping (Theorem 15.1.6), each $F(\beta_i)$ is an open subset of B whose boundary lies partly in S and partly in the set (11). In particular, V contains no boundary point of $F(\beta_i)$. Since $F(\beta_i)$ contains points arbitrarily close to η, $F(\beta_i)$ intersects V. Since V is connected, it follows that $V \subset F(\beta_i)$.

Put

$$(13) \qquad \Omega_i = \beta_i \cap F^{-1}(V) \qquad (1 \le i \le m).$$

Then $F(\Omega_i) = V$.

Since the m sets β_i are pairwise disjoint, and since no point of B has more than m inverse images under F, we conclude that F is one-to-one in each Ω_i.

Thus F is a biholomorphic map of Ω_i onto V.

To apply Theorem 15.3.4 to $F: \Omega_i \to V$, one more thing has to be checked, namely, that there is an $\varepsilon > 0$ such that Ω_i contains all $z \in B$ with $|z - \zeta_i| < \varepsilon$. If this were false, there would be a sequence $\{z_k\}$ in $\beta_i \setminus \Omega_i$, converging to ζ_i. For large k, $F(z_k) \in V$, since F is continuous on $\overline{B}$ and $F(\zeta_i) = \eta$. But if $F(z_k) \in V$ then $z_k \in \Omega_i$.

Theorem 15.3.4 shows now that $F \in \text{Aut}(B)$.

15.4.3. The preceding proof is basically that given by Alexander [3], except that at the very end he refers to a theorem of Fefferman [1] to conclude that the biholomorphic map $F: \Omega_i \to V$ has a C^∞-extension to the closure of $\bar{\Omega}_i$, and then uses his C^∞-version of Theorem 15.3.8(c) (Alexander [2]). Fefferman's proof of his theorem (asserting the C^∞-extendibility of bi-holomorphic maps between bounded strictly pseudo-convex domains with C^∞-boundaries) involved a difficult analysis of the boundary behavior of Bergman kernels, as well as a detailed study of the geodesics with respect to the Bergman metric.

Fefferman's theorem has recently been simplified by Ligocka [1], [2] and Bell–Ligocka [1]. The most elementary proof so far (though it is by no means simple) is apparently that of Nirenberg–Webster–Yang [1].

For related results, we refer to Bedford–Fornaess [1] and to Diedrich–Fornaess [1], [2], [3]. All of these papers contain numerous other references.

15.5. A Characterization of *B*

Some of the preceding proofs, especially that of Theorem 15.3.4, made very strong use of the transitivity of the group Aut(*B*). This transitivity is a very special property of *B*. In fact, as we shall now see, the only smoothly bounded domains in $\mathbb{C}^n$ that share it are the ones that are biholomorphically equi-valent to *B*. This is a corollary of Theorem 15.5.10.

The more precise formulation of this result gives us an opportunity to introduce the concept of strict pseudoconvexity; we have already alluded to this a few times, but informally.

15.5.1. Definitions. If W is an open set in $\mathbb{C}^n$ and ρ is a real-valued C^2-function with domain W, we define, for $w \in W$,

$$(1) \qquad N(w) = (\bar{D}_1\rho(w), \ldots, \bar{D}_n\rho(w))$$

$$(2) \qquad P_w(a) = \sum_{j,k=1}^{n} (D_j D_k \rho)(w) a_j a_k \qquad (\alpha \in \mathbb{C}^n)$$

$$(3) \qquad \langle H_w a, b \rangle = \sum_{j,k=1}^{n} (D_j \bar{D}_k \rho)(w) a_j \bar{b}_k \qquad (a, b \in \mathbb{C}^n)$$

and

$$(4) \qquad Q_w(a) = P_w(a) + \langle H_w a, a \rangle \qquad (a \in \mathbb{C}^n).$$

The "normal" vector $N(w)$ is perpendicular to the level surface of ρ through w. For instance, when $\rho(w) = |w|^2$, then $N(w) = w$.

P_w is a homogeneous polynomial of degree 2, H_w is a hermitian operator on $\mathbb{C}^n$ (the so-called *complex Hessian* of ρ at w; see §1.3.4), and the Taylor expansion of ρ, about any $w \in W$, can be written in the form

$$(5) \quad \rho(z) = \rho(w) + 2\,\mathrm{Re}\langle z - w, N(w)\rangle + \mathrm{Re}\,Q_w(z - w) + |z - w|^2 \varepsilon(z, w)$$

where ε is continuous on $W \times W$, and $\varepsilon(w, w) = 0$.

A region $\Omega \subset \mathbb{C}^n$ is said to have C^2-*boundary at a point* $\zeta_0 \in \partial\Omega$ if ζ_0 has a neighborhood W in which a function ρ is defined, as above, such that

$$(6) \qquad\qquad N(\zeta) \neq 0 \quad \text{for all} \quad \zeta \in W \cap \partial\Omega$$

and

$$(7) \qquad\qquad \Omega \cap W = \{z \in W : \rho(z) < 0\}.$$

Any such ρ is a *local defining function for* Ω at ζ_0. If the domain of W of ρ contains all of $\partial\Omega$, then ρ is *a defining function for* Ω. For example, $|z|^2 - 1$ is a defining function for B. If Ω has C^2-boundary at each of its boundary points, we say that Ω *has* C^2-*boundary*.

Let Ω have C^2-boundary at ζ. Then Ω is said to be *strictly pseudoconvex at* ζ if there exists a local defining function ρ for Ω at ζ whose complex Hessian H_ζ is strictly positive; more explicitly, there should exist $c > 0$ such that

$$(8) \qquad\qquad \langle H_\zeta a, a\rangle \geq c|a|^2 \qquad (a \in \mathbb{C}^n),$$

where H_ζ is defined by (3).

If Ω is strictly pseudoconvex at each of its boundary points, then Ω is said to be a *strictly pseudoconvex region*.

For example, when $\rho(z) = |z|^2 - 1$, then H_ζ is the identity operator. Thus B is strictly pseudoconvex.

The continuity of the second derivatives of ρ shows that the set of points at which Ω is strictly pseudoconvex is an open subset of $\partial\Omega$. For bounded Ω with C^2-boundary, this set cannot be empty, because of the following proposition.

15.5.2. Proposition. *Suppose* Ω *is a region with* C^2-*boundary, and* $\Omega \subset B$. *Then* Ω *is strictly pseudoconvex at every* $\zeta \in S \cap \partial\Omega$.

Proof. Let $\zeta = e_1$, without loss of generality. Near e_1, B has a defining function ρ_B of the form

$$(1) \qquad\qquad \rho_B(z) = x_1 - \left\{1 - y_1^2 - \sum_2^n |z_j|^2\right\}^{1/2},$$

and there is a C^2-function $\gamma = \gamma(y_1, z_2, \ldots, z_n)$ with the following properties: $\gamma \geq 0$, $\gamma = 0$ when $y_1 = z_2 = \cdots = z_n = 0$, and the boundary of Ω is defined, in some neighborhood V of e_1, by the equation

$$(2) \qquad\qquad \rho_B + \gamma = 0.$$

Put $\rho_\Omega = \rho_B + \gamma$, in V. Then

$$(3) \qquad \rho_\Omega(z) = \gamma + \tfrac{1}{2}(y_1^2 + |z_2|^2 + \cdots + |z_n|^2) + \cdots$$

where the missing terms are those whose second derivatives are 0 at e_1.

Since γ has a local minimum at e_1, its Hessian is ≥ 0 at e_1. This follows from the interpretation of the Hessian as a Laplacian that is mentioned in §1.3.4. The Hessian of the other term on the right of (3) is strictly positive at e_1.

The Hessian of ρ_Ω is thus strictly positive at e_1, and hence Ω is strictly pseudoconvex at e_1.

15.5.3. A Change of Variables. Let Ω be strictly pseudoconvex at ζ, with local defining function ρ. We shall see that there is a constant $c > 0$ and a biholomorphic map Φ, carrying a neighborhood W of ζ onto a neighborhood $\tilde{W}$ of 0, such that the transferred defining function $\tilde{\rho} = c\rho \circ \Phi^{-1}$ has the simple form

$$(1) \qquad\qquad \tilde{\rho}(z) = z_1 + \bar{z}_1 + |z|^2 + |z|^2 \varepsilon(z)$$

for $z \in \tilde{W}$, where ε is continuous, and $\varepsilon(0) = 0$. The set $\Phi(\Omega \cap W)$ consists then of those $z \in \tilde{W}$ for which $\tilde{\rho}(z) < 0$.

To do this, begin with ρ as in 15.5.1(5), with ζ in place of w. A translation makes $\zeta = 0$. Follow this by a unitary transformation and a dilation ($z \to tz$ for some $t > 0$), to achieve $N(0) = e_1$. Now ρ has the form

$$(2) \qquad \rho(z) = z_1 + \bar{z}_1 + P(z) + \overline{P(z)} + \langle Hz, z \rangle + |z|^2 \varepsilon(z)$$

where $P(z) = \tfrac{1}{2} \sum (D_j D_k \rho)(0) z_j z_k$ and H is a strictly positive hermitian operator. Make the change of variables

$$(3) \qquad w_1 = z_1 + P(z), \qquad w_2 = z_2, \ldots, w_n = z_n.$$

This is one-to-one in a neighborhood of the origin.

Put $\rho_0(w) = \rho(z)$. Then

$$(4) \qquad\qquad \rho_0(w) = w_1 + \bar{w}_1 + \langle Hw, w \rangle + |w|^2 \tilde{\varepsilon}(w)$$

with the same H as in (2).

Since H is a strictly positive hermitian operator on $\mathbb{C}^n$, a unitary transformation of $\mathbb{C}^n$ will transform $\langle Hw, w \rangle$ to $\sum \lambda_i |w_i|^2$, with $\lambda_i > 0$ for $i = 1, \ldots, n$. Replace $\lambda_i^{1/2} w_i$ by new variables; another dilation finally gives (1).

15.5.4. Remark. If $\tilde{\rho}$, of the form 15.5.3(1), is a local defining function for Ω, then $\operatorname{Re} z_1 < 0$ for all $z \in \Omega$ that are sufficiently close to 0. The function $\exp(z_1)$ shows therefore that 0 is a local peak point for $A(\Omega)$.

Every point at which Ω is strictly pseudoconvex is thus a local peak point for $A(\Omega)$.

Theorem 15.2.4 holds therefore for every strictly pseudoconvex bounded region Ω.

15.5.5. Lemma. *Suppose that ρ is a local defining function for Ω, that the domain W of ρ is convex, and that there is constant $t > 0$ such that*

$$\text{(1)} \qquad\qquad \operatorname{Re} Q_w(u) \geq t|u|^2 \qquad (u \in \mathbb{C}^n)$$

for all $w \in W$. Then

$$\text{(2)} \qquad\qquad 2\operatorname{Re}\langle \zeta - z, N(\zeta) \rangle \geq t|\zeta - z|^2$$

for all $\zeta \in W \cap \partial\Omega, z \in W \cap \overline{\Omega}$.

Proof. Put $h(s) = \rho((1 - s)\zeta + sz), 0 \leq s \leq 1$. Then $h(0) = \rho(\zeta) = 0, h(1) = \rho(z) \leq 0$. By the chain rule,

$$\text{(3)} \qquad\qquad h'(0) = 2\operatorname{Re}\langle z - \zeta, N(\zeta) \rangle$$

and

$$\text{(4)} \qquad\qquad h''(s) = 2 \operatorname{Re} Q_w(z - \zeta) \geq 2t|z - \zeta|^2,$$

where $w = (1 - s)\zeta + sz$. If (3) and (4) are inserted into the Taylor formula

$$\text{(5)} \qquad\qquad h(1) = h(0) + h'(0) + \tfrac{1}{2}h''(s)$$

which holds for some $s \in (0, 1)$, the result is (2).

Note. The real part of the inner product in (2) is the same as the dot-product $(\zeta - z) \cdot N(\zeta)$ (see §5.4.2). The positivity of this dot-product shows that $W \cap \Omega$ is entirely on one side of any hyperplane that is tangent to Ω at a point $\zeta \in W \cap \partial\Omega$. Thus $W \cap \Omega$ is a convex region.

In view of §15.5.3, it follows that strictly pseudoconvex regions can be made "locally convex" at any boundary point, by means of a local biholomorphic change of variables.

15.5.6. Spheres and Ellipsoids Tangent to $\partial\Omega$. Let Ω be a region with C^2-boundary at $\zeta \in \partial\Omega$, and let ρ be a local defining function for Ω at ζ. For any $t > 0$, the open ball with center

$$(1) \qquad\qquad a = \zeta - tN(\zeta)$$

and radius $t\,|N(\zeta)|$ will be denoted by $\beta_{\zeta,t}$.

Thus $z \in \beta_{\zeta,t}$ provided that

$$(2) \qquad\qquad |z - \zeta + tN(\zeta)| < t\,|N(\zeta)|.$$

One checks easily that $\beta_{\zeta,t}$ is tangent to $\partial\Omega$ at ζ.

Next, suppose $0 < c < 1$, and let E_c be the ellipsoid consisting of all $z = (z_1, z')$ such that

$$(3) \qquad\qquad \frac{|z_1 - 1 + c|^2}{c^2} + \frac{|z'|^2}{c} < 1.$$

(We have met such ellipsoids earlier, in connection with the Julia–Carathéodory theorem.) A simple computation shows that $E_c \subset B$. Clearly, $e_1 \in \partial E_c$.

Now let h be the composition of a dilation, a translation, and a unitary transformation of $\mathbb{C}^n$, such that $h(e_1) = \zeta$ and $h(0) = a$, where a is given by (1). Define

$$(4) \qquad\qquad E_{\zeta,t,c} = h(E_c).$$

Thus $E_{\zeta,t,c}$ is an ellipsoid in $\beta_{\zeta,t}$, having ζ in its boundary.

15.5.7. Proposition. *Suppose that Ω is a region with C^2-boundary at 0 that has*

$$(1) \qquad\qquad \rho(z) = z_1 + \bar{z}_1 + |z|^2 + |z|^2\varepsilon(z)$$

as a local defining function at 0.

Suppose that $0 < r < 1 < R < \infty$.

Then there is a neighborhood V of 0 in $\mathbb{C}^n$, and a constant $c > 0$, such that

$$(2) \qquad\qquad\qquad\qquad V \cap \Omega \subset \beta_{\zeta,R}$$

and

$$(3) \qquad\qquad\qquad\qquad E_{\zeta,r,c} \subset \Omega$$

for every $\zeta \in V \cap \partial\Omega$.

Proof. Choose s and t so that

$$(4) \qquad\qquad\qquad\qquad \frac{1}{R} < t < 1 < s < \frac{1}{r}.$$

Our assumption (1) says that $Q_0(u) = |u|^2$, in the notation of §15.5.1. Since $\rho \in C^2$, there is a convex neighborhood V of 0 in $\mathbb{C}^n$ such that $V \subset W$ (the domain of ρ) and

$$(5) \qquad\qquad t|u|^2 \leq \operatorname{Re} Q_w(u) \leq s|u|^2 \qquad (w \in V, u \in \mathbb{C}^n).$$

Suppose $\zeta \in V \cap \partial\Omega$, $z \in V \cap \Omega$. By (5) and Lemma 15.5.5,

$$(6) \qquad\qquad 2\operatorname{Re}\langle z - \zeta, N(\zeta)\rangle \leq -t|\zeta - z|^2$$

Hence

$$|z - \zeta + RN(\zeta)|^2 \leq |RN(\zeta)|^2 - (tR - 1)|z - \zeta|^2 < |RN(\zeta)|^2$$

so that $z \in \beta_{\zeta,R}$. This proves (2).

Next, choose $\delta > 0$, so small that the term $\varepsilon(z, w)$ in the Taylor expansion 15.5.1(5) satisfies

$$(7) \qquad\qquad \varepsilon(z, \zeta) < \frac{1}{r} - s \quad \text{if} \quad \zeta \in V \cap \partial\Omega, \qquad |z - \zeta| < \delta.$$

Assume this, and assume also that $z \in \beta_{\zeta,r}$. This means that $|z - \zeta + rN(\zeta)| < r|N(\zeta)|$. Hence there are $\lambda, \mu \in \mathbb{C}$, with $|\lambda|^2 + |\mu|^2 \leq 1$, and there is a vector $M(\zeta) \perp N(\zeta)$, with $|M(\zeta)| = |N(\zeta)|$, such that

$$(8) \qquad\qquad z - \zeta = r\{(\lambda - 1)N(\zeta) + \mu M(\zeta)\}.$$

This implies

(9) $$\operatorname{Re}\langle z - \zeta, N(\zeta)\rangle = -r|N(\zeta)|^2 \operatorname{Re}(1 - \lambda)$$

and

(10) $$|z - \zeta|^2 = |rN(\zeta)|^2(|1 - \lambda|^2 + |\mu|^2)$$
$$\leq 2|rN(\zeta)|^2 \operatorname{Re}(1 - \lambda).$$

By (9) and (10),

(11) $$2 \operatorname{Re}\langle z - \zeta, N(\zeta)\rangle \leq -\frac{1}{r}|z - \zeta|^2.$$

Finally, (5), (7), and (11) show that

$$\rho(z) = 2 \operatorname{Re}\langle z - \zeta, N(\zeta)\rangle + \operatorname{Re} Q_\zeta(z - \zeta) + |z - \zeta|^2 \varepsilon(z, \zeta)$$

$$< \left(-\frac{1}{r} + s + \frac{1}{r} - s\right)|z - \zeta|^2 = 0.$$

This proves that $z \in \Omega$ for all $z \in \beta_{\zeta, r}$ that satisfy $|z - \zeta| < \delta$, if $\zeta \in V \cap \partial\Omega$. Hence (3) holds if c is sufficiently small (depending only on δ and r).

15.5.8. Computation of a Jacobian. Suppose $0 < r < R < \infty$, $0 < c < 1$, $1 - rc < x < 1$. Let $E_{r,c}$ and β_R be the ellipsoid and ball defined by

(1) $$|z_1 - 1 + rc|^2 + c|z'|^2 < r^2 c^2$$

and

(3) $$|z_1 - 1 + R|^2 + |z'|^2 < R^2,$$

respectively. Put $p = xe_1$, and let ψ be a biholomorphic map of $E_{r,c}$ onto β_R that fixes p. We claim that the Jacobian of ψ satisfies

(3) $$|(J\psi)(p)|^2 = \left(\frac{R}{r}\right)^{n-1} \left\{ \frac{2 + \dfrac{1 - x}{R}}{2 + \dfrac{1 - x}{rc}} \right\}^{n+1}.$$

The point of (3) is that if R, r, c are held fixed but $x \nearrow 1$, the right side of (3) tends to $(R/r)^{n-1}$, which is independent of c.

To prove (3), write $\psi = f_3 \circ f_2 \circ f_1$, where

(4)
$$f_1(z_1, z') = \left(\frac{z_1 - 1 + rc}{rc}, \frac{z'}{r\sqrt{c}} \right)$$

maps $E_{r,c}$ onto B,

(5)
$$f_3(z_1, z') = (Rz_1 + 1 - R, Rz')$$

maps B onto β_R, and $f_2 \in \text{Aut}(B)$ takes $f_1(p)$ to $f_3^{-1}(p)$. The Jacobians of f_1 and f_3 can be read off from (4) and (5); the Jacobian of f_2 at $f_1(p)$ is given by 2.2.6(6); multiplying them, one obtains (3).

The following lemma, in which no smoothness assumptions are imposed on $\partial\Omega$, is similar to one that occurs in Rosay [2]. Combined with the preceding material, it will prove Theorem 15.5.10, the main result of this section.

15.5.9. Lemma. *Suppose Ω is a bounded region in $\mathbb{C}^n$, $w_0 \in \Omega$, and $\{F_i\}$, $\{G_i\}$, $\{D_i\}$ are sequences such that*

(a) $F_i: B \to \Omega$ *is holomorphic, $F_i(0) = w_0$,*
(b) D_i *is a region in Ω, $w_0 \in D_i$, and every compact subset of Ω lies in all but finitely many D_i,*
(c) $G_i: D_i \to B$ *is holomorphic and one-to-one, $G_i(w_0) = 0$, and*
(d) $\lim_{i \to \infty} |(JG_i)(w_0)(JF_i)(0)| = 1$.

Then Ω is biholomorphically equivalent to B.

Proof. Pass to subsequences, if necessary, so that the limits

(1)
$$G = \lim_{i \to \infty} G_i \quad \text{and} \quad H = \lim_{i \to \infty} G \circ F_i$$

exist, uniformly on compact subsets of Ω and B, respectively. Since $G_i(w_0) = 0$ for all i, and since each boundary point of B is a peak-point for $A(B)$, Proposition 15.2.2 implies that G maps Ω into B. The same argument shows that H maps B into B (and not merely into $\bar{B}$).

We claim that G is a biholomorphic map of Ω onto B.

Assume, to reach a contradiction, that G is not one-to-one in Ω. Then there exist $a \neq b$ in Ω, z_0 in B, with $G(a) = G(b) = z_0$. Since $\{(JF_i)(0)\}$ is bounded, (d) shows that $(JG)(w_0) \neq 0$. But JG is the limit, uniformly on compact subsets of Ω, of the zero-free functions JG_i. It follows (as in one variable) that JG has no zero in Ω, so that G is *locally* biholomorphic.

Consequently, there is a ball $V = z_0 + rB$, with $\bar{V} \subset B$, and there are disjoint neighborhoods V_a of a, V_b of b, such that G is a homeomorphism of

$\bar{V}_a$ onto $\bar{V}$, and of $\bar{V}_b$ onto $\bar{V}$. Let $h_a\colon \bar{V} \to \bar{V}_a$ invert G. Choose i so large that $|G_i - G| < r/2$ on $V_a \cup V_b$. Then $G_i \circ h_a$ moves no boundary point of the ball $\bar{V}$ by more than $r/2$. It follows that

$$(2) \qquad\qquad z_0 \in (G_i \circ h_a)(V) = G_i(V_a).$$

By the same argument, $z_0 \in G_i(V_b)$. This contradicts the fact that G_i is one-to-one.

We have now proved that G is one-to-one.

It remains to be proved that $G(\Omega) = B$.

To do this, let $H_i = G \circ F_i$. Then

$$(3) \qquad H_i'(0) = [G'(w_0) - G_i'(w_0)]F_i'(0) + G_i'(w_0)F_i'(0).$$

Since $F_i'(0)$ is bounded, $G'(w_0) - G_i'(w_0) \to 0$, and $H_i'(0) \to H'(0)$, by (1), it follows from (d) and (3) that $|(JH)(0)| = 1$. By Lemma 15.3.2, H is therefore a *unitary operator*.

Fix $z \in B$. Choose t so that $|z| < t < 1$. Since $H_i \circ H^{-1}$ converges to the identity map, uniformly on compact subsets of B, there is an index i such that $H_i \circ H^{-1}$ moves no point of tS by as much as $t - |z|$. Hence

$$(4) \qquad\qquad z \in H_i(H^{-1}(tB)) = G(F_i(H^{-1}(tB))).$$

In particular, $z \in G(\Omega)$, and the proof is complete.

We now come to the biholomorphic characterization of B that was found by Rosay [2]. For strictly pseudoconvex domains, the result was proved earlier by Wong [1].

15.5.10. Theorem. *If Ω is a bounded region in $\mathbb{C}^n$ that is strictly pseudoconvex at some point $\zeta_0 \in \partial\Omega$, and if there are automorphisms T_i of Ω such that*

$$(1) \qquad\qquad \lim_{i \to \infty} T_i(w_0) = \zeta_0$$

for some $w_0 \in \Omega$, then Ω is biholomorphically equivalent to B.

An automorphism of Ω is of course understood to be a biholomorphic map of Ω onto Ω.

Proof. By §15.5.3, there is a neighborhood W of ζ_0, and a biholomorphic map Φ of W onto a neighborhood $\tilde{W}$ of 0, with $\Phi(\zeta_0) = 0$, such that Φ maps

$\Omega \cap W$ onto $\tilde{\Omega}$, the set of all $z \in \tilde{W}$ that satisfy

$$(2) \qquad\qquad z_1 + \bar{z}_1 + |z|^2 + \eta(z) < 0,$$

where η is a C^2-function in $\tilde{W}$ that vanishes at 0, together with its first and second derivatives.

Choose r, R, so that $r < 1 < R$, and R/r is close to 1. Let V and c be given by Proposition 15.5.7, with $\tilde{\Omega}$ in place of Ω.

Put $w_i = T_i(w_0)$, $p_i = \Phi(w_i)$. (Note that $w_i \in W$ for all large i.) Then $w_i \to \zeta_0$ and $p_i \to 0$, as $i \to \infty$. For all sufficiently large i, the point ζ_i that is closer to p_i than any other point of $\partial\tilde{\Omega}$ will then lie in V, p_i will lie in the ellipsoid

$$(3) \qquad\qquad E_i = E_{\zeta_i, r, c} \subset \tilde{\Omega},$$

and the inclusion

$$(4) \qquad\qquad V \cap \tilde{\Omega} \subset \beta_{\zeta_i, R} = \beta_i$$

will hold. [(3) and (4) define E_i and β_i.]

Define

$$(5) \qquad\qquad D_i = T_i^{-1}(\Phi^{-1}(V \cap \tilde{\Omega})) \qquad (i = 1, 2, 3, \ldots).$$

Since ζ_0 is a local peak point for $A(\Omega)$ (§15.5.4), Proposition 15.2.2 implies that $T_i(w) \to \zeta_0$ uniformly on compact subsets of Ω. If $K \subset \Omega$ is compact, it follows that $T_i(K) \subset \Phi^{-1}(V \cap \tilde{\Omega})$, hence

$$(6) \qquad\qquad K \subset D_i$$

for all sufficiently large i.

Let f_i be a biholomorphic map of B onto E_i, such that $f_i(0) = p_i$.

Let g_i be a biholomorphic map of β_i onto B such that $g_i(p_i) = 0$.

Define $F_i = T_i^{-1} \circ \Phi^{-1} \circ f_i$. Then F_i maps B into Ω, and $F_i(0) = w_0$.

Define $G_i = g_i \circ \Phi \circ T_i$. Then G_i is a biholomorphic map from D_i into B, with $G_i(w_0) = 0$.

Note that $G_i \circ F_i = g_i \circ f_i$, and that $f_i \circ g_i$ is a biholomorphic map of the ball β_i onto the ellipsoid E_i that fixes p_i. Thus

$$|(JG_i)(w_o)(JF_i)(0)| = |J(g_i \circ f_i)(0)|$$
$$= |(Jf_i)(0)(Jg_i)(p_i)| = |J(f_i \circ g_i)(p_i)|,$$

and the latter tends to $(r/R)^{(n-1)/2}$ as $i \to \infty$, by §15.5.8.

Since r/R can be taken arbitrarily close to 1, the hypotheses of Lemma 15.5.9 are satisfied. This proves the theorem.

Corollary. *If Ω is a bounded region in $\mathbb{C}^n$, with C^2-boundary and if $\mathrm{Aut}(\Omega)$ is transitive on Ω, then Ω is biholomorphically equivalent to B.*

This follows from Theorem 15.5.10, because of Proposition 15.5.2.

Chapter 16

The $\bar{\partial}$-Problem

16.1. Differential Forms

Differential forms are often introduced purely algebraically (Spivak [1], Gunning–Rossi [1]), as members of a graded ring, or simply as "formal sums" that are to be manipulated according to the rules of "exterior algebra," but they can also be defined as complex-valued *functions* whose domain is the collection of all suitably differentiable surfaces of the appropriate dimension. We shall sketch this second approach, omitting all proofs; they are elementary, but long-winded and repetitious. Details may be found in Rudin [16]. The main purpose of this introductory section is to recall the basic facts and to establish notation.

16.1.1. Forms and Surfaces. Let $\Omega \subset R^N$ be open. A *k-surface* in Ω is a C'-map Φ from a parameter domain $D \subset R^k$ into Ω. A *differential form of order* $k \geq 1$ in Ω (briefly, a *k-form in Ω*) is a function α, symbolically represented by a sum

(1) $$\alpha = \sum a_{i_1 \ldots i_k}(x) dx_{i_1} \wedge \cdots \wedge dx_{i_k}$$

(the indices $i_1, \ldots, i_k$ range independently from 1 to n) which assigns to each k-surface Φ in Ω a number, called *the integral of α over Φ*, defined by

(2) $$\int_\Phi \alpha = \int_D \sum a_{i_1 \ldots i_k}(\Phi(u)) \frac{\partial(x_{i_1}, \ldots, x_{i_k})}{\partial(u_1, \ldots, u_k)} \, du.$$

Here du denotes Lebesgue measure in D. The Jacobians are the ones determined by the maps

(3) $$(u_1, \ldots, u_k) \to (\varphi_{i_1}(u), \ldots, \varphi_{i_k}(u))$$

where $\varphi_1, \ldots, \varphi_N$ are the components of Φ.

The coefficients $a_{i_1 \ldots i_k}$ in (1) are assumed to lie in $C(\Omega)$.

A 0-form is, by definition, a function in $C(\Omega)$.

330

Two k-forms α and β are defined to be equal if $\int_\Phi \alpha = \int_\Phi \beta$ for every eligible Φ.

The fact that determinants change sign when two rows are switched leads to the crucial *anti-commutative law*

(4) $$dx_i \wedge dx_j = -dx_j \wedge dx_i.$$

In particular,

(5) $$dx_i \wedge dx_i = 0 \qquad (1 \le i \le N).$$

Any term in (1) that has $i_r = i_s$ for some $r \ne s$ is thus 0. If we discard these terms, the others can be rewritten (using (4) repeatedly) so that the subscripts on the dx's are in increasing order. Collecting terms with the same subscripts leads then to the *standard presentation* of α,

(6) $$\alpha = \sum_I A_I(x)dx_I$$

in which each I is an increasing k-index (i.e., $I = (i_1, \ldots, i_k)$ with $i_1 < \cdots < i_k$), and

(7) $$dx_I = dx_{i_1} \wedge \cdots \wedge dx_{i_k}.$$

We call dx_I a *basic k-form*.

Each k-form α has a *unique* standard presentation. In particular, $\alpha = 0$ if and only if $A_I = 0$ in (6) for every I.

If $k > N$, all k-forms in R^N are 0, because of (5).

Addition and scalar multiplication of k-forms in Ω are defined in the obvious way.

16.1.2. Multiplication. If dx_I and dx_J are basic k-forms and l-forms, respectively, their product is defined to be the $(k + l)$-form

(1) $$dx_I \wedge dx_J = dx_{i_1} \wedge \cdots \wedge dx_{i_k} \wedge dx_{j_1} \wedge \cdots \wedge dx_{j_l}.$$

By 16.1.1.(4) and (5), this is 0 when I and J have an element in common. In the other case, let M be the $(k + l)$-index obtained by rearranging the members of I and J in increasing order; it follows that then

(2) $$dx_I \wedge dx_J = \varepsilon \, dx_M$$

where ε is the parity of the permutation

(3) $$(i_1, \ldots, i_k, j_1, \ldots, j_l) \to (m_1, \ldots, m_{k+1}).$$

If $\alpha = \sum A_I(x)dx_I$ and $\beta = \sum B_J(x)dx_J$ are standard presentations of a k-form α and an l-form β, one defines

(4)
$$\alpha \wedge \beta = \sum_{I, J} A_I(x)B_J(x)dx_I \wedge dx_J.$$

The resulting multiplication can be verified to be associative and distributive; as regards commutativity, it satisfies

(5)
$$\alpha \wedge \beta = (-1)^{kl}\beta \wedge \alpha.$$

We have tacitly assumed $k \geq 1$ and $l \geq 1$. The product of a 0-form f and a k-form α as above is defined to be the k-form

(6)
$$f\alpha = \alpha f = \sum f(x)A_I(x)dx_I.$$

It is customary to write $f\alpha$, rather than $f \wedge \alpha$, when f is a 0-form.

16.1.3. Differentiation. There is a differentiation operator d which associates a $(k + 1)$-form $d\alpha$ to every k-form α whose coefficients are in C^1.
For 0-forms f,

(1)
$$df = \sum_{i=1}^{N} \left(\frac{\partial f}{\partial x_i}\right)dx_i.$$

If $\alpha = \sum A_I(x)dx_I$ is the standard presentation of a k-form α, then

(2)
$$d\alpha = \sum_I dA_I \wedge dx_I$$

where each dA_I is defined by (1).
The relation $\partial^2 f/\partial x_i \, \partial x_j = \partial^2 f/\partial x_j \, \partial x_i$, combined with $dx_i \wedge dx_j = -dx_j \wedge dx_i$, leads to the very important fact that

(3)
$$d^2 = 0,$$

i.e., that $d(d\alpha) = 0$ (provided, of course, that the coefficients of α are in C^2).
The Leibnitz rule for differentiation of products becomes

(4)
$$d(\alpha \wedge \beta) = (d\alpha) \wedge \beta + (-1)^k\alpha \wedge d\beta$$

if α is a k-form.

16.1.4. Pull-Backs. Let Ω_1 and Ω_2 be open sets in R^N and R^M, respectively, and let

(1)
$$T: \Omega_1 \to \Omega_2$$

be a C^1-map with components $t_1, \ldots, t_M$. Denote points of Ω_1 and Ω_2 by x and y, respectively.

If $\alpha = \sum A_I(y)dy_I$ is a k-form in Ω_2, its *pull-back* α_T is the k-form in Ω_1 given by

(2)
$$\alpha_T = \sum A_I(T(x))dt_I$$

where $dt_I = dt_{i_1} \wedge \cdots \wedge dt_{i_k}$ when $I = (i_1, \ldots, i_k)$, and each dt_i is a 1-form, as in 16.1.3(1),

(3)
$$dt_i = \sum_{j=1}^{N} \left(\frac{\partial t_i}{\partial x_j} \right) dx_j \qquad (1 \le i \le M).$$

Pull-backs of forms have the following important properties: Let α and β be k- and l-forms in Ω_2. Then

(i) $(\alpha + \beta)_T = \alpha_T + \beta_T$ if $k = l$,
(ii) $(\alpha \wedge \beta)_T = \alpha_T \wedge \beta_T$,
(iii) $(\alpha_T)_S = \alpha_{TS}$ if $S: \Omega_0 \to \Omega_1$ is of class C^1,
(iv) $\int_{T\Phi} \alpha = \int_\Phi \alpha_T$ if Φ is a k-surface in Ω_1,
(v) $d(\alpha_T) = (d\alpha)_T$ if $\alpha \in C^1$ and $T \in C^2$.

Of these, (i) and (ii) are almost obvious; the others are consequences of the chain rule. The proof of (v) uses $d^2 t_i = 0$; hence the assumption $T \in C^2$. A very important special case of (v) is

(vi) $d(\alpha_T) = 0$ if $d\alpha = 0$.

Formula (iv) states an important invariance property.

16.1.5. Chains and Boundaries. We defined k-forms as functions that assign numbers to k-surfaces. This can be turned around: Each k-surface Φ in Ω (of class C^1) can be viewed as a function that assigns the numbers $\int_\Phi \alpha$ to k-forms α. Numerical functions can be added. It therefore makes sense to talk about finite "sums" of k-surfaces. These are usually called k-*chains*.

If $\Psi = \Phi_1 + \cdots + \Phi_r$, each Φ_i being a k-surface, then the integral of a k-form α over the chain Ψ is defined as

(1)
$$\int_\Psi \alpha = \sum_{i=1}^{r} \int_{\Phi_i} \alpha.$$

Here is an important example of this.

Let the parameter domain D be the compact unit cube in R^k; thus $u = (u_1, \ldots, u_k) \in D$ when $0 \le u_i \le 1$ for $i = 1, \ldots, k$. Let $I: D \to R^k$ be the identity map, a k-surface. The *boundary* ∂I of I is the following $(k-1)$-chain: For $i = 1, \ldots, k$, define the "faces"

(2) $$F_{i,0}(v) = (v_1, \ldots, v_{i-1}, 0, v_i, \ldots, v_{k-1})$$

(3) $$F_{i,1}(v) = (v_1, \ldots, v_{i-1}, 1, v_i, \ldots, v_{k-1})$$

where v ranges over the unit cube in R^{k-1}; they are $(k-1)$-surfaces in R^k, and, by definition

(4) $$\partial I = \sum_{i=1}^{k} (-1)^i (F_{i,0} - F_{i,1}).$$

An elementary computation, whose basic ingredient is the fundamental theorem of calculus in one variable, shows that the preceding definitions are made in such a way that

(5) $$\int_{\partial I} \alpha = \int_I d\alpha$$

for every $(k-1)$-form α whose coefficients are in $C'(D)$. (The details of this computation are to be found on pp. 102–3 of Spivak [1].) Formula (5) is a very special case of *Stokes' theorem*.

To state the general case, let $\Phi: D \to \Omega$ be a k-surface of class C^2; here D is the k-cube, as above, and Ω is open in R^N. The (oriented) *boundary* $\partial\Phi$ of Φ is defined to be the chain

(6) $$\partial\Phi = \sum_{i=1}^{k} (-1)^i (\Phi \circ F_{i,0} - \Phi \circ F_{i,1}).$$

Briefly, we may write this as: $\partial\Phi = \Phi(\partial I)$.

Let α be a $(k-1)$-form in Ω, with C'-coefficients.

16.1.6. Stokes' Theorem. *For such Φ and α,*

(1) $$\int_{\partial\Phi} \alpha = \int_\Phi d\alpha.$$

Since $\Phi = \Phi \circ I$, this follows from the previously mentioned special case by the pull-back machinery described in §16.1.4:

(2) $$\int_{\partial\Phi} \alpha = \int_{\Phi(\partial I)} \alpha = \int_{\partial I} \alpha_\Phi.$$

and

$$(3) \qquad \int_\Phi d\alpha = \int_{\Phi \circ I} d\alpha = \int_I (d\alpha)_\Phi = \int_I d(\alpha_\Phi).$$

The last integrals in (2) and (3) are equal, by 16.1.5(5).

Of course, (1) can be extended from surfaces Φ to chains. One needs only to define

$$(4) \qquad \partial(\Phi_1 + \cdots + \Phi_r) = \partial\Phi_1 + \cdots + \partial\Phi_r.$$

We shall almost exclusively encounter the case $k = N$ of Stokes' theorem. In that case, Φ will be a region Ω (parametrized in some way that will never be specified) with oriented C^2-boundary $\partial\Omega$, and α will be an $(N - 1)$-form on $\bar\Omega$, of class C^1; Stokes' formula is then

$$(5) \qquad \int_{\partial\Omega} \alpha = \int_\Omega d\alpha.$$

This case is also known as the divergence theorem.

16.2. Differential Forms in $\mathbb{C}^n$

16.2.1. Everything said in the preceding section applies when R^N is replaced by $\mathbb{C}^n = R^{2n}$, but the complex structure gives rise to further properties of differential forms.

As real coordinates in $\mathbb{C}^n$ we take

$$(1) \qquad x_1, y_1, \ldots, x_n, y_n.$$

The operator d, applied to the functions $z_j = x_j + iy_j$ and $\bar z_j = x_j - iy_j$, gives the 1-forms

$$(2) \qquad dz_j = dx_j + idy_j, \qquad d\bar z_j = dx_j - idy_j,$$

for $j = 1, \ldots, n$. Thus

$$(3) \qquad dx_j = \frac{1}{2}(dz_j + d\bar z_j), \qquad dy_j = \frac{1}{2i}(dz_j - d\bar z_j).$$

It follows from (3) that every k-form α in $\mathbb{C}^n$ (or in an open subset of $\mathbb{C}^n$), has a (unique) representation

$$(4) \qquad \alpha = \sum_{I, J} A_{I, J}(z)dz_I \wedge d\bar z_J.$$

The sum extends over all (I, J) such that $I = (i_1, \ldots, i_p)$ and $J = (j_1, \ldots, j_q)$ are increasing p- and q-indices, respectively, with $p + q = k$, each $A_{I,J}$ is a function, and

$$(5) \qquad dz_I = dz_{i_1} \wedge \cdots \wedge dz_{i_p}, \qquad d\bar{z}_J = d\bar{z}_{j_1} \wedge \cdots \wedge d\bar{z}_{j_q}.$$

Let p and q be given. A sum (4) in which every I is a p-index and every J is a q-index is said to be a form of *bidegree* (p, q), or of *type* (p, q), or simply to be a (p, q)-*form*.

Every k-form is thus a (unique) sum of forms of bidegree $(p, k - p)$, $p = 0, \ldots, k$.

16.2.2. Differentiation. If f is a function, then, in accordance with §16.1.3,

$$(1) \qquad df = \sum_{j=1}^{n} \left(\frac{\partial f}{\partial x_j} dx_j + \frac{\partial f}{\partial y_j} dy_j \right)$$

in our present setting. Recall the differential operators D_j and $\bar{D}_j$ defined in §1.3.1. If we use these, together with 16.2.1(2), we see that (1) is the same as

$$(2) \qquad df = \sum_{j=1}^{n} \{(D_j f)dz_j + (\bar{D}_j f)d\bar{z}_j\}.$$

This suggests a split of d into a sum

$$(3) \qquad\qquad\qquad\qquad d = \partial + \bar{\partial}$$

where

$$(4) \qquad \partial f = \sum_{j=1}^{n} (D_j f)dz_j, \qquad \bar{\partial} f = \sum_{j=1}^{n} (\bar{D}_j f)d\bar{z}_j.$$

If α is given by 16.2.1(4), then

$$(5) \qquad\qquad\qquad d\alpha = \sum_{I,J} (dA_{I,J}) \wedge dz_I \wedge d\bar{z}_J$$

so that

$$(6) \qquad\qquad\qquad d\alpha = \partial\alpha + \bar{\partial}\alpha,$$

where

$$(7) \qquad\qquad\qquad \partial\alpha = \sum (\partial A_{I,J}) \wedge dz_I \wedge d\bar{z}_J$$

and

(8)
$$\bar\partial\alpha = \sum (\bar\partial A_{I,J}) \wedge dz_I \wedge d\bar z_J.$$

Note that ∂f is a $(1, 0)$-form, $\bar\partial f$ is a $(0, 1)$-form.

Thus ∂ and $\bar\partial$ carry (p, q)-forms to forms of bidegree $(p + 1, q)$ and $(p, q + 1)$, respectively.

By (3), the equation $d^2 = 0$ becomes

(9)
$$\partial^2 + (\partial\bar\partial + \bar\partial\partial) + \bar\partial^2 = 0.$$

If α is a (p, q)-form, then $\partial^2\alpha$, $(\partial\bar\partial + \bar\partial\partial)\alpha$, and $\bar\partial^2\alpha$ have bidegrees $(p + 2, q)$, $(p + 1, q + 1)$, and $(p, q + 2)$, respectively. By (9), their sum is 0; hence they are individually 0. Consequently,

(10)
$$\partial^2 = 0, \qquad \partial\bar\partial = -\bar\partial\partial, \qquad \bar\partial^2 = 0.$$

16.2.3. The $\bar\partial$-Problem. Let Ω be open in $\mathbb{C}^n$. The definition of $\bar\partial$ (see 16.2.2(4)) shows that a function $h \in C^1(\Omega)$ is holomorphic if and only if $\bar\partial h = 0$.

Thus $\bar\partial h = 0$ is the Cauchy–Riemann equation. The $\bar\partial$-problem is concerned with the solution of the *inhomogeneous* Cauchy–Riemann equation: Given a $(0, 1)$-form

(1)
$$f = \sum_{j=1}^{n} f_j \, d\bar z_j$$

the problem is to find a function u such that

(2)
$$\bar\partial u = f.$$

Since $\bar\partial^2 = 0$, a *necessary* condition for solvability is

(3)
$$\bar\partial f = 0.$$

If f is given by (1), then

(4)
$$\bar\partial f = \sum_{k<j} (\bar D_k f_j - \bar D_j f_k) d\bar z_k \wedge d\bar z_j,$$

since $d\bar z_j \wedge d\bar z_k = -d\bar z_k \wedge d\bar z_j$. Thus (3) is a brief way of saying that the components of f have to satisfy the system of differential equations

(5)
$$\bar D_k f_j = \bar D_j f_k \qquad (j, k = 1, \dots, n).$$

Of course, one is not only interested in the mere existence of solutions, but one often wants to find solutions that satisfy good growth conditions or that have good smoothness properties.

There is a more general version of the $\bar{\partial}$-problem: Given a (p, q)-form α with $\bar{\partial}\alpha = 0$, find a $(p, q - 1)$-form β such that $\bar{\partial}\beta = \alpha$. Although this is not much harder than the case $p = 0$, $q = 1$, we shall not deal with it. The $(0, 1)$-case suffices for the applications that we shall make.

16.3. The $\bar{\partial}$-Problem with Compact Support

When the data of the $\bar{\partial}$-problem have compact support, its solution is particularly easy, and it exhibits an interesting difference between $n = 1$ and $n > 1$. The solution depends on two familiar one-variable facts, whose proof we include for the sake of completeness.

16.3.1. Proposition. *Let Ω be a bounded region in $\mathbb{C}$, with smooth oriented boundary $\partial\Omega$, as in §16.1.6. If $u \in C^1(\bar{\Omega})$, then*

$$u(a) = \frac{1}{2\pi i} \int_{\partial\Omega} \frac{u(\lambda)d\lambda}{\lambda - a} - \frac{1}{2\pi i} \int_{\Omega} \frac{(\bar{D}u)(\lambda)}{\lambda - a} d\bar{\lambda} \wedge d\lambda$$

for every $a \in \Omega$.

Note that this specializes to Cauchy's formula when u is holomorphic. Also

$$d\bar{\lambda} \wedge d\lambda = 2i\, dx \wedge dy,$$

setting $\lambda = x + iy$, so that the second integral is an ordinary Lebesgue integral over Ω, since $dx \wedge dy$ is Lebesgue measure in R^2.

Proof. Let $D_\varepsilon = \{\lambda : |\lambda - a| \leq \varepsilon\}$, choose ε so small that $D_\varepsilon \subset \Omega$, put $\Omega_\varepsilon = \Omega \backslash D_\varepsilon$, and apply Stokes' theorem to the $(1, 0)$-form

$$\beta = (\lambda - a)^{-1}u(\lambda)d\lambda$$

in Ω_ε. Since $d\lambda$ occurs in β, and since $(\lambda - a)^{-1}$ is holomorphic in Ω_ε, we have

$$d\beta = \bar{\partial}\beta = (\lambda - a)^{-1}(\bar{D}u)(\lambda)d\bar{\lambda} \wedge d\lambda,$$

so that

$$\int_{\partial\Omega} \beta - \int_{\partial D_\varepsilon} \beta = \int_{\Omega_\varepsilon} \frac{(\bar{D}u)(\lambda)}{\lambda - a} d\bar{\lambda} \wedge d\lambda.$$

As $\varepsilon \to 0$, the integral over ∂D_ε converges to $2\pi i u(a)$, and the integral over Ω_ε converges to that over Ω, since the integrand is in L^1.

16.3.2. Proposition. *Let $\Omega \subset \mathbb{C}$ be a bounded open set. Suppose $f \in C^1(\Omega)$, f is bounded, and*

(1)
$$u(z) = \frac{1}{2\pi i} \int_\Omega \frac{f(\lambda)}{\lambda - z} \, d\lambda \wedge d\bar{\lambda} \qquad (z \in \Omega).$$

Then $u \in C^1(\Omega)$ and $\bar{D}u = f$.

Proof. Extend f to $\mathbb{C}$, setting $f = 0$ outside Ω. Then (1) can be written in the form

(2)
$$u(z) = \frac{1}{2\pi i} \int_\mathbb{C} f(z + \lambda) \frac{d\lambda \wedge d\bar{\lambda}}{\lambda}.$$

Since this can be differentiated under the integral sign, $u \in C^1(\Omega)$.

Fix $a \in \Omega$. Choose $\psi \in C^1(\mathbb{C})$, with support in Ω, so that $\psi \equiv 1$ in a neighborhood V of a. If f is replaced by $(1 - \psi)f$ in (1), the resulting integral is holomorphic in V. Hence f can be replaced by ψf in the computation of $(\bar{D}u)(a)$, and we obtain

$$(\bar{D}u)(a) = \frac{1}{2\pi i} \int_\mathbb{C} \bar{D}(\psi f)(a + \lambda) \frac{d\lambda \wedge d\bar{\lambda}}{\lambda}$$

$$= \frac{1}{2\pi i} \int_\mathbb{C} \frac{\bar{D}(\psi f)(\lambda)}{\lambda - a} \, d\lambda \wedge d\bar{\lambda} = (\psi f)(a).$$

The last equality follows from Proposition 16.3.1, since $\psi f \in C^1(\mathbb{C})$, with compact support.

But $(\psi f)(a) = f(a)$, so that $(\bar{D}u)(a) = f(a)$.

Note: Since $\bar{\partial}u = (\bar{D}u)d\bar{\lambda} = f \, d\bar{\lambda}$, this proposition solves the $\bar{\partial}$-problem for every given $(0, 1)$-form $f d\bar{\lambda}$, when $n = 1$. The necessary condition $\bar{\partial}(f \, d\bar{\lambda}) = 0$ (see §16.2.3) is vacuously satisfied when $n = 1$, since 0 is then the only form of bidegree $(0, 2)$.

16.3.3. Definition. If $f = \sum f_j \, d\bar{z}_j$ is a $(0, 1)$-form in $\mathbb{C}^n$, the *support* of f is the smallest closed set $K \subset \mathbb{C}^n$ such that each of the functions $f_1, \ldots, f_N$ vanishes in the complement of K. In other words, the support of f is the union of the supports of $f_1, \ldots, f_n$.

16.3.4. Theorem. *Assume $n > 1$. Let f be a $(0, 1)$-form in $\mathbb{C}^n$, with C^1-coefficients and compact support K, such that*

(1)
$$\bar{\partial}f = 0.$$

Let Ω_0 be the unbounded component of $\mathbb{C}^n \backslash K$. There exists then a unique function $u \in C^1(\mathbb{C}^n)$ that satisfies

(2) $$\bar{\partial} u = f$$

as well as $u(z) = 0$ for every $z \in \Omega_0$.

Proof. Let $f = \sum f_j(z) d\bar{z}_j$. Define

(3) $$u(z) = \frac{1}{2\pi i} \int_C f_1(\lambda, z_2, \ldots, z_n) \frac{d\lambda \wedge d\bar{\lambda}}{\lambda - z_1}$$

for $z \in \mathbb{C}^n$. This can also be written as

(4) $$u(z) = \frac{1}{2\pi i} \int_C f_1(z_1 + \lambda, z_2, \ldots, z_n) \frac{d\lambda \wedge d\bar{\lambda}}{\lambda}.$$

If follows from (4) that $u \in C^1(\mathbb{C}^n)$, and that $\bar{D}_1 u = f_1$, by Proposition 16.3.2. For $2 \le j \le n$, (1) implies that $\bar{D}_j f_1 = \bar{D}_1 f_j$, so that

$$
\begin{aligned}
(\bar{D}_j u)(z) &= \frac{1}{2\pi i} \int_C (\bar{D}_j f_1)(\lambda, z_2, \ldots, z_n) \frac{d\lambda \wedge d\bar{\lambda}}{\lambda - z_1} \\
&= \frac{1}{2\pi i} \int_C (\bar{D}_1 f_j)(\lambda, z_2, \ldots, z_n) \frac{d\lambda \wedge d\bar{\lambda}}{\lambda - z_1} \\
&= f_j(z)
\end{aligned}
$$

by Proposition 16.3.1, applied to f_j, with $z_2, \ldots, z_n$ being fixed.

Thus $\bar{D}_j u = f_j$ for $1 \le j \le n$, which is the same as (2).

In particular, u is holomorphic in Ω_0. Since (3) shows that $u(z) = 0$ when $|z_2|$ is sufficiently large, the connectedness of Ω_0 implies that $u \equiv 0$ in Ω_0.

If u_1 is any other solution, then $\bar{\partial}(u - u_1) = 0$, thus $u - u_1$ is an entire function in $\mathbb{C}^n$. If $u_1 \equiv 0$ in Ω_0, it follows that $u - u_1 \equiv 0$ in Ω_0, hence $u - u_1 \equiv 0$ in $\mathbb{C}^n$. This proves the asserted uniqueness.

16.3.5. Remark. If $u \in C^1(\mathbb{C}^n)$ has compact support, it is clear that the Lebesgue integral of $\bar{D}_j u$ over $\mathbb{C}^n$ is 0, for $1 \le j \le n$. The hypotheses made in Theorem 16.3.4 imply therefore that each coefficient f_j of f has integral 0 over $\mathbb{C}^n$.

This shows that the theorem fails when $n = 1$: Simply choose an $f \in C^1(\mathbb{C})$, with compact support, whose integral is not 0. Theorem 16.3.2 yields $u \in C^1(\mathbb{C})$ such that $\bar{D} u = f$, but no such u can have compact support.

Theorems 16.3.4 can be used to prove the following theorem of Hartogs which says, roughly speaking, that domains of holomorphy "have no holes" when $n > 1$.

16.3.6. Theorem. *Suppose* $n > 1$, Ω *is open in* $\mathbb{C}^n$, $K \subset \Omega$, K *is compact, and* $\Omega \backslash K$ *is connected. Every* $g \in H(\Omega \backslash K)$ *has then an extension that is holomorphic in all of* Ω.

Proof. There is a function $\varphi \in C^\infty(\mathbb{C}^n)$ such that $\varphi \equiv 1$ in a neighborhood V of K, and such that φ has compact support $K_0 \subset \Omega$. Define

$$f = \begin{cases} g\bar{\partial}\varphi & \text{in } \Omega \backslash K, \\ 0 & \text{on the rest of } \mathbb{C}^n. \end{cases}$$

Since $\bar{\partial}\varphi = 0$ in V and outside K_0, f is a $(0, 1)$-form in $\mathbb{C}^n$, with C^∞-coefficients, whose support lies in K_0.

Let Ω_0 be the unbounded component of $\mathbb{C}^n \backslash K_0$. Let u be the solution of $\bar{\partial}u = f$ that vanishes in Ω_0 (Theorem 16.3.4), and define

$$G = \begin{cases} u + (1 - \varphi)g & \text{in } \Omega \backslash K, \\ u & \text{in } V. \end{cases}$$

Since $\varphi \equiv 1$ in V, G is well defined, and $G \in C^\infty(\Omega)$.
In V, $\bar{\partial}G = \bar{\partial}u = f = 0$.
In $\Omega \backslash K$, $\bar{\partial}G = \bar{\partial}u - g\bar{\partial}\varphi = f - f = 0$, since $\bar{\partial}g = 0$.
Thus $G \in H(\Omega)$.
Finally, in $\Omega_0 \cap (\Omega \backslash K)$ we have $\varphi = 0$ and $u = 0$, hence $G = g$. Since $\Omega_0 \cap (\Omega \backslash K)$ is not empty, and $\Omega \backslash K$ is connected, the holomorphic functions G and g must agree in all of $\Omega \backslash K$.

16.3.7. As an example to illustrate Hartog's theorem, let $n > 1, 0 < r < 1$, put

$$\Omega = \{z \in \mathbb{C}^n : r < |z| < 1\}$$

and let $g \in H(\Omega)$. Then g extends to a holomorphic function in B.

16.4. Some Computations

To facilitate the proofs of the integral formulas in the next section, which will lead to a solution of the $\bar{\partial}$-problem in convex regions, we now introduce some special forms and collect some of their properties.

16.4.1. Definitions. We fix $n > 1$ and define the following forms in $\mathbb{C}^n$:

(1) $$\omega(z) = dz_1 \wedge \cdots \wedge dz_n,$$

(2) $$\omega_j(z) = (-1)^{j-1}dz_1 \wedge \cdots [j] \cdots \wedge dz_n$$

for $j = 1, \ldots, n$, where the symbol $[j]$ indicates that the jth term dz_j is omitted, and

$$(3) \qquad \qquad \omega'(z) = \sum_{j=1}^{n} z_j \omega_j(z).$$

These are customary notations. It may be well to point out explicitly that $\omega'(z)$ is in no way a derivative of $\omega(z)$. The bidegree of $\omega(z)$ is $(n, 0)$, that of $\omega'(z)$ is $(n - 1, 0)$.

If $s = (s_1, \ldots, s_n)$ is a C^1-map of some region into $\mathbb{C}^n$, the corresponding pull-backs will be denoted by $\omega(s)$, $\omega_j(s)$, $\omega'(s)$. For example,

$$(4) \qquad \qquad \omega_j(s) = (-1)^{j-1} ds_1 \wedge \cdots [j] \cdots \wedge ds_n,$$

$\omega_j(\bar{z})$ is obtained from (2) by replacing each z_i by $\bar{z}_i$, and

$$(5) \qquad \qquad \omega'(\bar{z}) = \sum_{j=1}^{n} \bar{z}_j \omega_j(\bar{z})$$

is a form of bidegree $(0, n - 1)$.

16.4.2. Proposition. *With these notations,*

$$(1) \qquad \qquad d\bar{z}_j \wedge \omega_j(\bar{z}) = \omega(\bar{z}) \qquad (1 \le j \le n),$$

$$(2) \qquad \qquad d\bar{z}_k \wedge \omega'(\bar{z}) = \bar{z}_k \omega(\bar{z}) \qquad (1 \le k \le n),$$

$$(3) \qquad \qquad \bar{\partial}\omega'(\bar{z}) = n\omega(\bar{z}),$$

and

$$(4) \qquad \qquad z_1^{-n}\omega'(z) = d\left(\frac{z_2}{z_1}\right) \wedge \cdots \wedge d\left(\frac{z_n}{z_1}\right)$$

in the set $\{z \in \mathbb{C}^n : z_1 \neq 0\}$.

Proof. (1) is clear from the definitions. Since

$$(5) \qquad \qquad d\bar{z}_k \wedge \omega_j(\bar{z}) = 0 \quad \text{when} \quad k \neq j,$$

(1) implies (2). Since

$$(6) \qquad \qquad \bar{\partial}[\bar{z}_j \omega_j(\bar{z})] = d\bar{z}_j \wedge \omega_j(\bar{z}),$$

(3) is another consequence of (1).

The form on the right side of (4) is the product of $n - 1$ factors

(7) $$d\left(\frac{z_k}{z_1}\right) = z_1^{-1}\, dz_k - z_1^{-2}\, z_k\, dz_1 \qquad (2 \le k \le n),$$

in their natural order. Since $dz_1 \wedge dz_1 = 0$, the right side of (4) is thus the sum of

(8) $$z_1^{-n+1}\, dz_2 \wedge \cdots \wedge dz_n = z_1^{-n} z_1 \omega_1(z)$$

plus $n - 1$ terms

(9) $$-z_1^{-n} z_k\, dz_2 \wedge \cdots \wedge dz_{k-1} \wedge dz_1 \wedge dz_{k+1} \wedge \cdots \wedge dz_n$$
$$= (-1)^{k-1} z_1^{-n} z_k\, dz_1 \wedge \cdots [k] \cdots \wedge dz_n = z_1^{-n} z_k \omega_k(z),$$

where $k = 2, \ldots, n$. Adding (8) and (9) gives (4).

Here is a consequence of (4):

16.4.3. Proposition. *Let s and t be C^1-maps of an open set $\Omega \subset \mathbb{C}^n$ into $\mathbb{C}^n \backslash \{0\}$. If there is a function $g: \Omega \to \mathbb{C}$ such that $t = gs$, then*

(1) $$\omega'(t) = g^n \omega'(s).$$

Proof. It is clear that g has no zero in Ω. In the subset of Ω where $s_1 \ne 0$ we thus also have $t_1 \ne 0$, so that

(2) $$\omega'(s) = s_1^n d\left(\frac{s_2}{s_1}\right) \wedge \cdots \wedge \left(\frac{s_n}{s_1}\right)$$

and

(3) $$\omega'(t) = t_1^n\, d\left(\frac{t_2}{t_1}\right) \wedge \cdots \wedge d\left(\frac{t_n}{t_1}\right).$$

Since $s_k/s_1 = t_k/t_1$ and $t_1 = gs_1$, (1) follows from (2) and (3) wherever $s_1 \ne 0$.

At points of Ω where $s_1 = 0$, some other component s_j of s is $\ne 0$, and the preceding argument can be used with s_j in place of s_1.

Next, we wish to relate integrals of forms over the oriented boundary ∂B of the ball B to integrals of functions over the set S, with respect to the rotation-invariant measure σ.

16.4.4. Proposition. *There is a constant c_n, namely*

$$(1) \qquad c_n = \frac{(-1)^{n(n-1)/2}(2\pi i)^n}{n!}$$

such that

$$(2) \qquad \omega(\bar{z}) \wedge \omega(z) = c_n dv,$$

where v is Lebesgue measure on $\mathbb{C}^n$, normalized so that $v(B) = 1$. If $f \in C(S)$, then

$$(3) \qquad \int_{\partial B} f(\zeta)\omega_j(\bar{\zeta}) \wedge \omega(\zeta) = nc_n \int_S f(\zeta)\bar{\zeta}_j \, d\sigma(\zeta)$$

and

$$(4) \qquad \int_{\partial B} f(\zeta)\omega'(\bar{\zeta}) \wedge \omega(\zeta) = nc_n \int_S f \, d\sigma.$$

If h is a function that is continuous in a neighborhood of the point $a \in \mathbb{C}^n$, then

$$(5) \qquad \lim_{\varepsilon \searrow 0} \varepsilon^{-2n} \int_{\partial B(a;\varepsilon)} h(\zeta)\omega'(\bar{\zeta} - \bar{a}) \wedge \omega(\zeta) = nc_n h(a)$$

where $B(a; \varepsilon)$ is the ball of radius ε with center a.

Proof. Since $d\bar{z}_j \wedge dz_j = 2i \, dx_j \wedge dy_j$, a careful count of the number of transpositions needed to convert $d\bar{z}_1 \wedge dz_1 \wedge \cdots \wedge d\bar{z}_n \wedge dz_n$ to $\omega(\bar{z}) \wedge \omega(z)$ shows that $\omega(\bar{z}) \wedge \omega(z)$ equals

$$(-1)^{n(n-1)/2}(2i)^n \, dx_1 \wedge dy_1 \wedge \cdots \wedge dx_n \wedge dy_n.$$

Since $dx_1 \wedge dy_1 \wedge \cdots \wedge dx_n \wedge dy_n = dm_{2n}$, and since (see the proof of Proposition 1.4.9) $m_{2n}(B) = \pi^n/n!$, (2) is proved.

In proving (3) we may restrict ourselves to functions $f \in C^1(\mathbb{C}^n)$. Then

$$(6) \qquad \alpha = f(\zeta)\omega_j(\bar{\zeta}) \wedge \omega(\zeta)$$

if a form of bidegree $(n, n-1)$ on $\mathbb{C}^n$. The presence of $\omega(\zeta)$ shows that $\partial\alpha = 0$. Hence

$$(7) \qquad d\alpha = \bar{\partial}\alpha = (D_j f)\omega(\bar{\zeta}) \wedge \omega(\zeta)$$

by 16.4.2(1), so that Stokes' theorem gives

(8)
$$\int_{\partial B} \alpha = c_n \int_B (\bar{D}_j f) dv,$$

by (2). Consequently, (3) amounts to

(9)
$$\int_B (\bar{D}_j f) dv = n \int_S f(\zeta) \zeta_j \, d\sigma(\zeta).$$

When $f(z) = z^\beta \bar{z}^\beta \bar{z}_j$ for some multi-index β, Proposition 1.4.9 shows that (9) holds. For other monomials, both integrals are 0. Thus (9) holds for all polynomials in z and $\bar{z}$. This proves (3).

If we apply (3) to $f(\zeta)\zeta_j$ in place of $f(\zeta)$, for $j = 1, \ldots, n$, and add, we obtain (4).

Finally, (5) follows from (3) after a change of scale.

16.4.5. Proposition. *Fix* $z \in \mathbb{C}^n$. *Let* E *be the set of all* $(\zeta, \eta) \in \mathbb{C}^n \times \mathbb{C}^n$ *at which*

$$\langle \zeta - z, \eta \rangle = 0.$$

The $(n, n - 1)$-*form*

$$\gamma = \langle \zeta - z, \eta \rangle^{-n} \omega'(\bar{\eta}) \wedge \omega(\zeta)$$

satisfies $d\gamma = 0$ *in* $\mathbb{C}^n \times \mathbb{C}^n \backslash E$.

Proof. Using self-explanatory notation, the operator d splits into

$$d = d_\zeta + d_\eta = \partial_\zeta + \bar{\partial}_\zeta + \partial_\eta + \bar{\partial}_\eta.$$

The presence of $\omega(\zeta)$ implies $\partial_\zeta \gamma = 0$.

The inner product is a holomorphic function of ζ, so that $\bar{\partial}_\zeta \gamma = 0$, and is a conjugate-holomorphic function of η, so that $\partial_\eta \gamma = 0$.

It remains to be proved that $\bar{\partial}_\eta \gamma = 0$.

If g is any smooth function of η, Proposition 16.4.2 shows that

$$\bar{\partial}[g^{-n}\omega'(\bar{\eta})] = -ng^{-n-1}(\bar{\partial}_\eta g) \wedge \omega'(\bar{\eta}) + ng^{-n}\omega(\bar{\eta})$$

$$= ng^{-n-1}\left\{ g - \sum_{j=1}^n (\bar{D}_j g) \cdot \bar{\eta}_j \right\} \omega(\bar{\eta}).$$

If $g(\eta) = \langle \zeta - z, \eta \rangle$, then $\bar{D}_j g = \zeta_j - z_j$, hence the sum is g, and consequently

$$\bar{\partial}_\eta [\langle \zeta - z, \eta \rangle^{-n} \omega'(\bar{\eta})] = 0.$$

16.5. Koppelman's Cauchy Formula

16.5.1. In the proof of Proposition 16.3.1, Cauchy's formula (in one variable) came up as a special case of Stoke's theorem. As we shall now see, similar methods can be used in several variables as well.

Let Ω be a bounded region in $\mathbb{C}^n$, with C^2-boundary. Fix a point $z \in \Omega$, and assume that $s: \overline{\Omega} \to \mathbb{C}^n$ is a C^1-map such that

    (a) $s(\zeta) = \zeta - z$ in some neighborhood of z, and
    (b) $\langle \zeta - z, s(\zeta) \rangle \neq 0$ if $\zeta \in \overline{\Omega} \setminus \{z\}$.

To every such map s corresponds a differential form $K_s = K_s(z, \zeta)$, of bidegree $(n, n-1)$, given by

$$(1) \qquad K_s(z, \zeta) = \langle \zeta - z, s(\zeta) \rangle^{-n} \omega'(\bar{s}(\zeta)) \wedge \omega(\zeta),$$

for $\zeta \in \overline{\Omega} \setminus \{z\}$.

Comparison of (1) with the form γ that was featured in Proposition 16.4.5 shows that $K_s(z, \zeta)$ is a pull-back of γ, via the map that sends $\zeta \in \overline{\Omega} \setminus \{z\}$ to $(\zeta, s(\zeta))$. Since $d\gamma = 0$, it follows that

$$(2) \qquad dK_s(z, \zeta) = 0,$$

by one of the general properties of pull-backs. Let us stress that $d = d_\zeta$ in (2), since z is fixed.

If $f \in C^1(\overline{\Omega})$, note that

$$(3) \qquad d[f(\zeta)K_s(z, \zeta)] = df \wedge K_s(z, \zeta) = \bar\partial f \wedge K_s(z, \zeta).$$

The first equation holds because of (2), the second because $\partial f \wedge \omega(\zeta) = 0$.

We now apply Stokes' theorem to fK_s, in the region Ω_ε that consists of all $\zeta \in \Omega$ for which $|\zeta - z| > \varepsilon$. For sufficiently small ε, one obtains

$$(4) \qquad \int_{\partial\Omega} fK_s - \int_{\partial B(z;\varepsilon)} fK_s = \int_{\Omega_\varepsilon} \bar\partial f \wedge K_s$$

because of (3). If ε is small, $s(\zeta) = \zeta - z$ on $\bar{B}(z; \varepsilon)$, hence

$$(5) \qquad K_s(z, \zeta) = |\zeta - z|^{-2n} \omega'(\bar\zeta - \bar z) \wedge \omega(\zeta).$$

The integral over $\partial B(z; \varepsilon)$ is then equal to

$$(6) \qquad \varepsilon^{-2n} \int_{\partial B(z;\varepsilon)} f(\zeta)\omega'(\bar\zeta - \bar z) \wedge \omega(\zeta)$$

which converges to $nc_n f(z)$, as $\varepsilon \to 0$, by Proposition 16.4.4. The integral over Ω_ε converges to that over Ω, since (5) shows that the integrand is in L^1.

We summarize the result:

16.5.2. Proposition. *If Ω is a bounded region in $\mathbb{C}^n$, with C^2-boundary, if s satisfies* (a) *and* (b) *relative to a point $z \in \Omega$, and if $f \in C^1(\overline{\Omega})$, then*

$$nc_n f(z) = \int_{\partial\Omega} f(\zeta)K_s(z, \zeta) - \int_\Omega (\bar{\partial}f) \wedge K_s(z, \zeta).$$

In particular,

$$f(z) = \frac{1}{nc_n} \int_{\partial\Omega} f(\zeta)K_s(z, \zeta)$$

if $f \in C^1(\overline{\Omega}) \cap H(\Omega)$.

16.5.3. The easiest way to satisfy conditions (a) and (b) in §16.5.1 is to put $s(\zeta) = \zeta - z$. The corresponding form K_s is then called the *Bochner–Martinelli kernel*, and is usually denoted by K_b. Explicitly.

$$K_b(z, \zeta) = |\zeta - z|^{-2n} \sum_{j=1}^{n} (\bar{\zeta}_j - \bar{z}_j)\omega_j(\bar{\zeta}) \wedge \omega(\zeta).$$

The formulas in Proposition 16.5.2, with K_b in place of K_s, are the *Bochner–Martinelli formulas*.

The kernel K_b has the advantage of being quite explicit and universal (independent of Ω). It expresses holomorphic functions in terms of their boundary values, but it has the disadvantage of not depending holomorphically on z. Thus

$$\int_{\partial\Omega} f(\zeta)K_b(z, \zeta)$$

need not be a holomorphic function in Ω, for arbitrary f. In this way, K_b acts quite differently from the Cauchy kernel.

Theorem 16.5.6 will remedy this defect; however, the kernel obtained there will not be universal. The theorem will be an easy consequence of the following result due (in even greater generality) to Koppelman [1].

16.5.4. Theorem. *Suppose Ω is a bounded region in $\mathbb{C}^n$, with C^2-boundary, $z \in \Omega$, and*

(1) $\varphi: \partial\Omega \to \mathbb{C}^n$

is a C^1-map that satisfies

(2) $$\langle \zeta - z, \varphi(\zeta) \rangle \neq 0$$

for every $\zeta \in \partial\Omega$. Then

(3) $$f(z) = \frac{1}{nc_n} \int_{\partial\Omega} \frac{f(\zeta)}{\langle \zeta - z, \varphi(\zeta) \rangle^n} \, \omega'(\bar{\varphi}(\zeta)) \wedge \omega(\zeta)$$

for every $f \in C^1(\bar{\Omega}) \cap H(\Omega)$.

The astonishing feature of this theorem is that (2) alone is enough to give (3). The available choices for φ still depend on z, however. It may not be possible to find one φ that will do for every $z \in \Omega$; this will be the point of Theorem 16.5.6.

Proof. Let V be an open set in $\mathbb{C}^n$ that contains $\partial\Omega$ and to which φ can be extended so that the extension is of class C^1 and so that (2) holds for every $\zeta \in V$. Define

(4) $$g(\zeta) = \langle \zeta - z, \varphi(\zeta) \rangle \qquad (\zeta \in V).$$

Then

(5) $$\langle \zeta - z, g(\zeta)\varphi(\zeta) \rangle = |\langle \zeta - z, \varphi(\zeta) \rangle|^2 > 0 \qquad (\zeta \in V).$$

Choose $\psi \in C^\infty(\Omega)$, $0 \leq \psi \leq 1$, with support in Ω, so that $\psi \equiv 1$ in a neighborhood of $\Omega \setminus V$, and define

(6) $$s(\zeta) = \psi(\zeta)(\zeta - z) + [1 - \psi(\zeta)]g(\zeta)\varphi(\zeta)$$

for $\zeta \in \bar{\Omega}$. (The second summand is defined to be 0 where $\psi(\zeta) = 1$.)
It is then clear that $s(\zeta) = \zeta - z$ for all ζ near z; by (5),

(7) $$\langle \zeta - z, s(\zeta) \rangle = \psi(\zeta)|\zeta - z|^2 + [1 - \psi(\zeta)]|\langle \zeta - z, \varphi(\zeta) \rangle|^2$$

which is positive for every $\zeta \in \bar{\Omega} \setminus \{z\}$.
For ζ near $\partial\Omega$, $s(\zeta) = g(\zeta)\varphi(\zeta)$. Proposition 16.5.2 implies therefore that

(8) $$f(z) = \frac{1}{nc_n} \int_{\partial\Omega} \frac{f(\zeta)\omega'(\bar{g}\bar{\varphi}) \wedge \omega(\zeta)}{\langle \zeta - z, g(\zeta)\varphi(\zeta) \rangle^n}.$$

We now use Proposition 16.4.3:

(9) $$\omega'(\bar{g}\bar{\varphi}) = \bar{g}^n \omega'(\bar{\varphi}).$$

Hence (8) reduces to (3).

15.5.5. Let us now specialize Ω to a bounded *convex* region in $\mathbb{C}^n$, with C^2-boundary and defining function ρ. This means (see §15.5.1) that $\rho \in C^2(\mathbb{C}^n)$, that Ω is the set where $\rho < 0$, and that the gradient of ρ vanishes at no point of $\partial\Omega$; i.e., the vector

(1) $$N(\zeta) = (\bar{D}_1\rho(\zeta), \ldots, \bar{D}_n\rho(\zeta)) \qquad (\zeta \in \mathbb{C}^n)$$

satisfies $N(\zeta) \neq 0$ for every $\zeta \in \partial\Omega$.

If $\zeta \in \partial\Omega$, the convexity of Ω shows that the real $(2n-1)$-dimensional hyperplane that is tangent to Ω at ζ does not intersect Ω. Thus

(2) $$(\zeta - z) \cdot N(\zeta) > 0 \qquad (z \in \Omega, \zeta \in \partial\Omega).$$

This real dot-product is the real part of the complex inner product $\langle \zeta - z, N(\zeta) \rangle$. (See §5.4.2.) We conclude that $N(\zeta)$ can play the role of $\varphi(\zeta)$ in Theorem 16.5.4, and thus obtain the following Cauchy formula:

16.5.6. Theorem. *Let Ω be a bounded convex region in $\mathbb{C}^n$, with C^2-boundary, defining function ρ, and gradient vector*

(1) $$N(\zeta) = (\bar{D}_1\rho(\zeta), \ldots, \bar{D}_n\rho(\zeta)).$$

Then, for every $f \in A(\Omega)$ and every $z \in \Omega$,

(2) $$f(z) = \frac{1}{nc_n} \int_{\partial\Omega} \frac{f(\zeta)\omega'(\bar{N}(\zeta)) \wedge \omega(\zeta)}{\langle \zeta - z, N(\zeta) \rangle^n}.$$

Note that this is stated for all $f \in A(\Omega)$, without requiring that $f \in C^1(\bar{\Omega})$. The reason is that convexity of Ω allows us to use dilates of f to approximate f uniformly on $\bar{\Omega}$ by functions that are holomorphic in neighborhoods of $\bar{\Omega}$.

In the special case $\Omega = B$, we can take $\rho(z) = |z|^2 - 1$, in which case $N(\zeta) = \zeta$, and then (2) turns into the familiar formula

$$f(z) = \int_S \frac{f(\zeta)d\sigma(\zeta)}{(1 - \langle z, \zeta \rangle)^n}$$

because of 16.4.4(4).

16.5.7. The literature devoted to integral representation formulas in several variables is quite large. To quote just a few, there are papers by Bochner [1], Gleason [2], [3], Aizenberg [1] (where Theorem 16.5.6 is proved), Koppelman [1] (a brief note, but one that was quite influential), Henkin [2], [3], and Ramirez [1] (they obtained explicit formulas in strictly pseudo-convex regions). Many other references can be found in Aizenberg [2] and in Kerzman [2].

For our present purpose, we shall use a kernel suggested by Theorem 16.5.6 to solve the $\bar{\partial}$-problem in convex regions.

16.5.8. But first let us record the explicit form that the Bochner–Martinelli formula assumes in the ball: If $u \in C^1\bar{B}$ and $z \in B$, then, setting

$$f_j = \bar{D}_j u \qquad (1 \le j \le n)$$

it follows from 16.5.3, 16.5.2, and 16.4.4 that

$$u(z) = \int_S \frac{1 - \langle \zeta, z \rangle}{|\zeta - z|^{2n}} u(\zeta) d\sigma(\zeta) - \frac{1}{n} \int_B \frac{\langle f(w), w - z \rangle}{|w - z|^{2n}} dv(w)$$

where $\langle f(w), w - z \rangle = \sum_1^n f_j(w)(\bar{w}_j - \bar{z}_j)$.

16.6. The $\bar{\partial}$-Problem in Convex Regions

16.6.1. Construction of a Kernel. Let Ω be a bounded convex open set in $\mathbb{C}^n$, with C^2-boundary, defining function ρ, and gradient vector $N(\zeta)$, as in §16.5.5, so that

$$(1) \qquad \qquad \mathrm{Re}\langle \zeta - z, N(\zeta) \rangle > 0 \qquad (z \in \Omega, \zeta \in \partial\Omega).$$

Let $\Delta = \{(z, z) : z \in \Omega\}$ be the diagonal of $\Omega \times \Omega$. Then Δ and $\Omega \times \partial\Omega$ are disjoint closed subsets of $\Omega \times \bar{\Omega}$. Hence there is an infinitely differentiable function

$$(2) \qquad \qquad \psi : \Omega \times \bar{\Omega} \to [0, 1]$$

such that

(a) $\psi \equiv 1$ in a neighborhood of Δ,
(b) $\psi \equiv 0$ in a neighborhood of $\Omega \times \partial\Omega$,
(c) the inequality (1) holds for all (z, ζ) in $\Omega \times \bar{\Omega}$ at which $\psi(z, \zeta) < 1$.

The term "neighborhood" in (a) and (b) refers to the space $\Omega \times \bar{\Omega}$. Define

$$(3) \qquad \qquad s : \Omega \times \bar{\Omega} \to \mathbb{C}^n$$

by

$$(4) \qquad \qquad s(z, \zeta) = \psi(z, \zeta)(\zeta - z) + [1 - \psi(z, \zeta)]N(\zeta).$$

Then $s(z, \zeta) = \zeta - z$ in the neighborhood of Δ in which $\psi \equiv 1$, and $\langle \zeta - z, s(z, \zeta) \rangle$ has positive real part at all points of $\Omega \times \bar{\Omega} \backslash \Delta$. For each $z \in \Omega, s(z, \cdot)$ thus satisfies the requirements imposed in §16.5.1, so that

Proposition 16.5.2 holds for the kernel

(5) $$K_s(z, \zeta) = \langle \zeta - z, s(z, \zeta) \rangle^{-n} \omega'(\bar{s}(z, \zeta)) \wedge \omega(\zeta).$$

To avoid misunderstanding, let us point out that $\omega'(\bar{s}(z, \zeta))$ is to be understood as

$$\sum_{j=1}^{n} (-1)^{j-1} \bar{s}_j(z, \zeta) d_\zeta \bar{s}_1(z, \zeta) \wedge \cdots [j] \cdots \wedge d_\zeta \bar{s}_n(z, \zeta),$$

i.e., that z is held fixed in the differentiations. The presence of $\omega(\zeta)$ shows, incidentally, that d_ζ can be replaced by $\bar{\partial}_\zeta$ in this sum, without affecting $K_s(z, \zeta)$.

Note that K_s is constructed here just as in the proof of Theorem 16.5.4, except that its dependence on z must now be taken into account.

Suppose now that $u \in C^1(\bar{\Omega})$. Since $s(z, \zeta) = N(\zeta)$ when $\zeta \in \partial\Omega$, for *every* $z \in \Omega$, Proposition 16.5.2 can be stated in the form

(6) $$u(z) = h(z) + \frac{1}{nc_n} \int_\Omega K_s(z, \zeta) \wedge (\bar{\partial}u)(\zeta) \qquad (z \in \Omega)$$

where h is the *holomorphic* function

(7) $$h(z) = \frac{1}{nc_n} \int_{\partial\Omega} \frac{u(\zeta)\omega'(\bar{N}(\zeta)) \wedge \omega(\zeta)}{\langle \zeta - z, N(\zeta) \rangle^n} \qquad (z \in \Omega).$$

The preceding formulas (6) and (7) suggest a way to solve the $\bar{\partial}$-problem: Let f be a $(0, 1)$-form in Ω that is $\bar{\partial}$-exact; by definition, this means that there exists u with $\bar{\partial}u = f$. Assume $u \in C^1(\bar{\Omega})$, so that (6) holds. Put

(8) $$u^*(z) = \frac{1}{nc_n} \int_\Omega K_s(z, \zeta) \wedge f(\zeta) \qquad (z \in \Omega).$$

Then (6) becomes $u = h + u^*$. Since $\bar{\partial}u = f$ and $\bar{\partial}h = 0$, we see that $\bar{\partial}u^* = f$. Thus u^* solves the $\bar{\partial}$-problem, if there is a solution at all.

It is therefore reasonable to conjecture that (8) furnishes a solution of the $\bar{\partial}$-problem whenever f satisfies the necessary condition $\bar{\partial}f = 0$ (plus integrability, of course). The following theorem confirms this.

16.6.2. Theorem. *Suppose that Ω is a bounded convex region in $\mathbb{C}^n$, with C^2-boundary, and that*

(i)   *K_s is as constructed in §16.6.1,*
(ii) *f is a $(0, 1)$-form with coefficients in $C^1(\bar{\Omega})$,*
(iii) *$\bar{\partial}f = 0$, and*
(iv) *$u(z) = (1/nc_n)\int_\Omega K_s(z, \zeta) \wedge f(\zeta) \qquad (z \in \Omega).$*

Then $u \in C^1(\Omega)$ and $\bar{\partial}u = f$ in Ω.

Our proof will follow Øvrelid [1]. (He used this method in strictly pseudo-convex domains.)

Proof. Away from Δ, K_s is bounded. Near Δ, K_s coincides with the Bochner–Martinelli kernel K_b. The latter is translation-invariant and locally L^1. If K_s is replaced by K_b in (iv), the integral becomes a convolution. This implies that u inherits the differentiability of f. Thus $u \in C^1(\Omega)$.

The more difficult assertion $\bar{\partial} u = f$ will be proved by Stokes' theorem. We need some preparation before we can apply it. Define

$$(1) \qquad \gamma = \langle \zeta - z, \eta \rangle \omega'(\bar{\eta}) \wedge \omega(\zeta) \wedge \omega(z),$$

a differential form in $\mathbb{C}^{3n} \backslash E$, where E is the set on which $\langle \zeta - z, \eta \rangle = 0$. Letting

$$(2) \qquad d = d_z + d_\zeta + d_\eta = \partial_z + \bar{\partial}_z + \partial_\zeta + \bar{\partial}_\zeta + \partial_\eta + \bar{\partial}_\eta$$

one verifies, exactly as in Proposition 16.4.5, that

$$(3) \qquad\qquad\qquad d\gamma = 0.$$

Next, define a form Q in $(\Omega \times \bar{\Omega}) \backslash \Delta$ by

$$(4) \qquad\qquad Q(z, \zeta) = \langle \zeta - z, s(z, \zeta) \rangle^{-n} \beta(z, \zeta) \wedge \omega(\zeta)$$

where

$$(5) \qquad\qquad \beta(z, \zeta) = \sum_{i=1}^{n} (-1)^{i-1} \bar{s}_i(z, \zeta) \bigwedge_{j \neq i} (\bar{\partial}_z + \bar{\partial}_\zeta) \bar{s}_j(z, \zeta).$$

The definition of Q is so made that, first of all, the sum of those terms of Q that are of type $(0, 0)$ in z and $(n, n-1)$ in ζ is exactly $K_s(z, \zeta)$ (see 16.4.1(5)), and secondly

$$(6) \qquad Q(z, \zeta) \wedge \omega(z) = \langle \zeta - z, s(z, \zeta) \rangle^{-n} \omega'(\bar{s}(z, \zeta)) \wedge \omega(\zeta) \wedge \omega(z),$$

a pull-back of γ (see (1)) via the map that sends (z, ζ) to $(z, \zeta, s(z, \zeta))$. Hence (3) implies

$$(7) \qquad\qquad dQ \wedge \omega(z) = d(Q \wedge \omega(z)) = 0,$$

where $d = d_z + d_\zeta$.

Let φ be a smooth $(n, n - 1)$-form with compact support in Ω. If we can show that then

$$(8) \qquad \int_\Omega (\bar{\partial} u) \wedge \varphi = \int_\Omega f \wedge \varphi,$$

the arbitrariness of φ will imply that $\bar{\partial} u = f$.

Since $d(u\varphi) = \bar{\partial} u \wedge \varphi + u \, \bar{\partial}\varphi$, and since φ has compact support in Ω, Stokes' theorem gives

$$(9) \qquad \int_\Omega (\bar{\partial} u) \wedge \varphi = - \int_\Omega u \, \bar{\partial}\varphi.$$

Substitution of (iv) into (9) shows that we have to prove

$$(10) \qquad \int_{\Omega \times \Omega} f(\zeta) \wedge K_s(z, \zeta) \wedge (\bar{\partial}\varphi)(z) = nc_n \int_\Omega f \wedge \varphi.$$

Let U be a smoothly bounded convex region that contains the support of φ, such that $\bar{U} \subset \Omega$. For $\varepsilon > 0$, let H_ε be the "hole" in $\Omega \times \Omega$ given by

$$(11) \qquad H_\varepsilon = \{(z, \zeta) \in U \times \Omega : |\zeta - z| < \varepsilon\}.$$

We restrict ε to be so small that $\bar{H}_\varepsilon \subset \Omega \times \Omega$, and

$$(12) \qquad s(z, \zeta) = \zeta - z \quad \text{on} \quad \bar{H}_\varepsilon.$$

Put $(\Omega \times \Omega)_\varepsilon = (\Omega \times \Omega) \backslash \bar{H}_\varepsilon$.

The integral over $\Omega \times \Omega$ in (10) is the limit, as $\varepsilon \to 0$, of the corresponding integrals over $(\Omega \times \Omega)_\varepsilon$. Since $(\bar{\partial}\varphi)(z)$ has bidegree (n, n), K_s can be replaced by Q in these integrals. Put

$$(13) \qquad \alpha = f(\zeta) \wedge Q(z, \zeta) \wedge \varphi(z).$$

Since φ is of type $(n, n - 1)$, $\partial\varphi = 0$ and φ contains $\omega(z)$, so that

$$(14) \qquad dQ \wedge \varphi(z) = 0,$$

by (7). Since Q contains $\omega(\zeta)$, $\partial f \wedge Q = 0$. Also $\bar{\partial} f = 0$, by hypothesis. It follows that

$$(15) \qquad d\alpha = f(\zeta) \wedge Q(z, \zeta) \wedge (\bar{\partial}\varphi)(z).$$

The left side of (10) is thus the limit, as $\varepsilon \to 0$, of the integrals of $d\alpha$ over $(\Omega \times \Omega)_\varepsilon$; by Stokes' theorem, these are equal to the integrals of α over $\partial((\Omega \times \Omega)_\varepsilon)$. This boundary consists of $\partial(\Omega \times \Omega)$ and ∂H_ε. We claim that

$$\text{(16)} \qquad \int_{\partial(\Omega \times \Omega)} \alpha = 0$$

and

$$\text{(17)} \qquad -\int_{\partial H_\varepsilon} \alpha \to nc_n \int_\Omega f \wedge \varphi \quad \text{as} \quad \varepsilon \to 0.$$

The preceding discussion shows that (16) and (17) will prove (10), hence $\bar{\partial} u = f$.

To prove (16), recall that $s(z, \zeta) = N(\zeta)$ in the neighborhood of $\Omega \times \partial\Omega$ in which $\psi \equiv 0$ (§16.6.1). Choose $h \in C^\infty(\mathbb{C}^n)$ so that $h(\zeta) = 1$ for $\zeta \in \partial\Omega$, and whose support is so small that $s(z, \zeta) = N(\zeta)$ at all points $(z, \zeta) \in \Omega \times \bar{\Omega}$ with $z \in \text{supp}(\varphi)$, $\zeta \in \text{supp}(h)$. Since $\varphi = 0$ on $\partial\Omega$,

$$\text{(18)} \qquad \alpha = h(\zeta)\alpha \quad \text{on} \quad \partial(\Omega \times \Omega).$$

On the subset of $\Omega \times \bar{\Omega}$ on which $h(\zeta)\alpha \neq 0$, $s(z, \zeta) = N(\zeta)$ (independent of z), so that Q has type $(0, 0)$ in z, $(n, n-1)$ in ζ. It follows from (13) that $h(\zeta)\alpha$ has type $(n, n-1)$ in z, (n, n) in ζ. Hence

$$\int_{\partial(\Omega \times \Omega)} \alpha = \int_{\partial(\Omega \times \Omega)} h(\zeta)\alpha = \int_{\Omega \times \Omega} d(h(\zeta)\alpha) = \int_{\Omega \times \Omega} h(\zeta)d_z\alpha.$$

To evaluate the last integral, fix ζ and integrate first over z. The result is 0, since $\varphi(z) = 0$ on $\partial\Omega$. This proves (16).

We turn to (17). Recall that $s(z, \zeta) = \zeta - z$ on $\bar{H}_\varepsilon$. Thus

$$\text{(19)} \qquad \int_{\partial H_\varepsilon} \alpha = \varepsilon^{-2n} \int_{\partial H_\varepsilon} f(\zeta) \wedge \omega'(\bar{\zeta} - \bar{z}) \wedge \omega(\zeta) \wedge \varphi(z).$$

Since $d(\omega'(\bar{\zeta} - \bar{z})) = n\omega(\bar{\zeta} - \bar{z})$, Stokes' theorem converts (19) to

$$\text{(20)} \qquad -\int_{\partial H_\varepsilon} \alpha = -\varepsilon^{-2n} \int_{H_\varepsilon} f(\zeta) \wedge \omega'(\bar{\zeta} - \bar{z}) \wedge \omega(\zeta) \wedge (\bar{\partial}\varphi)(z)$$

$$+ n\varepsilon^{-2n} \int_{H_\varepsilon} f(\zeta) \wedge \omega(\bar{\zeta} - \bar{z}) \wedge \omega(\zeta) \wedge \varphi(z).$$

The first term on the right tends to 0, as $\varepsilon \to 0$, because of the factors $\bar{\zeta}_i - \bar{z}_i$ in $\omega'(\bar{\zeta} - \bar{z})$. (Note that the volume of H_ε is $\approx \text{const. } \varepsilon^{2n}$.) To compute the limit of the second term, write

$$f = \sum_{i=1}^{n} f_i(\zeta)d\bar{\zeta}_i \tag{21}$$

and note that

$$d\bar{\zeta}_i \wedge (d\bar{\zeta}_i - d\bar{z}_i) = d\bar{z}_i \wedge (d\bar{\zeta}_i - d\bar{z}_i). \tag{22}$$

The second term in (20) is therefore equal to

$$n\varepsilon^{-2n} \int_{H_\varepsilon} \sum_i f_i(\zeta)\omega(\bar{\zeta} - \bar{z}) \wedge \omega(\zeta) \wedge d\bar{z}_i \wedge \varphi(z)$$

$$= n \int_\Omega \sum_i \left\{ \varepsilon^{-2n} \int_{B(z;\,\varepsilon)} f_i(\zeta)\omega(\bar{\zeta}) \wedge \omega(\zeta) \right\} \wedge d\bar{z}_i \wedge \varphi(z)$$

by Fubini's theorem. The replacement of $\omega(\bar{\zeta} - \bar{z})$ by $\omega(\bar{\zeta})$ was legitimate, since $d\bar{z}_i \wedge \varphi(z)$ is of type (n, n). As $\varepsilon \to 0$, the expression in braces converges to $c_n f_i(z)$. (See Proposition 16.4.4.) Because of (21), it follows that (17) holds.

The proof is complete.

16.6.3. The $\bar{\partial}$-problem can be solved in arbitrary domains of holomorphy by Hilbert space methods (Hörmander [2], Wermer [1]). Under suitable assumptions about Ω and f, Kohn [1] proved the existence of solutions $u \in C^\infty(\bar{\Omega})$. Since 1970, the emphasis has been on integral formulas, which have led to very specific information about solutions. Among the many papers devoted to this subject, we cite Henkin [3], Øvrelid [1], Romanov–Henkin [1], Krantz [1], and Greiner–Stein [1], where further references may be found.

The kernel used in Theorem 16.6.2 is not canonically associated to Ω; it depends on the choice of the function ψ that is used to "connect" $\zeta - z$ on Δ to $N(\zeta)$ on $\partial\Omega$. In the next section we return to the ball and exhibit a canonical solution. It provides Lipschitz estimates, among other things, and a slight modification of it will be used in Chapter 17, in the proof of the Henkin–Skoda theorem.

16.6.4. Example. Here is a simple $\bar{\partial}$-problem that is not globally solvable.

Take $n = 2$. Let Ω_0, Ω_1 be the regions in $\mathbb{C}^2$ defined by $w \neq 0$, $z \neq 0$, respectively. Define g_i in Ω_i by

$$g_0(z, w) = \frac{\bar{z}}{w(z\bar{z} + w\bar{w})}, \qquad g_1(z, w) = \frac{-\bar{w}}{z(z\bar{z} + w\bar{w})},$$

and define f in $\Omega = \Omega_0 \cup \Omega_1$ by

$$f = \frac{\bar{w}\, d\bar{z} - \bar{z}\, d\bar{w}}{(z\bar{z} + w\bar{w})^2}.$$

Then $\bar{\partial}g_0 = f$ in Ω_0, $\bar{\partial}g_1 = f$ in Ω_1, so that

$$\bar{\partial}f = 0 \quad \text{in} \quad \Omega.$$

Assume, to reach a contradiction, that $\bar{\partial}u = f$ for some function u in Ω. Then there exist functions $h_i \in H(\Omega_i)$ such that $u = g_i + h_i$ in Ω_i, $i = 1, 2$. Hence

$$h_1 - h_0 = g_0 - g_1 = \frac{1}{zw} \quad \text{in} \quad \Omega_0 \cap \Omega_1.$$

Setting

$$\Phi_0(z, w) = zh_0(z, w) + \frac{1}{w}, \qquad \Phi_1(z, w) = zh_1(z, w),$$

we have $\Phi_i \in H(\Omega_i)$ and $\Phi_0 = \Phi_1$ in $\Omega_1 \cap \Omega_2$. Each Φ_i extends thus to a holomorphic function in $\Omega = \mathbb{C}^2 \backslash \{(0, 0)\}$. Since isolated singularities are removable (Theorem 16.3.6), Φ_0 extends to an entire function on $\mathbb{C}^2$.

But $\Phi_0(0, w) = 1/w$ in Ω_0, hence Φ_0 has no entire extension.

16.7. An Explicit Solution in B

16.7.1. The present section features an explicitly defined linear integral operator T that associates to each (suitably well-behaved) $(0, 1)$-form f in B, with $\bar{\partial}f = 0$, a function $u = Tf$ that satisfies

(1) $\bar{\partial}Tf = f.$

Theorem 16.6.2 will be involved only insofar as it proves the *existence* of some u with $\bar{\partial}u = f$. The particular way in which this existence was proved will be irrelevant.

The following facts will play a role: If u is one solution of $\bar{\partial}u = f$ and h is a holomorphic function, then $u + h$ is also a solution. Moreover, the difference of any two solutions is holomorphic. Thus, every solution is determined in B by its boundary values on S. In fact, once we know u on S, we have the data needed to compute u in B by the Bochner–Martinelli formula, since $\bar{\partial}u = f$ is given.

To illustrate the utility of an explicit solution, a Lipschitz estimate is included in the following theorem.

16.7.2. Theorem. *Suppose that f is a $(0,1)$-form in B, with coefficients in $(C^1 \cap L^\infty)(B)$, such that $\bar{\partial} f = 0$. Define Tf on S by*

$$(1) \qquad (Tf)(\zeta) = \frac{1}{n} \int_B \frac{\langle f(w), \zeta - w \rangle dv(w)}{(1 - \langle \zeta, w \rangle)^n (1 - \langle w, \zeta \rangle)}$$

and extend the definition of Tf to B by setting

$$(2) \qquad (Tf)(z) = J_1(z) - J_2(z)$$

where

$$(3) \qquad J_1(z) = \int_S \frac{1 - \langle \zeta, z \rangle}{|\zeta - z|^{2n}} (Tf)(\zeta) d\sigma(\zeta)$$

and

$$(4) \qquad J_2(z) = \frac{1}{n} \int_B \frac{\langle f(w), w - z \rangle}{|w - z|^{2n}} dv(w).$$

Then $Tf \in C^1(B) \cap (\text{Lip } \frac{1}{2})(\bar{B})$ and

$$(5) \qquad \bar{\partial} Tf = f \quad in \quad B.$$

Proof. Let $f = \sum f_j(z) d\bar{z}_j$ and put

$$(6) \qquad \|f\|_\infty = \sum_{j=1}^n \|f_j\|_\infty$$

where $\|f_j\|_\infty$ denotes the usual norm in $L^\infty(B)$. The inner products that occur in (1) and (4) are as in §16.5.8.

We assume, to begin with, that $f_j \in C^1(\Omega)$ for $1 \le j \le n$, where Ω is some region (a ball, for instance) that contains $\bar{B}$. This extra assumption will be removed at the end of the proof.

Theorem 16.6.2 furnishes a function $u \in C^1(\Omega)$ such that $\bar{\partial} u = f$. Let $C[u]$ be the Cauchy integral (as in Chapter 3) of u, or, more precisely, of the restriction of u to S. Define

$$(7) \qquad u_f(z) = u(z) - C[u](z) \qquad (z \in B).$$

The notation u_f calls for comment: If $v \in C^1(\Omega)$ satisfies $\bar{\partial} v = f$, then $v - u$ is holomorphic in Ω, hence $C[v - u] = v - u$ in B. The right side of

(7) is thus unchanged if u is replaced by v. Thus u_f depends only on f. The definition of $C[u]$ shows that

$$(8) \qquad u_f(z) = \int_S \frac{u(z) - u(\zeta)}{(1 - \langle z, \zeta \rangle)^n} \, d\sigma(\zeta) \qquad (z \in B).$$

The relation $\bar{\partial}u = f$ will now be used to convert (8) into a formula that no longer contains the arbitrarily chosen u.

Fix $z \in B$, for the moment. For $1 \le j \le n$, define

$$(9) \qquad h_j(w) = \frac{u(z) - u(w)}{(1 - \langle z, w \rangle)^n} \cdot \frac{\bar{w}_j - \bar{z}_j}{1 - \langle w, z \rangle} \qquad (w \in \bar{B}).$$

Then

$$(10) \qquad u_f(z) = \int_S \sum_{j=1}^{n} h_j(\zeta) \zeta_j \, d\sigma(\zeta)$$

which, by 16.4.4(9), is the same as

$$(11) \qquad u_f(z) = \int_B \frac{1}{n} \sum_{j=1}^{n} (\bar{D}_j h_j)(w) dv(w).$$

Compute $\bar{D}_j h_j$ from (9), keeping in mind that $\bar{D}_j u = f_j$ since $\bar{\partial}u = f$. Insert the result into (11). Appropriate grouping of terms leads to

$$(12) \qquad u_f(z) = g(z) + \int_B R(z, w)[u(z) - u(w)] dv(w)$$

where

$$(13) \qquad g(z) = \frac{1}{n} \int_B \frac{\langle f(w), z - w \rangle dv(w)}{(1 - \langle z, w \rangle)^n (1 - \langle w, z \rangle)}$$

and

$$(14) \qquad R(z, w) = \frac{1 - |z|^2}{(1 - \langle z, w \rangle)^{n+1}(1 - \langle w, z \rangle)}.$$

The following two facts will be proved presently:

Fact I *The integral (13) exists, as a Lebesgue integral, for every $z \in \bar{B}$, and $g \in (\mathrm{Lip} \frac{1}{2})(\bar{B})$.*

Fact II. *The integral in* (12) *converges to* 0 *when* $z \in B$ *converges to any* $\zeta \in S$.

Since $u \in C^1(S)$, its Cauchy integral extends continuously to $\bar{B}$; in fact, it lies in (Lip α)($\bar{B}$) for all $\alpha < 1$. Thus $u_f \in C(\bar{B})$, and Facts I and II, combined with (12), show that $u_f(\zeta)$ is equal to the integral (1) when $\zeta \in S$. This was the reason for defining $(Tf)(\zeta)$ by (1). Since (2) is just the Bochner–Martinelli formula, we conclude that $Tf = u_f$.

The formulas (1), (3), (4) make it clear that $Tf \in C^1(B)$.

So far, modulo Facts I and II, we have thus proved that Tf, as defined by (1) *and* (2), *satisfies* $\bar{\partial} Tf = f$, *under the additional assumption that* $f \in C^1(\bar{B})$.

The Lipschitz assertion will depend on a closer study of g, J_1, and J_2.

Proof of Fact I. Since $|z - w|^2 \le 2|1 - \langle z, w \rangle|$ for $z, w \in \bar{B}$, the absolute value of the integrand in (13) is dominated by

$$(15) \qquad A\|f\|_\infty |1 - \langle z, w \rangle|^{-n-1/2}.$$

Here, and in the remainder of this proof, A is a constant that depends only on the dimension n, but which may vary from one occurrence to the other.

Proposition 1.4.10 shows now that the integral (13), with the integrand replaced by its absolutely value, is a bounded function of z on $\bar{B}$.

Since $|1 - \langle r\zeta, w \rangle|^{-1} \le 2|1 - \langle \zeta, w \rangle|^{-1}$, the dominated convergence theorem implies that

$$(16) \qquad \lim_{r \nearrow 1} g(r\zeta) = g(\zeta) \qquad (\zeta \in S).$$

Computing $D_j g$ and $\bar{D}_j g$ in B from (13) shows, by another application of 1.4.10, that

$$(17) \qquad |(\text{grad } g)(z)| \le A\|f\|_\infty (1 - |z|)^{-1/2}.$$

By (16), (17), and Lemma 6.4.8,

$$(18) \qquad |g(z) - g(w)| \le A\|f\|_\infty |z - w|^{1/2} \qquad (z, w \in \bar{B}).$$

This proves Fact I.

Proof of Fact II. By Proposition 1.4.10 (with $t = 0$, $c = 1$),

$$(19) \qquad \sup_{z \in B} \int_B |R(z, w)| \, dv(w) < \infty.$$

If $\zeta \in S$ and V is any neighborhood of ζ (in $\bar{B}$), then (14) shows that

$$(20) \qquad \lim_{z \to \zeta} R(z, w) = 0$$

uniformly for $w \in B \setminus V$.

Now, given $\zeta \in S$ and $\varepsilon > 0$, there is a neighborhood of V of ζ such that $|u(\zeta) - u(w)| < \varepsilon$ for all $w \in V$. Apply (19) to the integral over V and (20) to the integral over $B \setminus V$ to conclude that the integral in (12) is $< A\varepsilon$ when z is sufficiently close to ζ.

This proves Fact II.

We now turn to the integrals J_1 and J_2 defined by (3) and (4).

The Integral $J_1(z)$. *We claim that*

$$(21) \qquad |(\operatorname{grad} J_1)(z)| \le A\|f\|_\infty (1 - |z|)^{-1/2},$$

so that

$$(22) \qquad |J_1(z) - J_1(w)| \le A\|f\|_\infty |z - w|^{1/2} \qquad (z, w \in B).$$

Since $|1 - \langle z, \zeta \rangle| = |\langle \zeta - z, \zeta \rangle| \le |\zeta - z|$, the z-gradient of the kernel

$$(23) \qquad |\zeta - z|^{-2n}(1 - \langle \zeta, z \rangle)$$

is $\le A|\zeta - z|^{-2n}$. The integral of this kernel over S is 1, for each $z \in B$; to see this, take $u \equiv 1$ in §16.5.8. It is enough to prove (21) when $z = re_1$. Then

$$(24) \qquad (D_k J_1)(z) = \int_S D_k \left\{ \frac{1 - \langle \zeta, z \rangle}{|\zeta - z|^{2n}} \right\} [(Tf)(\zeta) - (Tf)(e_1)]d\sigma(\zeta).$$

Since $Tf = g$ on S, (18) gives

$$(25) \qquad |(Tf)(\zeta) - (Tf)(e_1)| \le A\|f\|_\infty |\zeta - e_1|^{1/2}.$$

Our estimate for the gradient of (23) shows therefore that

$$(26) \qquad |(D_k J_1)(re_1)| \le A\|f\|_\infty \int_S \frac{|\zeta - e_1|^{1/2}}{|\zeta - re_1|^{2n}} d\sigma(\zeta).$$

Since $|\zeta - e_1|^2 = 2\,\mathrm{Re}(1 - \zeta_1)$, it follows from §1.4.5 that the integral in (26) is

$$\frac{2^{1/4}(n-1)}{\pi} \int_U \frac{(1-x)^{1/4}}{(1-2rx+r^2)^n} (1 - x^2 - y^2)^{n-2}\, dx\, dy$$

$$= A \int_{-1}^{1} \frac{(1-x)^{1/4}(1-x^2)^{n-3/2}}{(1-2rx+r^2)^n}\, dx$$

$$< A(1-r)^{-1/2} \int_0^\infty \frac{t^{n-5/4}}{(1+t)^n}\, dt.$$

To obtain the last inequality, use $1 - x^2 \le 2(1 - x)$,

$$3(1 - 2rx + r^2) \ge 1 - x + (1 - r)^2,$$

and put $1 - x = (1 - r)^2 t$.

Thus $|(D_k J_1)(z)| \le A\|f\|_\infty (1 - |z|)^{-1/2}$. The same estimate holds with $\bar{D}_k$ in place of D_k. This proves (21).

The Integral $J_2(z)$. First of all, J_2 is a convolution (extend f to $\mathbb{C}^n$, 0 outside B), hence J_2 inherits the differentiability of f. Thus $J_2 \in C^1(B)$.

Next, pick $x \in B$, $y \in B$, let $\delta = |x - y|$.

Then $|J_2(x) - J_2(y)|$ is at most

$$\frac{\|f\|_\infty}{n} \sum_{j=1}^{n} \int_B \left| \frac{x_j - w_i}{|x-w|^{2n}} - \frac{y_j - w_j}{|y-w|^{2n}} \right| dv(w).$$

Let $z = \frac{1}{2}(x + y)$. When $|w - z| > 2\delta$, these integrands are $\le A\delta|w - z|^{-2n}$, so that the integrals over $|w - z| > 2\delta$ are $< A\delta \log(1/\delta)$. The integrals over $|w - z| \le 2\delta$ are $< A\delta$. The modulus of continuity of J_2 is thus $\le A\delta \log(1/\delta)$, which is much better than $A\delta^{1/2}$.

In particular, J_2 also satisfies the uniform Lipschitz condition (22).

Conclusion of Proof. All that remains to be done is to remove the additional smoothness assumption that we imposed on f.

So, let f be a $(0, 1)$-form in B, with coefficients in $(C^1 \cap L^\infty)(B)$, and with $\bar\partial f = 0$. For $0 < r < 1$, let f_r be the dilate of f, given by

$$(27) \qquad\qquad f_r = \sum_{j=1}^{n} f_j(rz)\,d\bar z_j$$

and define Tf_r by (1) to (4), with f_r in place of f. Since $\|f_r\|_\infty \le \|f\|_\infty$, the integrals (1), (3), (4) show that $\{Tf_r\}$ is uniformly bounded on $\bar B$, and since

$f_r \in C^1(\Omega_r)$ for some $\Omega_r \supset \bar{B}$, what we have proved so far shows that

(28) $$|(Tf_r(z) - (Tf_r)(w)| \leq A\|f\|_\infty |z - w|^{1/2}$$

for all $z, w \in \bar{B}, 0 < r < 1$.

As $r \nearrow 1$, (1) shows that $(Tf_r)(\zeta) \to (Tf)(\zeta)$, for $\zeta \in S$, then (3) and (4) show that the same holds for all $z \in B$.

Clearly, (28) thus holds, with f in place of f_r.

Finally, the equation $\bar{\partial} Tf = f$ holds in B if and only if

(29) $$\int_B f \wedge \varphi = -\int_B (Tf)\bar{\partial}\varphi$$

holds for every smooth $(n, n - 1)$-form φ with compact support in B. Since each f_r satisfies (29), so does f, by the dominated convergence theorem.

16.7.3. Remark. The Cauchy transform projects $L^2(\sigma)$ orthogonally onto $H^2(S)$ (Theorem 5.6.9). Formula (7) of §16.7.2 exhibits u_f therefore as the orthogonal projection of u into $H^2(S)^\perp$. Since the null-space of $\bar{\partial}$ consists precisely of the holomorphic functions, and since $Tf = u_f$, we conclude:

Tf is that solution of $\bar{\partial}u = f$ that has the smallest $L^2(\sigma)$-norm.

16.7.4. Example. If $\alpha > \frac{1}{2}$, then Lip $\frac{1}{2}$ cannot be replaced by Lip α in the conclusion of Theorem 16.7.2.

Let $n = 2$. Let $f = \bar{\partial}u_0$, where

$$u_0(z, w) = (1 - z)^{-1/2}w\bar{w}.$$

Then $f = (1 - z)^{-1/2} w \, d\bar{w}$ is a $(0, 1)$-form in B, $\bar{\partial}f = 0$ is obvious, and $\|f\|_\infty = \sqrt{2}$.

Let u be any solution of $\bar{\partial}u = f$ in B.

Suppose $0 < x_0 < 1, x = \frac{1}{2}(1 + x_0), y_0^2 = \frac{1}{2}(1 - x_0^2)$. The points $(x, y_0 e^{i\theta})$ and $(x_0, y_0 e^{i\theta})$ are then in B, for $-\pi \leq \theta \leq \pi$. Since $u - u_0$ is holomorphic in B, and $u_0(z, 0) = 0$,

$$\frac{1}{2\pi} \int_{-\pi}^{\pi} u(x, y_0 e^{i\theta})d\theta - u(x, 0) = (1 - x)^{-1/2}y_0^2.$$

The same holds with x_0 in place of x.

If $u \in$ Lip α, the difference between the two left sides is $O(|x - x_0|^\alpha)$. The difference between the two right sides is

$$[(1 - x)^{-1/2} - (1 - x_0)^{-1/2}]y_0^2$$
$$= (1 + x_0)(1 - 2^{-1/2})|x - x_0|^{1/2} > \frac{1}{4}|x - x_0|^{1/2}.$$

Letting $x_0 \to 1$, we reach a contradiction, unless $\alpha \le \frac{1}{2}$.

Note also that $u_0 \in \text{Lip } \frac{1}{2}$.

Of course, if $\bar{\partial}u = f$, then the derivatives $\bar{D}_j u = f_j$ are bounded, but the derivatives $D_j u$ need not be. The present example shows this quite explicitly.

16.7.5. The Lip $\frac{1}{2}$ assertion of Theorem 16.7.2 was first proved (in strictly pesudoconvex domains) by Romanov–Henkin [1]. An example similar to the preceding one occurs in Kerzman [1], where it is ascribed to Stein. See Krantz [1] and Greiner–Stein [1] for further results in this direction.

The idea, used in Theorem 16.7.2, to obtain the boundary values of u directly from f occurs in Skoda [1] and Henkin [7].

Charpentier [1] has found other explicit solutions of the $\bar{\partial}$-problem in B; they minimize certain weighted L^2-norms on B.

Chapter 17

The Zeros of Nevanlinna Functions

17.1. The Henkin–Skoda Theorem

17.1.1. The Blaschke Condition. In Theorem 7.3.3 we saw that the zero-variety $Z(f)$ of a function $f \in N(B)$ satisfies the Blaschke condition. The Henkin–Skoda theorem asserts the converse: if a zero-variety V in B satisfies the Blaschke condition, then $V = Z(f)$ for some $f \in N(B)$. (Actually, both Henkin and Skoda proved this in strictly pseudoconvex domains.)

These statements are not quite precise, because the Blaschke condition is not a property of the set $Z(f)$ alone; multiplicities have to be taken into account. This problem can be avoided by defining the Blaschke condition in terms of counting functions, as was done in Section 7.3. The Jensen formula will enable us to state the condition in yet another way, which will be the most convenient for our present purpose. A more geometric description, in terms of the growth of the area of $Z(f) \cap rB$ as $r \to 1$, will be given in Section 17.3.

Let us recall the definition of N_f, for a function $f \in H(B)$ with $f(0) \neq 0$: For $\zeta \in S$, $0 < r \leq 1$, $n_f(\zeta, r)$ is the number of zeros (counted with multiplicities) of the slice function f_ζ in the disc rU, and

$$(1) \qquad N_f(\zeta, r) = \int_0^r n_f(\zeta, t)\, \frac{dt}{t}.$$

Assuming $f(0) = 1$, without loss of generality, Jensen's formula shows, when $r < 1$, that

$$(2) \qquad N_f(\zeta, r) = \frac{1}{2\pi} \int_{-\pi}^{\pi} \log|f(re^{i\theta}\zeta)|\, d\theta.$$

Integrate this over S. Since both sides of (2) are nondecreasing functions of r,

$$(3) \qquad \int_S N_f(\zeta, 1)\, d\sigma(\zeta) = \sup_{0 < r < 1} \int_S \log|f_r|\, d\sigma.$$

364

In Theorem 7.3.3, the Blaschke condition

(4) $$\int_S N_f(\zeta, 1)d\sigma(\zeta) < \infty$$

played a role. This is equivalent to the finiteness of the supremum in (3). Since $\log|f|$ is subharmonic, the latter condition is in turn equivalent to

(*) $\log|f|$ *has a harmonic majorant in B.*

It follows from Theorem 5.6.2 that it is also equivalent to

(**) $\log|f|$ *has an $\mathcal{M}$-harmonic majorant in B.*

The advantage of (**) is that it is an $\mathcal{M}$-invariant condition: if f satisfies it, and if $\psi \in \text{Aut}(B)$, then $f \circ \psi$ satisfies it. The assumption $f(0) \neq 0$ is therefore irrelevant as far as (**) is concerned. Hence it is also irrelevant to

(5) $$\sup_{0<r<1} \int_S \log|f_r|d\sigma < \infty,$$

which we shall adopt as the Blaschke condition.
By definition, every $f \in N(B)$ satisfies

(6) $$\sup_{0<r<1} \int_S \log^+ |f_r|d\sigma < \infty.$$

The proof of Theorem 5.6.4 showed that (6) implies

(7) $$\sup_{0<r<1} \int_S |\log|f_r||d\sigma < \infty.$$

The Henkin–Skoda theorem can therefore be stated in the following way:

17.1.2. Theorem. *If $f \in H(B)$ and*

$$\sup_{0<r<1} \int_S \log|f_r|d\sigma < \infty$$

then there exists $g \in H(B)$, with the same zeros as f, such that

$$\sup_{0<r<1} \int_S |\log|g_r||d\sigma < \infty.$$

Since $\log|f|$ is plurisubharmonic (§7.2.1) this can be deduced from the following result:

17.1.3. Theorem. *Assume that*

(a) *u is plurisubharmonic in B,*
(b) *u is pluriharmonic in εB, for some $\varepsilon > 0$, and*
(c) *$\sup_{0 < r < 1} \int_S u_r \, d\sigma < \infty$.*

Then there is a real-valued pluriharmonic function h in B such that

$$\sup_{0 < r < 1} \int_S |(u + h)_r| \, d\sigma < \infty.$$

The proof of this will occupy Section 17.2; it will be completed in §17.2.10.

Assumption (b) can be removed (see §17.2.13) but its presence makes the proof quite a bit easier, and it does no harm as far as the application to Theorem 17.1.2 is concerned:

17.1.4. Proof that 17.1.3 implies 17.1.2. The $\mathcal{M}$-invariance of the concepts involved in Theorem 17.1.2 shows that there is no loss of generality in assuming $f(0) \neq 0$. Then $u = \log|f|$ satisfies the assumptions of Theorem 17.1.3. Choose h as in the conclusion of 17.1.3, and choose $\varphi \in H(B)$ so that $h = \text{Re } \varphi$. Put $g = f \cdot \exp \varphi$. Then g has the same zeros as f, and

$$\log|g| = \log|f| + \text{Re } \varphi = u + h.$$

Hence g satisfies the conclusion of 17.1.2.

17.2. Plurisubharmonic Functions

17.2.1. Let Ω be open in $\mathbb{C}^n$. As in §7.2.1, an upper semi-continuous function $u: \Omega \to [-\infty, \infty)$ is said to be *plurisubharmonic* if the functions

(1) $$\lambda \to u(a + \lambda b)$$

are subharmonic in neighborhoods of the origin in $\mathbb{C}$, for every $a \in \Omega$, $b \in \mathbb{C}^n$.

If, in addition, $u \in C^2(\Omega)$, then the Laplacian of (1) is nonnegative at $\lambda = 0$. Comparison with §1.3.4 shows that this amounts to having

(2) $$\sum_{j,k=1}^n (D_j \bar{D}_k u)(a) b_j \bar{b}_k \geq 0 \qquad (a \in \Omega, b \in \mathbb{C}^n).$$

In other words, the complex Hessian $H_u(a)$ is a *positive* hermitian operator on $\mathbb{C}^n$:

(3) $\langle H_u(a)b, b \rangle \geq 0$ $(a \in \Omega, b \in \mathbb{C}^n)$.

This positivity will have important consequences concerning the complex-tangential behavior of the Laplacian of u.

Throughout this section, $n > 1$.

17.2.2. Definition. Suppose $u \in C^2(B)$. If $\zeta \in S$, $0 < r < 1$, $z = r\zeta$, define the "complex-radial" Laplacian $(\Delta_{\mathrm{rad}} u)(z)$ to be the Laplacian of the function

(1) $\lambda \to u(z + \lambda\zeta)$

at $\lambda = 0$. For example,

(2) $(\Delta_{\mathrm{rad}} u)(re_1) = 4(D_1 \bar{D}_1 u)(re_1)$.

The "complex-tangential" Laplacian of u is then defined by

(3) $\Delta_{\tan} u = \Delta u - \Delta_{\mathrm{rad}} u$.

Letting $[z] = \{\lambda z : \lambda \in \mathbb{C}\}$, $(\Delta_{\tan} u)(z)$ may thus be regarded as the Laplacian, at z, of the restriction of u to the $(n-1)$-dimensional affine set $z + [z]^{\perp}$.

17.2.3. Proposition. *If $u \in C^2(\bar{B})$ then*

(1) $\int_B r^2 (\Delta_{\tan} u) dv = (n-1) \int_B (1 - r^2) \Delta u \, dv$

and

(2) $\int_S u \, d\sigma - \int_B u \, dv = \frac{1}{4n} \int_B (1 - r^2) \Delta u \, dv$.

Here $r = |z|$, of course.

Proof. The radial Laplacian commutes with the action of the unitary group. Letting $u^{\#}$ denote the radialization of u (see §4.2.1), it follows that

(3) $(\Delta_{\mathrm{rad}} u)^{\#} = \Delta_{\mathrm{rad}}(u^{\#})$,

and that the same holds for $\Delta_{\tan}$.

It is therefore enough to prove (1) and (2) for radial functions u. In that case,

(4) $$u(z) = \varphi(r), \qquad r = |z|,$$

and

(5) $$\Delta u = \varphi'' + \frac{2n-1}{r}\varphi'$$

(6) $$\Delta_{\mathrm{rad}}\, u = \varphi'' + \frac{1}{r}\varphi'$$

so that

(7) $$\Delta_{\tan} u = \frac{2(n-1)}{r}\varphi'.$$

Setting $J = \int_0^1 r^{2n}\varphi'(r)\,dr$, integration in polar coordinates gives therefore

(8) $$\int_B r^2(\Delta_{\tan} u)\,dv = 4n(n-1)J,$$

(9) $$\int_S u\, d\sigma - \int_B u\, dv = 2n\int_0^1 [\varphi(1) - \varphi(r)]r^{2n-1}\, dr = J,$$

and

(10) $$\int_B (1-r^2)\Delta u\, dv = 2n\int_0^1 (1-r^2)\frac{d}{dr}[r^{2n-1}\varphi'(r)]\,dr = 4nJ.$$

These three relations prove (1) and (2).

17.2.4. Definition. Proposition 17.2.3 suggests the introduction of the abbreviation

(1) $$M(u) = \int_B (1-r^2)\Delta u\, dv.$$

Note that $-(1-r^2)$ is a defining function for B.

17.2.5. Proposition. *Suppose $u \in C^2(\bar{B})$ is plurisubharmonic in B. Let*

(1) $$\beta : B \to B \quad and \quad \tau : B \to B$$

be continuous maps, and suppose that τ *satisfies*

(2) $\qquad \langle \tau(z), z \rangle = 0, \qquad \tau(tz) = t\tau(z) \qquad (z \in B, 0 \leq t \leq 1).$

　　Then

(3) $$\int_B (1 - r^2)\langle H_u \beta, \beta \rangle dv \leq \tfrac{1}{4} M(u),$$

(4) $$\int_B \langle H_u \tau, \tau \rangle dv \leq \frac{n-1}{4} M(u),$$

and

(5) $$\int_B (1 - r^2)^{1/2} |\langle H_u \beta, \tau \rangle| dv \leq \frac{n}{8} M(u).$$

　　Here H_u is the Hessian of u, as in §17.2.2, and $\langle H_u \beta, \beta \rangle$ is short for $\langle H_u(z)\beta(z), \beta(z) \rangle$, etc.

　　Note that $\tau(z) \perp z$. Thus, for z near S, $\tau(z)$ is "complex-tangential." The most significant aspect of this proposition is that $1 - r^2$ occurs with exponent 0 in (4) and with exponent $\tfrac{1}{2}$ in (5).

Proof. Let us write H in place of H_u. Fix $z \in B$, for the moment, and let $\{\eta_1, \ldots, \eta_n\}$ be an orthonormal basis of $\mathbb{C}^n$ such that $\beta(z) = \lambda \eta_1$, $\lambda \in \mathbb{C}$. Since $|\lambda| \leq 1$ and $H(z)$ is a positive operator,

(6) $$\langle H\beta, \beta \rangle \leq \langle H\eta_1, \eta_1 \rangle \leq \sum_{j=1}^n \langle H\eta_j, \eta_j \rangle = \tfrac{1}{4}\Delta u.$$

The last equality depends on the fact that the Laplacian Δu can be computed with respect to *any* orthonormal basis.

　　The definition of $M(u)$ in §17.2.4 shows that (3) follows from (6).

　　The same reasoning, with $\{\eta_j\}$ chosen so that $\tau(z) = |\tau(z)|\eta_1$, shows, since $|\tau(z)| \leq |z|$ and $\tau(z) \perp z$, that

(7) $$\langle H(z)\tau(z), \tau(z) \rangle \leq \tfrac{1}{4}|z|^2 (\Delta_{\tan} u)(z).$$

Thus (4) follows from Proposition 17.2.3.

　　If $x \in \mathbb{C}^n$ and $y \in \mathbb{C}^n$, the positivity of H shows that

$$0 \leq \langle H(x + e^{i\theta}y), x + e^{i\theta}y \rangle = \langle Hx, x \rangle + \langle Hy, y \rangle + 2 \operatorname{Re}[e^{-i\theta}\langle Hx, y \rangle]$$

for every real θ. Hence

(8) $$2|\langle Hx, y \rangle| \leq \langle Hx, x \rangle + \langle Hy, y \rangle.$$

With $x = (1 - r^2)^{1/2}\beta$ and $y = \tau$, (8) shows that (5) follows from (3) and (4).

17.2.6. Lemma. *Suppose* $0 < \varepsilon < 1$, $a \geq \frac{1}{2}$, $-1 \leq b \leq 1$, $p: B \to [0, \infty)$ *is measurable and vanishes in* εB, *and*

(1) $$P(z) = \int_0^1 t^b p(tz)\, dt \qquad (z \in B).$$

Then

(2) $$\int_B (1 - |z|^2)^{a-1} P(z)\, dv(z) \leq \varepsilon^{-2n} \int_B (1 - |z|^2)^a p(z)\, dv(z).$$

Proof. Insert (1) into the left side of (2), use Fubini's theorem, put $tz = w$, and use Fubini's theorem again. The left side of (2) is then seen to equal

(3) $$\int_{\varepsilon < |w| < 1} p(w)\, dv(w) \int_{|w|}^1 (1 - t^{-2}|w|^2)^{a-1} t^{b-2n}\, dt.$$

Since $t^{-1-2n} < \varepsilon^{-2n}|w|^2 t^{-3}$ if $|w| > \varepsilon$ and $t > \varepsilon$, the inner integral in (3) is less than

$$\varepsilon^{-2n} \int_{|w|}^1 (1 - t^{-2}|w|^2)^{a-1} 2a|w|^2 t^{-3}\, dt = \varepsilon^{-2n}(1 - |w|^2)^a.$$

This proves (2).

17.2.7. Theorem. *If* $u \in C^3(\bar{B})$ *is real and*

(1) $$f = \sum_{k=1}^n f_k(z)\, d\bar{z}_k$$

is the (0, 1)-*form defined on* $\bar{B}$ *by*

(2) $$f_k(z) = \sum_{j=1}^n z_j \int_0^1 t(D_j \bar{D}_k u)(tz)\, dt$$

then $\bar{\partial} f = 0$ *and* $\partial f - \bar{\partial} f = \partial \bar{\partial} u$ *in* B.
   If, in addition, u is plurisubharmonic in B and pluriharmonic in εB *for some* $\varepsilon > 0$, *then*

(3) $$\int_B |f_k|\, dv \leq n\varepsilon^{-2n} M(u),$$

and

(4)
$$\int_B (1 - r^2)^{-1/2} |\langle f, \tau \rangle| \, dv \le n\varepsilon^{-2n} M(u)$$

for all τ that satisfy the hypotheses of Proposition 17.2.5.

Note that $\tilde{f} = \sum \tilde{f}_k(z) dz_k$, by definition.

Proof. To simplify the notation, we put

(5)
$$\mu_{jk} = D_j \bar{D}_k u \qquad (j, k = 1, \dots, n).$$

Since $\bar{D}_m \mu_{jk} = \bar{D}_k \mu_{jm}$ for all j, k, m (this is why it is assumed that $u \in C^3$), (2) shows directly that $\bar{D}_m f_k = \bar{D}_k f_m$. Thus $\bar{\partial} f = 0$.
Since $\tilde{f} = \sum \tilde{f}_i \, dz_i$ we have

(6)
$$\partial f - \bar{\partial} \tilde{f} = \sum_{i,k} (D_i f_k + \bar{D}_k \tilde{f}_i) dz_i \wedge d\bar{z}_k$$

whereas

(7)
$$\partial\bar{\partial} u = \sum_{i,k} \mu_{ik} dz_i \wedge d\bar{z}_k.$$

We have to show that (6) and (7) are equal.
By (2)

(8)
$$(D_i f_k)(z) = \int_0^1 \left\{ t\mu_{ik}(tz) + \sum_j t^2 z_j (D_i \mu_{jk})(tz) \right\} dt$$

and

(9)
$$(\bar{D}_k \tilde{f}_i)(z) = \int_0^1 \left\{ t\bar{\mu}_{ki}(tz) + \sum_j t^2 \bar{z}_j (\bar{D}_k \bar{\mu}_{ji})(tz) \right\} dt.$$

Note that $\bar{\mu}_{ki} = \mu_{ik}$, $D_i \mu_{jk} = D_j \mu_{ik}$, and

(10)
$$\bar{D}_k \bar{\mu}_{ji} = \bar{D}_k \mu_{ij} = \bar{D}_j \mu_{ik}.$$

Addition of (8) and (9) gives therefore

$$(D_i f_k)(z) + (\bar{D}_k \tilde{f}_i)(z) = \int_0^1 \left\{ 2t\mu_{ik}(tz) + t^2 \sum_j z_j (D_j \mu_{ik})(tz) + \bar{z}_j (\bar{D}_j \mu_{ik})(tz) \right\} dt$$

$$= \int_0^1 \frac{\partial}{\partial t} [t^2 \mu_{ik}(tz)] dt = \mu_{ik}(z).$$

In conjunction with (6) and (7) this implies

(11) $$\partial f - \bar{\partial}\bar{f} = \partial\bar{\partial}u,$$

and the first half of the theorem is proved.

We turn to the estimates (3) and (4).

If u is plurisubharmonic, the Hessian H_u is a positive operator, so that $\mu_{jj} \geq 0$ and

(12) $$2|\mu_{jk}| \leq \mu_{jj} + \mu_{kk}.$$

(This is the case $x = e_j$, $y = e_k$, of 17.2.5(8).) If, furthermore, u is pluri-harmonic in εB, then $\mu_{jk} = 0$ in εB, for all j and k.

Hence (2), (12), Lemma 17.2.6, and Definition 17.2.4 give

$$\int_B |f_k|\,dv \leq \sum_j \int_B dv(z) \int_0^1 t|\mu_{jk}(tz)|\,dt$$

$$\leq n \int_B dv(z) \int_0^1 t(\Delta u)(tz)\,dt$$

$$\leq n\varepsilon^{-2n} \int_B (1 - r^2)(\Delta u)\,dv$$

$$= n\varepsilon^{-2n}M(u)$$

which is (3). Next,

$$\langle f, \tau\rangle(z) = \sum_k f_k(z)\bar{\tau}_k(z)$$

$$= \int_0^1 \sum_{j,k} \mu_{jk}(tz)z_j\bar{\tau}_k(z)t\,dt$$

$$= \int_0^1 t^{-1}\langle H(tz)tz, \tau(tz)\rangle\,dt.$$

If we now apply Lemma 17.2.6 with

$$p(z) = \langle H(z)z, \tau(z)\rangle,$$

$b = -1$, $a = \frac{1}{2}$, we obtain

$$\int_B (1 - r^2)^{-1/2}|\langle f, \tau\rangle|\,dv \leq \varepsilon^{-2n} \int_B (1 - |z|^2)^{1/2} \langle H(z)z, \tau(z)\rangle\,dv(z)$$

$$\leq n\varepsilon^{-2n}M(u)$$

by Proposition 17.2.5. This is (4).

The theorem just proved is the first of the two major steps that constitute the proof of Theorem 17.1.3. The second one is the construction of a solution to the $\bar\partial$-problem, with data as furnished by 17.2.7, whose norm in $L^1(\sigma)$ is under control.

17.2.8. Theorem. *Suppose* $f = \sum f_k(z)d\bar z_k$ *is a* $(0, 1)$-*form in B, such that*

    (i) $\bar\partial f = 0$ *in B, and*

    (ii) $f_k \in (C^1 \cap L^\infty)(B)$ *for* $1 \le k \le n$.

        Let $M < \infty$ *be so chosen that*

(iii)
$$\int_B | f_k | dv \le M \qquad (1 \le k \le n)$$

    and

(iv)
$$\int_B (1 - r^2)^{-1/2} |\langle f, \tau \rangle| dv \le M$$

for every map τ *that satisfies the hypotheses of Proposition 17.2.5.*

Let Tf be the solution of $\bar\partial u = f$ *given by Theorem 16.7.2, and define*

(1)
$$V(z) = (Tf)(z) + \frac{1}{n} \int_B \frac{\langle f(w), w \rangle}{(1 - \langle z, w \rangle)^n} dv(w) \qquad (z \in \bar B).$$

Then $\bar\partial V = f$ *in B,* $V \in (\text{Lip } \tfrac{1}{2})(\bar B)$, *and*

(2)
$$\int_S | V | d\sigma \le AM$$

where A is a constant that depends only on the dimension n of B.

Proof. The integral in (1) is holomorphic in B. Hence $\bar\partial V = \bar\partial Tf = f$. Proposition 1.4.10 shows that the gradient of the integral is dominated by $\log((1 - |z|)^{-1})$. The Lipschitz assertion follows from this and Theorem 16.7.2.

We turn to the proof of (2), the main point of the theorem. Recall that

(3)
$$(Tf)(\zeta) = \frac{1}{n} \int_B \frac{\langle f(w), \zeta - w \rangle dv(w)}{(1 - \langle \zeta, w \rangle)^n (1 - \langle w, \zeta \rangle)} \qquad (\zeta \in S).$$

Hence (1) can be rewritten in the form

(4)
$$V(\zeta) = I_1(\zeta) + I_2(\zeta) \qquad (\zeta \in S)$$

where

(5) $$I_1(\zeta) = \frac{1}{n} \int_B \frac{(1 - |w|^2)\langle f(w), \zeta \rangle}{(1 - \langle \zeta, w \rangle)^n (1 - \langle w, \zeta \rangle)} \, dv(w)$$

and

(6) $$I_2(\zeta) = \frac{1}{n} \int_B \frac{\langle f(w), \langle w, w \rangle \zeta - \langle \zeta, w \rangle w \rangle}{(1 - \langle \zeta, w \rangle)^n (1 - \langle w, \zeta \rangle)} \, dv(w).$$

By 1.4.10,

$$\int_S \frac{1 - |w|^2}{|1 - \langle \zeta, w \rangle|^{n+1}} \, d\sigma(\zeta)$$

is a bounded function of w in B. Hence (5) and (iii) give

(7) $$\int_S |I_1| \, d\sigma \leq A_1 M.$$

To estimate the integral of $|I_2|$, note that

$$\langle w, w \rangle \zeta - \langle \zeta, w \rangle w = \sum_{i,k} (w_i \bar{w}_i \zeta_k e_k - \zeta_k \bar{w}_k w_i e_i)$$

$$= \sum_{i,k} w_i \zeta_k (\bar{w}_i e_k - \bar{w}_k e_i)$$

$$= \sum_{i<k} c_{ik}(\zeta, w) \tau_{ik}(w)$$

where

(8) $$c_{ik}(\zeta, w) = w_i \zeta_k - w_k \zeta_i$$

and

(9) $$\tau_{ik}(w) = \bar{w}_i e_k - \bar{w}_k e_i$$

Thus

(10) $$\langle f(w), \langle w, w \rangle \zeta - \langle \zeta, w \rangle w \rangle = \sum_{i<k} \overline{c_{ik}(\zeta, w)} \langle f(w), \tau_{ik}(w) \rangle.$$

Each τ_{ik} satisfies the hypotheses of Proposition 17.2.5, and, for $\zeta \in S, w \in B$,

(11) $$|c_{ik}(\zeta, w)| \leq |\zeta - w| \leq 2|1 - \langle \zeta, w \rangle|^{1/2}.$$

Another application of 1.4.10 gives

(12) $$\int_S |1 - \langle \zeta, w \rangle|^{-n-1/2} \, d\sigma(\zeta) = O((1 - |w|^2)^{-1/2}).$$

It now follows from (10), (11), (12), (6), and our hypothesis (iv) that

(13) $$\int_S |I_2| \, d\sigma \le A_2 M.$$

Finally, (4), (7), and (13) give (2). This completes the proof.

The following smooth version of Theorem 17.1.3 is now easy:

17.2.9. Theorem. *Assume that*

(a) *is plurisubharmonic in B,*
(b) *u is pluriharmonic in εB, for some $\varepsilon > 0$,*
(c) *$u \in C^3(\bar{B})$, and $u(0) \ge 0$.*
Then there is an $h \in C(\bar{B})$, pluriharmonic in B, such that

(1) $$\int_S |u + h| \, d\sigma \le A(\varepsilon) \int_S u \, d\sigma.$$

As usual, the constant $A(\varepsilon)$ depends also on the dimension n, and may not be the same at each occurrence.

Proof. Being plurisubharmonic, u is also subharmonic in B, hence

(2) $$\int_B u \, dv \ge u(0) \ge 0$$

which gives

(3) $$M(u) \le 4n \int_S u \, d\sigma$$

by 17.2.3, 17.2.4.

Let f be the $(0, 1)$-form furnished by Theorem 17.2.7, that satisfies

(4) $$\bar{\partial} f = 0 \quad \text{and} \quad \partial f - \bar{\partial} f = \partial \bar{\partial} u.$$

The function V associated to f by Theorem 17.2.8 satisfies then $\bar{\partial} V = f$ and

(5) $$\int_S |V| \, d\sigma \le A(\varepsilon) M(u) \le A(\varepsilon) \int_S u \, d\sigma.$$

Observe now that

(6) $$\partial\bar\partial V = \partial f$$

and therefore (since $\bar\partial V = f$ implies $\partial\bar V = \bar f$)

(7) $$\partial\bar\partial\bar V = -\bar\partial\partial\bar V = -\bar\partial\bar f,$$

so that

(8) $$\partial\bar\partial(V + \bar V) = \partial f - \bar\partial\bar f = \partial\bar\partial u.$$

Put $h = V + \bar V - u$. Then $\partial\bar\partial h = 0$, by (8). But

(9) $$\partial\bar\partial h = \sum_{j,k} (D_j\bar D_k h)dz_j \wedge d\bar z_k.$$

Thus $D_j\bar D_k h = 0$ for all j, k. Consequently, h is pluriharmonic in B. Since $|u + h| \le 2|V|$, (1) follows from (5).

17.2.10. Proof of Theorem 17.1.3. We are given that u is plurisubharmonic in B, pluriharmonic in εB for some $\varepsilon > 0$, and that there is an $M < \infty$ such that

(1) $$\int_S u(r\zeta)d\sigma(\zeta) \le M \qquad (0 < r < 1).$$

We assume that $u(0) = 0$, without loss of generality.

Choose $0 < r_1 < r_2 < \cdots$, $\lim r_i = 1$. Choose δ_i so that $0 < \delta_i < (1 - r_i)\varepsilon$. Let $\chi_i \in C^\infty(\mathbb{C}^n)$ be radial, positive in $\delta_i B$, 0 outside $\delta_i B$, with $\int \chi_i \, dv = 1$, and define

(2) $$u_i(z) = \int u(z - w)\chi_i(w)dv(w) = \int u(w)\chi_i(z - w)dv(w),$$

for $|z| < 1 - \delta_i$.

The subharmonicity of u shows that $u_i \ge u$ in $(1 - \delta_i)B$, and that $u_i(z) \searrow u(z)$ as $i \to \infty$, for every $z \in B$.

The first integral in (2) shows that u_i is plurisubharmonic in $(1 - \delta_i)B$ (being a convex combination of translates of u), and that $u_i(z) = u(z)$ if $|z| < \varepsilon - \delta_i$, by the harmonicity of u in εB.

That $u_i \in C^\infty((1 - \delta_i)B)$ is clear from the second integral in (2).

Assumption (1) is equivalent to the existence of a harmonic function $H \ge u$ in B, with $H(0) = M$. Using $u \le H$ in the first integral (2), we conclude that $u_i(z) \le H(z)$ if $|z| < 1 - \delta_i$. Thus u_i satisfies (1) if $r < 1 - \delta_i$.

Note that $r_i < 1 - \delta_i$.

The functions $\tilde{u}_i$ given by $\tilde{u}_i(z) = u_i(r_i z)$ thus satisfy the hypotheses of Theorem 17.2.9. Hence there are pluriharmonic functions h_i in $r_i B$, continuous on $r_i \bar{B}$, such that, setting $\tilde{h}_i(z) = h_i(r_i z)$, we have

$$(3) \qquad \int_S |\tilde{u}_i + \tilde{h}_i| d\sigma \leq A(\varepsilon) M.$$

Put $v_i = u_i + h_i$. Then (3) becomes

$$(4) \qquad \int_S |v_i(r_i \zeta)| d\sigma(\zeta) \leq C \qquad (i = 1, 2, 3, \ldots)$$

where $C = A(\varepsilon) M$.

Now fix r, $0 < r < 1$, choose i so large that $r_i > r$. We want to obtain an analogue of (4), but with r in place of r_i.

Since $h_i \geq -|v_i| - u_i$, and h_i is harmonic,

$$(5) \qquad \int_S h_i(r\zeta) d\sigma(\zeta) = \int_S h_i(r_i \zeta) d\sigma(\zeta) \geq -C - M.$$

Since u_i is subharmonic,

$$(6) \qquad \int_S u_i(r\zeta) d\sigma(\zeta) \geq u_i(0) = u(0) = 0.$$

Since v_i is subharmonic, so is v_i^+. Thus (5) and (6) give

$$\int_S |v_i(r\zeta)| d\sigma(\zeta) = 2 \int_S v_i^+(r\zeta) d\sigma(\zeta) - \int_S v_i(r\zeta) d\sigma(\zeta)$$

$$\leq 2 \int_S v_i^+(r_i \zeta) d\sigma(\zeta) + C + M$$

$$\leq 2 \int_S |v_i(r_i \zeta)| d\sigma(\zeta) + C + M.$$

Hence, using (4),

$$(7) \qquad \int_S (|u_i(r\zeta) + h_i(r\zeta)|) d\sigma(\zeta) \leq M + 3C \quad \text{if} \quad r_i > r.$$

Still keeping r fixed, the monotone convergence theorem shows that

$$(8) \qquad \lim_{i \to \infty} \int_S |u_i(r\zeta) - u(r\zeta)| d\sigma(\zeta) = 0.$$

The integrals $\int_S |u_i(r\zeta)| d\sigma(\zeta)$ form thus a bounded sequence (this bound depends on r). By (7), the same is true of $\int_S |h_i(r\zeta)| d\sigma(\zeta)$. Since the h_i's are pluriharmonic, they are real parts of holomorphic functions, and boundedness of the above integrals implies that they form a normal family in rB.

This is true for every $r < 1$. Hence $\{h_i\}$ is a normal family in B, and some subsequence of $\{h_i\}$ converges to a pluriharmonic function h in B which, because of (7), satisfies

$$(9) \qquad \int_S |u(r\zeta) + h(r\zeta)| d\sigma(\zeta) \le M + 3C \qquad (0 < r < 1).$$

Theorem 17.1.3 is now proved.

As we saw in §17.1.4, our principal objective, namely the fact that every zero-variety in B that satisfies the Blaschke condition is the zero-variety of some $f \in N(B)$, is an easy consequence of Theorem 17.1.3.

Nevertheless, it may be of some interest to see how one can remove the assumption that u be pluriharmonic in some neighborhood of the origin. This will be sketched in §17.2.13. The following one-variable lemma will be used in the proof of Proposition 17.2.12.

17.2.11. Lemma. *If $u \in C^2(r\bar{U})$, $|z| < r < 1$, and*

$$g_r(z, \lambda) = \frac{1}{\lambda - z} - \frac{\bar{\lambda}}{r^2 - \bar{\lambda}z}$$

then

$$2\pi i(Du)(z) = \int_{\partial(rU)} \frac{u(\lambda)d\lambda}{(\lambda - z)^2} + \int_{rU} g_r(z, \lambda)(D\bar{D}u)d\lambda \wedge d\bar{\lambda}.$$

Proof. Put $\Gamma = \partial(rU)$, let γ_ε be the (positively oriented) boundary of a disc with center z, radius $\varepsilon < r - |z|$, and let Ω_ε consist of all λ such that $|\lambda| < r$ and $|\lambda - z| > \varepsilon$. Write $g(\lambda)$ in place of $g_r(z, \lambda)$. Since $D\bar{D}g = 0$ in Ω_ε, Stokes' theorem gives

$$\left(\int_\Gamma - \int_{\gamma_\varepsilon}\right)(gDu \, d\lambda + u\bar{D}g \, d\bar{\lambda}) = \int_{\Omega_\varepsilon} (\bar{D}Du)g \, d\bar{\lambda} \wedge d\lambda.$$

Note that $g(\lambda) = (\lambda - z)^{-1}(r^2 - \bar{\lambda}z)^{-1}(r^2 - |\lambda|^2)$, which is 0 when $|\lambda| = r$.

The lemma follows therefore from the three relations

$$\lim_{\varepsilon \to 0} \int_{\gamma_\varepsilon} gDu \, d\lambda = \lim_{\varepsilon \to 0} \int_{\gamma_\varepsilon} (Du)(\lambda) \, \frac{d\lambda}{\lambda - z} = 2\pi i (Du)(z),$$

$$\lim_{\varepsilon \to 0} \int_{\gamma_\varepsilon} u\bar{D}g \, d\bar{\lambda} = 0,$$

and

$$\int_\Gamma u\bar{D}g \, d\bar{\lambda} = -\int_\Gamma \frac{r^2 u(\lambda)}{(r^2 - \bar{\lambda}z)^2} \, d\bar{\lambda} = \int_\Gamma \frac{u(\lambda)}{(\lambda - z)^2} \, d\lambda.$$

The first two of these are clear. In the third, one uses $r^2 = \lambda\bar{\lambda}$ and the fact that therefore $\lambda \, d\bar{\lambda} + \bar{\lambda} \, d\lambda = 0$ on Γ.

17.2.12. Proposition. *If $u \in C^2(B)$ is plurisubharmonic and $0 < t < \frac{1}{2}$, then*

$$\int_{tB} |\operatorname{grad} u| \, dv \leq At^{2n-1} \int_S |u_{2t}| \, d\sigma + At \int_{2tB} \Delta u \, dv.$$

Here A depends only on the dimension n of B.

Proof. Write $z' = (z_1, \ldots, z_{n-1})$, $z = (z', z_n)$. For every z' with $|z'| < t$, apply Lemma 17.2.11, with $r^2 = 4t^2 - |z'|^2$, to the function $u(z', \cdot)$: if $|z_n| < r$, then

$$2\pi i (D_n u)(z) = \int_{\partial(rU)} \frac{u(z', \lambda)}{(\lambda - z_n)^2} \, d\lambda + \int_{rU} g_r(z_n, \lambda)(D_n \bar{D}_n u)(z', \lambda) d\lambda \wedge d\bar{\lambda}.$$

Note that $r^2 > 3t^2$. If $|\lambda| = r$ and $|z_n| < t$, it follows that $|\lambda - z_n| > (\sqrt{3} - 1)t$. Also

$$\int_{|z_n| < t} |g_r(z_n, \lambda)| \, dm_2(z_n) < 4\pi t.$$

Hence

$$\int_{|z_n| < t} |(D_n u)(z', z_n)| \, dm_2(z_n)$$

$$\leq At \int_{-\pi}^{\pi} |u(z', re^{i\theta})| \, d\theta + At \int_{rU} (D_n \bar{D}_n u)(z', \lambda) \, dm_2(\lambda).$$

Now integrate this over $|z'| < t$, using 1.4.7(2). Do the same with $D_1, \ldots,$ D_{n-1} in place of D_n, and add the resulting inequalities.

17.2.13. We shall now sketch how the proofs of Theorems 17.1.3 and 17.2.9 have to be changed if (b) is dropped from their hypotheses.

Begin with 17.2.9, assuming just (a) and (c). Fix a small t, say $t = \frac{1}{10}$, so that Proposition 17.2.12 can be applied. Define

$$M^*(u) = \int_S u \, d\sigma + \int_S |u_{2t}| \, d\sigma + \int_{2tB} \Delta u \, dv + \int_{tB} |u| \, dv.$$

Fix a function $\varphi \in C^\infty(\mathbb{C}^n)$, $\varphi \geq 0$, such that $\varphi \equiv 0$ in some neighborhood of the origin, and $\varphi \equiv 1$ outside tB. Compute $(D_j \bar{D}_k)(\varphi u)$ by the product rule. Since $\varphi u = u$ outside tB, it follows from 17.2.5 and 17.2.12 that

$$\int_B (1 - r^2)|(D_j \bar{D}_k)(\varphi u)| \, dv \leq A M^*(u).$$

The constant A depends on t and φ, but these are fixed.

Now replace u by φu in Theorem 17.2.7, obtaining f so that $\bar{\partial} f = 0$, $\partial f - \bar{\partial} \bar{f} = \partial \bar{\partial}(\varphi u)$, and so that $\varepsilon^{-2n} M(u)$ is replaced by $A M^*(u)$. Use this f in Theorem 17.2.8, obtaining a function V that satisfies the differential equation

$$\partial \bar{\partial}(V + \bar{V} - \varphi u) = 0$$

and the estimate $\int_S |V| \, d\sigma \leq A M^*(u)$, as in the proof of Theorem 17.2.9. If $h = V + \bar{V} - \varphi u$, then

$$\int_S |u + h| \, d\sigma \leq A M^*(u),$$

since $\varphi = 1$ on S.

This gives the improved version of Theorem 17.2.9. The derivation of Theorem 17.1.3, without (b), proceeds now as in §17.2.10, using the same smooth approximations u_i, since the quantities $M^*(u_i)$ are bounded as $i \to \infty$.

17.2.14. Example. We shall now show, following Skoda, that (iv) cannot be dropped from the assumptions in Theorem 17.2.8.

Let $n = 2$. Let h be holomorphic on $\bar{U}$. Put $f_1 = 0$, $f_2(z_1, z_2) = h(z_1)$, $f = f_2 \, d\bar{z}_2$. Then $\bar{\partial} f = 0$, and

(1) $$\int_B |f_2| \, dv = \frac{2}{\pi} \int_U (1 - |\lambda|^2)|h(\lambda)| \, dm_2(\lambda).$$

If $\bar{\partial} V = f$, then $V - \bar{z}_2 f_2$ is holomorphic, hence so is $z_2 \bar{z}_2 f_2 - z_2 V$. If we fix z_1, put $z_2 = r(z_1)e^{i\theta}$, where $r(z_1) = (1 - |z_1|^2)^{1/2}$, and integrate over θ, we obtain

$$r^2(z_1)h(z_1) = \frac{1}{2\pi} \int_{-\pi}^{\pi} r(z_1)V(z_1, r(z_1)e^{i\theta})e^{i\theta} \, d\theta.$$

Divide by $r(z_1)$, take absolute values, and integrate over z_1. This gives

(2) $$\frac{1}{\pi} \int_U (1 - |\lambda|^2)^{1/2}|h(\lambda)| \, dm_2(\lambda) \leq \int_S |V| \, d\sigma.$$

All we have to do now is to choose h so that the integral over U in (1) is small, whereas the one on the left of (2) is very large.

17.2.15. Remark. Both Henkin [7], [8], [9] and Skoda [1] solve a "tangential" $\bar{\partial}$-problem, in place of our Theorem 17.2.8; this refers to the tangential Cauchy–Riemann equation. We shall take this up in Chapter 18, but will not consider the corresponding $\bar{\partial}$-problem.

They work directly with an arbitrary plurisubharmonic function (thus avoiding the approximation argument in §17.2.10), rather than with smooth ones. This requires the use of differential forms whose coefficients are measures, rather than smooth functions. The point is that the Laplacian of any subharmonic function is a positive measure, in the sense of distribution theory. In Theorem 17.2.7, the data μ_{jk} are measures, the output f is a $(0, 1)$-form with measure coefficients, but when this f is used in Theorem 17.2.8, the resulting V is in $L^1(\sigma)$.

These papers contain a variety of other interesting results.

17.3. Areas of Zero-Varieties

17.3.1. In Section 17.1, the Blaschke condition was formulated in terms of the integrated counting function N_f, and, via Jensen's formula, in terms of the integral of $\log|f_r|$ over S. The purpose of the present section is to explain how this can also be viewed as a condition on the area (i.e., the $(2n - 2)$-dimensional volume) of $V \cap rB$, V being the zero-variety of some $f \in H(B)$.

Throughout this section, $n > 1$, $\{e_1, \ldots, e_n\}$ is the standard basis for $\mathbb{C}^n$, $\pi_1, \ldots, \pi_n$ are the linear projections that satisfy $\pi_i e_i = 0$, $\pi_i e_k = e_k$ when $i \neq k$, and

$$Y_i = \{z \in \mathbb{C}^n : z_i = 0\}$$

is the range of π_i.

If E is a polyhedron that lies in some $(n - 1)$-dimensional subspace of $\mathbb{C}^n$, its $(2n - 2)$-dimensional volume will be denoted by $A(E)$.

To motivate the definition of $A(V)$, we begin with the simplest case, letting f be a linear function ($\neq 0$). Then $Z(f)$ is the range of a linear operator that maps $\mathbb{C}^{n-1}$ into $\mathbb{C}^n$.

17.3.2. Proposition. *If* $T: \mathbb{C}^{n-1} \to \mathbb{C}^n$ *is linear, and* E *is the unit cube in* $\mathbb{C}^{n-1} = R^{2n-2}$, *then*

$$(1) \qquad\qquad A(T(E)) = \sum_{i=1}^{n} A(\pi_i T(E)).$$

Proof. The left side of (1) is obviously unchanged if T is replaced by UT, where U is any unitary operator on $\mathbb{C}^n$. We claim that the same is true of the sum on the right.

Identify T with its matrix, relative to $\{e_1, \ldots, e_n\}$, so that T has n rows, $n - 1$ columns. Let $J_i(T)$ be the determinant of the matrix obtained by deleting the ith row of T. Then (see Lemma 1.3.5),

$$(2) \qquad\qquad A(\pi_i T(E)) = |J_i(T)|^2 \qquad (1 \leq i \leq n).$$

For any $z \in \mathbb{C}^n$, let $[z \mid T]$ be the square matrix obtained from T be adjoining the column vector z on the left. Let U be unitary. Then

$$(3) \qquad\qquad U[z \mid T] = [Uz \mid UT]$$

so that $|\det[z \mid T]| = |\det[Uz \mid UT]|$. Setting $w = Uz$, this becomes

$$\left| \sum_{i=1}^{n} (-1)^i z_i J_i(T) \right|^2 = \left| \sum_{i=1}^{n} (-1)^i w_i J_i(UT) \right|^2 \leq |z|^2 \sum_{i=1}^{n} |J_i(UT)|^2$$

since $|w| = |z|$. If we maximize the left side, letting z run over the unit vectors in $\mathbb{C}^n$, we obtain

$$(4) \qquad\qquad \sum_{i=1}^{n} |J_i(T)|^2 \leq \sum_{i=1}^{n} |J_i(UT)|^2.$$

Replace T and U by UT and U^{-1} in (4), to obtain the opposite inequality. Equality holds therefore in (4). Now it follows from (2) that *the right side of* (1) *is unchanged if* T *is replaced by* UT, as claimed.

Choose U so that U maps the range of T into Y_n. For $i < n$, $\pi_i UT$ has rank $< n - 1$, hence $A(\pi_i UT(E)) = 0$. Also, $\pi_n UT = UT$. The sum on the right of (1) reduces therefore to $A(UT(E))$. This proves (1). ∎

17.3.3. If $\Omega \subset \mathbb{C}^n$ is a bounded region, f is the restriction to Ω of a linear (or affine) functional ($\neq 0$), and $V = Z(f)$, then it follows from Proposition

17.3.2 that

(1) $$A(V) = \sum_{i=1}^{n} A(\pi_i V).$$

The area of V is thus the sum of the areas of its orthogonal projections into $Y_1, \ldots, Y_n$.

Note that $A(\pi_i V)$ is the Lebesgue integral (identifying Y_i with $\mathbb{C}^{n-1}$) of the characteristic function of $\pi_i V$, and that this characteristic function counts the zeros of f in a certain way. This suggests that a reasonable definition of $A(V)$ might be obtained, for arbitrary zero-varieties V in Ω, by the following procedure.

For $w \in Y_i$, put

(2) $$\Omega_i(w) = \{\lambda \in \mathbb{C} : (w_1, \ldots, w_{i-1}, \lambda, w_{i+1}, \ldots, w_n) \in \Omega\}.$$

For $f \in H(\Omega)$, define $(\#_i f)(w)$ to be the number of zeros, counted according to their multiplicities, that the function

(3) $$\lambda \to f(w_1, \ldots, w_{i-1}, \lambda, w_{i+1}, \ldots, w_n)$$

has in $\Omega_i(w)$. Note that $(\#_i f)(w)$ may well be ∞. If $\Omega_i(w)$ is empty, then $(\#_i f)(w) = 0$.

We now define

(4) $$A(V) = \sum_{i=1}^{n} \int_{Y_i} (\#_i f) dv_{n-1}$$

to be the "area" of the zero-variety $V = V_f$ of $f \in H(\Omega)$.

Here v_{n-1} is Lebesgue measure on $\mathbb{C}^{n-1} = R^{2n-2}$, normalized so that the unit ball has measure 1.

When $\Omega = B$, Theorem 17.3.5 will show that $A(V)$ is closely related to the counting functions and the integrals of $\log|f|$ that we met in §17.1.1.

17.3.4. Lemma. *If $\varphi \in C^1$ in some neighborhood of a point $z \in \mathbb{C}^n$, then*

$$\frac{\partial}{\partial t}\bigg|_{t=1} \varphi(tz) = \sum_{i=1}^{n} \frac{\partial}{\partial t}\bigg|_{t=1} \varphi(z_1, \ldots, z_{i-1}, tz_i, z_{i+1}, \ldots, z_n).$$

Proof. It suffices to prove this for monomials $\varphi(z) = z^\alpha \bar{z}^\beta$; for these, each side is $(|\alpha| + |\beta|)\varphi(z)$.

17.3.5. Theorem. *If $f \in H(B)$, V is the zero-variety of f, $0 < r < 1$, and*

(1) $$V(r) = V \cap rB,$$

then

(2)
$$\frac{A(V(r))}{r^{2n-2}} = r\frac{\partial}{\partial r}\int_S \log|f_r|d\sigma,$$

(3)
$$\frac{A(V(r))}{r^{2n-2}} = \int_S n_f(\zeta, r)d\sigma(\zeta),$$

and

(4)
$$\int_\varepsilon^r \frac{A(V(t))}{t^{2n-1}}\,dt = \int_S \log|f_r|d\sigma - \int_S \log|f_\varepsilon|d\sigma,$$

for $0 < \varepsilon < r$.

Proof. Recall Jensen's formula in the unit disc: If $g \in H(U)$ and $n_g(r)$ is the number of zeros of g in rU $(0 < r < 1)$ then

(5)
$$n_g(r) = r\frac{d}{dr}\frac{1}{2\pi}\int_{-\pi}^\pi \log|g(re^{i\theta})|d\theta,$$

at least when g has no zero on $\partial(rU)$; if d/dr is interpreted as a left-hand derivative, then (5) holds for every $r \in (0, 1)$. Note that this form of Jensen's theorem is true regardless of whether $g(0)$ is or is not 0.

It will be advantageous to rewrite (5) as

(6)
$$n_g(r) = \frac{\partial}{\partial t}\Big|_{t=1}\frac{1}{2\pi}\int_{-\pi}^\pi \log|g(tre^{i\theta})|d\theta.$$

Assume next that $f \in H(\Omega)$, for some $\Omega \supset \bar{B}$. Apply Lemma 17.3.4 to $\log|f|$ at any point $\zeta \in S$ where $f(\zeta) \neq 0$:

(7)
$$\frac{\partial}{\partial t}\Big|_{t=1}\log|f(t\zeta)| = \sum_{i=1}^n \frac{\partial}{\partial t}\Big|_{t=1}\log|f(\zeta_1, \ldots, \zeta_{i-1}, t\zeta_i, \zeta_{i+1}, \ldots, \zeta_n)|.$$

By (6) and the definition of $\#_i$ in §17.3.3, the ith summand in (7) gives

(8)
$$(\#_i f)(\zeta_1, \ldots, \zeta_{i-1}, 0, \zeta_{i+1}, \ldots, \zeta_n).$$

(Note that we only count zeros in B. The assumption that $f \in H(\Omega)$ was made merely to avoid complications on S.) The integral of (8) over B_{n-1} is the same as its integral over S, since (8) is independent of ζ_i (§1.4.5). On the other hand, the sum of these n integrals is $A(V)$. Thus (7) implies that

$$(9) \qquad A(V) = \int_S \frac{\partial}{\partial t}\bigg|_{t=1} \log|f(t\zeta)| \, d\sigma(\zeta) = \frac{\partial}{\partial t}\bigg|_{t=1} \int_S \log|f_t| \, d\sigma.$$

Denote V by V_f, for the moment. We can apply (9) to f_r in place of f. Observe that

$$(10) \qquad A(V_f(r)) = r^{2n-2} A(V_{f_r}).$$

This follows from the definition of $A(V)$ in §17.3.3: In computing $A(V_f(r))$, one integrates over rB_{n-1} insteadof B_{n-1}.

By (9) and (10),

$$(11) \qquad A(V(r)) = r^{2n-2} \frac{\partial}{\partial t}\bigg|_{t=1} \int_S \log|f_{rt}| \, d\sigma,$$

which is equivalent to (2).

The Jensen formula (5), applied to the slice functions of f, leads from (2) to (3), and (4) is an obvious consequence of (2).

This completes the proof.

Note that ε can be replaced by 0 in (4) if $f(0) \neq 0$. In that case (4) becomes

$$(12) \qquad \int_0^r \frac{A(V(t))}{t^{2n-1}} \, dt = \int_S \log|f_r| \, d\sigma - \log|f(0)|.$$

17.3.6. Remarks. (a) Conclusion (4) of Theorem 17.3.5, with some fixed ε, shows that

$$(1) \qquad \sup_r \int_S \log|f_r| \, d\sigma < \infty$$

if and only if

$$(2) \qquad \int_0^1 A(V(t)) \, dt < \infty.$$

The zero-variety V of f thus satisfies the Blaschke condition if and only if (2) holds.

(b) Let f be a linear function in Theorem 17.3.5, say $f(z) = z_1$. The quotient on the left of (3) is then 1, for every r. If $f(\zeta) \neq 0$, then $n_f(\zeta, r) = 1$, because of the simple zero that f_ζ has at 0. If $f(\zeta) = 0$ (this happens only on a set of measure 0), then $n_f(\zeta, r) = \infty$. Thus (3) is verified for linear f.

The point of this is that, among the functions $f \in H(B)$ with $f(0) = 0$, $n_f(\zeta, r) \geq 1$, so that the ratio

$$\frac{A(V(r))}{r^{2n-2}}$$

is minimized when f is linear. Conclusion (3) of Theorem 17.3.5 shows also that this ratio is a nondecreasing function of r.

(c) If $f \in H(\Omega)$ then the Laplacian $\Delta \log |f|$ is a positive measure (in the sense of distribution theory) which is obviously concentrated on the zero-variety V of f. In fact, this measure turns out to be exactly the area measure defined in §17.3.3. We shall not go into the details of proving this, but refer to Lelong [1]. The finiteness of $M(u)$ (see Definition 17.2.4) which was crucial in Section 17.2 thus becomes

$$(3) \qquad\qquad\qquad \int_B \rho \, dA < \infty$$

where dA is area measure on V and $-\rho(z) = |z|^2 - 1$ is a defining function for B. The necessity of this condition occurs in Chee [1], [2].

It is in this form that the Blaschke condition is defined in general regions.

Chapter 18

Tangential Cauchy–Riemann Operators

18.1. Extensions from the Boundary

The theme of this section is that holomorphic functions in a region $\Omega \subset \mathbb{C}^n$ that are smooth on $\overline{\Omega}$ satisfy "tangential Cauchy–Riemann equations" on $\partial\Omega$ when $n > 1$, and that, conversely, all functions defined on a portion M of the boundary that satisfy these equations extend to be holomorphic on one or even on both sides of M, provided certain geometric conditions hold.

Throughout this chapter, $n > 1$.

18.1.1. Definitions. Let W be a region in $\mathbb{C}^n$, let ρ be a real-valued C^2-function with domain W, and put

$$(1) \qquad N(w) = (\bar{D}_1\rho(w), \dots, \bar{D}_n\rho(w)) \qquad (w \in W),$$

as in §15.5.1. Let M be the set where $\rho = 0$, and assume that

$$(2) \qquad N(\zeta) \neq 0 \qquad \text{for all} \quad \zeta \in M.$$

Let $a: W \to \mathbb{C}^n \setminus \{0\}$ be continuous, and consider the corresponding linear first-order differential operator $L \ (= L_a)$ defined in W by

$$(3) \qquad L = \sum_{j=1}^{n} \bar{a}_j(z)\bar{D}_j.$$

Obviously, $Lf = 0$ for every $f \in H(W)$; for this reason, L is called a Cauchy–Riemann operator. If L is applied to the defining function ρ, one obtains

$$(4) \qquad (L\rho)(z) = \langle N(z), a(z) \rangle \qquad (z \in W).$$

We say that L is *tangential* to M if

$$(5) \qquad (L\rho)(\zeta) = 0 \qquad (\zeta \in M),$$

or, equivalently, if $a \perp N$ at every point of M.

18.1.2. Remark. Let L be tangential to M, as above. Suppose $u_1, u_2 \in C^1(W)$, and $u_1(\zeta) = u_2(\zeta)$ for all $\zeta \in M$. Put $u = u_1 - u_2$. Then $u = 0$ on M, so that the gradient of u is a scalar multiple of the gradient of ρ at all points of M; i.e., there is a function $h: M \to \mathbb{C}$ such that

$$(1) \qquad\qquad (\bar{D}_j u)(\zeta) = h(\zeta)(\bar{D}_j \rho(\zeta)) \qquad (\zeta \in M).$$

Hence

$$(2) \qquad\qquad (Lu)(\zeta) = h(\zeta)(L\rho)(\zeta) = 0 \qquad (\zeta \in M).$$

Consequently, $Lu_1 = Lu_2$ on M.

In other words, if $f \in C^1(W)$ and $\zeta \in M$, then $(Lf)(\zeta)$ depends only on the restriction of f to M.

We may therefore regard L as acting on the space $C^1(M)$: if $f \in C^1(M)$, then Lf is independent of whatever C^1-extension we use to compute the derivatives of f.

18.1.3. Proposition. *Let $u \in C^1(W)$. Then*

$$(1) \qquad\qquad \bar{\partial} u \wedge \bar{\partial} \rho = 0 \quad on \quad M$$

if and only if $Lu = 0$ on M for every Cauchy–Riemann operator L that is tangential to M.

Proof. Note that

$$(2) \qquad\qquad \bar{\partial} u \wedge \bar{\partial} \rho = \sum_{j<k} (\bar{D}_k \rho \, \bar{D}_j u - \bar{D}_j \rho \bar{D}_k u) d\bar{z}_j \wedge d\bar{z}_k.$$

Setting

$$(3) \qquad\qquad L_{jk} = \bar{D}_k \rho \bar{D}_j - \bar{D}_j \rho \bar{D}_k \qquad (1 \le j < k \le n)$$

it follows that L_{jk} is tangential to M, and that u satisfies (1) if and only if

$$(4) \qquad\qquad (L_{jk} u)(\zeta) = 0 \qquad (j < k, \zeta \in M).$$

In terms of the notation used in §18.1.1, L_{jk} corresponds to the vector

$$(5) \qquad\qquad a = a_{jk} = (D_k \rho)e_j - (D_j \rho)e_k.$$

If $\zeta \in M$, then $(D_m \rho)(\zeta) \ne 0$ for some m. The $n - 1$ vectors a_{jm} and a_{mk} $(1 \le j < m < k \le n)$ are then linearly independent. Therefore they span the complex tangent space to M at ζ. Hence (4) implies that $Lu = 0$ on M for *every* tangential Cauchy–Riemann operator L.

18.1.4. Example. Let $W = \mathbb{C}^n$, $\rho(z) = |z|^2 - 1$, so that $M = S$. The preceding operators L_{jk} are then given by

(1)
$$L_{jk} = \zeta_k \bar{D}_j - \zeta_j \bar{D}_k.$$

If Γ is an open subset of S, and $u \in C^1(\Gamma)$, we accordingly say that "u satisfies the tangential Cauchy–Riemann equations" provided that

(2)
$$\zeta_k(\bar{D}_j u)(\zeta) = \zeta_j(\bar{D}_k u)(\zeta) \qquad (1 \le j < k \le n, \zeta \in \Gamma).$$

Note that the system (2) reduces to just one equation when $n = 2$.

Proposition 18.1.6 will show that the vanishing of $\bar{\partial} u \wedge \bar{\partial} \rho$ (see 18.1.3) is equivalent to the vanishing of certain integrals over M. The proof uses the following computation.

18.1.5. A Pull-back. In addition to the other assumptions made in §18.1.1, let us now assume that ρ is given by an equation

(1)
$$\rho(z) = x_1 - \Phi(y_1, z_2, \ldots, z_n)$$

where $\Phi \in C^2(W)$. [If we restrict our attention to a sufficiently small neighborhood of any point of M, we can achieve (1) by means of a complex-linear change of variables.]

Let M_0 be the set of all (y_1, z'), where $z' = (z_2, \ldots, z_n)$, such that

(2)
$$T(y_1, z') = (\Phi(y_1, z') + iy_1, z')$$

lies in M. Pull-backs of functions and forms from M to M_0, via T, will be indicated by using T as a subscript.

Recall the definitions of $\omega(z)$, $\omega_k(z)$ given in §16.4.1. *We claim that*

(3)
$$[\omega_k(\bar{z}) \wedge \omega(z)]_T = (\bar{D}_k \rho)_T \omega_1(\bar{z}) \wedge 2i\, dy_1 \wedge \omega_1(z)$$

for $1 \le k \le n$.

Note that $\omega_1(\bar{z}) \wedge 2i\, dy_1 \wedge \omega_1(z) = c\, dm_{2n-1}$, where c depends only on n, and m_{2n-1} is ordinary Lebesgue measure on R^{2n-1}. Hence (3) amounts to the assertion (for appropriate u) that

(4)
$$\int_M u(\zeta) \omega_k(\bar{\zeta}) \wedge \omega(\zeta) = c \int_{M_0} (u\bar{D}_k \rho)_T\, dm_{2n-1}.$$

We turn to the proof of (3).

The pull-back is induced by the equations

(5)
$$\operatorname{Re} \zeta_1 = \Phi(y_1, z'), \qquad \operatorname{Im} \zeta_1 = y_1, \qquad \zeta_j = z_j \qquad (2 \le j \le n).$$

Thus

(6) $$(d\zeta_1)_T = d\Phi_T + i\, dy_1, \quad (d\bar{\zeta}_1)_T = d\Phi_T - i\, dy_1.$$

Also, $(\bar{D}_1\rho)(\zeta) = \tfrac{1}{2}(1 - i\, \partial\Phi/\partial\eta_1)$, where $\eta_1 = \operatorname{Im}\zeta_1$, so that

(7) $$2i(\bar{D}_1\rho)_T(y_1, z') = i + \frac{\partial\Phi_T}{\partial y_1}.$$

Hence $\omega_1(\zeta) \wedge \omega(\zeta) = \omega_1(\zeta) \wedge d\zeta_1 \wedge \omega_1(\zeta)$ pulls back to

$$\omega_1(\bar{z}) \wedge (d\Phi_T + i\, dy_1) \wedge \omega_1(z)$$

$$= \omega_1(\bar{z}) \wedge \left(\frac{\partial\Phi_T}{\partial y_1} + i\right) dy_1 \wedge \omega_1(z)$$

$$= (\bar{D}_1\rho)_T(y_1, z')\omega_1(\bar{z}) \wedge 2i\, dy_1 \wedge \omega_1(z),$$

which is the case $k = 1$ of (3).

When $1 < k \le n$, then

(8) $$\omega_k(\zeta) \wedge \omega(\zeta) = (-1)^{k-1}d\bar{\zeta}_1 \wedge \cdots [k] \cdots \wedge d\bar{\zeta}_n \wedge d\zeta_1 \wedge \omega_1(\zeta).$$

By (6), $(d\bar{\zeta}_1)_T \wedge (d\zeta_1)_T = 2i\, d\Phi_T \wedge dy_1$. Hence (8) shows that the pull-back of the left side is

(9) $$(-1)^{k-1}d\Phi_T \wedge d\bar{z}_2 \wedge \cdots [k] \cdots \wedge d\bar{z}_n \wedge 2i\, dy_1 \wedge \omega_1(z),$$

which is unchanged if $d\Phi_T$ is replaced by

(10) $$(\bar{D}_k\Phi_T)d\bar{z}_k = -(\bar{D}_k\rho)_T\, d\bar{z}_k.$$

This proves (3) when $k > 1$.

18.1.6. Proposition. *Let W, ρ, M be as in §18.1.1. For $u \in C^1(M)$, the following are equivalent:*

    (i) $\bar{\partial}u \wedge \bar{\partial}\rho = 0.$

    (ii) $\bar{D}_j\rho\bar{D}_ku = \bar{D}_k\rho\bar{D}_ju$ *for all j, k.*

    (iii) $\int_M \bar{\partial}u \wedge \alpha = 0$ *for every $(n, n-2)$-form α with C^1-coefficients and compact support in W.*

    (iv) $\int_M u\, \bar{\partial}\alpha = 0$ *for every α as in (iii).*

Proof. The equivalence of (i) and (ii) was noted in proving Proposition 18.1.3.

Every α described in (iii) is a sum of forms

$$(1) \qquad\qquad a(\zeta)\tau_{jk}(\bar\zeta) \wedge \omega(\zeta) \qquad (1 \le j < k \le n)$$

where $\tau_{jk}(\bar\zeta) = d\bar\zeta_1 \wedge \cdots [j] \cdots [k] \cdots \wedge d\bar\zeta_n$, and the support of a is so small that §18.1.5 can be applied after a linear change of variables. Since

$$(2) \qquad \bar\partial u \wedge \tau_{jk}(\bar\zeta) = (-1)^{j+k}[(\bar D_j u)\omega_k(\bar\zeta) - (\bar D_k u)\omega_j(\bar\zeta)],$$

we deduce from 18.1.5(4) that

$$(3) \quad \int_M \bar\partial u \wedge \alpha = (-1)^{j+k}c \int_{M_0} \{[\bar D_j u)(\bar D_k \rho) - (\bar D_k u)(\bar D_j \rho)]a\}_T \, dm_{2n-1}$$

if α is given by (1). Hence (ii) is equivalent to (iii).

Finally, let Γ be an open subset of M that contains $M \cap \operatorname{supp} \alpha$; thus $\alpha = 0$ on $\partial\Gamma$, and Stokes' theorem implies that

$$(4) \qquad\qquad 0 = \int_{\partial\Gamma} u\alpha = \int_\Gamma (u\,\bar\partial\alpha + \bar\partial u \wedge \alpha).$$

This shows the equivalence of (iii) and (iv).

Note: Condition (iv) makes sense without imposing any differentiability assumptions on u. *Any u that satisfies (iv) is therefore said to satisfy the tangential Cauchy–Riemann equations *in the weak sense.*

18.1.7. The extension theorems that are our objective depend on the possibility of solving a certain $\bar\partial$-problem. We will do this in the following setting.

Let $W = W' \times \mathbb{C}$, where W' is open in $\mathbb{C}^{n-1}$.

Let Ω be an open subset of W.

For $z' \in \mathbb{C}^{n-1}$, let $E(z')$ be the set of all $z_n \in \mathbb{C}$ such that $(z', z_n) \in \bar\Omega$. Assume that

(i) $W\backslash\bar\Omega$ is connected,

(ii) $E(z')$ is compact for every $z' \in W'$, and

(iii) $E(z')$ is empty for all z' in some nonempty open set $W_0' \subset W'$.

18.1.8. Proposition. *Suppose $n > 1$. Let W and Ω be as above. If f is a $(0,1)$-form with coefficients in $C^1(W)$, such that $\bar\partial f = 0$ in W and $f = 0$ in $W\backslash\Omega$, then there exists $v \in C^1(W)$ such that $\bar\partial v = f$ in W and $v = 0$ in $W\backslash\Omega$.*

Proof. If $f = \Sigma f_j(z)d\bar z_j$, define

$$v(z) = \frac{1}{2\pi i} \int_{\mathbb{C}} \frac{f_n(z', \lambda)}{\lambda - z_n} \, d\lambda \wedge d\bar\lambda.$$

The equation $\bar{\partial} v = f$ is verified as in Theorem 16.3.4. Thus $v \in H(W \setminus \bar{\Omega})$. Also, $v = 0$ in $W_0' \times \mathbb{C} \subset W \setminus \bar{\Omega}$. Hence $v = 0$ in $W \setminus \Omega$.

18.1.9. Theorem. *Suppose $n > 1$. Suppose that W and Ω are as in §18.1.7, and that there is a defining function $\rho \in C^4(W)$ such that $\Omega = \{\rho < 0\}$.*
If $u \in C^4(M)$ satisfies the tangential Cauchy–Riemann equation

$$\text{(1)} \qquad\qquad \bar{\partial} u \wedge \bar{\partial} \rho = 0,$$

then there is a function $U \in C^1(W)$ which is holomorphic in Ω, such that $U = u$ on M.

Proof (Hörmander [2]). Let $u \in C^4(W)$ be an extension of the given u on M. As noted in §18.1.2, (1) is independent of which extension is chosen. By (1) and 18.1.2(1),

$$\text{(2)} \qquad\qquad \bar{\partial} u = h_1 \, \bar{\partial} \rho + \rho \beta_1$$

where $h_1 \in C^3(W)$ and β_1 is a $(0, 1)$-form with coefficients in $C^2(W)$. Putting

$$\text{(3)} \qquad\qquad \beta_2 = \beta_1 - \bar{\partial} h_1$$

we have $\rho \beta_2 = \bar{\partial}(u - \rho h_1)$, so that $\beta_2 \wedge \bar{\partial} \rho = 0$ on M. Hence

$$\text{(4)} \qquad\qquad \beta_2 = h_2 \, \bar{\partial} \rho + \rho \beta_3$$

where $h_2 \in C^2(W)$ and β_3 is a $(0, 1)$-form with coefficients in $C^1(W)$. Put

$$\text{(5)} \qquad\qquad u_0 = u - h_1 \rho - \tfrac{1}{2} h_2 \rho^2.$$

It follows that $u_0 \in C^2(W)$ and that

$$\begin{aligned}
\bar{\partial} u_0 &= (\bar{\partial} u - h_1 \, \bar{\partial} \rho) - \rho \, \bar{\partial} h_1 - \tfrac{1}{2} \bar{\partial}(h_2 \rho^2) \\
&= \rho(\beta_1 - \bar{\partial} h_1 - h_2 \, \bar{\partial} \rho) - \tfrac{1}{2} \rho^2 \, \bar{\partial} h_2 \\
&= \rho^2(\beta_3 - \tfrac{1}{2} \bar{\partial} h_2).
\end{aligned}$$

Hence, setting

$$\text{(6)} \qquad\qquad f = \begin{cases} \bar{\partial} u_0 & \text{in} \quad \Omega \\ 0 & \text{in} \quad W \setminus \Omega, \end{cases}$$

the coefficients of the $(0, 1)$-form f are in $C^1(W)$, and Proposition 18.1.8 furnishes a $v \in C^1(W)$ such that $v = 0$ in $W \setminus \Omega$ and $\bar{\partial} v = \bar{\partial} u_0$ in Ω.
The function $U = u_0 - v$ has the desired properties.

We shall now give some special cases of this result. The first is a strengthened form of Theorem 16.3.6, due to Bochner [1].

18.1.10. Theorem. *Suppose* $n > 1$, Ω *is a bounded region in* $\mathbb{C}^n$, *with* C^4-*boundary, and* $\mathbb{C}^n \backslash \bar{\Omega}$ *is connected.*

Every $u \in C^4(\partial\Omega)$ *that satisfies the tangential Cauchy–Riemann equations extends then to a function* $U \in C^1(\bar{\Omega}) \cap H(\Omega)$.

This is an immediate consequence of 18.1.9.

Weinstock [1], [2], [3] has proved analogous theorems in which $u \in C(\partial\Omega)$ is only assumed to satisfy the tangential Cauchy–Riemann equations in the weak sense (as in Proposition 18.1.6).

18.1.11. Theorem. *Define* ρ *in* $\mathbb{C}^3$ *by*

$$(1) \qquad\qquad \rho(z) = z_1 \bar{z}_1 - z_2 \bar{z}_2 + \operatorname{Re} z_3.$$

Let $M = \{z \in \mathbb{C}^3 : \rho(z) = 0\}$.

Every $u \in C^4(M)$ *that satisfies the tangential Cauchy–Riemann equations extends then to an entire function on* $\mathbb{C}^3$.

Proof. Let Ω^+ and Ω^- be the sets where $\rho > 0$ and $\rho < 0$. We claim that both of these satisfy the conditions of §18.1.7, with $\mathbb{C}^3$ in place of W.

It is clear that both Ω^+ and Ω^- are connected.

If z_1 and z_3 are fixed, then the set of all z_2 for which $\rho(z) \geq 0$ is compact, and is empty when $z_1\bar{z}_1 + \operatorname{Re} z_3 < 0$. Hence Ω^+ satisfies §18.1.7.

If z_2 and z_3 are fixed, then the set of all z_1 for which $\rho(z) \leq 0$ is compact, and is empty when $\operatorname{Re} z_3 - z_2\bar{z}_2 > 0$. Hence Ω^- satisfies §18.1.7.

The given function u has therefore a C^1-extension to $\mathbb{C}^3$ which is holomorphic in $\Omega^+ \cup \Omega^-$. Complex lines parallel to the coordinate directions intersect M in circles and straight lines. Morera's theorem, applied in each variable separately, shows therefore that our extension is entire.

18.1.12. Theorem. *Suppose* $n > 1$, $\zeta_0 \in S$, $t < 1$. *Let*

$$\Gamma = \{\zeta \in S : t < \operatorname{Re}\langle \zeta, \zeta_0 \rangle\},$$
$$\Omega = \{z \in B : t < \operatorname{Re}\langle z, \zeta_0 \rangle\}.$$

If $g \in C(\Gamma)$ *satisfies*

$$(1) \qquad\qquad \int_\Gamma g\,\bar{\partial}\alpha = 0$$

for every $(n, n - 2)$-*form* α *with coefficients in* $C^1(\mathbb{C}^n)$ *such that* $\Gamma \cap (\operatorname{supp} \alpha)$ *is compact, then* g *has a continuous extension to* $\Gamma \cup \Omega$ *which is holomorphic in* Ω.

Note: Γ is a "spherical cap" whose convex hull is $\Gamma \cup \Omega$. If $\Gamma = S$, the theorem characterizes the members of $A(S)$ as those $g \in C(S)$ that satisfy the weak Cauchy–Riemann equations (1).

Proof. Take $\zeta_0 = e_1$, without loss of generality. If

$$W = \{z \in \mathbb{C}^n : t < \operatorname{Re} z_1\},$$

we are in the situation of §18.1.7, so that Theorem 18.1.9 can be applied. Proposition 18.1.6 shows therefore that our present theorem is true if $g \in C^4(\Gamma)$.

To go from $C^4(\Gamma)$ to $C(\Gamma)$, let $\Gamma = \Gamma_t$, choose s, $t < s < 1$, and consider (1) only for those forms α whose support intersects Γ_t within Γ_s. If $g \in C(\Gamma_t)$ satisfies (1) for these α, so does $g \circ U$ for all unitary operators U that lie in some sufficiently small neighborhood N of the identity element I of the group $\mathcal{U}$. Let χ be an approximate identity on $\mathcal{U}$, $\chi \in C^\infty$, with support in N, and put

$$(2) \qquad h(\zeta) = \int_{\mathcal{U}} g(U\zeta)\chi(U)dU \qquad (\zeta \in \Gamma_s).$$

Then (1) holds with h in place of g.

Let $\zeta \to U_\zeta$ be a C^∞-map of Γ_s into $\mathcal{U}$, such that $U_\zeta e_1 = \zeta$. Then

$$h(\zeta) = \int_{\mathcal{U}} g(UU_\zeta e_1)\chi(U)dU$$

$$= \int_{\mathcal{U}} g(Ue_1)\chi(UU_\zeta^{-1})dU,$$

so that $h \in C^\infty(\Gamma_s)$.

Thus h has a holomorphic extension to Ω_s, the set of all $z \in B$ with $\operatorname{Re} z_1 > s$.

As the support of χ shrinks to I, h converges to g, uniformly on Γ_s, hence the holomorphic extensions of h converge uniformly to a holomorphic extension of g, in Ω_s. Since $s > t$ was arbitrary, the proof is complete.

18.1.13. Remark. Suppose again that W, ρ, M are as in §18.1.1. Let $\Omega^+ = \{\rho > 0\}$, $\Omega^- = \{\rho < 0\}$. If the complex Hessian of ρ (also called the *Levi form* of ρ) has a positive eigenvalue at some point $\zeta \in M$, then there are local biholomorphic changes of coordinates (as in §15.5.3) that transform the situation into the one described in §18.1.7, with $V \cap \Omega^-$ in place of Ω, where V is some neighborhood of ζ. Every smooth function u on M that satisfies the tangential Cauchy–Riemann equations thus has a holomorphic extension to $V \cap \Omega^-$. This was first proved by H. Lewy [1]; see also Theorem

2.6.13 in Hörmander [2], and R. Nirenberg [1], where tangential Cauchy–Riemann systems are studied on surfaces in $\mathbb{C}^n$ whose dimension is less than $2n - 1$.

If there is a negative eigenvalue, the same holds with Ω^+ in place on Ω^-.

If, at some $\zeta \in M$, there is a positive eigenvalue and also a negative one, then u extends to a holomorphic function in a full neighborhood V of ζ. Theorem 18.1.11 was an explicit global example of this extension phenomenon.

Finally, note that these extension theorems fail to be true when all eigenvalues of the Levi form are 0. For example, let $W = \mathbb{C}^2$, $\rho(z_1, z_2) = z_1 + \bar{z}_1 = 2x_1$, so that

$$\bar{\partial} u \wedge \bar{\partial}\rho = (\bar{D}_2 u)d\bar{z}_2 \wedge d\bar{z}_1.$$

Thus $\bar{\partial} u \wedge \bar{\partial}\rho = 0$ amounts to having $\bar{D}_2 u = 0$. Every C^∞-function u of y_1 alone thus satisfies the tangential Cauchy–Riemann equations on $M = \{x_1 = 0\}$, and it is clear that not all of these functions have holomorphic extensions to either side of M.

18.2. Unsolvable Differential Equations

18.2.1. Hans Lewy [1], [2], was the first to discover a linear partial differential operator A, namely

(1) $$A = \frac{\partial}{\partial x} - i\frac{\partial}{\partial y} + i(x - iy)\frac{\partial}{\partial t},$$

such that the equation $Au = f$ is locally unsolvable for some $f \in C^\infty(R^3)$.

Following Greiner–Kohn–Stein [1], we shall now show that the same is true of all adjoints of Cauchy–Riemann operators that are tangential to S.

18.2.2. Adjoints. Let $L = \Sigma\,\bar{a}_j(z)d\bar{z}_j$ be tangential to S. Define its adjoint L^* by

(1) $$[Lu, v] = [u, L^*v],$$

say for all $u, v \in C^1(S)$, where $[f, g] = \int_S f\bar{g}\,d\sigma$ is the standard inner product in $L^2(\sigma)$.

When the coefficients a_j are in $C^1(S)$, then L^* is a differential operator with continuous coefficients. As an example, we compute L^* if

(2) $$L = L_{12} = \zeta_1\bar{D}_2 - \zeta_2\bar{D}_1.$$

The identity 16.4.4(9)

(3) $$\int_S h(\zeta)\bar\zeta_i \, d\sigma(\zeta) = \frac{1}{n}\int_B (\bar D_i h)dv$$

shows that

$$\int_S (\zeta_1\bar D_2 f - \zeta_2\bar D_1 f)\bar g \, d\sigma = \frac{1}{n}\int_B \{(\bar D_1 \bar g)(\bar D_2 f) - (\bar D_2 \bar g)(\bar D_1 f)\}dv$$

$$= -\int_S (\zeta_1\bar D_2\bar g - \zeta_2\bar D_1\bar g)f \, d\sigma.$$

Thus

(4) $$(L_{12})^* = -\bar\zeta_1 D_2 + \bar\zeta_2 D_1.$$

18.2.3. Theorem. *Suppose that*

   (i)   *L is a Cauchy–Riemann operator that is tangential to S,*
   (ii) *Γ is an open subset of S, $u \in C^1(\Gamma)$,*
   (iii) *$f \in C(S)$ and*

(1) $$L^*u = f \quad in \quad \Gamma.$$

The Cauchy integral $C[f]$ extends then holomorphically across Γ.

More precisely, there is a region $\Omega \supset B \cup \Gamma$ and a function $h \in H(\Omega)$ such that $h = C[f]$ in B.

For example, (1) is not solvable if f is the boundary function of some $F \in H(B)$ such that F has a singularity at some point of Γ; this can happen even if $f \in C^\infty(S)$.

Proof. Let $K \subset \Gamma$ be compact, with nonempty interior (relative to S). Choose $\psi \in C^\infty(S)$ with support in Γ, such that $\psi = 1$ on K.

For $z \in B$, recall that the Cauchy kernel is

(2) $$C(z, \zeta) = C_z(\zeta) = (1 - \langle z, \zeta \rangle)^{-n},$$

that $C[f](z) = [f, \bar C_z]$, and that $\bar C_z$ is a holomorphic function of ζ on $\bar B$. If ψu is defined to be 0 on $S\backslash\Gamma$, it follows that

(3) $$C[L^*(\psi u)](z) = [L^*(\psi u), \bar C_z] = [\psi u, L\bar C_z] = 0,$$

since L annihilates boundary values of holomorphic functions. Hence

(4) $$C[f] = C[f - L^*(\psi u)]$$

in B. Put $g = f - L^*(\psi u)$.

On $K, g = f - L^*u = 0$, by (1). Hence $C[g]$ extends holomorphically across the interior of K. By (4), the same is true of $C[f]$. Varying K, we obtain the desired extension across all of Γ.

18.2.4. Remark. Let $n = 2$, and put

$$(1) \qquad \qquad \rho(z, w) = z\bar{z} - \frac{w - \bar{w}}{2i} = z\bar{z} - t$$

where $w = s + it$. Then $M = \{\rho = 0\}$ is the Heisenberg group, the boundary of the Cayley transform of B. (See Section 2.3). The operator

$$(2) \qquad \qquad L = 2iz \frac{\partial}{\partial \bar{w}} - \frac{\partial}{\partial \bar{z}}$$

is tangential to M. If $h \in C^1(M)$, then Lh is independent of any particular extension of h to $\mathbb{C}^2$. Let us extend h so that $\partial h/\partial s = 0$. Then $\partial h/\partial \bar{w} = \frac{1}{2} \partial h/\partial t$, so that

$$(3) \qquad \qquad Lh = iz \frac{\partial h}{\partial t} - \frac{\partial h}{\partial \bar{z}}.$$

The analogue of Theorem 18.2.3 holds for this L, and gives Lewy's original example.

For further details on this topic, we refer to Chap. VI of Hörmander [1], and to Kohn [2].

18.3. Boundary Values of Pluriharmonic Functions

18.3.1. Let $\Gamma \subset S$ and $\Omega \subset B$ be as in Theorem 18.1.12, let $n > 1$. Define tangential operators

$$(1) \qquad \qquad L_{ij} = z_i \bar{D}_j - z_j \bar{D}_i \qquad \bar{L}_{ij} = \bar{z}_i D_j - \bar{z}_j D_i$$

for $i, j \in \{1, \dots, n\}$. Theorem 18.1.12 showed (when combined with Proposition 18.1.6) that a function $u \in C^1(\Gamma)$ has a continuous extension to $\Gamma \cup \Omega$ which is holomorphic in Ω if and only if $L_{ij}u = 0$ for all i, j. This characterization of the boundary values of holomorphic functions has an analogue, with pluriharmonic in place of holomorphic. The case $\Gamma = S$ was treated by Bedford [1] and by Bedford–Federbush [1]. The local case was studied by Audibert [1], [2] who showed, among other things, that the local extension theorem for pluriharmonic functions requires much stronger hypotheses than the global one, contrary to what is true for holomorphic functions.

Theorems 18.3.4 and 18.3.7 are the main results of this section.

18.3.2. Proposition. *With Ω and Γ as above, assume $u \in C^3(\overline{\Omega})$ and u is pluri-harmonic in Ω. Then*

$$L_{ij} L_{km} \bar{L}_{rs} u = 0 \quad and \quad \bar{L}_{ij} \bar{L}_{km} L_{rs} u = 0$$

for all $i, j, k, m, r, s \in \{1, \ldots, n\}$.

Proof. In Ω, $u = f + \bar{g}$, where $f, g \in H(\Omega)$. Hence $D_r u \in H(\Omega)$ for all r, and $\bar{D}_m D_r u = 0$ for all m. It follows that $L_{km} \bar{L}_{rs} u \in H(\Omega)$, hence $L_{ij} L_{km} \bar{L}_{rs} u = 0$. The other conclusion is proved in the same way.

In the proof of Theorem 18.3.4, the global converse of Proposition 18.3.2, we shall use the spaces $H(p, q)$ that were described in Section 12.2.

18.3.3. Proposition.

   (i)   *Both L_{ij} and $\bar{L}_{ij}$ commute with the Laplacian.*

   (ii) *L_{ij} maps $H(p, q)$ into $H(p + 1, q - 1)$ if $q \geq 1$, and L_{ij} annihilates $H(p, 0)$.*

  (iii) *$\bar{L}_{ij}$ maps $H(p, q)$ into $H(p - 1, q + 1)$ if $p \geq 1$, and $\bar{L}_{ij}$ annihilates $H(0, q)$.*

  (iv) *If $n = 2$ and $q \geq 1$, then L_{12} is a one-to-one map of $H(p, q)$ onto $H(p + 1, q - 1)$.*

   (v) *If $n = 2$ and $p \geq 1$, then $\bar{L}_{12}$ is a one-to-one map of $H(p, q)$ onto $H(p - 1, q + 1)$.*

Proof. For any i, j, k,

$$(1) \qquad\qquad D_k \bar{D}_k L_{ij} - L_{ij} D_k \bar{D}_k = \delta_{ik} \bar{D}_k D_j - \delta_{jk} \bar{D}_k D_i.$$

Add these equations, for $k = 1, \ldots, n$. On the right, the sum is $\bar{D}_i D_j - \bar{D}_j D_i = 0$. Hence

$$(2) \qquad\qquad\qquad \Delta L_{ij} = L_{ij} \Delta.$$

The same is true with $\bar{L}_{ij}$ in place of L_{ij}.

In particular, L_{ij} and $\bar{L}_{ij}$ preserve harmonicity. Since L_{ij} converts bi-degree (p, q) to $(p + 1, q - 1)$, (ii) is proved. The proof of (iii) goes the same way.

Suppose now that $n = 2$. Let us write L in place of L_{12}. The system of tangential Cauchy–Riemann equations reduces now to just $Lu = 0$. If $q \geq 1$, $u \in H(p, q)$, and $u \neq 0$, then u is not holomorphic, hence $Lu \neq 0$ (Theorem 18.1.2). Thus

$$(2) \qquad\qquad L: H(p, q) \to H(p + 1, q - 1)$$

is one-to-one. The same is true of

(3) $$\bar{L}: H(p + 1, q - 1) \to H(p, q).$$

The spaces $H(p, q)$ and $H(p + 1, q - 1)$ have therefore the same dimension (which one can easily check to be $p + q + 1$, by taking $q = 0$). This proves (iv) and (v).

18.3.4. Theorem (Audibert [1]). *If $u \in C^3(S)$ satisfies*

(1) $$\bar{L}_{ij} \bar{L}_{ij} L_{ij} u = 0$$

for all $i, j \in \{1, \ldots, n\}$, then the Poisson integral of u is pluriharmonic in B.

Proof. First, let $n = 2$, write L for L_{12}, choose $h \in H(p, q)$ with $p > 0, q > 0$. By Proposition 18.3.3, $\bar{L}LL$ maps $H(p - 1, q + 1)$ onto $H(p, q)$. Hence $h = \bar{L}LLg$ for some $g \in H(p - 1, q + 1)$. Since $\bar{L} = -L^*$ (see §18.2.2), (1) gives

$$[u, h] = [u, \bar{L}LLg] = -[L\bar{L}Lu, g] = 0.$$

Thus $u \perp H(p, q)$ whenever $p > 0$ and $q > 0$. The Poisson integral of u is therefore pluriharmonic in B.

Assume next that $n \geq 3$ and make the induction hypothesis that the theorem is true in dimension $n - 1$. Fix $k, 1 \leq k \leq n$, and use the equation (1) with $i \neq k, j \neq k$. The induction hypothesis, applied in $(n - 1)$-balls of varying radii, yields an extension f_k of u which is pluriharmonic in B as a function of the variables $(z_1, \ldots, z_{k-1}, z_{k+1}, \ldots, z_n)$. Since being pluriharmonic is a 2-variable property ($D_i \bar{D}_j u = 0$ for all i, j), it suffices to prove that the extensions $f_1, \ldots, f_n$ coincide in B.

Fix a point $a = (a_1, \ldots, a_n) \in B$. Let j, k, m be distinct. The functions

$$\lambda \to f_k(a_1, \ldots, a_{j-1}, \lambda, a_{j+1}, \ldots, a_n)$$
$$\lambda \to f_m(a_1, \ldots, a_{j-1}, \lambda, a_{j+1}, \ldots, a_n)$$

are then harmonic in the disc defined by

$$|\lambda|^2 < 1 - \sum_{i \neq j} |a_i|^2$$

and have the same boundary values. Hence they coincide in this disc, and in particular at $\lambda = a_j$. Thus $f_k = f_m$. This completes the proof.

One rather remarkable feature of this theorem is that the hypothesis (1) is much weaker than the conclusion of Proposition 18.3.2. This is due to the fact that (1) is assumed to hold on all of S. There are examples that show that (1) is not enough for local theorems:

18.3.5. Examples (Audibert [1])

(a) Take $n = 2$, put $u(z) = \bar{z}_1/z_2$ on the part of S where $z_2 \neq 0$. With $L_{12} = L$, one has $Lu = -1$, hence $\bar{L}LLu = 0$, so that 18.3.4(1) holds. But $LL\bar{L}u = -2 \neq 0$, so that u has no pluriharmonic extension, because of Proposition 18.3.2.

(b) Take $n = 3$, put $u(z) = \bar{z}_1/z_2 + \bar{z}_2/z_1$ on the part of S where $z_1 z_2 \neq 0$. Then

$$L_{ij} L_{ij} \bar{L}_{ij} u = 0 = \bar{L}_{ij} \bar{L}_{ij} L_{ij} u$$

for all $i, j \in \{1, 2, 3\}$. But $\bar{L}_{12} \bar{L}_{13} L_{13} u \neq 0$, so that u again has no pluriharmonic extension.

The following facts will be used in the proof of the local extension theorem 18.3.7.

18.3.6. Proposition. Let $n = 2$, put $L_{12} = L$, define

(1) $$\Lambda_1 = \bar{z}_1 L\bar{L} + z_2 \, \bar{L}, \qquad \Lambda_2 = \bar{z}_2 L\bar{L} - z_1 \bar{L}.$$

Then

(2) $$L\Lambda_1 = \bar{z}_1 LL\bar{L}, \qquad L\Lambda_2 = \bar{z}_2 LL\bar{L}$$

and

(3) $$\Lambda_1 \Lambda_2 - \Lambda_2 \Lambda_1 = |z|^2 \bar{L}LL.$$

If f is pluriharmonic, then

(4) $$\Lambda_1 f = -|z|^2 D_1 f, \qquad \Lambda_2 f = -|z|^2 D_2 f.$$

Proof. Since $L\bar{z}_1 = -z_2$ and $Lz_2 = 0$,

(5) $$L\Lambda_1 = \bar{z}_1 LL\bar{L} + (L\bar{z}_1)L\bar{L} + z_2 L\bar{L} = \bar{z}_1 LL\bar{L}.$$

The second part of (2) follows in the same way from $L\bar{z}_2 = z_1$, $Lz_1 = 0$.
A straightforward but rather laborious computation leads from (1) to

(6) $$\Lambda_1 \Lambda_2 - \Lambda_2 \Lambda_1 = |z|^2 \{2\bar{L}L\bar{L} + 2\bar{L} - L\bar{L}\bar{L}\}.$$

Since

(7) $$\bar{L}L - L\bar{L} = z_1 D_1 + z_2 D_2 - \bar{z}_1 \bar{D}_1 - \bar{z}_2 \bar{D}_2,$$

one obtains

$$(8) \qquad \bar{L}\bar{L}L - 2\bar{L}L\bar{L} + L\bar{L}\bar{L} = \bar{L}(\bar{L}L - L\bar{L}) - (\bar{L}L - L\bar{L})\bar{L} = 2\bar{L}.$$

Substitution of (8) into (6) gives (3).
For pluriharmonic f,

$$\begin{aligned}
\Lambda_1 f &= (\bar{z}_1 L + z_2)(\bar{z}_1 D_2 f - \bar{z}_2 D_1 f)\\
&= (\bar{z}_1 z_1 \bar{D}_2 - \bar{z}_1 z_2 \bar{D}_1 + z_2)(\bar{z}_1 D_2 f - \bar{z}_2 D_1 f)\\
&= -(\bar{z}_1 z_1 + \bar{z}_2 z_2)D_1 f
\end{aligned}$$

since $D_1 f$ and $D_2 f$ are holomorphic. The second half of (4) is proved in the same way.

18.3.7. Theorem (Audibert [1]). *Suppose* $\zeta_0 \in S, t < 1$,

$$\Gamma = \{\zeta \in S : t < \operatorname{Re}\langle \zeta, \zeta_0 \rangle\},$$
$$\Omega = \{z \in B : t < \operatorname{Re}\langle z, \zeta_0 \rangle\},$$

and $u \in C^3(\Gamma)$ *satisfies the equations*

$$(1) \qquad\qquad L_{ij} L_{km} \bar{L}_{rs} u = 0 = \bar{L}_{ij} \bar{L}_{km} L_{rs} u$$

for all $i, j, k, m, r, s \in \{1, \dots, n\}$.
Then u *has a continuous extension to* $\Omega \cup \Gamma$ *that is pluriharmonic in* Ω.

Proof. We first handle the case $n = 2$. Put $L_{12} = L$, as before. Since $LL\bar{L}u = 0$, 18.3.6(2) shows that $L\Lambda_1 u = 0$ on Γ and $L\Lambda_2 u = 0$ on Γ. By Theorem 18.1.12, there are functions $g_i \in C(\Omega \cup \Gamma) \cap H(\Omega)$ such that

$$(2) \qquad\qquad g_i = \Lambda_i u \quad \text{on } \Gamma \qquad (i = 1, 2).$$

By 18.3.6(4),

$$\Lambda_1 g_2 = -|z|^2 D_1 g_2, \qquad \Lambda_2 g_1 = -|z|^2 D_2 g_1$$

in Ω, so that

$$(3) \qquad\qquad \Lambda_1 g_2 - \Lambda_2 g_1 = -|z|^2(D_1 g_2 - D_2 g_1)$$

in Ω. But on Γ, 18.3.6(3) shows that

$$(4) \qquad\qquad \Lambda_1 g_2 - \Lambda_2 g_1 = (\Lambda_1 \Lambda_2 - \Lambda_2 \Lambda_1)u = \bar{L}\bar{L}Lu = 0,$$

by the second half of (1). Since $D_1 g_2 - D_2 g_1$ is holomorphic, (3) and (4) imply that

$$(5) \qquad\qquad\qquad D_1 g_2 = D_2 g_1.$$

Fix $a \in \Omega$. Since Ω is convex, one can define

$$\varphi(z) = \int_0^1 \sum_{k=1}^2 (z_k - a_k) g_k(a + t(z - a)) dt$$

for $z \in \Omega \cup \Gamma$. Clearly, $\varphi \in H(\Omega)$. Differentiation under the integral sign leads to

$$(D_1 \varphi)(z) = \int_0^1 \{ g_1(a + t(z - a)) + t \frac{\partial}{\partial t} g_1(a + t(z - a)) \} dt,$$

because of (5). An integration by parts now gives the first of the equations

$$(6) \qquad\qquad\qquad D_i \varphi = g_i \qquad (i = 1, 2).$$

The second one is proved in exactly the same way.

Hence, on Γ, (6) and (2) give

$$\bar{L}\varphi = \bar{z}_1 D_2 \varphi - \bar{z}_2 D_1 \varphi = \bar{z}_1 g_2 - \bar{z}_2 g_1 = (\bar{z}_1 \Lambda_2 - \bar{z}_2 \Lambda_1) u = -\bar{L}u.$$

The last equality follows directly from the definition of Λ_1 and Λ_2.

Thus $\bar{L}(u + \varphi) = 0$ on Γ. By Theorem 18.1.12, $u + \varphi$ has a conjugate-holomorphic extension ψ, and thus $\psi - \varphi$ is the desired pluriharmonic extension of u.

This proves the theorem when $n = 2$.

For the general case, we again proceed by induction, assuming that $n \geq 3$ and that the theorem is true in dimension $n - 1$. It is easy to check that the hypothesis (1) is preserved by unitary changes of variables. It follows that u extends pluriharmonically to every $(n - 1)$-ball that is the intersection of an $(n - 1)$-dimensional affine set with Ω. The proof that these extensions coincide is similar to the one given at the end of Theorem 18.3.4.

Chapter 19

Open Problems

19.1. The Inner Function Conjecture

19.1.1. We define an *inner function* in B to be a *nonconstant* $f \in H^\infty(B)$ whose radial limits f^* satisfy $|f^*(\zeta)| = 1$ for almost all $\zeta \in S$.

When $n = 1$, i.e., when B is the unit disc in $\mathbb{C}$, inner functions play a well-known very important role in factorization theorems involving H^p-functions, in the classification of invariant subspaces of H^2, in the complete description of the closed ideals of the disc algebra, and so on. Theorem 7.3.8 destroys any possibility of finding such H^p-applications when $n > 1$. In any case, no one has ever seen an inner function in B when $n > 1$, and there is strong evidence in favor of the following.

Conjecture. *There are no inner functions in B_n when $n > 1$.*

This conjecture goes back at least to 1966. It has turned out to be "curiously resistant," to borrow a phrase from Littlewood.

19.1.2. The inner function conjecture would be proved for arbitrary $n > 1$ if it were proved for $n = 2$, and it would be enough to prove it under some additional hypotheses. For if there were an inner function f in B, then there would also exist

(a) a zero-free inner function, namely

$$\exp\left\{\frac{f+1}{f-1}\right\},$$

(b) an inner function g with

$$\lim_{r \nearrow 1} \int_S \log|g_r|\, d\sigma = 0,$$

via Frostman's theorem (Rudin [1], [6]); the least harmonic majorant of $\log|g|$ would be 0, and almost all slice functions of g would be Blaschke products (as in Theorem 5.3.2 of Rudin [1]);

403

(c)   an inner function h that satisfies (b) and is not a product of two inner functions (i.e., h is *irreducible* in the unit ball of H^∞, using the terminology of Ahern–Rudin [1]).

If an inner function did exist in B, it would have to be extremely oscillatory near S. The following local result (whose hypothesis is probably vacuous) shows this.

19.1.3. Proposition. *Assume $n > 1$, $f \in H^\infty(B)$, f is not constant, Γ is a nonempty open subset of S, and*

$$(1) \qquad \lim_{r \to 1} |f(r\zeta)| = 1 \qquad \textit{a.e. on } \Gamma.$$

Then Γ contains a dense G_δ-subset H such that f maps every radius of B that ends at a point of H onto a dense subset of the unit disc U in $\mathbb{C}$.

(Sadullaev [1] drew a somewhat weaker conclusion.)
In particular, there is no $\zeta \in \Gamma$ such that f has a continuous extension to $B \cup \{\zeta\}$.

Proof. Note first that $f(B \cap V)$ is dense in U if V is any open set in $\mathbb{C}^n$ that intersects Γ; otherwise, V would contain one-dimensional analytic discs D, with $\partial D \subset S$, such that $f|_D$ is a one-variable inner function whose range is not dense in U, an impossibility. (The Lindelöf–Čirka theorem was tacitly used here.)
It follows that each of the closed sets

$$(2) \qquad E_{ik} = \left\{ \zeta \in \Gamma : |f(r\zeta)| \geq \frac{1}{k} \text{ if } 1 - \frac{1}{i} \leq r < 1 \right\}$$

$(i, k = 1, 2, 3, \ldots)$ has empty interior. For $\alpha \in U$, put

$$(3) \qquad H_\alpha = \left\{ \zeta \in \Gamma : \liminf_{r \to 1} |f(r\zeta) - \alpha| = 0 \right\}.$$

Since $H_0 = \Gamma \setminus \bigcup E_{ik}$, Baire's theorem shows that H_0 is a dense G_δ in Γ. The same is true of each H_α, since $(f - \alpha)/(1 - \bar\alpha f)$ satisfies the hypothesis.
To finish, let α range over a countable dense subset of U, and let H be the intersection of the corresponding sets H_α.
As a corollary, we note that, at every $\zeta \in \Gamma$, the cluster set of f is the whole closed unit disc, a fact which was also proved by Range [1].

19.1.4. Eric Bedford and B. A. Taylor [1] have observed that the gradient of an inner function f in B cannot lie in $L^2(v)$ if $n > 1$. Here is a simple proof:

Almost all slice functions of f would map U onto a Riemann surface of finite area, hence would be finite Blaschke products, and this would force f to be rational (Rudin [1], Theorem 5.2.2), contradicting 19.1.3.

19.1.5. If f were inner in B, with $f(0) = 0$, then $\{1, f, f^2, \ldots\}$ would be an orthonormal set in $H^2(B)$ that is bounded in $H^\infty(B)$.

(Quite recently, P. Wojtaszczyk has proved the existence of polynomials F_k on $\mathbb{C}^n$ that are homogeneous of degree k and satisfy $|F_k| \leq 2^n$ on B for $k = 0, 1, 2, \ldots$. Every $H^2(B)$ thus contains an infinite uniformly bounded orthonormal set.)

Also, $f^*: S \to T$ would be a measure-preserving map. To see this, note that

$$\int_S (h \circ f^*)d\sigma = \frac{1}{2\pi} \int_{-\pi}^{\pi} h(e^{i\theta})d\theta,$$

first for trigonometric polynomials h, hence for all bounded Borel functions h on T, and in particular for characteristic functions of Borel sets $E \subset T$.

If φ is a conformal map of U onto the strip $0 < x < 1$, with $\varphi(0) = \frac{1}{2}$, then $\text{Re}[\varphi \circ f]$ would be a bounded pluriharmonic function in B whose radial limits are 0 and 1 at almost all points of S. There would exist a set E, $\sigma(E) = \frac{1}{2}$, whose characteristic function has a pluriharmonic Poisson integral.

19.1.6. Let W be the set of all $F \in H(B)$ such that $\text{Re } F > 0$ in B and $F(0) = 1$. The formula

$$F = \frac{1 + f}{1 - f}$$

sets up a one-to-one correspondence between the inner functions f in B that have $f(0) = 0$ and those $F \in W$ whose real parts have radial limit 0 a.e. on S. The inner function conjecture can therefore be reformulated as follows.

Conjecture. *If $n > 1$, $F \in W$, and $u = \text{Re } F$, then u^* cannot be 0 a.e. on S.*

Let $\tilde{W}$ be the set of all probability measures μ on S whose Poisson integrals $P[\mu]$ are pluriharmonic. Since each $F \in W$ is uniquely determined by its real part u, there is a natural affine one-to-one correspondence between W and $\tilde{W}$, given by $u = P[\mu]$. In terms of $\tilde{W}$, the preceding conjecture becomes:

No $\mu \in \tilde{W}$ is singular with respect to σ.

Probably, more is true:

19.1.7. Conjecture. *If* $\mu \in \tilde{W}$ *then* $\mu \ll \sigma$.

It would be enough to prove this for the extreme points of $\tilde{W}$. (Note that $\tilde{W}$ is convex and weak*-compact.) For if $E \subset S$ is compact then

$$\mu(E) = \inf\left\{\int_S f\,d\mu : f \geq \chi_E, f \in C(S)\right\}$$

for every $\mu \in \tilde{W}$. The linear function that takes μ to $\mu(E)$ is thus upper semi-continuous on $\tilde{W}$, hence attains its maximum (relative to $\tilde{W}$) at some extreme point of $\tilde{W}$. If $\mu(E) = 0$ whenever μ is an extreme point of $\tilde{W}$, it follows therefore that $\mu(E) = 0$ for every $\mu \in \tilde{W}$.

This reduction to extreme points is Theorem 1.4.1 in Forelli [6].

Section 19.2 contains some of the results (all due to Forelli) that are known about $\tilde{W}$.

19.1.8. The inner function conjecture is closely related to some problems concerning $H^1(B)$. For example 19.1.7 would obviously be established by a proof of the following.

Conjecture. *If* $n > 1$, $F \in H(B)$, *and* $\mathrm{Re}\, F > 0$, *then* $F \in H^1(B)$.

[It is trivial that $F \in H^p(B)$ for all $p < 1$, even when $n = 1$.]

To illustrate, take $F_0(z, w) = 1/(1 - z^2 - w^2)$. This seems to be about as large as any $F \in W$ can be when $n = 2$. (See §19.2.8.)

A computation shows that $F \in H^p(B)$ for all $p < \frac{3}{2}$.

In fact, if $n > 1$, no $F \in W$ seems to be known that is not in $H^p(B)$ for all $p < (n + 1)/2$.

In any case, the following inequality would prove that $F \in H^1(B)$:

19.1.9. Conjecture. *If* $n > 1$, *there is a constant* $c(n) < \infty$ *such that*

$$\int_S |g|\,d\sigma \leq c(n) \int_S |\mathrm{Re}\, g|\,d\sigma$$

for every $g \in A(B)$.

Note that this is true for those g that depend only on $n - 1$ variables (i.e., those that satisfy $g = g \circ P$ for some orthogonal projection P taking $\mathbb{C}^n$ onto a lower-dimensional subspace) since the integrals over $S = \partial B_n$ reduce then to integrals over B_{n-1}, so that Theorem 7.1.5 can be applied.

19.1.10. *If there were an inner function in B, then there would be a* $g \in H^1(B)$, $\|g\|_1 = 1$, *which is not an extreme point of the closed unit ball* X *of* $H^1(B)$.

The proof of this is as in de Leeuw–Rudin [1] or Hoffman [1], pp. 140–141.

Assume $n > 1$. If $h \in H^1(B)$, $\|h\|_1 = 1$, and if some $\zeta \in S$ has a neighborhood V in $\mathbb{C}^n$ such that h is bounded from 0 in $B \cap V$, then h is an extreme point of X.

For if h is not extreme, there is a nonconstant $\varphi: S \to [-1, 1]$ and a $g \in H^1(B)$ such that $g^* = \varphi h^*$ a.e. on S. A contradiction is reached by noting that $h|_D$ is bounded from 0, hence is an outer function in $H^1(D)$, for every analytic disc $D \subset V$ that has its boundary in S.

The set of extreme points of the closed unit ball of $H^1(B)$ is thus norm-dense in the unit sphere of $H^1(B)$, if $n > 1$.

This lends some credibility to the following conjecture, whose truth would prove the nonexistence of inner functions:

Conjecture. *If $n > 1$, then every function in the unit sphere of $H^1(B)$ is an extreme point of the closed unit ball of $H^1(B)$.*

19.1.11. The Nevanlinna class $N(B)$ has a subclass $N_*(B)$ (often called the Smirnov class, in the unit disc), consisting of all $f \in N(B)$ for which the function $\log^+ |f_r|$, $0 < r < 1$, are *uniformly integrable*. This means that to each $\varepsilon > 0$ corresponds a $\delta > 0$ such that

$$\int_E \log^+ |f_r| d\sigma < \varepsilon \qquad (0 < r < 1)$$

for all $E \subset S$ with $\sigma(E) < \delta$.

In U, every $f \in N \setminus N^*$ has the form $f = g/h$, where $g \in N^*$ and h is a zero-free inner function.

Conjecture. *When $n > 1$, then $N(B) = N_*(B)$.*

Assume this to be true, and let $F \in W$ (see §19.1.6), $u = \mathrm{Re}\, F$, $G = e^F$. Then $\log^+ |G| = u$, so that $G \in N(B)$, hence (by the present conjecture) $G \in N_*(B)$, so that $\{u_r: 0 < r < 1\}$ is uniformly integrable. This implies that $u = P[u^*]$ and $u^* \in L^1(S)$.

The truth of $N(B) = N_*(B)$ would thus imply the conjecture made in §19.1.6.

19.1.12. Let $n = 2$. For $0 < \alpha < \pi/2$, let U_α^2 be the polydisc in $\mathbb{C}^2$ defined by $|z| < \cos \alpha$, $|w| < \sin \alpha$.

Any inner function f in B_2 would be inner in almost every U_α^2.

More explicitly, for almost all α it would be true that

$$\lim_{r \to 1} |f(re^{i\theta} \cos \alpha, re^{i\varphi} \sin \alpha)| = 1$$

for almost all (θ, φ).

Quite a bit is known about inner functions in polydiscs, but apparently not enough to answer the following:

Question. *Do there exist $\alpha < \beta < \gamma$ in $(0, \pi/2)$, such that some nonconstant f is holomorphic in the union of $U_\alpha^2, U_\beta^2, U_\gamma^2$ and is inner in each of these three polydiscs?*

No rational f can do this.

In fact, if a nonconstant rational function f is inner in U_α^2 and in U_β^2, the explicitly known form of rational inner functions in U^2 (Rudin–Stout [1]; Rudin [1], Theorem 5.2.5) implies that there are relatively prime positive integers p and q such that

$$\frac{\log(\sin \beta/\sin \alpha)}{\log(\cos \alpha/\cos \beta)} = \frac{p}{q}$$

and that f is a function of the single variable $z^p w^q$. Hence f cannot be inner in U_γ^2.

19.1.13. Instead of trying to prove that there are no inner functions in B when $n > 1$, one might shoot for the full analogue of the fact that $f(B) \subset f(S)$ for every $f \in A(B)$:

Conjecture. *If $n > 1$, $f \in H^\infty(B)$, V is open in $\mathbb{C}$, and $V \subset f(B)$, then there is a set $E \subset S$, with $\sigma(E) > 0$, such that $f^*(E) \subset V$.*

In other words, the conjecture is that the essential range of f^* on S is the closure of $f(B)$.

19.2. RP-Measures

19.2.1. Those real Borel measures on S whose Poisson integrals are pluriharmonic (and thus are real parts of holomorphic functions in B) will be called RP-measures. They may also be characterized as being those that are orthogonal to all spaces $H(p, q)$ with both $p > 0$ and $q > 0$.

As stated in §19.1.7, it seems likely that $\mu(E) = 0$ whenever $\sigma(E) = 0$ and μ is an RP-measure. Theorem 19.2.3 will prove this for certain classes of sets E.

19.2.2. Let $a = (a_1, \ldots, a_n)$ be an n-tuple of positive real numbers. Define

$$(1) \qquad\qquad g_\lambda(z) = (e^{i\lambda a_1} z_1, \ldots, e^{i\lambda a_n} z_n)$$

for $\lambda \in \mathbb{C}$, $z \in \mathbb{C}^n$. As t runs through the real numbers, $\{g_t\}$ is a one-parameter group G_a of unitary operators on $\mathbb{C}^n$. A set $E \subset S$ is said to be G_a-*invariant* if $g_t(E) = E$ for $-\infty < t < \infty$.

19.2.3. Theorem (Forelli [4]). *If μ is an* RP-*measure on S and G_a is as above, then $\mu(E) = 0$ for every G_a-invariant set E that has $\sigma(E) = 0$.*

Proof. Fix $w \in \mathbb{C}$ with $v = \text{Im } w > 0$. The Poisson formula

$$(1) \qquad\qquad h(w) = \frac{1}{\pi} \int_{-\infty}^{\infty} \frac{v}{|w - t|^2} h(t) dt$$

holds then for all harmonic functions h that are bounded in the upper half-plane of $\mathbb{C}$.

Let $\tilde{\mu}$ be the measure determined by requiring that

$$(2) \qquad\qquad \int_S \varphi \, d\tilde{\mu} = \frac{1}{\pi} \int_{-\infty}^{\infty} \frac{v \, dt}{|w - t|^2} \int_S \varphi(g_{-t}(\zeta)) d\mu(\zeta)$$

for all $\varphi \in C(S)$. Then (2) holds equally well for every bounded Borel function φ on S. If $E \subset S$ is a G_a-invariant Borel set and φ is the characteristic function of E, then the inner integral on the right of (2) is $\mu(E)$ for every t. We conclude that

$$(3) \qquad\qquad \tilde{\mu}(E) = \mu(E)$$

if E is G_a-invariant.

To find out more about $\tilde{\mu}$, let us compute $P[\tilde{\mu}]$. Recall that $u = P[\mu]$ is pluriharmonic in B, by hypothesis. The identity

$$(4) \qquad\qquad P(z, g_{-t}(\zeta)) = P(g_t(z), \zeta) \qquad (z \in B, \zeta \in S, t \in R)$$

leads to

$$(5) \qquad\qquad P[\tilde{\mu}](z) = \frac{1}{\pi} \int_{-\infty}^{\infty} \frac{v \, dt}{|w - t|^2} u(g_t(z)) dt = u(g_w(z)).$$

The first equality in (5) follows from (2) and (4), the second from (1), since $\lambda \to u(g_\lambda(z))$ is harmonic for each fixed $z \in B$. Note also that g_w maps $\bar{B}$ into B (since $v > 0$). Thus, setting $f(z) = u(g_w(z))$, we find that $f \in C(\bar{B})$ and $f = P[\tilde{\mu}]$. Thus $d\tilde{\mu} = f \, d\sigma$, or

$$(6) \qquad\qquad \tilde{\mu}(E) = \int_E f \, d\sigma$$

for every Borel set $E \subset S$.

The theorem follows from (3) and (6).

Forelli proved this under a weaker assumption: $a_j \geq 0$ for all j, and $a_j > 0$ for at least one j. It is then no longer true that $g_w(\bar{B}) \subset B$, but one still obtains (6), with $f \in L^1(\sigma)$, by a more delicate argument.

The case $a_1 = \cdots = a_n = 1$ shows that $\mu(E) = 0$ for all RP-measures μ and for all circular sets $E \subset S$ with $\sigma(E) = 0$.

19.2.4. The discussion in §19.1.6 to 19.1.8 showed that the inner function conjecture would be proved if one could show that all extreme points of the convex set W (defined in §19.1.6) lie in $H^1(B)$ when $n > 1$. The rest of this section is devoted to these extreme points. Very little seems to be known about them. We begin with a necessary condition.

19.2.5. Theorem (Forelli [8]). *If*

$$F = \frac{1+f}{1-f}$$

is an extreme point of W, then f is irreducible: there is no factorization $f = uv$, with $u, v \in H^\infty(B)$, $|u| < 1$, $|v| < 1$.

Proof. Assume $f = uv$, $u, v \in H^\infty(B)$, $|u| \leq 1$, $|v| \leq 1$, and $u(0) = 0$. (Note that $f(0) = 0$, since $F(0) = 1$.) We shall prove that then v is a constant of absolute value 1.

Replacing v by $e^{i\alpha}v$, hence u by $e^{-i\alpha}u$, we can assume that $v(0) = it$, $0 \leq t \leq 1$. Put

$$F_1 = \frac{(1-u)(1-v)}{1-uv} + it, \qquad F_2 = \frac{(1+u)(1+v)}{1-uv} - it.$$

Since $2(1 - uv)/(1 - u)(1 - v) = (1 + u)/(1 - u) + (1 + v)/(1 - v)$, we see that Re $F_j > 0$ in B. Also, $F_j(0) = 1$. Thus $F_j \in W$. Since $F_1 + F_2 = 2F$ and F is extreme, $F_1 = F_2$, which leads to

$$v = i\frac{t + iu}{1 + itu}.$$

If this is substituted into $F = (1 + uv)/(1 - uv)$, one obtains

$$F = \frac{1+t}{2}\frac{1+iu}{1-iu} + \frac{1-t}{2}\frac{1-iu}{1+iu},$$

a convex combination of two members of W that are distinct, since $u \neq 0$. Since F is extreme, it follows that $t = 1$, so that $v = i$.

19.2.6. The function $(1 + z)/(1 - z)$ shows that the preceding necessary condition is not sufficient when $n > 1$, since

$$\frac{1 + z}{1 - z} + w^2 h(z, w)$$

has positive real part in B for every h such that $|h| \leq \frac{1}{4}$.

Every extreme point of W gives rise to others, by means of the automorphisms of B:

19.2.7. Proposition (Forelli [5], [8]). *If $\psi \in \mathrm{Aut}(B)$ and Λ_ψ is defined by*

$$(\Lambda_\psi F)(z) = \frac{F(\psi(z)) - i \operatorname{Im} F(\psi(0))}{\operatorname{Re} F(\psi(0))}$$

then Λ_ψ is a map of W onto W that carries extreme points to extreme points.

Proof. Clearly, $\Lambda_\psi(W) \subset W$. Also, Λ_ψ is the identity map when ψ is the identity. A simple calculation shows that

$$\Lambda_{\psi_1 \psi_2} F = \Lambda_{\psi_2} \Lambda_{\psi_1} F.$$

Hence $\Lambda_{\psi^{-1}} = (\Lambda_\psi)^{-1}$, and $\Lambda_\psi(W) = W$.

Now write Λ for Λ_ψ. Assume F is extreme in W, and $\Lambda F = sG + tH$, where $G, H \in W, s \geq 0, t \geq 0, s + t = 1$. Then

$$F = \Lambda^{-1}(sG + tH) = s'\Lambda^{-1}G + t'\Lambda^{-1}H,$$

where, setting $a = \psi^{-1}(0)$,

$$s' = \frac{\operatorname{Re}[sG(a)]}{\operatorname{Re}[sG(a) + tH(a)]}, \qquad t' = \frac{\operatorname{Re}[tH(a)]}{\operatorname{Re}[sG(a) + tH(a)]}.$$

Since F is extreme, $s' = 0$ or $t' = 0$, hence $s = 0$ or $t = 0$.

19.2.8. Theorem (Forelli [5]). *If $n > 1$ and*

$$f(z) = z_1^2 + \cdots + z_n^2$$

then $F = (1 + f)/(1 - f)$ is an extreme point of W.

Proof. First, consider one-variable functions

(1) $\varphi(\lambda) = 1 + c_1 \lambda + c_2 \lambda^2 + \cdots$ $(|\lambda| < 1)$

such that Re $\varphi > 0$. To every φ of this type corresponds a probability measure μ on the circle T such that

(2)
$$\varphi(\lambda) = \int_T \frac{\alpha + \lambda}{\alpha - \lambda} \, d\mu(\alpha).$$

Since $(\alpha + \lambda)/(\alpha - \lambda) = 1 + 2\Sigma(\lambda/\alpha)^k$, it follows that $|c_k| \le 2$ for $k = 1, 2, 3, \ldots$. Moreover, if $c_2 = 2$, then μ is concentrated on $\{1, -1\}$, so that $c_{2m} = c_2$ and $c_{2m-1} = c_1$, for $m = 1, 2, 3, \ldots$; also, c_1 is real in this case.

Now let $H = H_1 + H_2 + \cdots$ be the homogeneous expansion of some holomorphic function in B that satisfies

(3)
$$\text{Re}[F \pm H] > 0 \text{ in } B.$$

We have to prove that $H = 0$.

Fix $x \in S \cap R^n$. (In other words, x is any point on S whose coordinates are real; these are exactly the points of S at which $f = 1$.) Define

(4) $\varphi_1(\lambda) = F(\lambda x) + H(\lambda x), \qquad \varphi_2(\lambda) = F(\lambda x) - H(\lambda x),$

for $|\lambda| < 1$. Writing φ for φ_1 or φ_2,

(5)
$$\varphi(\lambda) = 1 + 2\lambda^2 + 2\lambda^4 + + \cdots \pm \sum_{k=1}^{\infty} H_k(x)\lambda^k,$$

where $+$ refers to φ_1, $-$ to φ_2. The coefficient of λ^2 is $2 \pm H_2(x)$. Its absolute value cannot exceed 2. Hence $H_2(x) = 0$.

The first paragraph of this proof implies now that

(6) $H_{2m}(x) = 0, \qquad H_{2m+1}(x) = H_1(x) \qquad (x \in S \cap R^n)$

for $m = 1, 2, 3, \ldots$. Since H_{2m} is homogeneous, it follows from (6) that $H_{2m} = 0$ on R^n, hence on C^n. Similarly, $H_{2m+1} - H_1 f^m$ is homogeneous, vanishes on $S \cap R^n$ by (6), hence vanishes on C^n. Thus $H_{2m+1} = H_1 f^m$. Also, $H_1(x)$ is real when $x \in R^n$. Thus

(7) $H_1(z) = a_1 z_1 + \cdots + a_n z_n$

with real coefficients a_j.

Collecting all this information, we obtain

(8)
$$F \pm H = \frac{1+f}{1-f} \pm \sum_0^{\infty} H_1 f^m = \frac{1+f \pm H_1}{1-f}.$$

Use (8) at $z = (s, it, 0, \ldots, 0)$, where $s > 0, t > 0, s^2 + t^2 = 1$. Then $1 - f = 2t^2, 1 + f = 2s^2$, and the positivity of $\text{Re}[F \pm H]$ implies that

$$(9) \qquad\qquad \text{Re}[2s^2 \pm (a_1 s + ia_2 t)] \geq 0$$

or $2s \pm a_1 \geq 0$. Letting $s \to 0$, we find $a_1 = 0$.

In the same way one proves that $a_j = 0$ for $j = 2, \ldots, n$. Thus $H_1 = 0$, hence $H = 0$.

19.2.9. Here is one more result of this type: Let

$$f(z) = cz_1^{m_1} \cdots z_n^{m_n}$$

be a *monomial* with $m_j > 0$ for all j, where c is so chosen that the maximum of $|f|$ on S is 1.

Then $F = (1 + f)/(1 - f)$ is an extreme point of W if and only if the greatest common divisor of $\{m_1, \ldots, m_n\}$ is 1.

For the proof we refer to Forelli [9].

19.3. Miscellaneous Problems

This section contains brief descriptions of several problems related to the unit ball of $\mathbb{C}^n$, $n > 1$. Some of them have already been mentioned in earlier chapters.

19.3.1. Factorization *Does there exist an $f \in H^1(B)$ that is not a product of two members of $H^2(B)$?*

The answer is undoubtedly yes. By analogy with the same problem in the polydisc context, it seems in fact likely that the set of factorable functions is of the first category in $H^1(B)$. (Rosay [1].)

It is quite easy to prove, by the same device that established the lemma in Rosay [1], that the bounded bilinear map that sends $(g, h) \in H^2(B) \times H^2(B)$ to $gh \in H^1(B)$ is not open at the origin if $n > 1$. However, in contrast to the linear case, such bilinear maps may be surjective. This was first proved by Paul Cohen [1]. A much simpler *finite-dimensional* example, namely a bilinear map of $\mathbb{C}^3 \times \mathbb{C}^3$ onto $\mathbb{C}^4$ that is not open at $(0, 0)$, was found by Horowitz [2].

Coifman–Rochberg–Weiss [1] (see also Coifman–Weiss [2]) have developed a substitute for factorization by showing that every $f \in H^1(B)$ is an infinite sum of products of H^2-functions:

$$f = \sum_{i=1}^{\infty} g_i h_i \quad \text{and} \quad \sum_{i=1}^{\infty} \|g_i\|_2 \|h_i\|_2 \leq c\|f\|_1.$$

The constant c depends only on the dimension n. It is not known whether every $f \in H^1(B)$ is a finite sum of this type.

19.3.2. Zero-Varieties. *Is there an $f \in H^\infty(B_3)$ ($f \not\equiv 0$) whose zero-variety has infinite area? (See §7.3.6.)*

Using the Henkin–Skoda technique, Berndtsson [1] has proved the following: Let V be a zero-variety in B, and let $A(r)$ be the area of $V \cap (rB)$. Then $V = Z(f)$ for some $f \in H^\infty(B)$

    (a)   if $n = 2$ and $A(1) < \infty$,
    (b)   if $n = 3$ and dA/dr is bounded.

Another type of problem concerns the way in which zero-varieties approach the boundary. For example:

    If $f \in A(B)$ and $f(0) = 0$, must $Z(f)$ contain a path that approaches some point of S nontangentially?

The same question can of course be asked with H^∞, H^p, $N, \ldots$, in place of A.

As indicated in Section 7.3, a characterization of the zero-varieties of $H^p(B)$ seems out of reach. Varopoulos [3], [4] formulated a *uniform* Blaschke condition which, when satisfied by V, ensures that V is an H^p-zero-variety for *some $p > 0$*.

19.3.3. Radial Limits. *If $f \in H^\infty(B)$ and $f(0) = 0$, must there be a $\zeta \in S$ such that*

$$\lim_{r \to 1} f(r\zeta) = 0?$$

Because of the Lindelöf–Cirka theorem, this is related to the preceding question, and it too can be asked in the context of other function spaces.

    If $f \in H^\infty(B)$ and if E_f is the exceptional set consisting of all $\zeta \in S$ where $\lim f(r\zeta)$ does not exist, must E_f be totally null? (See §11.3.5.)

One major difficulty with this question is that we do not know nearly enough about the totality of the representing measures.

The question has a negative answer if $H^\infty(B)$ is replaced by $H^p(B)$, $p < \infty$. To see this, take $n = 2$, refer to Theorem 7.2.4 which says that the restriction of $H^p(B)$ to any complex line through 0 is the Bergman space $(L^p \cap H)(U)$, and use the known fact that there are functions in $(L^p \cap H)(U)$ that have radial limits at almost no point of T.

19.3.4. Radial Pathology. Do the following holomorphic functions exist in B:

    (a)   An f with $\lim_{r \to 1} f(r\zeta) = \infty$ a.e. on S?
    (b)   An $f \not\equiv 0$ with $\lim_{r \to 1} f(r\zeta) = 0$ a.e. on S?
    (c)   An f whose radial limits exist at almost no point of S?
    (d)   An $f \in H^\infty(B)$ (or in $H(B)$) such that the image of almost every radius is nonrectifiable?

In the disc, the answers are: yes.

In (a) and (b), the answers become no if radial limits are replaced by K-limits. (See Theorems 5.5.8 and 5.5.9.)

19.3.5. Natural Boundaries. There is an $f \in N(B)$ with the following property (Globevnik–Stout [1]):

If $\Phi: \mathbb{C} \to \mathbb{C}^n$ is any affine map that sends T into S, then $f \circ \Phi$ (which is holomorphic in U) has T as its natural boundary.

Is the same true for some $f \in A(B)$, or even for some $f \in A^\infty(B)$?

19.3.6. The Closed Ideals of $A(B)$. A complete description of these, of the sort that is known for the disc algebra (Rudin [17], Hoffman [1]), is probably impossible. But one might be able to answer more specific questions.

If J is a closed ideal of $A(B)$, define

$$Z(J) = \{z \in \bar{B}: f(z) = 0 \text{ for all } f \in J\}.$$

It is known (the proof of Theorem 4.4.2 in Rudin [1] works in balls just as in polydiscs) that $A(B)$ contains closed ideals J that are not the closure of any finitely generated ideal.

Can this happen if $Z(J)$ is a single point of S?

To each compact $K \subset S$ corresponds the ideal

$$J_K = \{f \in A(B): f \,|_K = 0\}.$$

Clearly, $K \subset Z(J_K)$; equality holds when K is a (Z)-set for $A(B)$.

Can it happen that $K = Z(J_K)$ although K is not a (Z)-set?

This question was discussed in §10.5.10.

19.3.7. The Corona Problem. *Suppose $f_1, \ldots, f_k \in H^\infty(B)$, $\delta > 0$, and $\Sigma | f_j(z)| > \delta$ for all $z \in B$. Do there exist $g_1, \ldots, g_k \in H^\infty(B)$ such that $\Sigma f_j g_j = 1$?*

Equivalently, is B dense in the maximal ideal space of $H^\infty(B)$?

When $n = 1$, the answer is affirmative. This is the famous corona theorem of Carleson [4]. When $n > 1$, attempts have been made to attack the problem with solutions of the $\bar{\partial}$-equation (Varopoulos [2]), but without success so far.

19.3.8. H^p Isometries. *If $p \neq 2$, are there any isometries of $H^p(B)$ into $H^p(B)$ whose range is not all of $H^p(B)$?*

Since multiplication by an inner function would be such an isometry, a negative answer would show that there are no inner functions in B.

(Theorem 7.5.6 described the surjective isometries of $H^p(B)$. Isometries of the Nevanlinna class were discussed by Stephenson [1].)

With U in place of B, there are other isometries that are not surjective, for example $f(z) \to f(z^2)$. This suggests the next question.

19.3.9. Inner Maps. *Suppose that* $F: B \to B$ *is holomorphic, and*

$$\lim_{r \to 1} |F(r\zeta)| = 1$$

for almost all $\zeta \in S$. *Does it follow that* $F \in \mathrm{Aut}\,(B)$?

An affirmative answer would represent a considerable strengthening of the corresponding theorem about proper maps (Theorem 15.4.2) and would imply the inner function conjecture.

19.3.10. Proper Maps. *When* $1 < n < p$, *what are the proper holomorphic maps of* B_n *into* B_p?

The case $p = n + 1$ has been settled by Webster [3], under additional boundary smoothness assumptions:

If $n \geq 3$ *then every map of* $\bar{B}_n$ *into* $\bar{B}_{n+1}$ *whose boundary values yield a* C^3-*immersion of* ∂B_n *into* ∂B_{n+1} *and which is holomorphic in* B_n *has its range in an affine set of (complex) dimension* n.

By Theorem 15.4.2, these maps are thus completely known.

Alexander's example $(z, w) \to (z^2, \sqrt{2}zw, w^2)$ shows that Webster's theorem does not extend to $n = 2$.

19.3.11. Multiplier Theorems. Every bounded sequence $\{\lambda_i\}$ of complex numbers induces a multiplier transformation T of $H(B)$ into $H(B)$: If $f = \Sigma F_i$ is the homogeneous expansion of f, define $Tf = \Sigma \lambda_i F_i$.

Since these transformations act equally on all slices, their effect on $H(B)$ can be deduced from the way they act on $H(U)$. For example, if $\{\lambda_i\}$ takes $H^p(U)$ into $H^p(U)$, for some p, then, for the same p, $\{\lambda_i\}$ takes $H^p(B)$ into $H^p(B)$, for all n.

The multiplier transformation that occurs in §6.6.3 is not of this type. The problem is to find others like it.

19.3.12. Interpolation Manifolds. *Which compact connected smooth* $(n - 1)$-*manifolds, other than* T^{n-1} *and* S^{n-1}, *can occur as* (PI)-*sets in* ∂B_n?

Since this involves complex-tangential embeddings (see §10.5.7), there may be some interesting connections with differential topology.

19.3.13. Interpolation Sets. *If* $K \subset S$ *is an* (I)-*set, is there an integral operator that produces* $A(B)$-*extensions of any* $f \in C(K)$?

Do such operators exist at least when K is a manifold? The integrals that define the functions g_δ in the proof of Theorem 10.5.4 come close, but don't quite do it.

If $K \subset S$ is compact, and if no C^1-curve γ with $\langle \gamma', \gamma \rangle \neq 0$ intersects K in a set of positive (one-dimensional) measure, does it follow that K is a (PI)-set?
An affirmative answer would be a converse of Theorem 11.2.5.

19.3.14. Peak Sets. *Let $K \subset S$ be compact, $\sigma(K) = 0$. Does there exist an $f \in C(S)$ that peaks on K and that is orthogonal to all RP-measures?*

To require $\int f \, d\mu = 0$ for all RP-measures μ is the same as to say that f is to lie in the closed linear span of the $H(p, q)$-spaces with both $p > 0$ and $q > 0$. This span is not an algebra, but nevertheless it should be of some interest to study its peak sets, interpolation sets, etc.

19.3.15. Boundary Values of $|f|$. *Suppose $f \in A(B)$, $g \in A(B)$, E is the set of all $f \in S$ where $|f(\zeta)| = |g(\zeta)|$, and $\sigma(E) > 0$. Does it follow that $f = cg$, for some constant c?*

This is a slight generalization of the problem discussed in Section 11.4.
The answer is yes when $g = 0$, and is also yes if "$\sigma(E) > 0$" is replaced by "E has nonempty interior."
If $f \in A(B)$ and $|f| < 1$ in B, is

$$\int_S \log(1 - |f|) d\sigma > -\infty?$$

19.3.16. The Invariant Laplacian. In Theorem 4.1.3, computation established the formula

$$(\tilde{\Delta}f)(a) = (1 - |a|^2)[(\Delta f)(a) - (\Delta f_a)(1)].$$

Is there a more intuitive (geometric?, group-theoretic?) way of seeing why, except for the factor $1 - |a|^2$, $\tilde{\Delta}f$ is a difference of two ordinary Laplacians?

19.3.17. Moebius Invariance. *Does the Fréchet algebra $C(B)$ have any non-trivial closed $\mathcal{M}$-subalgebras other than $H(B)$ and conj $H(B)$?*

See §13.4.6. This problem is open even in one variable, as is the following one:
Let Y be a closed $\mathcal{M}$-invariant subspace of $C(\bar{B})$, put $Y_0 = Y \cap C_0(B)$. Is $Y|_S$ closed in $C(S)$?
Is Y the direct sum of Y_0 and one of the spaces of Poisson integrals listed in Theorem 13.3.2?

Bibliography

Kenzo Adachi [1] Extending bounded holomorphic functions from certain subvarieties of a strongly pseudoconvex domain, Bull. Fac. Sci. Ibaraki Univ., Math., No. 8, 1976.

M. L. Agranovskii [1] Invariant algebras on the boundaries of symmetric domains, Dokl. Akad. Nauk SSSR **197**, 9–11 (1971), Soviet Math. Dokl. **12**, 371–374 (1971).

[2] Invariant algebras on noncompact Riemannian symmetric spaces, Dokl. Akad. Nauk SSSR **207**, 513–516 (1972), Soviet Math. Dokl. **13**, 1538–1542 (1972).

M. L. Agranovskii and R. E. Valskii [1] Maximality of invariant algebras of functions, Sib. Mat. Zh. **12**, 3–12 (1971), Sib. Mat. J. **12**, 1–7 (1971).

P. R. Ahern [1] On the generalized F. and M. Riesz Theorem, Pacific J. Math. **15**, 373–376 (1965).

P. R. Ahern and D. N. Clark [1] Radial Nth derivatives of Blaschke products, Math. Scand. **28**, 189–201 (1971).

P. R. Ahern and W. Rudin [1] Factorization of bounded holomorphic functions, Duke Math. J. **39**, 767–777 (1972).

P. R. Ahern and Robert Schneider [1] Isometries of H^∞, Duke Math. J. **42**, 321–326 (1975).

[2] The boundary behavior of Henkin's kernel, Pacific J. Math. **66**, 9–14 (1976).

[3] Estimates of solutions of $\bar{\partial}u = f$ using covering techniques (Preprint).

[4] Holomorphic Lipschitz functions in pseudoconvex domains, Amer. J. Math. **101**, 543–565 (1979).

[5] A smoothing property of the Henkin and Szegö projections, Duke Math. J. **47**, 135–143 (1980).

L. V. Ahlfors and A. Beurling [1] Conformal invariants and function-theoretic null sets, Acta Math. **83**, 101–129 (1950).

L. A. Aizenberg [1] Integral representations of functions which are holomorphic in convex regions of $\mathbb{C}^n$-space, Dokl. Akad. Nauk SSSR **151**, 1247–1249 (1963), Soviet Math. Dokl. **4**, 1149–1152 (1963).

[2] Integral representations of holomorphic functions of several complex variables, Trudy Moskov. Mat. Obsc. **21**, 3–26 (1970), Trans. Moscow Math. Soc. **21**, 1–29 (1970).

[3] On a formula for the generalized multidimensional logarithmic residue and the solution of systems of nonlinear equations, Dokl. Akad. Nauk SSSR **234**, 505–508 (1977), Soviet Math. Dokl. **18**, 691–695 (1977).

H. Alexander [1] Extending bounded holomorphic functions from certain subvarieties of a polydisc, Pacific J. Math. **29**, 485–490 (1969).

[2] Holomorphic mappings from the ball and polydisc, Math. Ann. **209**, 249–256 (1974).

[3] Proper holomorphic mappings in $\mathbb{C}^n$, Indiana Univ. Math. J. **26**, 137–146 (1977).

[4] Extremal holomorphic imbeddings between the ball and polydisc, Proc. Amer. Math. Soc. **68**, 200–202 (1978).

H. Alexander, B. A. Taylor, and D. L. Williams [1] The interpolating sets for A^∞, J. Math. Anal. Appl. **36**, 556–566 (1971).

Eric Amar [1] Suites d'interpolation pour les classes de Bergman de la boule et du polydisque de $\mathbb{C}^n$, Can. J. Math. **30**, 711–737 (1978).

[2] Représentation des fonctions BMO et solutions de l'équation $\bar{\partial}_b$, Math. Ann. **239**, 21–33 (1979).

[3] Extensions de fonctions analytiques avec estimations, Ankiv för Mat. **17**, 123–138 (1979).

E. Amar and A. Bonami [1] Mesures de Carleson d'ordre α et estimations de solutions du $\bar{\partial}$, Bull. Soc. Math. France **107**, 23–48 (1979).

T. Audibert [1] Opérateurs différentielles sur la sphère de $\mathbb{C}^n$ caractérisant les restrictions des fonctions pluriharmoniques, Thèse, Université de Provence, 1977.

[2] Caractérisation locale par des opérateurs differentielles des restrictions à la sphere de $\mathbb{C}^n$ des fonctions pluriharmoniques, C. R. Acad. Sci. Paris **284**, A1029–A1031 (1977).

A. Aytuna and A.-M. Chollet [1] Une extension d'un résultat de W. Rudin, Bull. Soc. Math. France **104**, 383–388 (1976).

Eric Bedford [1] The Dirichlet problem for some overdetermined systems on the unit ball in $\mathbb{C}^n$, Pacific J. Math. **51**, 19–25 (1974).

Eric Bedford and Paul Federbush [1] Pluriharmonic boundary values, Tohoku Math. J. **26**, 505–511 (1974).

Eric Bedford and John E. Fornaess [1] Biholomorphic maps of weakly pseudoconvex domains, Duke Math. J. **45**, 711–719 (1978).

Eric Bedford and B. A. Taylor [1] Two applications of a nonlinear integral formula to analytic functions, Indiana Math. J. **29**, 463–465 (1980).

Steve Bell and Ewa Ligocka [1] A simplification and extension of Fefferman's theorem on biholomorphic mappings, Invent. Math. **57**, 283–289 (1980).

Carlos A. Berenstein and Lawrence Zalcman [1] Pompeiu's problem on spaces of constant curvature, J. d'Analyse Math. **30**, 113–130 (1976).

Stefan Bergman [1] The Kernel Function and Conformal Mapping, Amer. Math. Soc. Surveys 5, 1950.

B. Berndtsson [1] Integral formulas for the $\partial\bar{\partial}$-equation and zeros of bounded holomorphic functions in the unit ball, Math. Annalen (to appear).

Errett Bishop [1] A general Rudin–Carleson theorem, Proc. Amer. Math. Soc. **13**, 140–143 (1962).

[2] Differentiable manifolds in complex euclidean space, Duke Math. J. **32**, 1–22 (1965).

J. Bochnak and J. Siciak [1] Analytic functions in topological vector spaces, Studia Math. **39**, 77–112 (1971).

Salomon Bochner [1] Analytic and meromorphic continuation by means of Green's formula, Ann. Math. **44**, 652–673 (1943).

[2] Group invariance of Cauchy's formula in several variables, Ann. Math. **45**, 686–707 (1944).

[3] Classes of holomorphic functions of several variables in circular domains, Proc. Nat. Acad. Sci. **46**, 721–723 (1960).

S. Bochner and W. T. Martin [1] *Several Complex Variables*, Princeton University Press, Princeton, NJ, 1948.

L. Boutet De Monvel [1] On the index of Toeplitz operators of several complex variables, Inventiones Math. **50**, 249–272 (1979).

L. Bungart [1] Boundary kernel functions for domains on complex manifolds, Pacific J. Math. **14**, 1151–1164 (1964).

Daniel Burns and E. L. Stout [1] Extending functions from submanifolds of the boundary, Duke Math. J. **43**, 391–404 (1976).

A. P. Calderón [1] The behavior of harmonic functions at the boundary, Trans. Amer. Math. Soc. **68**, 47–54 (1950).

A. P. Calderón and A. Zygmund [1] On the existence of certain singular integrals, Acta Math. **88**, 85–139 (1952).

C. Carathéodory [1] Über das Schwarzsche Lemma bei analytischen Funktionen von zwei komplexen Veränderlichen, Math. Ann. **97**, 76–98 (1927).

[2] Über die Abbildungen, die durch Systeme von analytischen Funktionen von meheren Veränderlichen erzeugt werden, Math. Z. **34**, 758–792 (1932).

[3] *Theory of Functions of a Complex Variable*, Chelsea, 1954.

Lennart Carleson [1] Sets of uniqueness for functions regular in the unit circle, Acta Math. **87**, 325–345 (1952).

[2] Representations of continuous functions, Math. Z. **66**, 447–451 (1957).

[3] *Selected Problems on Exceptional Sets*, Van Nostrand, Princeton, NJ, 1967.

[4] Interpolation by bounded analytic functions and the corona problem, Ann. Math. **76**, 547–559 (1962).

Henri Cartan [1] Les fonctions de deux variables complexes et le problème de la représentation analytique, J. de Math. Pures et Appl. **96**, 1–114 (1931).

[2] Sur les fonctions de plusieurs variables complexes. L'itération des transformations intérieures d'un domaine borné, Math. Z. **35**, 760–773 (1932).

[3] Sur les transformations analytiques des domains cerclés et semi-cerclés bornés, Math. Ann. **106**, 540–573 (1932).

Philippe Charpentier [1] Formules explicites pur les solutions minimales de l'equation $\bar{\partial}u = f$ dans la boule et dans le polydisque de $\mathbb{C}^n$ (Preprint, Orsay).

J. Chaumat and A.-M. Chollet [1] Ensembles pics pour $A^\infty(D)$, Ann. Inst. Fourier **29** (3), 171–200 (1979).

[2] Quelques propriétés des ensembles pics de $A^\infty(D)$, C. R. Acad. Sci. Paris **288**, 611–613 (1979).

[3] Caractérisation et propriétés des ensembles localement pics de $A^\infty(D)$ (Preprint, Orsay).

Pak Soong Chee [1] The Blaschke condition for bounded holomorphic functions, Trans. Amer. Math. Soc. **148**, 249–263 (1970).

[2] On the generalized Blaschke condition, Trans. Amer. Math. Soc. **152**, 227–231 (1970).

[3] Universal functions in several complex variables, J. Austral. Math. Soc. **28**, 189–196 (1979).

Anne-Marie Chollet [1] Ensembles de zéros à la frontière de fonctions analytiques dans des domaines strictement pseudo-convexes, Ann. Inst. Fourier **26**, 51–80 (1976).

[2] Ensembles de zéros, ensembles pics et d'interpolation pour $A(D)$ (Preprint, Orsay).

E. M. Čirka [1] The Lindelöf and Fatou theorems in $\mathbb{C}^n$, Mat. Sb. **92**, 622–644 (1973). Math. U.S.S.R. Sb. **21**, 619–641 (1973).

E. M. Čirka and G. M. Henkin [1] Boundary properties of holomorphic functions of several complex variables, J. Soviet Math. **5**, 612–687 (1976).

L. A. Coburn [1] Singular integral operators and Toeplitz operators on odd spheres, Indiana Univ. Math. J. **23**, 433–439 (1973).

P. J. Cohen [1] A counterexample to the closed graph theorem for bilinear maps, J. Functional Anal. **16**, 235–239 (1974).

R. R. Coifman and Guido Weiss [1] Analyse harmonique noncommutative sur certains espaces homogènes, Lecture Notes in Mathematics, No. 242, Springer-Verlag, Heidelberg, 1971.

[2] Extensions of Hardy spaces and their use in analysis, Bull. Amer. Math. Soc. **83**, 569–645 (1977).

R. R. Coifman, R. Rochberg, and G. Weiss [1] Factorization theorems for Hardy spaces in several variables, Ann. Math. **103**, 611–635 (1976).

Brian Cole and R. Michael Range [1] A-measures on complex manifolds and some applications, J. Functional Anal. **11**, 393–400 (1972).

Anne Cumenge [1] Extensions dans des classes de Hardy de fonctions holomorphes, C. R. Acad. Sci. Paris **289**, 385–388 (1979).

A. M. Davie and N. P. Jewell [1] Toeplitz operators in several complex variables, J. Functional Anal. **26**, 356–368 (1977).

A. M. Davie and B. Øksendal [1] Peak interpolation sets for some algebras of analytic functions, Pacific J. Math. **41**, 81–87 (1972).

Jacqueline Detraz [1] Restrictions à la diagonale des classes de Hardy du bidisque, C. R. Acad. Sci. Paris **287**, 997–999 (1978).

Klas Diederich and John E. Fornaess [1] Proper holomorphic maps onto pseudoconvex domains with real-analytic boundary, Ann. Math. **110**, 575–592 (1979).

[2] Biholomorphic mappings between certain real analytic domains in $\mathbb{C}^2$ (Preprint).

[3] Biholomorphic mappings between two-dimensional Hartogs domains with real-analytic boundaries (Preprint).

N. Dunford and J. T. Schwartz [1] *Linear Operators, Part I*, Wiley-Interscience, New York, 1958.

P. L. Duren [1] *Theory of H^p-Spaces*, Academic Press, New York, 1970.

P. L. Duren and A. L. Shields [1] Restrictions of H^p functions to the diagonal of the polydisc, Duke Math. J. **42**, 751–753 (1975).

Manuel Elgueta [1] Extensions to strictly pseudoconvex domains of functions holomorphic in a submanifold in general position and C^∞ up to the boundary, Ill. J. Math. **24**, 1–17 (1980).

P. Fatou [1] Séries trigonométriques et séries de Taylor, Acta Math. **30**, 335–400 (1906).

Charles Fefferman [1] The Bergman kernel and biholomorphic mappings of pseudo-convex domains, Invent. Math. **26**, 1–65 (1974).

G. B. Folland [1] The tangential Cauchy–Riemann complex on spheres, Trans. Amer. Math. Soc. **171**, 83–133 (1972).

Frank Forelli [1] Analytic measures, Pacific J. Math. **13**, 571–578 (1963).

[2] The isometries of H^p, Can. J. Math. **16**, 721–728 (1964).

[3] A theorem on isometries and the application of it to the isometries of $H^p(S)$ for $2 < p < \infty$, Can. J. Math. **25**, 284–289 (1973).

[4] Measures whose Poisson integrals are pluriharmonic, Ill. J. Math. **18**, 373–388 (1974).

[5] Measures whose Poisson integrals are pluriharmonic II, Ill. J. Math. **19**, 584–592 (1975).

[6] Measures whose Poisson integrals are pluriharmonic III (Preprint).

[7] Pluriharmonicity in terms of harmonic slices, Math. Scand. **41**, 358–364 (1977).

[8] A necessary condition on the extreme points of a class of holomorphic functions, Pac. J. Math. **73**, 81–86 (1977).

[9] Some extreme rays of the positive pluriharmonic functions, Can. J. Math. **31**, 9–16 (1979).

Frank Forelli and W. Rudin [1] Projections on spaces of holomorphic functions in balls, Indiana Univ. Math. J. **24**, 593–602 (1974).

John Erik Fornaess [1] Embedding strictly pseudoconvex domains in convex domains, Amer. J. Math. **98**, 529–569 (1976).

John Erik Fornaess and Edgar Lee Stout [1] Spreading polydiscs on complex manifolds, Amer. J. Math. **99**, 933–960 (1977).

[2] Polydiscs in complex manifolds, Math. Ann. **227**, 145–153 (1977).

H. Fürstenberg [1] A Poisson formula for semi-simple Lie groups, Ann. Math. **77**, 335–386 (1963).

T. W. Gamelin [1] *Uniform Algebras*, Prentice-Hall, Englewood Cliffs, NJ, 1969.

A. M. Gleason [1] Finitely generated ideals in Banach algebras, J. of Math. and Mechanics **13**, 125–132 (1964).

[2] The abstract theorem of Cauchy–Weil, Pacific J. Math. **12**, 511–525 (1962).

[3] The Cauchy–Weil theorem, J. Math. and Mech. **12**, 429–444 (1963).

Irving Glicksberg [1] The abstract F. and M. Riesz theorem, J. Functional Anal. **1**, 109–122 (1967).

[2] Recent Results on Function Algebras, Amer. Math. Soc. Regional Conference Series No. 11, 1972.

Josip Globevnik and Edgar Lee Stout [1] Highly noncontinuable functions on convex domains (Preprint).

Hans Grauert and Ingo Lieb [1] Das Ramirezsche Integral und die Lösung der Gleichung $\bar{\partial}f = \alpha$ im Bereich der beschränkten Formen, Rice Univ. Studies **56**, No. 2, 29–50 (1970).

P. C. Greiner, J. J. Kohn, and E. M. Stein [1] Necessary and sufficient conditions for solvability of the Lewy equation, Proc. Nat. Acad. Sci. USA **72**, 3287–3289 (1975).

P. C. Greiner and E. M. Stein [1] Estimates for the $\bar{\partial}$-Neumann problem, *Mathematics Notes*, Princeton University Press, Princeton, NJ, 1977.

R. C. Gunning and H. Rossi [1] *Analytic Functions of Several Complex Variables*, Prentice Hall, Englewood Cliffs, NJ, 1965.

K. T. Hahn and J. Mitchell [1] H^p spaces on bounded symmetric domains, Trans. Amer. Math. Soc. **146**, 521–531 (1969).

Monique Hakim and Nessim Sibony [1] Quelques conditions pour l'éxistence de fonctions pics dans des domaines pseudoconvexes, Duke Math. J. **44**, 399–406 (1977).

[2] Ensembles pics dans des domaines strictement pseudoconvexes, Duke Math. J. **45**, 601–617 (1978).

G. H. Hardy and J. E. Littlewood [1] A maximal theorem with function-theoretic applications, Acta. Math. **54**, 81–116 (1930).

Lawrence A. Harris [1] Schwarz's lemma in normed linear spaces, Proc. Nat. Acad. Sci. **62**, 1014–1017 (1969).

[2] Banach algebras with involution and Möbius transformations, J. Functional Analysis **11**, 1–16 (1972).

[3] Bounded symmetric homogeneous domains in infinite dimensional spaces, *Lecture Notes in Mathematics 364*, 13–40, Springer-Verlag, New York, 1973.

[4] Operator Siegel domains, Proc. Royal Soc. Edinburgh **79A**, 137–156 (1977).

Fritz Hartogs [1] Zur Theorie der analytischen Funktionen mehrerer unabhängiger Veränderlichen, insbesondere über die Darstellung derselben durch Reihen, welche nach Potenzen einer Veränderlichen fortschreiten, Math. Ann. **62**, 1–88 (1906).

T. L. Hayden and T. J. Suffridge [1] Biholomorphic maps in Hilbert space have a fixed point, Pacific J. Math. **38**, 419–422 (1971).

Sigurdur Helgason [1] A duality for symmetric spaces with applications to group representations, Advances Math. **5**, 1–154 (1970).

[2] Eigenspaces of the Laplacian; integral representations and irreducibility, J. Functional Anal. **17**, 328–353 (1974).

G. M. Henkin [1] Banach spaces of analytic function on the ball and on the bicylinder are not isomorphic, Funkcional Anal. i Prilozen **2**, No. 4, 82–91 (1968), Functional Anal. Appl. **2**, 334–341 (1968).

[2] Integral representation of functions holomorphic in strictly pseudoconvex domains and some applications, Mat. Sb. **78**, 611–632 (1969), Math. USSR Sb. **7**, 597–616 (1969).

[3] Integral representation of functions in strictly pseudoconvex domains and applications to the $\bar{\partial}$-problem, Mat. Sb. **82**, 300–308 (1970), Math. USSR Sb. **11**, 273–281 (1970).

[4] The approximation of functions in pseudoconvex domains and a theorem of A. L. Leibenson, Bull. Acad. Polon. Sci. Ser. Sci. Math. Astron. Phys. **19**, 37–42 (1971).

[5] Continuation of bounded holomorphic functions from submanifolds in general position to strictly pseudoconvex domains, Izv. Akad. Nauk SSSR Ser. Mat. **36**, 540–567 (1972), Math. USSR Izv. **6**, 536–563 (1972).

[6] An analytic polyhedron is not holomorphically equivalent to a strictly pseudoconvex domain, Dokl. Akad. Nauk SSSR **210**, 1026–1029 (1973). Soviet Math. Dokl. **14**, 858–862 (1973).

[7] Solutions with estimates of the H. Lewy and Poincaré–Lelong equations. Construction of functions of the Nevanlinna class with prescribed zeros in strictly pseudoconvex domains, Dokl. Akad. Nauk SSSR **225**, 771–774 (1975), Soviet Math. Dokl. **16**, 1310–1314 (1975).

[8] H. Lewy's equation and analysis on a pseudoconvex manifold I, Uspehi Mat. Nauk **32**, No. 3 (195), 57–118 (1977), Russian Math. Surveys **32**, 59–130 (1977).

[9] H. Lewy's equation and analysis on a pseudoconvex manifold II, Mat. Sbornik **102**, No. 1, 71–108 (1944), Math USSR Sbornik **31**, 63–94 (1977).

G. M. Henkin and E. M. Čirka [1] Boundary properties of holomorphic functions of several complex variables, Sovrem. Probl. Mat. **4**, 13–142 (1975). J. Soviet Math. **5**, 612–687 (1976).

Kenneth Hoffman [1] *Banach Spaces of Analytic Functions*, Prentice-Hall, Englewood Cliffs, NJ, 1962.

Lars Hörmander [1] *Linear Partial Differential Operators*, Springer-Verlag, New York, 1963.

[2] *An Introduction to Complex Analysis in Several Variables*, Van Nostrand, Princeton, NJ, 1966.

[3] L^p estimates for (pluri-) subharmonic functions, Math. Scand. **20**, 65–78 (1967).

Charles Horowitz [1] Zeros of functions in the Bergman spaces, Duke Math. J. **41**, 693–710 (1974).

[2] An elementary counterexample to the open mapping principle for bilinear maps, Proc. Amer. Math. Soc. **53**, 293–294 (1975).

[3] Factorization theorems for functions in the Bergman spaces, Duke Math. J. **44**, 201–213 (1977).

Charles Horowitz and Daniel M. Oberlin [1] Restrictions of H^p functions to the diagonal of U^n, Indiana Math. J. **24**, 767–772 (1975).

L. K. Hua [1] *Harmonic Analysis of Functions of Several Complex Variables in the Classical Domains*, Science Press, Peking, 1958; Amer. Math. Soc. Transl., Math. Monograph 6, 1963.

Jan Janas [1] An application of the theorem of Rudin on the Toeplitz operators on odd spheres, Math. Z. **150**, 185–187 (1976).

Nicholas P. Jewell [1] Multiplication by the coordinate functions on the Hardy space of the unit sphere in $\mathbb{C}^n$, Duke Math. J. **44**, 839–851 (1977).

S. Kakutani [1] Concrete representation of abstract (L)-spaces and the mean ergodic theorem, Ann. of Math. **42**, 523–537 (1941).

N. Kerzman [1] Hölder and L^p-estimates for solutions of $\bar{\partial}u = f$ in strongly pseudoconvex domains, Comm. Pure Appl. Math. **24**, 301–379 (1971).

[2] Singular integrals in complex analysis, Proceedings of Symposia in Pure Math. A.M.S., Vol. 35, Part 2 (1979), pp. 3–41.

N. Kerzman and A. Nagel [1] Finitely generated ideals in certain function algebras, J. Functional Anal. **7**, 212–215 (1971).

N. Kerzman and E. Stein [1] The Szegö kernel in terms of Cauchy–Fantappiè kernels, Duke Math. J. **45**, 197–224 (1978).

J. J. Kohn [1] Harmonic integrals on strongly pseudoconvex manifolds I, Ann. Math. **78**, 206–213 (1963), II, Ann. Math. **79**, 450–472 (1964).

[2] Methods of partial differential equations in complex analysis, Proc. of A.M.S. Symposia in Pure Mathematics, Vol. 30, Part 1, 215–237.

Clinton J. Kolaski [1] An F. and M. Riesz type theorem for the unit ball in complex N-space, Proc. Amer. Math. Soc. **61**, 19–25 (1976).

[2] Measures whose integral transforms are pluriharmonic, Proc. Amer. Math. Soc. **75**, 75–80 (1979).

[3] A new look at a theorem of Forelli and Rudin, Indiana U. Math. J. **28**, 495–499 (1979).

H. König and G. Seever [1] The abstract F. and M. Riesz theorem, Duke Math. J. **36**, 791–797 (1969).

Walter Koppelman [1] The Cauchy integral for functions of several complex variables, Bull. Amer. Math. Soc. **73**, 373–377 (1967).

A. Korányi [1] The Poisson integral for generalized half-planes and bounded symmetric domains, Ann. Math. **82**, 332–350 (1965).

[2] Harmonic functions on Hermitian hyperbolic space, Trans. Amer. Math. Soc. **135**, 507–516 (1969).

[3] A remark on boundary values of functions of several complex variables, *Lecture Notes in Mathematics*, 155, Springer-Verlag, Heidelberg, 1970.

A. Korányi and S. Vagi [1] Singular integrals in homogeneous spaces and some problems of classical analysis, Ann. Scuola Normale Superiore Pisa **25**, 575–648 (1971).

[2] Isometries of H^p spaces of bounded symmetric domains, Can. Math. J. **28**, 334–340 (1976).

S. G. Krantz [1] Structure and interpolation theorems for certain Lipschitz spaces and estimates for the $\bar{\partial}$-equation, Duke Math. J. **43**, 417–439 (1976).

[2] Optimal Lipschitz and L^p regularity for the equation $\bar{\partial}u = f$ on strongly pseudoconvex domains, Math. Ann. **219**, 233–260 (1976).

[3] Boundary values and estimates for holomorphic functions of several complex variables, Duke Math. J. **47**, 81–98 (1980).

Serge Lang [1] $SL_2(R)$, Addison-Wesley, Reading, MA, 1975.

Guy Laville [1] Resolution du $\partial\bar{\partial}$ avec croissance dans des ouverts pseudoconvexes étoilés de $\mathbb{C}^n$, C. R. Acad. Sci. Paris **274**, A 554–A 556 (1972).

[2] Sur les diviseurs de la classe de Nevanlinna dans la boule de $\mathbb{C}^2$, C. R. Acad. Sci. Paris **281**, 145–147 (1975).

Karel de Leeuw and Walter Rudin [1] Extreme points and extremum problems in H^1, Pacific J. Math. **8**, 467–485 (1958).

K. de Leeuw, W. Rudin, and J. Wermer [1] The isometries of some function spaces, Proc. Amer. Math. Soc. **11**, 694–698 (1960).

Pierre Lelong [1] Propriétés métriques des variétés analytiques complexes définies parune équation, Ann. Sci. Ec. Norm. Sup. **67**, 393–419 (1950).

[2] Intégration sur un ensemble analytique complexe, Bull. Soc. Math. France **85**, 239–262 (1957).

[3] *Plurisubharmonic Functions and Positive Differential Forms*, Gordon and Breach, New York, 1968.

Hans Lewy [1] On the local character of the solutions of an atypical linear differential equation in three variables and a related theorem for regular functions of two complex variables, Ann. Math. **64**, 514–522 (1956).

[2] An example of a smooth linear partial differential equation without solution, Ann. Math. **66**, 155–158 (1957).

[3] On hulls of holomorphy, Comm. Pure Appl. Math. **13**, 587–591 (1960).

Ewa Ligocka [1] A proof of Fefferman's theorem on biholomorphic mappings without use of differential geometry, Institute of Mathematics, Polish Academy of Sciences (Preprint) No. 169, 1979.

[2] Some remarks on extension of biholomorphic mappings (Preprint).

E. Lindelöf [1] Sur un principe générale de l'analyse et ses applications à la théorie de la représentation conforme, Acta Soc. Sci. Fennicae **46**, 1–35 (1915).

G. Lumer [1] Espaces de Hardy en plusieurs variables complexes, C. R. Acad. Sci. Paris **273**, 151–154 (1971).

Arne Magnus [1] On polynomial solutions of a differential equation, Math. Scand. **3**, 255–260 (1955).

Paul Malliavin [1] Travaux de H. Skoda sur la classe de Nevanlinna, Séminaire Bourbaki, Exposé No. 504, 1977.

E. Mantinelli [1] Sopra una dimonstrazione di R. Fueter per un teorema di Hartogs, Comment. Math. Helv. **15**, 340–349 (1942).

Gerard McDonald [1] The maximal ideal space of $H^\infty + C$ on odd spheres (Preprint).

[2] Fredholm properties of a class of Toeplitz operators on the ball, Indiana Math. J. **26**, 567–576 (1977).

[3] Toeplitz operators on the ball with piecewise continuous symbol. Ill. J. Math. **23**, 286–294 (1979).

Joseph Miles [1] Zero sets in $H^p(U^m)$, Illinois J. Math. **17**, 458–464 (1973).

B. Moulin and J. P. Rosay [1] Sur la restriction des fonctions plurisousharmoniques à la diagonale du polydisque, Indiana Math. J. **26**, 869–873 (1977).

Alexander Nagel [1] Smooth zero-sets and interpolation sets for some algebras of holomorphic functions on strictly pseudoconvex domains, Duke Math. J. **43**, 323–348 (1976).

[2] Cauchy transforms of measures, and a characterization of smooth peak interpolation sets for the ball algebra, Rocky Mountain J. Math. **9**, 299–305 (1979).

Alexander Nagel and W. Rudin [1] Moebius-invariant function spaces on balls and spheres, Duke Math. J. **43**, 841–865 (1976).

[2] Local boundary behavior of bounded holomorphic functions, Can. J. Math. **30**, 583–592 (1978).

Alexander Nagel and Stephen Wainger [1] Limits of bounded holomorphic functions along curves (Preprint).

R. Narasimhan [1] *Several Complex Variables*, University of Chicago Press, Chicago, 1971.

L. Nirenberg, S. Webster, and P. Yang [1] Local boundary regularity of holomorphic mappings, Comm. Pure Appl. Math. **33**, 305–338 (1980).

Ricardo Nirenberg [1] On the H. Lewy extension phenomenon, Trans. Amer. Math. Soc. **168**, 337–356 (1972).

Bernt K. Øksendal [1] A short proof of the F. and M. Riesz theorem, Proc. Amer. Math. Soc. **30**, 204 (1971).

Nils Øvrelid [1] Integral representation formulas and L^p-estimates for the $\bar{\partial}$-equation, Math. Scand. **29**, 137–160 (1971).

[2] Generators of the maximal ideals of $A(D)$, Pacific J. Math. **39**, 219–223 (1971).

Aleksander Pelczynski [1] Banach spaces of analytic functions and absolutely summing operators, Regional Conference Series in Mathematics, No. 30, A.M.S., 1977.

D. H. Phong and E. M. Stein [1] Estimates for the Bergman and Szegö projections on strongly pseudo-convex domains, Duke Math. J. **44**, 695–704 (1977).

S. I. Pinčuk [1] A boundary uniqueness theorem for holomorphic functions of several complex variables, Mat. Zametki **15**, 205–212 (1974), Math. Notes **15**, 116–120 (1974).

[2] On proper holomorphic mappings of strictly pseudoconvex domains, Sib. Mat. Zh. **15**, 909–917 (1974), Siberian Math. J. **15**, 644–649 (1974).

[3] On the analytic continuation of holomorphic mappings, Mat. Sb. **98**, 416–435 (1975), Math. USSR Sb. **27**, 375–392 (1975).

[4] Analytic continuation of mappings along strictly pseudoconvex hypersurfaces, Dokl. Akad. Nauk SSSR **236**, 544–547 (1977), Soviet Math. Dokl. **18**, 1237–1240 (1977).

[5] On holomorphic mappings of real-analytic hypersurfaces, Mat. Sbornik **105**, 574–593 (1978), Math. USSR Sbornik **34**, 503–519 (1978).

A. Plessner [1] Ueber das Verhalten analytischer Funktionen auf dem Rande des Definitionsbereiches, J. Reine Angew. Math. **158**, 219–227 (1928).

Henri Poincaré [1] Les fonctions analytiques de deux variables et la représentation conforme, Rend. Circ. Matem. Palermo **23**, 185–220 (1907).

I. Priwaloff [1] Sur une généralisation du théorème de Fatou, Rec. Math. (Mat. Sbornik) **31**, 232–235 (1923).

J. Rainwater [1] A note on the preceding paper, Duke Math. J. **36**, 798–800 (1969).

E. Ramirez De Arellano [1] Ein Divisionsproblem und Randintegraldarstellungen in der komplexen Analysis, Math. Ann. **184**, 172–187 (1970).

R. Michael Range [1] On the modulus of boundary values of holomorphic functions, Proc. Amer. Math. Soc. **65**, 282–286 (1977).

K. Reinhardt [1] Über die Abbildungen durch analytische Funktionen zweier Veränderlichen, Math. Ann. **83**, 211–255 (1921).

F. and M. Riesz [1] Über Randwerte einer analytischen Funktion, Quatrième Congrès des mathématiciens scandinaves, 27–44 (1916).

N. M. Rivière [1] Singular integrals and multiplier operators, Ark. för Mat. **9**, 243–278 (1971).

C. A. Rogers [1] *Hausdorff Measures*, Cambridge University Press, Cambridge, 1970.

A. V. Romanov and G. M. Henkin [1] Exact Hölder estimates for the solutions of the $\bar{\partial}$-equation, Izv. Akad. Nauk SSSR, Ser. Mat. **35**, 1171–1183 (1971). Math. USSR Izv. **5**, 1180–1192 (1971).

J. P. Rosay [1] Sur la non-factorization des élements de l'espace de Hardy $H^1(U^2)$, Illinois J. Math. **19**, 479–482 (1975).

[2] Sur une caractérisation de la boule parmi les domaines de $\mathbb{C}^n$ par son groupe d'automorphismes, Ann. Inst. Fourier **29** (4), 91–97 (1979).

Linda Preiss Rothschild and E. M. Stein [1] Hypoelliptic differential operators and nilpotent groups, Acta Math. **137**, 247–320 (1976).

Walter Rudin [1] *Function Theory in Polydiscs*, Benjamin, New York, 1969.

[2] *Functional Analysis*, McGraw-Hill, New York, 1973.

[3] *Real and Complex Analysis*, 2nd ed., McGraw-Hill, New York, 1974.

[4] Boundary values of continuous analytic functions, Proc. Amer. Math. Soc. **7**, 808–811 (1956).

[5] Projections on invariant subspaces, Proc. Amer. Math. Soc. **13**, 429–432 (1962).

[6] A generalization of a theorem of Frostman, Math. Scand. **21**, 136–143 (1967).

[7] Spaces of type $H^\infty + C$, Ann. Institut Fourier **25**, 99–125 (1975).

[8] L^p-isometries and equimeasurability, Indiana Univ. Math. J. **25**, 215–228 (1976).

[9] Zeros of holomorphic functions in balls, Indag. Math. **38**, 57–65 (1976).

[10] Pluriharmonic functions in balls, Proc. Amer. Math. Soc. **62**, 44–46 (1977).

[11] Lumer's Hardy spaces, Mich. Math. J. **24**, 1–5 (1977).

[12] Holomorphic Lipschitz functions in balls, Comment. Math. Helvetici **53**, 143–147 (1978).

[13] The fixed-point sets of some holomorphic maps, Bull. Malaysian Math. Soc. **1**, 25–28 (1978).

[14] Peak-interpolation sets of class C^1, Pacific J. Math. **75**, 267–279 (1978).

[15] Unitarily invariant algebras of continuous functions on spheres, Houston J. Math. **5**, 253–265 (1979).

[16] *Principles of Mathematical Analysis*, 3rd ed., McGraw-Hill, New York, 1976.

[17] The closed ideals in an algebra of analytic functions, Canadian J. Math. **9**, 426–434 (1957).

Walter Rudin and E. L. Stout [1] Boundary properties of functions of several complex variables, J. Math. Mech. **14**, 991–1006 (1965).

A. Sadullaev [1] Inner functions in $\mathbb{C}^n$, Mat. Zametki **19**, 63–66 (1976).

Donald Sarason [1] Generalized interpolation in H^∞, Trans. Amer. Math. Soc. **127**, 179–203 (1967).

[2] Algebras of functions on the unit circle, Bull. Amer. Math. Soc. **79**, 286–299 (1973).

Robert B. Schneider [1] Isometries of $H^p(U^n)$, Can. J. Math. **25**, 92–95 (1973).

[2] Unit preserving isometries are homomorphisms in certain L^p, Can. J. Math. **27**, 133–137 (1975).

H. S. Shapiro and A. L. Shields [1] On the zeros of functions with finite Dirichlet integral and some related function spaces, Math. Z. **80**, 217–229 (1962).

Joel H. Shapiro [1] Mackey topologies, reproducing kernels, and diagonal maps on the Hardy and Bergman spaces, Duke Math. J. **43**, 187–200 (1976).

[2] Zeros of functions in weighted Bergman spaces, Michigan Math. J. **24**, 243–256 (1977).

Nessim Sibony [1] Valeurs au bord de fonctions holomorphes et ensembles polynomialement convexes, *Lecture Notes in Mathematics*, No. 578, 300–313, Springer-Verlag, Heidelberg, 1977.

Henri Skoda [1] Valeurs au bord pour les solutions de l'opérateur d″, et caractérisation des zéros des fonctions de la classe de Nevanlinna, Bull. Soc. Math. France, **104**, 225–299 (1976).

Michael Spivak [1] Calculus on Manifolds, Benjamin, New York, 1965.

Charles M. Stanton [1] Embedding Riemann surfaces in polydiscs, Duke Math. J. **43**, 791–796 (1976).

E. M. Stein [1] *Singular Integrals and Differentiability Properties of Functions*, Princeton University Press, Princeton, NJ, 1970.

[2] Boundary behavior of holomorphic functions of several complex variables, *Mathematics Notes*, Princeton University Press, Princeton, NJ, 1972.

[3] Singular integrals and estimates for the Cauchy-Riemann equations, Bull. Amer. Math. Soc. **79**, 440–445 (1973).

[4] Note on the class $L \log L$, Studia Math. **32**, 305–310 (1969).

[5] On limits of sequences of operators, Ann. Math. **74**, 140–170 (1961).

E. M. Stein and G. Weiss [1] *Introduction to Fourier Analysis on Euclidean Spaces*, Princeton University Press, Princeton, NJ, 1971.

Kenneth Stephenson [1] Isometries of the Nevanlinna class, Indiana Univ. Math. J. **26**, 307–324 (1977).

Manfred Stoll [1] Harmonic majorants for plurisubharmonic functions on bounded symmetric domains with applications to the spaces H_Φ and N_*, J. Reine Angew. Math. **282**, 80–87 (1976).

[2] Mean value theorems for harmonic and holomorphic functions on bounded symmetric domains, J. Reine Angew. Math. **290**, 191–198 (1977).

Edgar Lee Stout [1] *The Theory of Uniform Algebras*, Bogden and Quigley, Tarrytown on Hudson, NY, 1971.

[2] On the multiplicative Cousin problem with bounded data, Scuola Normale Superiore di Pisa **27**, 1–17 (1973).

[3] An integral formula for holomorphic functions on strictly pseudoconvex hypersurfaces, Duke Math. J. **42**, 347–356 (1975).

[4] Bounded extensions. The case of discs in polydiscs. J. d'Analyse Math. **28**, 239–254 (1975).

[5] H^p-functions on strictly pseudoconvex domains, Amer. J. Math. **98**, 821–852 (1976).

[6] The boundary values of holomorphic functions of several complex variables, Duke Math. J. **44**, 105–108 (1977).

[7] Interpolation manifolds (Preprint).

[8] Cauchy integrals on strongly pseudoconvex domains, Scuola Normale Superiore di Pisa, Ser. 4, **6**, 685–702 (1979).

T. J. Suffridge [1] Common fixed points of commuting holomorphic maps of the hyperball, Mich. Math. J. **21**, 309–314 (1974).

B. A. Taylor and D. L. Williams [1] The peak sets of A^m, Proc. Amer. Math. Soc. **24**, 604–606 (1970).

[2] Zeros of Lipschitz functions analytic in the unit disc, Michigan Math. J. **18**, 129–139 (1971).

E. Thorp and R. Whitley [1] The strong maximum modulus theorem for analytic functions into a Banach space, Proc. Amer. Math. Soc. **18**, 640–646 (1967).

E. C. Titchmarsh [1] *The Theory of Functions*, Oxford University Press, Oxford, 1939.

A. E. Tumanov [1] A peak set for the disc algebra of metric dimension 2.5 in the three-dimensional unit sphere, Izv. Akad. Nauk SSSR **41**, No. 2 (1977), Math. USSR Izv. **11**, 353–359 (1977).

R. E. Valskii [1] On measures orthogonal to analytic functions in $\mathbb{C}^n$, Dokl. Akad. Nauk SSSR **198**, 502–505 (1971), Sov. Math. Dokl. **12**, 808–812 (1971).

N. Th. Varopoulos [1] Ensembles pics et ensembles d'interpolation pour les algèbres uniformes, C. R. Acad. Paris **272**, 866–867 (1971).

[2] BMO functions and the $\bar{\partial}$-equation, Pac. J. Math. **71**, 221–273 (1977).

[3] B.M.O. functions in complex analysis, Proceedings of Symposia in Pure Math. A.M.S., Vol. 35, Part 2, 43–61 (1979).

[4] Zeros of H^p-functions in several complex variables (Preprint; Orsay).

S. M. Webster [1] On the reflection principle in several complex variables, Proc. Amer. Math. Soc. **71**, 26–28 (1978).

[2] The rigidity of C-R hypersurfaces in a sphere, Indiana U. Math. J. **28**, 405–416 (1979).

[3] On mapping an n-ball into an $(n + 1)$-ball in complex space, Pacific J. Math. **81**, 267–272 (1979).

B. M. Weinstock [1] Continuous boundary values of analytic functions of several complex variables, Proc. Amer. Math. Soc. **21**, 463–466 (1969).

[2] An approximation theorem for $\bar{\partial}$-closed forms of type $(n, n - 1)$, Proc. Amer. Math. Soc. **26**, 625–628 (1970).

[3] Continuous boundary values of holomorphic functions on Kähler domains, Can. J. Math. **28**, 513–522 (1976).

[4] Zero-sets of continuous holomorphic functions on the boundary of a strongly pseudoconvex domain, J. London Math. Soc. **18**, 484–488 (1978).

John Wermer [1] *Banach Algebras and Several Complex Variables*, 2nd. ed., Springer-Verlag, New York, 1976.

Nicholas Weyland [1] A note on the zeros of H^p functions, Indiana U. Math. J. **28**, 507–510 (1979).

B. Wong [1] Characterization of the unit ball in $\mathbb{C}^n$ by its automorphism group, Inventiones Math. **41**, 253–257 (1977).

Lawrence Zalcman [1] Analyticity and the Pompeiu problem, Arch. Rat. Mech. Anal. **47**, 237–254 (1972).

A. Zygmund [1] A remark on functions of several complex variables, Acta Szeged. **12**, 66–68 (1950).

[2] On a theorem of Marcinkiewicz concerning interpolation of operations, J. de Math. **35**, 233–248 (1956).

[3] *Trigonometric Series*, 2nd ed., Cambridge University Press, Cambridge, 1958.

Index

absolute convergence   130, 171
absolutely continuous   68
Adachi, K.   167
admissible convergence   76
affine set   32
Agranovskii, M.L.   264, 287
Ahern, P.R.   101, 107, 115, 119, 184, 235, 404
Ahlfors, L.V.   62
Aizenberg, L.A.   349
Alexander, H.   229, 308, 313, 316, 319, 416
algebra pattern   264
Amar, E.   168
analytic
   cover   305
   measure   187
   subvariety   291
   variety   291
annihilating measure   186
anticommutative law   331
approach curve   169
   , restricted   170
   , special   170
approach region   72
asymptotic value   172
Audibert, T.   397, 399, 400, 401
automorphism   23, 25, 311, 313, 327
Aytuna, A.   114

balanced set   59, 161
ball   2, 65
ball algebra   39, 185
basic forms   331

Bedford, E.   319, 397, 404
Bell, S.   319
Berenstein, C.   58
Bergman
   formula   37
   kernel   36, 38
Bergman, S.   38, 40
Berndtsson, B.   414
Beurling, A.   62
bidegree   255, 336
Bieberbach, L.   25
biholomorphic map   11, 303
bi-invariance   54
bilinear map   413
Bishop, E.   205, 209, 216, 224
Blaschke condition   133, 365, 385
Bochnak, J.   20
Bochner−Martinelli formula   347, 350
Bochner, S.   15, 25, 349, 393
Bonami, A.   420
boundary   320, 333
Boutet de Monvel, L.   110
Bungart, L.   17, 40
Burns, D.   216

Calderón, A.P.   79, 91, 129
Carathéodory, C.   175, 177
Carleson, L.   205, 229, 235, 250, 415
Carleson set   250
Cartan, H.   23, 24
Cauchy
   formula   3, 39, 40, 349
   integral   39
   kernel   4, 38, 92

Cauchy [*cont.*]
  transform   39
Cauchy−Riemann
  equation   8, 252, 337
  operator   387
Cayley transform   31
chain   333
chain rule   8
Charpentier, P.   363
Chaumat, J.   230
Chee, P.S.   386
Chollet, A.-M.   114, 205, 230
circular
  measure   201
  set   24
Čirka's theorem   171, 174
Čirka, E.M.   168, 171, 216
Clark, D.N.   184
closed map   301
Coburn, L.A.   110
Cohen, P.J.   413
Coifman, R.R.   91, 413
Cole, B.   198
Cole−Range theorem   185, 198, 202, 205
commute   256
compact variety   292, 294
complex line   6
complex-tangential curve   101, 212, 214, 237
complex-tangential map   214, 216
complex tangent space   73, 74
composition   5
corona problem   415
counting functions   134
covering
  lemma   68, 94
  map   305
critical
  set   301, 303
  value   301, 303
Cumenge, A.   168
curve   102, 169
  , approach   169
  , complex-tangential   101, 212, 214, 237
  , restricted   170
  , special   170
Davie, A.M.   211
Davie−Øksendal theorem   211
defining function   320
de Leeuw, K.   153, 160, 406
derivative   7, 10
  of an automorphism   26

of a form   332, 336
of a measure   72, 78, 79
  , radial   103
determinant   9
determining set   133, 222
Detraz, J.   128
Diederich, K.   319
differential form   330
dilate   56
Dini function   110
divergence theorem   335
domain of holomorphy   7, 126, 340
dot-product   73
Dunford, N.   203
Duren, P.L.   103, 128, 133

Elgueta, M.   168
ellipsoid   29, 175, 323
elliptic operator   53
epsilon-cover   295
exceptional sets   247, 414
extension   127, 167, 312
extremal functions   164

factorization   413
Fatou's theorem   72, 235
Fatou, P.   25, 205
Federbush, P.   397
Fefferman, C.   17, 319
fixed-point set   33, 165
Folland, G.B.   17
Forelli, F.   14, 60, 122, 153, 196, 406, 409, 410, 411, 413
Fornaess, J.E.   309, 313, 319
F_σ -set   193
function
  , bi-invariant   54
  , defining   320
  , Dini   110
  , holomorphic   2
  , inner   403
  , K-bounded   82
  , Lipschitz   101, 222
  , $\mathscr{M}$-harmonic   49
  , pluriharmonic   9, 59, 63, 397
  , plurisubharmonic   126, 366, 375, 379
  , radial   4
  , real-analytic   52, 282, 310
  , slice   6, 61, 132, 134
  , spherical   55
  , subharmonic   20
function algebra   185
Fürstenberg, H.   56

Gamelin, T.W. 197
generic manifold 225
GKS decomposition 194, 197, 198, 202, 247
Gleason's problem 114
Gleason, A.M. 114, 349
Glicksberg, I. 191, 192, 194, 197, 207
Globevnik, J. 415
Green's theorem 235, 254
Greiner, P.C. 32, 355, 363, 395
Gunning, R.C. 305, 330

Haar measure 13
Hahn, K.T. 423
Hakim, M. 230, 233
Hardy, G.H. 86, 103
Hardy—Littlewood theorems 86, 103
Hardy space 84
harmonic conjugate 223
Harris, L.A. 30
Hartogs, F. 2, 20, 340, 341
Hartogs' theorem 2, 4, 341
Hausdorff dimension 220, 296
Hausdorff measure 248, 295
Hayden, T.L. 30, 33
Helgason, S. 55
Henkin, G.M. 115, 133, 167, 187, 189, 203, 216, 315, 349, 355, 363, 381
Henkin measure 187, 189, 198, 202, 244, 246
Henkin—Skoda theorem 133, 135, 365
Henkin's theorem 189, 198, 202
Hessian 9, 320, 367, 394
Hoffman, K. 119, 133, 160, 262, 406, 415
holomorphic
    function 2
    map 5
    monomial 3
    retract 166
homogeneous
    expansion 19
    polynomial 19
homomorphism 118
Hopf lemma 231, 312
Hörmander, L. 2, 52, 53, 126, 235, 355, 392, 395, 397
Horowitz, C. 128, 145, 413
Hua, L.K. 40

ideals 415
inhomogeneous Cauchy—Riemann
    equation 337
inner function 403

inner map 416
inner product 1, 38, 73, 254, 256
integration by slices 15
interpolation
    manifold 220, 416
    set 204, 416
invariant
    Laplacian 47
    mean value property 43
    Poisson kernel 40
inverse function theorem 11, 302
involution 26, 34
isometry 152
isomorphic 149

Jacobian 11, 28, 310, 330
Janas, J. 110
Jensen's formula 134, 384
Jewell, N.P. 424
Julia's theorem 175, 176
Julia—Carathéodory theorem 174, 177

Kakutani, S. 203
K-bounded function 82
kernel function 38, 257
Kerzman, N. 115, 116, 349, 363
K-limit 76, 315, 317
K-null set 191
Kohn, J.J. 355, 395, 397
Kolaski, C.J. 424
Kolmogorov, A.N. 101
König, H. 191, 194, 197
Koppelman, W. 347, 349
Korányi, A. 56, 65, 72, 75, 76, 85, 91, 99, 153
Korányi's theorem 75
Korányi—Vagi theorem 99, 125
Krantz, S.G. 355, 363
K-singular measure 191

Lang, S. 55
Laplacian 8
    , complex-radial 367
    , complex-tangential 367
    , invariant 47
Laville, G. 425
Lebesgue decomposition theorem 68, 191
Lebesgue point 70
Leibenson, A.L. 115
Leibnitz rule 332
Lelong, P. 386
Levi form 394
Lewy, H. 394, 395, 397

Ligocka, E. 319
Lindelöf–Čirka theorem 168, 171, 239, 414
Lindelöf, E. 168
Lipschitz
    condition 101
    function 101, 222
Littlewood, J.E. 86, 103, 403
local peak point 305, 322
local peak set 230
Lumer, G. 145, 146
Lumer's Hardy space 145, 198

Magnus, A. 25
majorant 84, 145, 198, 365
Malliavin, P. 425
map
    , biholomorphic 11, 303
    , bilinear 413
    , closed 301
    , complex-tangential 214, 216
    , holomorphic 5
    , inner 416
    , nonsingular 215
    , open 301
    , proper 300
Marcinkiewicz interpolation theorem 69, 88, 99, 100
Martinelli, E. 425
Martin, W.T. 25
maximal difference 93
maximal function 68, 74, 77, 236
maximal operator 69
maximal subalgebra 269
maximal theorem 69, 75, 86, 95
maximum modulus theorem 5, 291, 295
maximum principle 55
McDonald, G. 110
measure
    , absolutely contionuous 68
    , analytic 187
    , annihilating 186
    , circular 201
    , Hausdorff 248, 295
    , Henkin, 187, 189, 198, 202, 244, 246
    , K-singular 191
    , representing 185
    , RP 417
    , singular 68
    , totally singular 186, 202
metric 65
$\mathscr{M}$-harmonic
    function 49

majorant 84, 365
minimax theorem 192, 194, 196
$\mathscr{M}$-invariant 43, 278, 365
Mitchell, J. 423
modulus of continuity 110
Moebius
    group 23
    invariance 43
monic polynomial 290, 298
Montel sequence 187
Moulin, B. 128
multi-index 3
multiple power series 4, 6
multiplicity 13, 303
multiplier transformation 118, 262, 416

Nagel, A. 64, 115, 116, 172, 216, 230, 235, 242, 244, 253, 261, 287
Narasimhan, R. 2, 308
natural boundary 415
neighborhood 2
Nevanlinna class 83, 133, 153, 365, 407
Nirenberg, L. 319
Nirenberg, R. 395
nonsingular map 215
norm 1, 36, 56, 84, 146, 161, 199, 200, 223
normal family 5
normal vector 319
norm-preserving extension 166
null set 204

Oberlin, D.M. 128
Øksendal, B. 211
one-dimensionally removable 62
one-radius theorem 58
order of zero 288
oriented boundary 334
oscillation 236
Øvrelid, N. 352, 355

peak-interpolation set 204
peak set 204, 230, 237
Pelczynski, A. 187, 203
Phong, D.H. 426
Pinčuk, S.I. 225, 226, 228, 313
Plessner, A. 79
pluriharmonic
    function 9, 59, 63, 397
    majorant 145, 198
plurisubharmonic function 126, 366, 375, 379
Poincaré, H. 30

point
  of density   70
  of strict pseudoconvexity   320
Poisson
  kernel   40, 45
  integral   41
polar coordinates   13
polydisc   2, 168, 205
Privalov, I.   79
probability measure   12, 185
projection theorem   292
proper map   300, 416
pull-back   333

radial
  derivative   103
  function   14
radialization   49, 281
Radon−Nikodym theorem   68
Radó's theorem   302
Rainwater, J.   192
Rainwater's lemma   193
Ramirez de Arellano, E.   349
Range, R.M.   198, 404
rank theorem   301
real-analytic   52, 282, 310
region   23
regular point   299
regular value   301, 303
Reinhardt, K.   426
removable set   62
removable singularities   62
representing measure   185
restricted approach curve   170
restricted K-limit   170
restriction of measure   190
restriction operator   127
retraction   166
Riesz, F.   205
Riesz, M.   92, 101, 205
Riesz theorem   185, 189, 195, 197, 205,
    211
Rivière, N.   91
Rochberg, R.   413
Rogers, C.A.   426
Romanov, A.V.   355, 363
Rosay, J.-P.   128, 326, 327, 413
Rossi, H.   305, 330
rotation-invariance   32
Rothschild, L.P.   12
RP-measure   417
Rudin, W.   13, 64, 101, 113, 114, 122,
    128, 133, 145, 153, 160, 165, 166,

    168, 172, 193, 196, 205, 216, 221,
    235, 238, 253, 261, 262, 263, 287,
    403, 404, 405, 406, 408, 415

Sadullaev, A.   404
same zeros   133, 365
Sarason, D.   114
Schneider, R.B.   101, 107, 115, 119,
    153, 157
Schwartz, J.T.   203
Schwarz lemma   161, 163
Seever, G.   191, 194, 197
semigroup   267, 269
set
  , affine   32
  , balanced   59, 161
  , Carleson   250
  , circular   24
  , critical   301, 303
  , determining   133, 303
  , H^∞-removable   62
  , interpolation   204, 416
  , K-null   191
  , local peak   230
  , null   204
  of type F_σ 193, 249
  of type G_δ 404
  , one-dimensionally removable   62
  , peak   204, 230, 237
  , peak-interpolation   204
  , totally null   186, 204, 242
  , zero   204
Shapiro, H.S.   135
Shapiro, J.H.   128, 145
Shields, A.L.   128, 135
Sibony, N.   230, 233, 248, 252
Siciak, J.   20
singular
  measure   68
  point   299
Skoda, H.   133, 363, 380, 381
slice function   6, 61, 132, 134
slice integration formula   15
Smirnov class   407
special approach curve   170
sphere   2
spherical
  function   55
  harmonics   253
Spivak, M.   330, 334
standard orthonormal basis   2
standard presentation   331
Stanton, C.   167

Stein, E.M.   32, 74, 85, 88, 91, 100,
    101, 109, 129, 145, 236, 255, 355,
    363, 395
Stephenson, K.   153, 415
Stokes' theorem   334
Stoll, M.   428
strictly convex region   165
strictly pseudoconvex   114, 115, 168,
    205, 211, 216, 320, 327
Stout, E.L.   115, 145, 152, 197, 205,
    216, 219, 221, 244, 309, 408, 415
subharmonic function   20
subvariety   292, 303
Suffridge, T.J.   30, 33, 166
surface   330
symbol   110

tangential Cauchy−Riemann
    equations   252, 389, 391, 393
tangential Cauchy−Riemann
    operator   387
tangent space   74
Taylor, B.A.   229, 404
Thorp, E.   428
Titchmarsh, E.C.   59, 134
Toeplitz operator   110
totally null   186, 204, 242
totally real manifold   223
totally real vector space   218
totally singular measure   186, 202
transitivity   27, 319, 329
triangle inequality   66
Tumanov, A.E.   220
two-function lemma   288

$\mathcal{U}$-invariant   256
Ullrich, D.   33, 56

$\mathcal{U}$-minimal   259
uncomplemented subspace   262
uniform Blaschke condition   414
unitarily invariant   256
unitary group   15
unit ball   2
$\mathcal{U}$-space   256

Vagi, S.   91, 99, 153
Valskii, R.E.   187, 205, 264, 287
Valskii's decomposition   187, 198, 203
Varopoulos, N.T.   205, 207, 414, 415
von Neumann, J.   13, 192

Wainger, S.   238
weak L^1 69
weak type (1,1)   69
Webster, S.M.   319, 416
Weierstrass polynomial   290, 298
Weierstrass theorem   290
Weinstock, B.M.   393
Weiss, G.   88, 91, 236, 255, 413
Wermer, J.   153, 160, 355
Whitley, R.   428
Williams, D.L.   229
Wojtaszczyk, P.   405
Wong, B.   327

Yang, P.   319

Zalcman, L.   58, 287
zero set   204
zero variety   133, 291, 414
Zygmund, A.   83, 88, 91, 101, 110, 129,
    132

Grundlehren der mathematischen Wissenschaften

A Series of Comprehensive Studies in Mathematics

A Selection

114. Mac Lane: Homology
131. Hirzebruch: Topological Methods in Algebraic Geometry
144. Weil: Basic Number Theory
145. Butzer/Berens: Semi-Groups of Operators and Approximation
146. Treves: Locally Convex Spaces and Linear Partial Differential Equations
152. Hewitt/Ross: Abstract Harmonic Analysis. Vol. 2: Structure and Analysis for Compact Groups. Analysis on Locally Compact Abelian Groups
153. Federer: Geometric Measure Theory
154. Singer: Bases in Banach Spaces I
155. Müller: Foundations of the Mathematical Theory of Electromagnetic Waves
156. van der Waerden: Mathematical Statistics
157. Prohorov/Rozanov: Probability Theory. Basic Concepts. Limit Theorems. Random Processes
158. Constantinescu/Cornea: Potential Theory on Harmonic Spaces
159. Köthe: Topological Vector Spaces I
160. Agrest/Maksimov: Theory of Incomplete Cylindrical Functions and their Applications
161. Bhatia/Szegö: Stability of Dynamical Systems
162. Nevanlinna: Analytic Functions
163. Stoer/Witzgall: Convexity and Optimization in Finite Dimensions I
164. Sario/Nakai: Classification Theory of Riemann Surfaces
165. Mitrinovic/Vasic: Analytic Inequalities
166. Grothendieck/Dieudonné: Eléments de Géométrie Algébrique I
167. Chandrasekharan: Arithmetical Functions
168. Palamodov: Linear Differential Operators with Constant Coefficients
169. Rademacher: Topics in Analytic Number Theory
170. Lions: Optimal Control of Systems Governed by Partial Differential Equations
171. Singer: Best Approximation in Normed Linear Spaces by Elements of Linear Subspaces
172. Bühlmann: Mathematical Methods in Risk Theory
173. Maeda/Maeda: Theory of Symmetric Lattices
174. Stiefel/Scheifele: Linear and Regular Celestial Mechanic. Perturbed Two-body Motion—Numerical Methods—Canonical Theory
175. Larsen: An Introduction to the Theory of Multipliers
176. Grauert/Remmert: Analytische Stellenalgebren
177. Flügge: Practical Quantum Mechanics I
178. Flügge: Practical Quantum Mechanics II
179. Giraud: Cohomologie non abélienne
180. Landkof: Foundations of Modern Potential Theory
181. Lions/Magenes: Non-Homogeneous Boundary Value Problems and Applications I
182. Lions/Magenes: Non-Homogeneous Boundary Value Problems and Applications II
183. Lions/Magenes: Non-Homogeneous Boundary Value Problems and Applications III
184. Rosenblatt: Markov Processes. Structure and Asymptotic Behavior
185. Rubinowicz: Sommerfeldsche Polynommethode
186. Handbook for Automatic Computation. Vol. 2. Wilkinson/Reinsch: Linear Algebra

187. Siegel/Moser: Lectures on Celestial Mechanics
188. Warner: Harmonic Analysis on Semi-Simple Lie Groups I
189. Warner: Harmonic Analysis on Semi-Simple Lie Groups II
190. Faith: Algebra: Rings, Modules, and Categories I
191. Faith: Algebra II, Ring Theory
192. Mallcev: Algebraic Systems
193. Pólya/Szegö: Problems and Theorems in Analysis I
194. Igusa: Theta Functions
195. Berberian: Baer*-Rings
196. Athreya/Ney: Branching Processes
197. Benz: Vorlesungen über Geometric der Algebren
198. Gaal: Linear Analysis and Representation Theory
199. Nitsche: Vorlesungen über Minimalflächen
200. Dold: Lectures on Algebraic Topology
201. Beck: Continuous Flows in the Plane
202. Schmetterer: Introduction to Mathematical Statistics
203. Schoeneberg: Elliptic Modular Functions
204. Popov: Hyperstability of Control Systems
205. Nikollskii: Approximation of Functions of Several Variables and Imbedding Theorems
206. André: Homologie des Algèbres Commutatives
207. Donoghue: Monotone Matrix Functions and Analytic Continuation
208. Lacey: The Isometric Theory of Classical Banach Spaces
209. Ringel: Map Color Theorem
210. Gihman/Skorohod: The Theory of Stochastic Processes I
211. Comfort/Negrepontis: The Theory of Ultrafilters
212. Switzer: Algebraic Topology—Homotopy and Homology
213. Shafarevich: Basic Algebraic Geometry
214. van der Waerden: Group Theory and Quantum Mechanics
215. Schaefer: Banach Lattices and Positive Operators
216. Pólya/Szegö: Problems and Theorems in Analysis II
217. Stenström: Rings of Quotients
218. Gihman/Skorohod: The Theory of Stochastic Processes II
219. Duvaut/Lions: Inequalities in Mechanics and Physics
220. Kirillov: Elements of the Theory of Representations
221. Mumford: Algebraic Geometry I: Complex Projective Varieties
222. Lang: Introduction to Modular Forms
223. Bergh/Löfström: Interpolation Spaces. An Introduction
224. Gilbarg/Trudinger: Elliptic Partial Differential Equations of Second Order
225. Schütte: Proof Theory
226. Karoubi: K-Theory. An Introduction
227. Grauert/Remmert: Theorie der Steinschen Räume
228. Segal/Kunze: Integrals and Operators
229. Hasse: Number Theory
230. Klingenberg: Lectures on Closed Geodesics
231. Lang: Elliptic Curves: Diophantine Analysis
232. Gihman/Skorohod: The Theory of Stochastic Processes III
233. Stroock/Varadhan: Multi-dimensional Diffusion Processes
234. Aigner: Combinatorial Theory
235. Dynkin/Yushkevich: Markov Control Processes and Their Applications
236. Grauert/Remmert: Theory of Stein Spaces
237. Köthe: Topological Vector Spaces II
238. Graham/McGehee: Essays in Commutative Harmonic Analysis
239. Elliott: Probabilistic Number Theory I
240. Elliott: Probabilistic Number Theory II